2018 International Semiconductor Conference (CAS 2018)

Sinaia, Romania
10-12 October 2018

IEEE Catalog Number: CFP18CAS-POD
ISBN: 978-1-5386-4483-6

**Copyright © 2018 by the Institute of Electrical and Electronics Engineers, Inc.
All Rights Reserved**

Copyright and Reprint Permissions: Abstracting is permitted with credit to the source. Libraries are permitted to photocopy beyond the limit of U.S. copyright law for private use of patrons those articles in this volume that carry a code at the bottom of the first page, provided the per-copy fee indicated in the code is paid through Copyright Clearance Center, 222 Rosewood Drive, Danvers, MA 01923.

For other copying, reprint or republication permission, write to IEEE Copyrights Manager, IEEE Service Center, 445 Hoes Lane, Piscataway, NJ 08854. All rights reserved.

*** This is a print representation of what appears in the IEEE Digital Library. Some format issues inherent in the e-media version may also appear in this print version.

IEEE Catalog Number:	CFP18CAS-POD
ISBN (Print-On-Demand):	978-1-5386-4483-6
ISBN (Online):	978-1-5386-4482-9
ISSN:	1545-827X

Additional Copies of This Publication Are Available From:

Curran Associates, Inc
57 Morehouse Lane
Red Hook, NY 12571 USA

Phone:	(845) 758-0400
Fax:	(845) 758-2633
E-mail:	curran@proceedings.com
Web:	www.proceedings.com

2018 International Semiconductor Conference (CAS 2018)

Sinaia, Romania
10-12 October 2018

IEEE Catalog Number: CFP18CAS-POD
ISBN: 978-1-5386-4483-6

2018
INTERNATIONAL
SEMICONDUCTOR CONFERENCE

41st Edition
October 10-12, Sinaia, Romania
www.imt.ro/CAS

CAS 2018 PROCEEDINGS

- ◆ Plenary Sessions
- ◆ Nanoscience and Nanoengineering 1
- ◆ Nanoscience and Nanoengineering 2
- ◆ Microwave and Millimeter Wave Circuits and Systems
- ◆ Microsensors and Microsystems
- ◆ Modelling
- ◆ Semiconductor Devices
- ◆ Integrated Circuits 2
- ◆ Integrated Circuits 1 - Student papers
- ◆ Integrated Circuits 3 - Student papers
- ◆ Devices & Integrated Circuits - Student papers
- ◆ Nanoscience; Micro- and nanophotonics
- ◆ Semiconductor devices and Microsystems
- ◆ Workshop "Microsystems for Energy Harvesting and Environment Monitoring"

CAS 2018

organized by

**NATIONAL INSTITUTE FOR
RESEARCH AND DEVELOPMENT
IN MICROTECHNOLOGIES – IMT Bucharest**

CAS Office:

IMT Bucharest

126A (32B), Erou Iancu Nicolae Street, R-077190, Bucharest, Romania

Phone: +40.21-269.07.70   +40.21-269.07.74
Fax: +40.21-269.07.72   +40.21-269.07.76
E-mail: cas@imt.ro
http://www.imt.ro/CAS

INTERNATIONAL SEMICONDUCTOR CONFERENCE (CAS) 2018

organized by:

National Institute for Research and Development
in Microtehnologies – IMT Bucharest

co-sponsored by:

IEEE - Electron Devices Society	Ministry of Research and Innovation	University of Cambridge, Dept. of Engineering, UK

HISTERESIS S.R.L., Romania	NANOTEAM S.R.L., Romania	SCHAEFER SouthEast Europe S.R.L., Romania

SC. SITEX 45 S.R.L., Romania	S.C. New Style Trans Prest S.R.L., Romania	S.C. Marido Cafe Club S.R.L., Romania

FOREWORD

It is our great pleasure to welcome you to the 41st edition of the International Semiconductor Conference (CAS 2018), taking place October 10-12, 2018 in Sinaia, Romania!

CAS is an annual conference, with a long-standing tradition, taking place every year since 1978. The conference was organized by ICCE (Institute of Research for Electronic Components) until 1996 as an *Annual Semiconductor Conference* (in Romanian *"Conferinta Anuala de Semiconductoare – CAS"*). Since 1991, CAS was opened to foreign participants, becoming an international conference and changed its name accordingly to *International Semiconductor Conference*. Starting with the 1995 edition, the conference became **an IEEE (Institute of Electrical and Electronic Engineers) event**, being sponsored by the IEEE EDS (Electron Devices Society). Since 1997, the organizer of CAS has been **the National Institute for Research and Development in Microtechnologies – IMT Bucharest** (www.imt.ro).

The Conference profile has been extended from semiconductor device physics, technology and integrated circuits to **micro- and nanotechnologies** in general, including micro-nanoelectronics and micro-nanosystems technologies. Nanostructures and nanostructured materials are also considered.

CAS 2018 will deliver a rich technical program, including **60 oral papers** (12 invited, 48 regular) and **23 posters presentations**, with authors from 17 countries on 4 continents (13 countries from Europe). The quality of papers accepted for publication in the Proceedings was guaranteed by the evaluation made by referees, members of the Conference Paper Review Board and of the Technical Program Committee. We take this opportunity to thank all the reviewers for their devoted work.

For this conference edition, **12** distinguished researchers are joining us as **invited speakers**, originating from France, Germany, Italy, Ireland, Japan, Romania, Serbia, The Netherlands.

CAS 2018 is sponsored by IEEE Electron Devices Society, under whose authority the CAS 2018 Proceedings is published, by IMT Bucharest, the Ministry of Research and Innovation. The co-sponsors of this conference edition are: University of Cambridge, Dept. of Engineering, UK; HISTERESIS S.R.L.; NANOTEAM S.R.L.; Schaefer SouthEast Europe S.R.L.; S.C. SITEX 45 SRL; S.C. New Style Trans Prest S.R.L.; S.C. Marido Cafe Club S.R.L.

We would like to express our sincere appreciation to the members of the conference committees, the invited speakers, the experts chairing various sessions of CAS 2018, as well as to all the conference participants and sponsors, for their strong contributions to make this event successful!

It is an honour to have you all here and hope you will enjoy the conference sessions, as well as your staying in Sinaia!

Miron Adrian Dinescu
CAS General Chairman

ORGANIZING COMMITTEE

General Chairman: **Miron Adrian Dinescu** - IMT Bucharest, Romania
Technical Program Chair: **Mircea Dragoman** - IMT Bucharest, Romania

Technical Program Vice Chair **Gheorghe Brezeanu** - "Politehnica" Univ. of Bucharest, Romania
Manager: **Claudia Roman** - IMT Bucharest, Romania

HONORARY ADVISORY COMMITTEE

G.A.J. Amaratunga	- Cambridge Univ., UK
C. Anghel	- Inst. Superieur d'Electronique de Paris, France
N. Barsan	- Tuebingen Univ, Germany
S. Bellucci	- INFN Lab. Nazionali di Frascati, Roma, Italy
C. Bulucea*	- Texas Instruments, Santa Clara, CA, USA (retired)
S. Cristoloveanu	- CNRS-INPG-UJF, Grenoble, France
D. Dascalu**	- "Politehnica" Univ. of Bucharest, Romania
D. Dragomirescu	- Univ. of Toulouse; LAAS-CNRS; France
H. Hartnagel	- Techn. Univ. Darmstadt, Germany
A.M. Ionescu	- EPF, Lausanne, Switzerland
G. Konstantinidis	- FORTH Heraklion, Greece
R. Marcelli	- CNM, Rome, Italy
S. Melinte	- Univ. Catholique de Louvain, Belgium
A. Mocuta	- IMEC, Belgium
K. Mutamba	- Infineon Technologies, Germany
D. Planson	- Institut National des Sciences Appliquées de Lyon, France
A. Rusu	- KTH Royal Inst. of Technology, Sweden
K. Szaciłowski	- AGH Univ. of Science and Technology, Poland
F. Udrea	- Cambridge Univ., UK

TECHNICAL PROGRAM COMMITTEE

Gh. Brezeanu	- "Politehnica" Univ. of Bucharest, Romania
M.L. Ciurea	- National Inst. of Material Physics, Bucharest, Romania
D. Cristea	- IMT Bucharest, Romania
M.A. Dinescu	- IMT Bucharest, Romania
D. Dobrescu	- "Politehnica" Univ. of Bucharest, Romania
M. Dragoman	- IMT Bucharest, Romania
A. Müller	- IMT Bucharest, Romania
R. Müller	- IMT Bucharest, Romania
D. Neculoiu	- IMT Bucharest, Romania

---

*CAS Founder Chairman

** CAS Chairman (1998-2016)

PAPER REVIEW BOARD

V. Anghel	- ON Semiconductor, Bucharest, Romania
V. Banu	- IMB-CNM, Barcelona, Spain
G. Bartolucci	- Univ. of Rome Tor Vergata, Italy
S. Bellucci	- INFN Lab. Nazionali di Frascati, Roma, Italy
Gh. Brezeanu	- "Politehnica" Univ. of Bucharest, Romania
M. Brezeanu	- PwC Romania, Bucharest, Romania
O. Buiu	- IMT Bucharest, Romania
E. Ceuca	- "1 Decembrie 1918" University of Alba Iulia
M.L. Ciurea	- National Institute of Materials Physics, Bucharest, Romania
D. Cristea	- IMT Bucharest, Romania
M. Dinescu	- National Inst. for Laser, Plasma and Radiation Physics, Romania
D. Dobrescu	- "Politehnica" Univ. of Bucharest, Romania
L. Dobrescu	- "Politehnica" Univ. of Bucharest, Romania
F. Draghici	- "Politehnica" Univ. of Bucharest, Romania
D. Dragoman	- Univ. of Bucharest, Romania
M. Dragoman	- IMT Bucharest, Romania
L. Goras	- „Gh. Asachi" Technical Univ. of Iasi, Romania
H. Hartnagel	- Techn. Univ. Darmstadt, Germany
M. Kusko	- IMT Bucharest, Romania
S. Lazanu	- National Institute of Materials Physics, Bucharest, Romania
M. Lazar	- INSA Lyon, France
R. Marcelli	- CNR-IMM, Rome, Italy
S. Melinte	- Univ. Catholique de Louvain, Belgium
M. Mihaila	- IMT Bucharest, Romania
N. Militaru	- "Politehnica" Univ. of Bucharest, Romania
C. Moldovan	- IMT Bucharest, Romania
A. Müller	- IMT Bucharest, Romania
R. Müller	- IMT Bucharest, Romania
M. Neag	- Tehnical Univ. of Cluj-Napoca, Romania
D. Neculoiu	- IMT Bucharest, Romania
O. Nedelcu	- IMT Bucharest, Romania
R. Pascu	- IMT Bucharest, Romania
Gh. Pristavu	- "Politehnica" Univ. of Bucharest, Romania
M. Purica	- IMT Bucharest, Romania
P. Pursula	- VTT Technical Research Centre of Finland Ltd, Finland
M. Pustan	- Tehnical Univ. of Cluj-Napoca, Romania
B. Serban	- IMT Bucharest , Romania
S. Simion	- Technical Military Academy of Bucharest, Romania
I. Stavarache	- National Institute of Materials Physics, Bucharest, Romania
T. Stoica	- National Institute of Materials Physics, Bucharest, Romania
A. Tulbure	- "1 Decembrie 1918" Univ. Alba Iulia, Romania
F. Udrea	- Cambridge Univ., UK
T. Visan	- Infineon Technologies Romania, Bucharest, Romania
C. Wang	- Heriot-Watt University, UK

EDITORS

Gh. Brezeanu	- "Politehnica" Univ. of Bucharest, Romania
M.L. Ciurea	- National Inst. of Material Physics, Bucharest, Romania
D. Cristea	- IMT Bucharest, Romania
M.A. Dinescu	- IMT Bucharest, Romania
D. Dobrescu	- "Politehnica" Univ. of Bucharest, Romania
M. Dragoman	- IMT Bucharest, Romania
A. Müller	- IMT Bucharest, Romania
R. Müller	- IMT Bucharest, Romania
D. Neculoiu	- IMT Bucharest, Romania

CONTENTS

PLENARY SESSIONS

LOW-DIMENSIONAL-STRUCTURE DEVICES FOR FUTURE ELECTRONICS, *S. Oda, T. Kawanago, H. Wakabayashi*, Tokyo Inst. of Technology, Japan 3

PROCESSING ISSUES IN SiC AND GaN POWER DEVICES TECHNOLOGY: THE CASES OF 4H-SiC PLANAR MOSFET AND RECESSED HYBRID GaN MISHEMT, *F. Roccaforte, G. Greco, P. Fiorenza*, Istituto per la Microelettronica e Microsistemi (CNR-IMM), Catania, Italy....................... 7

DFT CALCULATIONS OF STRUCTURE AND OPTICAL PROPERTIES IN WIDE BAND-GAP SEMICONDUCTOR CLUSTERS FOR DYE-SENSITIZED SOLAR CELLS, *C.I. Oprea[1], P. Panait[1,2], Reda M. AbdelAal, M.A. Gîrțu[1]*, [1]Ovidius Univ. of Constanța, [2]Univ. of Bucharest, Romania, [3]Suez Univ., Suez, Egypt........................ 17

ANALYTICAL MODELLING APPROACH IN STUDY OF THE TRANSIENT RESPONSE OF THERMOPILE-BASED MEMS SENSORS APPLIED FOR SIMULTANEOUS DETECTION OF PRESSURE AND GAS COMPOSITION, *D.V. Randjelović*, Univ. of Belgrade, Belgrade, Serbia........... 27

SECOND HARMONIC GENERATION FOR NON-DESTRUCTIVE CHARACTERIZATION OF DIELECTRIC-SEMICONDUCTOR INTERFACES, *I. Ionica[1], D. Damianos[1], A. Kaminski-Cachopo[1], D. Blanc-Pelissier[2], M. Lei[3], J. Changala[3], A. Bouchard[1], X. Mescot[1], M. Gri[1], G. Grosa[1], S. Cristoloveanu[1], G. Vitrant[1]*, [1]Univ. Grenoble Alpes, CNRS, [2]INSA de Lyon, France, [3]FemtoMetrix, USA........................ 35

DESIGN AUTOMATION FOR MICRO-ELECTRO-MECHANICAL SYSTEMS, *D. Kriebel, H. Schmidt, M. Schiebold, M. Freitag, B. Arnold, M. Naumann, J.E. Mehner*, Univ. of Technology Chemnitz, Germany 43

ON EFFECTIVE GRAPHENE BASED COMPUTING, *N.C. Laurenciu, S.D. Cotofana*, Delft Univ. of Technology, The Netherlands 51

Session N&N1: NANOSCIENCE AND NANOENGINEERING 1 – Oral presentations

ENHANCED PHOTOCONDUCTIVITY OF SiGe-TRILAYER STACK BY RETRENCHING ANNEALING CONDITIONS, *M.T. Sultan[1], J.T. Gudmundsson[2,3], A. Manolescu[1], M.L. Ciurea[4,5], C. Palade[4], A.V. Maraloiu[4], H.G. Svavarsson[1]*, [1]Reykjavik Univ., Iceland, [2]KTH-Royal Inst. of Tech., Sweden, [3]Univ. of Iceland, Iceland, [4]National Inst. for R&D in Material Physics, Bucharest, [5]Academy of Romanian Scientists, Bucharest, Romania........................ 61

FROM PENTACENE THIN FILM TRANSISTOR TO NANOSTRUCTURED MATERIALS SYNTHESIS FOR GREEN ORGANIC-TFT, *C. Ravariu, D.E. Mihaiescu, D. Istrati, M. Stanca*, "Politehnica" Univ. of Bucharest, Romania........................ 65

ELECTRICAL PROPERTIES OF AS-DEPOSITED ALD HfO_2 FILMS RELATED TO SILICON SURFACE STATE, *C. Cobianu, F. Nastase, N. Dumbravescu, O. Buiu, A. Albu, B. Serban, M. Danila, C. Romanitan, O. Ionescu*, CENASIC - IMT Bucharest, Romania........................ 69

ENHANCED PHOTOCURRENT IN GeSi NCs/TiO_2 MULTILAYERS, *C. Palade[1], A. Slav[1], O. Cojocaru[1], V.S. Teodorescu[1], S. Lazanu[1], T. Stoica[1], M.T. Sultan[3], H.G. Svavarsson[3], M.L. Ciurea[1,2]*, [1]National Inst. for R&D in Material Physics, Bucharest, [2]Academy of Romanian Scientists, Romania, [3]Reykjavik Univ., Iceland........................ 73

XI

Session N&N2: NANOSCIENCE AND NANOENGINEERING 2 – Oral presentations

TiO$_2$ – GRAPHENE OXIDE THIN FILMS OBTAINED BY SPRAY PYROLYSIS DEPOSITION, *I. Tismanar, L. Isac, A.C. Obreja[*], O. Buiu[*], A. Duta*, Transilvania Univ. of Brasov, [*]IMT Bucharest, Romania.. 79

EFFECT OF THE DEPOSITION CONDITIONS ON TITANIUM OXIDE THIN FILMS PROPERTIES, *M. Pustan, C. Birleanu, A. Trif, S. Garabagiu[*], D. Marconi[*], L. Barbu-Tudoran[*]*, Technical Univ. of Cluj-Napoca, [*]National Inst. for Research and Development of Isotopic and Molecular Technologies, Cluj-Napoca, Romania.. 83

MOS DOSIMETER BASED ON Ge NANOCRYSTALS IN HfO$_2$, *C. Palade, A. Slav, A.M. Lepadatu, I. Stavarache, I. Dascalescu, O. Cojocaru, T. Stoica, M.L. Ciurea, S. Lazanu*, National Inst. for R&D in Material Physics, Bucharest, Romania... 87

Session MW: MICROWAVE AND MILLIMETER WAVE CIRCUITS AND SYSTEMS – Oral presentations

INVESTIGATION OF LIQUID METAL AND FDM 3D PRINTED MICROWAVE DEVICES, *K.Y. Chan, X. Li, R. Ramer*, The Univ. of New South Wales, Sydney, Australia............................ 93

WAFER LEVEL PACKAGING OF GaN/Si SAW BAND PASS FILTERS WITH OPERATING FREQUENCIES ABOVE 5 GH, *A.-C. Bunea[1], D. Neculoiu[1,2], A. Dinescu[1]*, [1]IMT Bucharest, [2]"Politehnica" Univ. of Bucharest, Romania.. 97

METAL-INSULATOR TRANSITION IN MONOLAYER MoS2 FOR TUNABLE AND RECONFIGURABLE DEVICES, *M. Aldrigo, M. Dragoman, D. Masotti[*]*, IMT Bucharest, Romania, [*]Univ. of Bologna, Italy... 101

DESIGN ASPECTS AND EXPERIMENTAL RESULTS ON BROADBAND MONOPOLE DIELECTRIC RESONATOR ANTENNA, *S. Simion, S. Iordanescu[*]*, MTA – Bucharest, Romania, [*]IMT Bucharest, Romania... 105

PERMITTIVITY CHARACTERIZATION USING A DOUBLE-SIDED PARALLEL-STRIP LINE RESONATOR, *D.A. Nesic, Ivana Radnovic[*]*, Univ. of Belgrade, [*]Inst. IMTEL, New Belgrade, Serbia... 109

Session M&M: MICROSENSORS AND MICROSYSTEMS – Oral presentations

LOW-TEMPERATURE PACKAGING METHODS AS A KEY ENABLERS FOR MICROSYSTEMS ASSEMBLY AND INTEGRATION, *S. Stoukatch, F. Dupont, M. Kraft[*]*, Liege Univ., [*]Univ. of Leuven (KUL), Belgium.. 115

SENSING APPLICATIONS BASED ON CAVITY PERTURBATION METHOD – A PROOF OF CONCEPT, *V. Buiculescu, R. Rebigan*, IMT Bucharest, Romania 119

CONTINUOUS-WAVE MM-WAVE WAVE-GUIDE-BASED PROBE FOR SKIN TISSUE CHARACTERISATION, *K.Y. Chan, X. Li, Y. Fu, R. Ramer*, The Univ. of New South Wales, Sydney, Australia... 123

Session M: MODELLING – Oral presentations

MULTI-SCALE FINITE ELEMENT MODELING OF CNT-POLYMER-COMPOSITES, *M. Schiebold, J. Mehner*, Chemnitz Univ. of Technology, Chemnitz, Germany............................ 129

MODELING OF HIGH TOTAL IONIZING DOSE (TID) EFFECTS FOR ENCLOSED LAYOUT TRANSISTORS IN 65 NM BULK CMOS, *A. Nikolaou[1], M. Bucher[1], N. Makris[1], A. Papadopoulou[1], L. Chevas[1], G. Borghello[2,4], H.D. Koch[3,4], F. Faccio[4]*, [1]Technical Univ. of Crete, Greece, [2]DPIA, Univ. degli Studi di Udine, Italy, [3]SEMi, Univ. de Mons, Belgium, [4]EP Dept., CERN, Switzerland............... 133

ANALYTICAL ANALYSIS OF THE PLASMONIC ENHANCEMENT OF RESONANCE ENERGY TRANSFER IN THE VICINITY OF A SPHERICAL NANOPARTICLE, *T. Sandu, C. Tibeica, O.T. Nedelcu, M. Gologanu*, IMT Bucharest, Romania 137

Session SD: SEMICONDUCTOR DEVICES – *Oral presentations*

Ψ-MOSFET CONFIGURATION FOR DNA DETECTION, *L. Benea[1], M. Banu[2], M. Bawedin[1], C. Delacour[3], M. Simion[2], M. Kusko[2], S. Cristoloveanu[1], I. Ionica[1]*, [1]Univ. Grenoble Alpes, CNRS, France, [2]IMT Bucharest, Romania, [3]Néel Inst., CNRS, France... 143

INTERFACE TRAP EFFECTS IN THE DESIGN OF A 4H-SiC MOSFET FOR LOW VOLTAGE APPLICATIONS, *G. De Martino, F. Pezzimenti, F.G. Della Corte*, Univ. of Reggio Calabria, Italy...... 147

HIGH PILLAR DOPING CONCENTRATION FOR SiC SUPERJUNCTION IGBTs, *H. Kang, F. Udrea*, Univ. of Cambridge, U.K... 151

SURFACE RECOMBINATION EVALUATION IN BIPOLAR JUNCTION TRANSISTORS BY COMBINED ELECTRO-OPTICAL METHOD, *V. Banu, J. Montserrat[*], X. Jorda[*], P. Godignon[*]*, D+T Microelectronica A.I.E., [*]IMB-CNM, CSIC, Spain... 155

Session IC2: INTEGRATED CIRCUITS 2 – *Oral presentations*

LOW POWER AND LOW AREA CMOS CAPACITANCE MULTIPLIER, *G. Bonteanu, A. Cracan*, "Gh. Asachi" Technical Univ. of Iasi, Romania 161

REGULATION MECHANISM FOR DICKSON CHARGE PUMPS USING CHARGE RECYCLING AND ADIABATIC CHARGING, *F. Bîzîitu[1], L. Goraş[2,3]*, [1]Infineon Technologies Romania, [2]"Gh. Asachi" Tech. Univ. of Iaşi, [3]Romanian Academy, Iaşi Branch, Romania.................................. 165

APPLICATION SPECIFIC INTEGRATED CIRCUIT (ASIC) FOR AN ENERGY EFFICIENT IMPULSE RADIO ULTRA-WIDEBAND TRANSCEIVER. TESTING AND STATISTIC ASSESSMENT, *N. Varachiu, B. Benamrouche[*], J.-L. Noullet[*], A. Rumeau[*], D. Dragomirescu[*]*, IMT Bucharest, [*]LAAS-CNRS, Univ. de Toulouse, France... 169

WIDE DYNAMIC RANGE CURRENT MIRROR, *A. Cracan, G. Bonteanu*, "Gh. Asachi" Tech. Univ. of Iasi, Romania... 173

Session IC1-S: INTEGRATED CIRCUITS 1 – STUDENT PAPERS – *Oral presentations*

A HIGH PERFORMANCE MIXED-VOLTAGE DIGITAL OUTPUT BUFFER, *A.M. Dragan[1,2], A. Enache[1,2], A. Negut[1], A.M. Tache[1], G. Brezeanu[2]*, [1]ON Semiconductor Romania, Bucharest, [2]"Politehnica" Univ. of Bucharest, Romania.. 179

DUTY CYCLE ADJUSTMENT FOR THE LOW COST HIGH FREQUENCY CHARGE/DISCHARGE CMOS OSCILLATOR, *A.M. Antonescu, L. Dobrescu, D. Dobrescu*, "Politehnica" Univ. of Bucharest, Romania.. 183

RESISTOR BASED TEMPERATURE SENSOR USING ACTIVE INDUCTOR OSCILLATOR, *V.-S. Savinescu, I.-A. Nica, L. Goras[2]*, "Gh. Asachi" Technical Univ. Iasi, [*]Inst. for Information Tech. Iasi, Romania.. 187

LDO WITH A DUAL COMPLEMENTARY BUFFER ARCHITECTURE, *M. Dicianu, V. Ionescu[*], C. Dan*, "Politehnica" Univ. of Bucharest, [*]Infineon Technologies Romania SCS, Bucharest, Romania.. 191

XIII

Session: IC3-S: INTEGRATED CIRCUITS 3 – STUDENT PAPERS – Oral presentations

I/O BLOCKS RELIABILITY FOR AN SRAM-BASED FPGA WHEN EXPOSED TO IONIZING RADIATION, V.M. Placinta[1,2], I.N. Cojocariu[1], C. Ravariu[2], [1]"Horia Hulubei" National Inst. for R&D in Physics and Nuclear Engineering, [2]"Politehnica" Univ. of Bucharest, Romania............................ 197

FAULT IMPACT ASSESSMENT FOR AUTO-MOTIVE SMART POWER PRODUCTS IN AN ELECTRIC POWER STEERING APPLICATION, J. Stricker, C. Kain[*], A. Buzo[*], J. Kirscher[*], L. Maurer, G. Pelz[*], Bundeswehr Univ. München, [*]Infineon Technologies AG, Germany...................... 201

MESSAGE RECOVERED: A ROBUST FAULT DETECTION AND REPORTING METHOD FOR GALVANICALLY ISOLATED IGBT GATE DRIVERS, I. Hurez[1,2], T. Chen[3], F. Vlădoianu[2], V. Anghel[2], G. Brezeanu[1], [1]"Politehnica" Univ. of Bucharest, [2]ON Semiconductor Romania, Bucharest, Romania, [3]ON Semiconductor USA.. 205

COMPARISON OF LEVEL SHIFTER ARCHITECTURES: APPLICATION TO I/O CELL, R.-V. Petrica[1,2], M.-D. Dobre[1,2], P. Coll[1], F. Draghici[2], G. Brezeanu[2], [1]Microchip Technology Inc., [2]Politehnica Univ., Bucharest, Romania.. 209

Session D&IC-S: DEVICES & INTEGRATED CIRCUITS – STUDENT PAPERS – Oral presentations

POWER SUPPLY DUTY CYCLING FOR HIGHLY CONSTRAINED IOT DEVICES, A. Monti, E. Alata, A. Takacs, D. Dragomirescu, LAAS-CNRS, Univ. of Toulouse, France..................... 215

OVER-TEMPERATURE PROTECTION FOR A SWITCHED-CAPACITOR DC-DC CONVERTER WITH CONTROLLED CHARGING CURRENT, C.-S. Plesa, M. Neag, C.M. Boianceanu[*], Technical Univ. of Cluj-Napoca, [*]Infineon Technologies, Bucharest, Romania.......................... 219

INFLUENCE OF PLATINUM-HYDROGEN COMPLEXES ON SILICON P+/N-DIODE CHARAC-TERISTICS, J. Prohinig[1,2], F. Rasinger[1], H.-J. Schulze[3], G. Pobegen[1], [1]KAI Kompetenzzentrum Automobil- u. Industrieelektronik GmbH, [2]Graz Univ. of Technology, Austria, [3]Infineon Technologies AG, Germany.. 223

NUMERICAL SIMULATIONS OF RADIATION DAMAGE EFFECTS IN ACTIVE-EDGE SILICON PIXEL SENSORS FOR HIGH-ENERGY PHYSICS EXPERIMENTS, D. Djamai[1] E. Leonidas Gkougkousis[2], M. Chahdi[3], A. Lounis[4], S. Oussalah[5], [1]Univ. Abbes Laghrour Khenchela, Algeria, [1]Inst. de Fisica d'Altes Energies (IFAE), Barcelona, Spain, [3]Univ. de Batna, Algeria, [4]Univ. Paris-Sud XI, France, [5]Centre de Développement des Technologies Avancées, Algeria.............................. 227

INVESTIGATION OF CARBON INTERSTITIALS IN THE VICINITY OF THE SiO_2/4H-SiC(0001) INTERFACE, H. Alsnani, J.P. Goss, O. Al-Ani, S.H. Olsen, P.R. Briddon, M.J. Rayson, A.B. Horsfall[*], Newcastle Univ., [*]Durham Univ., UK... 231

Session N&MN: NANOSCIENCE; MICRO- AND NANOPHOTONICS Poster Session

GRAPHENE AND TiO_2 – PVDF NANOCOMPOSITES FOR POTENTIAL APPLICATIONS IN TRIBO-ELECTRONICS, P. Pascariu[1], I.V. Tudose[2], C. Pachiu[3], M. Danila[3], O. Ionescu[3], M. Popescu[3], E. Koudoumas[2], M. Suchea[2,3], [1]"Petru Poni" Inst. of Macromolecular Chemistry, Iaşi, Romania, [2]Technological Educational Inst. of Crete, Greece, [3]IMT Bucharest, Romania.. 237

KINETICS OF LANTHANUM AND YTTRIUM DOPED ZIRCONIA CRYSTALLIZATION BY X-RAY POWDER DIFFRACTION, D.V. Drăguţ, V. Bădiliţă, R.R. Piticesccu, A. Motoc, National R&D Inst. For Non-Ferrous and Rare Metals – INCDMNR-IMNR, Bucharest, Romania........................ 241

COMPARATIVE STUDY OF Sm And La DOPED ZnO properties, I.V. Tudose[1], P. Pascariu[2], C. Pachiu[3], F. Comanescu[3], M. Danila[3], R. Gavrila[3], E. Koudoumas[1], M. Suchea[1,3], [1]Technological Educational Inst. of Crete, Heraklion, Greece, [2]"Petru Poni" Inst. of Macromolecular Chemistry, Iasi, [3]IMT Bucharest, Romania.. 245

CARBON NANOTUBE/POLYANILINE COMPOSITE FILMS PREPARED BY HYDRO-THERMAL-ELECTROCHEMICAL METHOD FOR BIOSENSOR APPLICATIONS, *L.M. Cursaru (Popescu), A.G. Plaiasu*, C.M. Ducu*, R.M. Piticescu, I.A. Tudor*, National R&D Inst. for Non-Ferrous and Rare Metals, Bucharest, **Univ. of Pitesti, Romania.. 249

GeSi NANOCRYSTALS IN SiO$_2$ MATRIX WITH EXTENDED PHOTORESPONSE IN NEAR INFRARED, *I. Stavarache[1], L. Nedelcu[1] V.S. Teodorescu[1], V.A. Maraloiu[1], I. Dascalescu[1], M.L. Ciurea[1,2]*, [1]National Inst. for R&D in Material Physics, Bucharest, [2]Academy of Romanian Scientists, Romania.. 253

THE EFFECT OF H$_2$/Ar PLASMA TREATMENT OVER PHOTOCONDUCTIVITY OF SiGe NANOPARTICLES SANDWICHED BETWEEN SILICON OXIDE MATRIX, *M.T. Sultan[1], J.T. Gudmundsson[2,3], A. Manolescu[1], M.L. Ciurea[4,5], H.G. Svavarsson[1]*, [1]Reykjavik Univ., Iceland, [2]KTH-Royal Inst. of Technology, Sweden, [3]Science Inst., Univ. of Iceland, Iceland, [4] National Inst. for R&D in Material Physics, Bucharest, [5]Academy of Romanian Scientists, Romania............................,............ 257

DIRECT WRITING PATTERNS FOR GOLD THIN FILM WITH DPN TECHNIQUE, *M. Carp, C. Pachiu, V. Dediu*, IMT Bucharest, Romania... 261

SUBSTRATE EFFECT ON THE MORPHOLOGY AND OPTICAL PROPERTIES OF ZnO NANORODS LAYERS GROWN BY MICROWAVE-ASSISTED HIDROTHERMAL METHOD, *A. Filip, V. Musat, N. Tigau, A. Cantaragiu, C. Romanitan*, M. Purica**, "Dunărea de Jos" Univ. of Galati, *IMT Bucharest, Romania.. 265

RAMAN INVESTIGATION OF CRITICAL STEPS IN MONOLAYER GRAPHENE TRANSFER FORM COPPER SUBSTRATE TO OXIDIZED SILICON BY MEANS OF ELECTROCHEMICAL DELAMINATION, *F. Comanescu, A. Istrate, M. Purica*, IMT Bucharest, Romania...................... 269

ELECTRON TRANSFER AND DYE REGENERATION IN DYE-SENSITIZED SOLAR CELLS, *C.I. Oprea[1], A. Ndiaye[2], A. Trandafir[1,3], F. Cimpoesu[4], M.A. Gîrțu[1]*, [1]Ovidius Univ. of Constanța, Romania, [2]Univ. of Dakar, Senegal, [3]Univ. of Bath, UK, [4]Inst. for Physical Chemistry, Romania..................... 273

Session SD&Ms: SEMICONDUCTOR DEVICES AND MICROSYSTEMS – Poster Session

EFFICIENCY AND TOTAL HARMONIC DISTORSION IN COMPOSITE RIGHT-/LEFT-HANDED DISTRIBUTED OSCILLATORS, *G. Bartolucci, L. Scucchia, S. Simion**, Univ. of Roma Tor Vergata, Italy, *MTA – Bucharest, Romania.. 279

EFFECT OF DEGENERATION ON A MILLIMETER WAVE LNA: APPLICATION OF MICROSTRIP TRANSMISSION LINES, *M. Fanoro, S.S. Olokede, S. Sinha*, Univ. of Johannesburg, South Africa.. 283

MILLIMETER WAVE AND TERAHERTZ INVESTIGATIONS ON SOME DIELECTRIC MATERIALS, *M.G. Banciu[1], T. Furuya[2], D.C. Geambasu[1], L. Nedelcu[1], D. Pantelica[3], M.-D. Dracea[3], P. Ionescu[3], A. Iuga[1], C. Chirila[1], L. Hrib[1], L. Trupina[1], M. Tani[2]*, [1]National Inst. for R&D in Material Physics, Bucharest, Romania, [2]Fukui Univ., Japan, [3]"Horia Hulubei" National Inst. for R&D in Physics and Nuclear Engineering, Bucharest, Romania.. 287

METHODS FOR ART PRESERVATION AND RESTAURATION. IDENTIFICATION OF PARAMETERS FOR POTENTIAL MONITORING THE TEMPORAL EVOLUTION OF PUTTIES, *I.-M. Giura[1], C. Pachiu[2], M. Popescu[2], B. Bita[2], O.N. Ionescu[2], M. Suchea[2,3]*, [1]Romanian Patriarchy, Bucharest, [2]IMT Bucharest, Romania, [3]Technological Educational Inst. of Crete, Greece.................... 291

TEMPERATURE MEASUREMENTS WITH FOUR-RESISTOR SENSOR PATTERNED ON GOLDEN LAYER, *M. Sarajlić, M. Frantlović, P. Poljak, K. Radulović, D.V. Radović*, Univ. of Belgrade, Serbia... 295

MANUFACTURE AND INVESTIGATION OF A VERTICAL MEMS SWITCH, *A. Baracu, R. Müller, R.C. Voicu, M. Pustan*, C. Birleanu*, A. Dinescu*, IMT Bucharest, *Technical Univ. of Cluj-Napoca, Romania... 299

XV

THE GATE CURRENT IN MOSFETs VERSUS PLANAR-NOI DEVICES, *C. Ravariu[1], E. Manea[2], C. Parvulescu[2], F. Babarada[1], A. Popescu[2], A. Srinivasulu[3]*, [1]"Politehnica" Univ. of Bucharest, [2]IMT Bucharest, Romania, [3]JECRC Univ., India... 303

IMPROVED Ti/Pt/Au – N-TYPE Si CONTACTS BY POST-METALLIZATION ANNEALING IN NITROGEN ATMOSPHERE, *R. Pascu, M. Danila, P. Varasteanu, M. Kusko, G. Pristavu, G. Brezeanu[*], F. Draghici[*]*, IMT Bucharest, [*]"Politehnica" Univ. of Bucharest, Romania 307

COMPARATIVE STUDY OF THE ELECTRICAL PROPERTIES OF CZTS-TiO$_2$ AND CZTS-ZnO HETEROJUNCTIONS FOR PV APPLICATIONS, *M. Covei, C. Bogatu, D. Perniu, S. Cisse[*], A. Duta*, Transilvania Univ. of Brasov, Romania, [*]Cheikh Anta Diop Univ. of Dakar, Sénégal...................... 311

THREE PHASE SYNCHRONOUS BOOST RECTIFIER, *V. Trifa, Gh. Brezeanu, E. Ceuca[*]*, "Politehnica" Univ. of Bucharest, [*]"1 December 1918" Univ. of Alba Iulia, Romania....................... 315

Workshop "Microsystems for Energy Harvesting and Environment Monitoring", organized in connection with the projects "PiezoMEMS", "WaterSafe" and "PiezoHARV"

POWER HARVESTING AND STORAGE CIRCUIT FOR A DOUBLE ARRAY OF LEAD-FREE PIEZOELECTRIC CANTILEVERS, *G. Muscalu[1,2], B. Firtat[1,2], S. Dinulescu[1], C. Moldovan[1], A. Anghelescu[1], I. Stan[3]*, [1]IMT Bucharest, [2]"Politehnica" Univ. of Bucharest, [3]ROMELGEN SRL, Bucharest, Romania... 321

DESIGN AND SIMULATION OF PIEZOELECTRIC ENERGY HARVESTER FOR AEROSPACE APPLICATIONS, *G. Muscalu[1,2], B. Firtat[1,2], S. Dinulescu[1], C. Moldovan[1], A. Anghelescu[1], C. Vasile[3,2], D. Ciobotaru[3], C. Hutanu[3]*, [1]IMT Bucharest, [2]"Politehnica" Univ. of Bucharest, [3]Advanced Technologies Inst., Romania.. 325

ELECTROCHEMICAL SENSORS FOR DETECTION OF DIFFERENT IONIC SPECIES (NITRITES/NITRATES AND HEAVY METALS) IN NATURAL WATER SOURCES, *M. Gartner[1], C. Lete[1], M. Chelu[1], H. Stroescu[1], M. Zaharescu[1], C. Moldovan[2], C. Brasoveanu[2], M. Gheorghe[3], S. Gheorghe[3], A. Duta[4], Z. Labadi[5], B. Kalas[5], A. Saftics[5], M. Fried[5,6], P. Petrik[5], E. Tóth[7], H. Jankovics[7], F. Vonderviszt[7]*, [1]"Ilie Murgulescu" Inst. of Physical Chemistry, [2]IMT Bucharest, [3]NANOM MEMS SRL, Rasnov, [4]"Transilvania" Univ. of Braşov, Romania, [5]Inst. for Technical Physics and Materials Science, Hungary, [6]Óbuda Univ., [7]Univ. of Pannonia, Hungary.................................. 329

CAS 2018 PROCEEDINGS

2018 INTERNATIONAL SEMICONDUCTOR CONFERENCE

October 10-12, Sinaia, ROMANIA

NATIONAL INSTITUTE FOR RESEARCH AND DEVELOPMENT IN MICROTECHNOLOGIES - IMT Bucharest

co-sponsored by
IEEE – Electron Devices Society

Ministry of Research and Innovation

Session PS

PLENARY SESSIONS

978-1-5386-4483-6/18 $31.00 © 2018 IEEE

Low-Dimensional-Structure Devices for Future Electronics*

Shunri Oda, Takamasa Kawanago, Hitoshi Wakabayashi
Tokyo Institute of Technology
soda@pe.titech.ac.jp

Abstract— Recent progress of nanotechnology has made possible observations of unique characteristic of nano-structure which are not possible in bulk semiconductors. In this talk, novel properties and possible device applications of quantum dots (0D), nanowires (1D) and atomic layer (2D) devices are discussed.

Keywords— quantum dots; nanowires; 2D materials; zero-power devices; qubit

1. Introduction

One of the major application targets for future electronics is zero-power (without the need of recharging battery) wearable communication tools [1]. Combination of low-power consumption devices and energy harvesting devices are necessary. Tunnel field-effect-transistors (TFET) are promising since extremely low-voltage operation of switching beyond the limitation of CMOS devices would be possible [2]. 2D materials and 1D nanowires attract attention not only because these materials would be suitable for the fabrication of TFETs, but also various novel application such as sensors, displays would be possible [3].

On the other hand, quantum computing is no longer a future technology. Recent advances in D-Wave computers based on quantum annealing [4] and superconducting devices, and the demonstration of long spin decoherence times in isotopically-enriched Si qubits [5], have accelerated the research and development of this technology. The remaining challenge is large scale integration of qubits. Physically-defined coupled quantum dots (QDs) on silicon-on-insulator substrates represent potential multiple scaled qubits.

*Partially supported by JST-CREST and JSPS KAKENHI Grant No. 26249048.

In this paper, we discuss recent progress of quantum dot (0D), nanowire (1D) and 2D atomic-layer material devices

2. Coupled Quantum Dots for Si Qubits

A. Device Fabrication

Physically-defined double quantum dots connected to source/drain electrodes, five side gate electrode to control the number of electrons in QDs and interaction between QDs and electrodes, are fabricated on SOI (silicon-on-insulator) substrates by electron-beam lithography and reactive ion etching [6].

B. Pauli Spin Blockade

Electron transport was measured at 250 mK. Current counter plot as functions of two control gate electrodes showed honeycomb-like structure, which suggests quantum interaction between two quantum dots. We also observed current rectification in coupled QDs due to Pauli spin blockade, which is a valuable tool for the initialization and readout of spin states during the operation of spin qubits [6, 7].

C. A Few Electron Regime

To implement quantum logic gates based on electron spin, it is necessary to reduce the number of electrons in individual QDs to only a few or even to a single electron, so as to create spin states that are energetically well defined and separate from other states. We prepared integrated charge sensors (CSs) composed of QDs in order to determine the absolute number of electrons in the individual QDs.

We have successfully determined the number of electrons in each QD and controlled a few electron regimes in coupled QDs as shown in Fig. 1; essential steps for qubit operation [7,8].

Recently, single-qubit gate fidelities exceeding 99.9% have been observed on an electron spin confined in a ^{28}Si/SiGe quantum dot [9]. This is quite promising for the realization of fault-tolerant universal quantum computation.

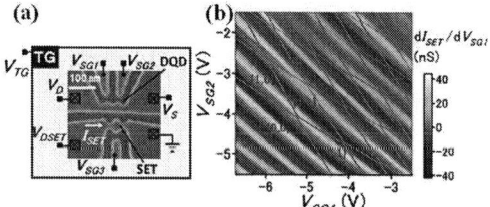

Fig. 1 (a) SEM image of a double-quantum dot (DQD) device and a schematic of the measurement setup. The DQD device was fabricated by the same process used to make the single QD device and measurements were performed at a base temperature of 300 mK. (b) Plots of the SET transconductance, dI_{SET}/dV_{SG1}, as functions of V_{SG1} and V_{SG2} in a DQD; V_{TG} = 5.8 V and V_{SG3} = 0 V, $V_D = V_S$ = 0.9 V, and V_{DSET} = 3 mV. A charge stability diagram of the few electron regimes in the DQD is clearly obtained. [7]

3. Ge/Si Core/Shell Nanowires

For the implementation of zero-power devices, energy harvesting systems from light, thermal, vibration, and RF energy are investigated. We propose thermoelectric devices using the temperature deference between human skin and ambient based on Ge/Si core/shell nanowires are promising for wearable power generators.

The figure of merit ZT of the thermoelectric system is described as

$$ZT = \frac{\sigma S^2}{\kappa} T \qquad (1)$$

where κ is the thermal conductivity, σ is the electrical conductivity, S is the Seebeck coefficient, and T is the absolute temperature.

Because of the unique density of states distribution of one-dimensionally quantized structure of nanowires, high Seebeck coefficient is expected. Since Ge core region is confined by Si potential well, high hole concentration is obtained even in undoped Ge, that results in high electrical conductivity. Surface phonon scattering of Si shell region suppresses thermal conductivity. All these effects result in large value of the figure of merit for thermoelectric energy conversion.

We have prepared high crystalline quality and narrow Si and Ge nanowires by vapor-liquid-solid chemical vapor deposition [10-12]. We have also fabricated Ge/Si core/shell nanowires. The problem we encountered was the formation of branch structure due to migrated Au particles on the surface of Ge nanowires. We solved the problem by suppressing Au migration using two step growth methods [13]. We have obtained very high quality interface between Ge core and Si shell, as shown in Fig. 2, and measured electron transport in core/shell structures [14]. We also prepared thermoelectric devices based on Ge/Si core/shell nanowires. The preliminary measurements show very promising properties for energy harvesting devices for zero-power wearable devices.

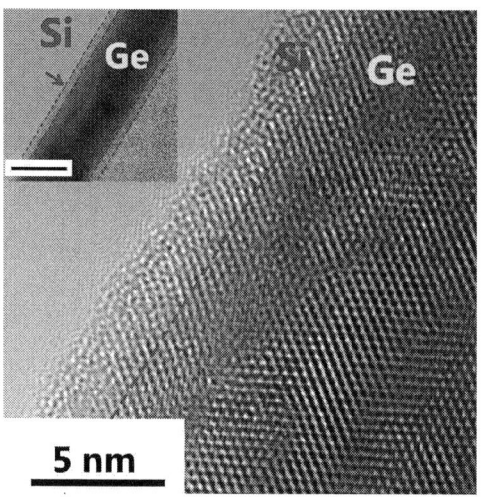

Fig. 2. TEM image of a Ge/Si core/shell nanowire. The inset shows the whole Ge/Si NW with a 20 nm scale bar. [14]

978-1-5386-4483-6/18 $31.00 © 2018 IEEE

4. Two-Dimensional Atomic Layer Transistors

Two-dimensional atomic layer materials are promising for wearable device application, because of high-electric and thermal conductivity, low-power consumption, flexibility and transparency [15]. Particularly, MoS_2 transistors are promising, because of low interface defect density due to free of dangling bonds, scalability to sub-5-nm region due to ultra-thin uniform layer and relatively large bandgap [16, 17].

A self-assembled monolayer (SAM), an organic molecular film that spontaneously forms on the surface of a substrate exactly one molecule thick, has ideal characteristics for gate dielectrics in FETs because of insulating properties at a very thin layer and easy fabrication process. We applied n-octadecylphosphonic acid (ODPA) and AlOx dual layer as gate dielectrics for MoS_2 transistors. The electrical properties are excellent with very steep subthreshold slope of 69 mV/dec and no hysteresis, which means free of interface defects [18]. A SAM layer is also used as a self-alignment mask to simplify the device fabrication process as shown in Fig. 3 [19, 20]. Adhesion lithography using SAM has been applied to various material systems including organic semiconductors [21]. Normally-off characteristics are achieved by controlling threshold voltage using proper gate metals [22].

Wafer scale MoS_2 films are commonly prepared by chemical vapor deposition. However, sometimes triangular structure deteriorates the uniformness of CVD grown films. In order to circumvent this problem, we deposited MoS_2 films by RF sputtering [23]. We clarified that surface flatness of the substrate is essential for high quality films [24]. To reduce the number of sulfur defects, annealing in the forming gas [25], H_2S [26] and sulfur vapor [27] was effective and low carrier density MoS_2 films were obtained. These processing technologies are important steps towards future high-performance wearable devices.

5. Conclusion

Low-dimensional materials are emerging. Novel properties, self-assembled fabrication are promising for next generation wearable devices and new architecture computations.

Acknowledgments. The authors thank Tetsuo Kodera, Tomohiro Noguchi, Marolop Simanullang and Kousuke Horibe for discussion and performing experiments.

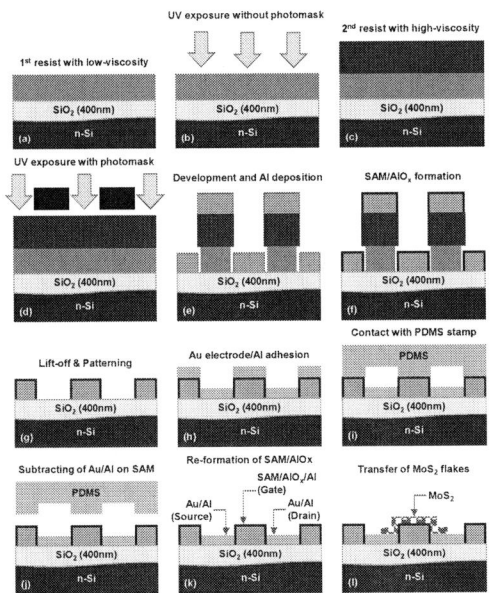

Fig. 3. Process flow for fabrication of MoS_2 transistors using SAM-based adhesion lithography. [19]

References

[1] A. M. Ionescu, *"Energy efficient computing and sensing in the zettabyte era: from silicon to the cloud,"* IEDM 2017, 1.2.

[2] A. M. Ionescu and H. Riel, *"Tunnel field-effect transistors as energy-efficient electronic switches,"* Nature, **479**, pp. 329-337, 2011.

[3] D. Sarkar et al., *"A Subthermionic Tunnel Field-Effect Transistor with an Atomically Thin Channel,"* Nature, **526**, pp. 91-95, 2015.

[4] M.W. Johnson et al., *"Quantum annealing with manufactured spins,"* Nature, **473**, 194-198 2011.

[5] M. Veldhorst et al., *"An addressable quantum dot qubit with fault-tolerant control-fidelity,"* Nature Nanotech. **9**, 981, 2014.

[6] G. Yamahata, T. Kodera, H. O. H. Churchill, K. Uchida, C. M. Marcus, and S. Oda, *"Magnetic field dependence of Pauli spin blockade: A window into the sources of spin relaxation in silicon quantum dots,"* Phys. Rev. B, **86**, 115322, 2012.

[7] S. Oda, G. Yamahata, K. Horibe and T. Kodera, *"Coupled quantum dots on SOI as highly integrated Si qubits,"* IEDM 2016, 13.3.

[8] K. Horibe, T. Kodera and S. Oda, *"Lithographically-defined few-electron silicon quantum dots based on a silicon-on-insulator substrate,"* Appl. Phys. Lett. **106**, 083111, 2015.

[9] J. Yoneda et al., *"A quantum-dot spin qubit with coherence limited by charge noise and fidelity higher than 99.9%,"* Nature Nanotechnology, **13**, pp. 102-106, 2017.

[10] S. Akhtar, K. Usami, Y. Tsuchiya, H. Mizuta, and S. Oda, *"Vapor-liquid-solid growth of small- and uniform-diameter silicon nanowires at low temperature from Si_2H_6",* Appl. Phys. Express, **1**, 014003, 2008.

[11] C. B. Li, K. Usami, T. Muraki, H. Mizuta, and S. Oda, *"The impacts of surface conditions on the vapor-liquid-solid growth of germanium nanowires on Si (100) substrates,"* Appl. Phys. Lett. **93**, 041917, 2008.

[12] M. Simanullang, K. Usami, T. Kodera, K. Uchida, and S. Oda, *"Germanium Nanowires with 3-nm-Diameter Prepared by Low Temperature Vapour–Liquid–Solid Chemical Vapour Deposition,"* J. Nanosci. Nanotechnol. **11**, 8163, 2011.

[13] T. Noguchi, Marolop Simanullang, Z. Y. Xu, K. Usami, T. Kodera and S. Oda, *"Synthesis of Ge/Si core/shell nanowires with suppression of branch formation,"* Appl. Phys. Express, **9**, 055504, 2016.

[14] T. Noguchi et al., *"Ge/Si core/shell nanowires with controlled low temperature grown Si shell thickness,"* Phys. Status Solidi A, **212**, 1578, 2015.

[15] P. Ajayan, P. Kim, and K. Banerjee, *"Two – dimensional van der Waals materials,"* Physics Today, **69**, 38-44, 2016

[16] W. Cao, J. Kang, D. Sarkar, W. Liu, and K. Banerjee, *"2D semiconductor FETs - Projections and design for sub-10 nm VLSI,"* IEEE Trans. Electron Dev., **62**, pp. 3459–3469, Nov. 2015.

[17] A. Pal, W. Cao, J. Kang, and K. Banerjee, *"How to Derive the Highest Mobility from 2D FETs – A First-Principle Study,"* IEDM 2017, 31.3.

[18] T. Kawanago and S. Oda, *"Utilizing self-assembled-monolayer-based gate dielectrics to fabricate molybdenum disulfide field-effect transistors,"* Applied Physics Letters, **108**, 041605, 2016.

[19] T. Kawanago, R. Ikoma, W.J. Du and S. Oda, *"Adhesion lithography to fabricate MoS_2 FETs with self-assembled monolayer-based gate dielectrics,"* IEEE ESSDERC 2016, pp. 251-254.

[20] W.J. Du, T. Kawanago and S. Oda, *"Use of self-assembled monolayers for selective metal removal and ultrathin gate dielectrics in MoS_2 field-effect transistors,"* Japanese Journal of Applied Physics, **56**, 04CP10, 2017.

[21] J. Semple et al., *"Large-area plastic nanogap electronics enabled by adhesion lithography,"* npj Flexible Electronics (online), **2**, 18, 2018.

[22] T. Kawanago and S. Oda, *"Control of threshold voltage by gate metal electrode in molybdenum disulfide field effect transistors,"* Applied Physics Letters, **110**, 133507, 2017.

[23] T. Ohashi, et al., *"Multi-layered MoS_2 film formed by high-temperature sputtering for enhancement-mode nMOSFETs,"* Japanese Journal of Applied Physics. **54**, 04DN08, 2015.

[24] T. Ohashi et al., *"Quantitative relationship between sputter-deposited- MoS_2 properties and underlying-SiO_2 surface roughness,"* Appl. Phys. Express, **10**, 041202, 2017.

[25] J. Shimizu et al., *"High-mobility and low-carrier-density sputtered MoS_2 film formed by introducing residual sulfur during low-temperature in 3%-H_2 annealing for three-dimensional ICs,"* Japanese Journal of Applied Physics, **56**, 04CP06, 2017.

[26] J. Shimizu et al., *"Low-carrier density sputtered-MoS_2 film by H_2S annealing for normally-off accumulation-mode FET,"* Electron Devices Technology and Manufacturing Conference (EDTM) 2017, P-22.

[27] K. Matsuura et al., *"Low-carrier-density sputtered MoS_2 film by vapor-phase sulfurization,"* Journal of Electronic Materials, **47**, pp. 3947 - 3501, 2018.

Processing issues in SiC and GaN power devices technology: the cases of 4H-SiC planar MOSFET and recessed hybrid GaN MISHEMT

F. Roccaforte *, G. Greco, P. Fiorenza

* Consiglio Nazionale delle Ricerche – Istituto per la Microelettronica e Microsistemi (CNR-IMM).
Strada VIII, n. 5 – Zona Industriale, 95121 Catania - Italy
E-mail: fabrizio.roccaforte@imm.cnr.it

Abstract—This paper aims to give a short overview on some relevant processing issues existing in SiC and GaN power devices technology. The main focus is put on the importance of the channel mobility in transistors, which is one of the keys to reduce R_{ON} and power dissipation. Specifically, in the case of the 4H-SiC planar MOSFETs the most common solutions and recent trends to improve the channel mobility are presented. In the case of GaN, the viable routes to achieve normally-off HEMTs operation are briefly introduced, giving emphasis to the case of the recessed hybrid MISHEMT.

Keywords—wide band gap semiconductors, SiC, GaN.

1. Introduction

The worldwide increasing need of electric energy is a serious concern in our society. In fact, the energy consumption in the world is estimated to increase of 40% in the next two decades [1] and the largest fraction (up to 60%) of the consumed energy will be electric energy. Hence, energy efficiency has become a challenge in modern semiconductor power devices technologies, to ultimately reduce the global energy consumption.

Currently, power electronics market is almost entirely based on Silicon (Si) devices [2]. However, Si-based power electronics has reached its performance limits, in terms of maximum power levels, frequency and operation temperatures. Hence, the only way to overcome the physical limits of Si is a radical innovation of the technology for discrete semiconductor power devices.

In this context, due to their excellent physical properties [3], the most popular wide band gap (WBG) semiconductors, silicon carbide (4H-SiC) and gallium nitride (GaN), are considered the best materials to replace Si in the future high efficient power electronics. *Fig. 1* shows a graphical comparison of some relevant physical properties of Si, SiC and GaN. As can be seen, the large values of energy gap and critical electric field allow these materials to operate at high breakdown voltages (B_V). The high saturated electron velocity enables superior performances under high frequency operation. Finally, the high thermal conductivity (in the case of SiC) is an important feature that guarantees an easy heat dissipation for operation at high temperature and high current levels.

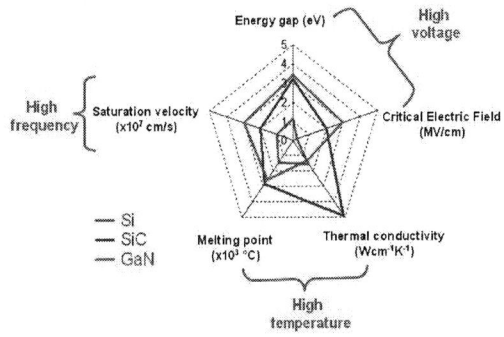

Fig. 1 Comparison of Si, SiC and GaN relevant properties for power devices applications.

These outstanding properties of SiC and GaN enable to design transistors with a smaller ON-resistance (R_{ON}) and smaller parasitic capacitances with respect to the Si counterparts for a fixed targeted maximum operation voltage. The direct impact of a lower R_{ON} is a reduction of the total power dissipation [4]. Hence, SiC and GaN devices can find several applications in power electronics in many important fields. To visualize the huge potential of these materials, *Fig. 2* depicts the major applications of WBG power devices in a power versus voltage chart. As can be seen the possible application areas enter our daily life, e.g., consumer electronics (PFC/power supply, audio amplifiers,…), EV/HEV automotive components (converters, battery chargers, ….), industrial applications (motor drives,..), renewable energies (PV-inverters,…), transportations, etc.

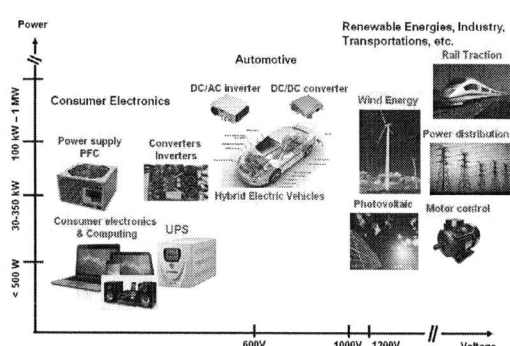

Fig. 2 Main application areas of SiC and GaN power devices. .

Today, while several 4H-SiC and GaN transistors with excellent performances have already reached the market, there are still some important physical problems related to the fabrication processes of these devices, which are still object of intensive investigation by the scientific community.

This paper aims to give a brief overview on some current processing issues encountered in SiC and GaN power devices, with a focus on transistors technology. In particular, the most common approaches to improve the MOS interface quality in 4H-SiC planar MOSFETs are presented, highlighting their advantages and limitations. Moreover, the feasible solutions to achieve normally-off operation in GaN HEMTs are presented, with special attention to the case of the recessed hybrid MISHEMT.

2. 4H-SiC MOSFET

One of the long standing problems in 4H-SiC planar MOSFETs technology is the low inversion channel mobility, especially below 1 kV, i.e., where the channel mobility can represent an important contribution to the total R_{ON}. This latter can be clearly seen in ***Fig. 3*** , reporting the specific R_{ON} as a function of the breakdown voltage B_V for different values of the inversion layer channel mobility [5].

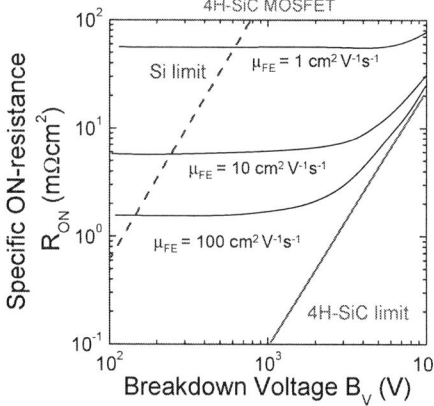

Fig. 3 Specific ON-resistance R_{ON} versus breakdown voltage B_V for 4H-SiC MOSFETs, estimated for different values of the channel mobility μ_{FE}.

The problem of the channel mobility in 4H-SiC MOSFETs has been recently reviewed by *Cabello et al.* [6]. In general, low values of the channel mobility (typically < 5-10 cm^{2}V^{-1}s^{-1}) are obtained with thermal SiO$_2$ gates, due to the high density of interface traps (D_{it}) near the conduction band edge [7,8], determining Coulombic scattering effects by charges trapped at the interface states and inside the oxide [9,10]. Hence, post deposition annealing or innovative gate oxide processes are mandatory to increase the channel mobility and decrease the $R_{ON.}$

Fig. 4 reports the values of the field effect mobility μ_{FE} of 4H-SiC planar MOSFETs for different treatments of the gate oxide. For a direct comparison of the data, the mobility curves are reported as a function of the difference between the gate voltage and the threshold voltage (V_g-V_{th}). The mobility curve μ_{FE} of an "untreated" dry oxide is also reported as a reference (< 5 cm^{2}V^{-1}s^{-1}). As can be seen, the experimental μ_{FE} mobility curves versus gate voltage typically exhibit a maximum (peak mobility) after the threshold voltage V_{th} is reached. Then, the channel mobility slightly decreases with increasing the gate voltage (i.e., with increasing the transversal electric field) due to the dominance of phonon and interface scattering mechanisms [11].

To improve the channel mobility, nitridation processes of the gate oxides, i.e., post-oxidation-annealing (POA) or post-deposition-annealing (PDA) in nitrogen-rich atmospheres (NO or N_2O) in the temperature range 1000-1300 °C, have been introduced at the end of the 90's [12,13,14,15].

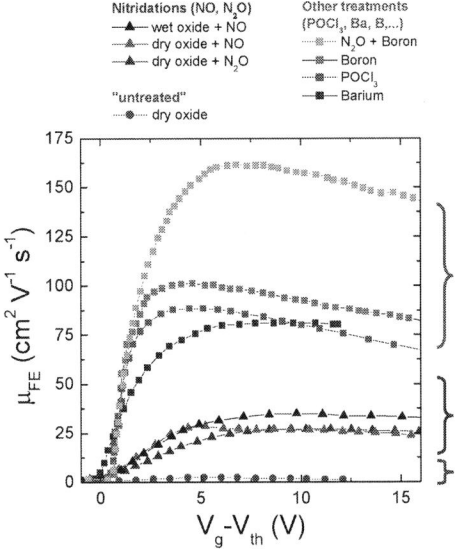

Fig. 4 Field effect mobility (μ_{FE}) as a function of the difference between the gate voltage and the threshold voltage (V_g-V_{th}) in 4H-SiC planar MOSFETs fabricated employing different gate oxides treatments. The data are from Ref. [2] and references therein.

The improvement of the mobility (up to 25-50 $cm^2V^{-1}s^{-1}$) obtained upon nitridation is typically accompanied by a reduction of the interface state density D_{it} down to the low 10^{12} $eV^{-1}cm^{-2}$ range.

As an alternative to the nitridations, the introduction of different species in the gate oxide has been considered to passivate the SiO_2/SiC interface states and increase the mobility μ_{FE}. *Okamoto et al.* [16] demonstrated that an annealing of the gate oxide in phosphoryl chloride ($POCl_3$) can significantly increase the 4H-SiC MOSFETs mobility (89 $cm^2V^{-1}s^{-1}$). Later, other authors

explored similar phosphorous-based passivation routes, obtaining mobility values higher than 100 $cm^2V^{-1}s^{-1}$, with D_{it} in the 10^{11} $cm^{-2}eV^{-1}$ range [17,18,19].

During nitridation (N_2O or NO) or $POCl_3$ processes, the presence of n-type dopant (i.e., nitrogen and phosphorous) in the annealing atmosphere determines notable electrical changes in the SiO_2/SiC interface. In fact, nitrogen and phosphorous atoms can be incorporated in the SiC substrate during annealing, and act as n-type shallow donors in the material [20,21]. Using scanning probe microscopy analyses at the SiO_2/SiC interface allowed to demonstrate the "counter doping effect" of the p-type implanted regions in the MOSFET channel [17,22]. These measurements also showed a higher electrically active phosphorous incorporation in $POCl_3$ with respect to the active nitrogen incorporated in N_2O [22].

In spite of the high channel mobility, the drawback of the $POCl_3$ annealing is the poor reliability of the gate oxides, caused by the large amount of charge traps in the SiO_2 network after a phosphorous incorporation [23]. Some research groups proposed other phosphorous-based processes ($POCl_3$ pre-annealings before oxide deposition, combination of N- and P-based annealings, P-ion-implantation), with promising results in terms of mobility and improvement of the V_{th} stability [18,19,24,25,26]. More recently, channel mobility values > 100 $cm^2V^{-1}s^{-1}$ have been obtained using other group-V elements (e.g., As, Sb), in conjunction with nitric oxide (NO) post-oxidation annealing [27]. However, the μ_{FE} curves of As- or Sb-doped 4H-MOSFETs channels exhibit pronounced maxima at low electric fields, but decrease rapiidly at high fields (e.g.>10 V). Hence, As- or Sb-counter-doping appears of limited effectiveness in real devices [6].

Another recent approach to increase the 4H-SiC MOSFET mobility is the use of Boron (B). *Okamoto et al.* [28] achieved a mobility of about 100 $cm^2V^{-1}s^{-1}$ using Boron thermal diffusion (by a planar BN diffusion source) into a dry oxide. Since B is an acceptor for SiC, "counter doping" does not occur and cannot explain the increased mobility. Hence, these results were attributed to a stress relaxation of the interface by the incorporation of B-atoms in the SiO_2 matrix [29]. This process was recently optimized, by combining the N_2O oxinitridation with B-diffusion [30,31]. In this way, a peak mobility of $160 \ cm^2V^{-1}s^{-1}$ has been obtained, while a stable threshold voltage V_{th} at least at room temperature [6].

Finally, the use of alkali or alkaline earth elements (Rb, Cs, Sr, Ba,...) has been proposed to passivate the SiO_2/4H-SiC interface states and increase the 4H-SiC MOSFET mobility. These processes typically consist in the deposition of a thin layer of alkali/alkaline-earth material on SiC, followed by the deposition and post-annealing (in O_2 or O_2/N_2 ambient)) of SiO_2 gate oxide. Among various elements the most promising results were achieved with Sr and Ba, with mobility values of μ_{FE} up to 65 and $110 \ cm^2V^{-1}s^{-1}$, respectively [32,33,34]. It has been also shown that Ba incorporation allows to obtain a threshold voltage stability under stress at 175 °C and 2 MV/cm gate bias. The beneficial role of Ba was explained in term of interface stress release using transmission electron microscopy analysis. In particular, the tensile strain of the SiC region close to the SiO_2/SiC interface is released in the presence of an oxidized Ba interlayer. Such an "unstrained" interface is the key factor for the increase of the channel mobility [35,36].

Despite the significant improvements of the channel mobility achievable with the aforementioned approaches, most of these processes are still far to be employed in "real" devices, since they are affected by threshold voltage V_{th} instability issues. Hence, nitridation of the gate oxide remains the process of choice in the fabrication of state-of-the-art 4H-SiC MOSFETs.

3. Normally-OFF GaN HEMTs

In principle, due to its higher critical electric field (*Fig. 1*) one may expect from GaN a better high voltage operation behavior than SiC. However, a large density of defects is still present in GaN-based materials, which hinders to reach the theoretical electric field strength. Moreover, the lack of high quality large diameter bulk GaN substrates does not allow the realization of power devices with vertical architectures, as needed for a high breakdown voltages at low R_{ON}. Consequently, lateral heterojunction devices are nowadays the preferred solution to fabricate GaN-based transistors. In particular, GaN high electron mobility transistors (HEMTs) are normally-ON devices, due to the presence of the two dimensional electron gas (2DEG) in AlGaN/GaN heterostructures. However, power electronics applications typically require normally-OFF devices, to guarantee fail-safe operation and gate drivers simplicity [37,38,39]. Hence, significant efforts have been devoted in the last decade to develop physical methods to control the 2DEG in the channel and obtain HEMT with a positive threshold voltage V_{th}.

The use of a p-GaN gate is currently the only commercial solution for normally-OFF GaN HEMTs [40]. *Greco et al.* [41] recently summarized in a review the most relevant processing issues in normally-OFF HEMTs with the p-GaN gate approach. Hence, this layout will be not subject of discussion in the present paper.

Another promising approach consists in the complete removal of the AlGaN barrier under the gate [42,43], creating a metal insulator semiconductor (MIS) recessed-gate hybrid HEMT (MISHEMT). The recessed-gate hybrid MISHEMT enables to have a positive threshold voltage V_{th} of the MIS channel, preserving a low on resistance R_{ON} in the access regions. The most important part of such a device is the recessed channel, in which the carriers mobility is influenced by several factors (roughness of the etched surface, defects, quality of the gate insulator, etc). Hence, characterizing the properties of

978-1-5386-4483-6/18 $31.00 © 2018 IEEE

insulator/GaN interface and understanding the mechanisms limiting the channel mobility are key aspects for the progress of the recessed-gate MISHEMTs technology.

Various dielectric materials have been proposed to fabricate recessed-gate normally-OFF hybrid GaN MISHEMTs (SiO_2, SiN, Al_2O_3, AlN/SiN....) [44,45]. As in the case of standard MOSFET, the field effect mobility μ_{FE} is an important parameter that must be optimized in order to reduce the total device R_{ON} [2].

Similarly to the case of a MOSFET, also in the MISHEMT the field effect mobility μ_{FE} increases with the gate bias V_g up to a maximum $\mu_{FE(peak)}$ and then decreases at high electric fields.

As can be seen in **Table 1** the values of peak mobility $\mu_{FE(peak)}$ reported in literature vary approximately in the range 30–250 $cm^2V^{-1}s^{-1}$, with threshold voltage values V_{th} of 1-2Volts. The specific on-resistance R_{ON} (taken at gate bias values of $V_g > 15V$) lies in the interval 7–20 Ωmm.

Table 1. Values of mobility μ_{FE} and threshold voltage V_{th} reported for normally-OFF recessed hybrid GaN MISHEMTs, employing different gate insulators.

Gate insulator and thickness	$\mu_{FE(peak)}$ ($cm^2V^{-1}s^{-1}$)	V_{th} (V)	Ref.
SiN (20nm)	120	5.2	[42]
Al_2O_3 (30nm)	225	2	[46]
Al_2O_3 (38nm)	55	3.5	[47]
SiO_2 (60nm)	166	3.7	[48]
SiO_2 (60nm)	94	2.4	[48]
Al_2O_3 (10nm)	251	1.7	[49]
Al_2O_3 (20nm)	148	2.9	[50]
Al_2O_3 (30nm)	170	3.5	[51]
$SiN_{(2nm)LT}/ SiN_{(15nm)HT}$	160	2.37	[52]
SiN (17nm) HT	38	1.28	[52]
SiN (20nm)	203	1.2	[53]
Al_2O_3 (18nm)	65	7.6	[54]
SiO_2 (50nm)	110	0.7	[55]
$AlN_{(7nm)}/ SiN_{(7nm)}$	180	1.2	[56]
Al_2O_3 (5nm)/SiN(25nm)	122	1.7	[57]

From these data, it is not simple to find a correlation between the values of $\mu_{FE(peak)}$ and R_{ON}, due to the fact that the reported devices are extremely different (in terms of geometry, recession processes to prepare the channel region, etc.).

However, besides its maximum, it is important to have high channel mobility values also at the operative electric field.

In order to predict the device behavior under operative conditions, it is very important to understand the dependence of the mobility on different parameters (surface roughness, interface traps, electric field, temperature, etc.).

Fiorenza et al. [55] investigated the temperature and field dependence of the channel mobility in recessed-gate hybrid GaN MISHEMTs using SiO_2 as gate insulator. ***Fig. 5*** reports the peak mobility $\mu_{FE(peak)}$ (the maxima of the μ_{FE} curves) as a function of the temperature for a recessed SiO_2/GaN MISHEMT [55]. From this figure, it is possible to see that the experimental $\mu_{FE(peak)}$ data slightly decrease with increasing the measurement temperature. Assuming a formalism analogous to a standard MOSFET, the channel mobility was expressed including in the Matthiessen's rule different scattering contributions, i.e., the bulk mobility factor (μ_B), the acoustic-phonon scattering (μ_{AC}), the surface roughness scattering (μ_{SR}), and the Coulomb scattering (μ_C) due to interface charges [55].

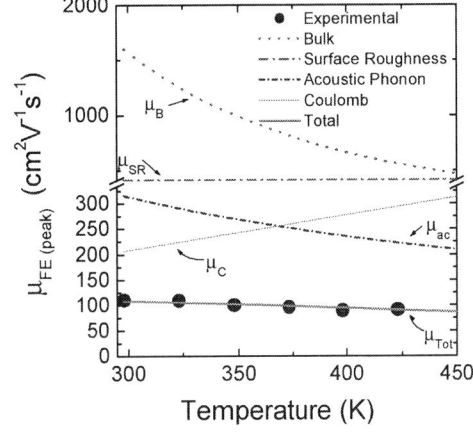

Fig. 5. Peak mobility values $\mu_{FE(peak)}$ as a function of the temperature for a recessed hybrid MISHEMT using SiO_2 as gate insulator. The experimental data were fitted with a total mobility curve (μ_{TOT}) including the different contributions in the Matthiessen's rule (μ_B, μ_{SR}, μ_{AC}, and μ_C). The data are from Ref. [55].

The single contributions to the total mobility depend on several physical features of the insulator/GaN interface (roughness, doping, interface traps, etc.). Some of these parameters can be determined by direct electrical and morphological analyses of the channel region [55]. In particular, using the experimental values of interface trapped charges ($Q_{trap} = 1.35 \times 10^{12}$ cm^{-2}) and surface roughness (RMS = 0.15 nm), determined by C-V and AFM measurements respectively, it was possible to extract the single contributions to the mobility. The total mobility μ_{TOT} and the single contributions are also reported in *Fig. 5*, and show a good agreement with the experimental data.

The temperature dependence of the peak mobility suggests that the main limiting factors to the carrier flow in the channel are the surface roughness (μ_{SR}), the acoustic phonon (μ_{AC}), and the Coulomb scattering (μ_C) contributions.

Hence, the optimization of the insulator/GaN interface in the recessed channel in terms of roughness and the interface trap density is a fundamental issue to improve the mobility.

In this context, the structural and electronic quality of the recessed interface could be improved by using an innovative AlN/SiN stack, grown by metal organic chemical vapour deposition (MOCVD), as gate insulating material [56]. In particular, in this case the overall the interface states D_{it} was reduced with respect to the SiO$_2$/GaN MISHEMT, as can be seen in the D_{it} versus energy plot in *Fig. 6a*. In fact, the total amount of trapped charge in AlN/SiN ($Q_{trap} = 6.4 \times 10^{11}$ cm^{-2}), i.e., the integral of the interface state density over the energy, is less than one half of the value obtained in SiO$_2$ (1.35×10^{12} cm^{-2}). This improvement allows the increase of the peak mobility from 110 cm^{2}V^{-1}s^{-1} (SiO$_2$ gate dielectric) up to 180 cm^{2}V^{-1}s^{-1} (AlN/SiN gate dielectric), shown in *Fig. 6b*.

The high on/off ratio observed in the case of the transistors employing AlN/SiN makes this system very promising for power switching applications [56].

As conclusive remark, it must be mentioned that channel mobility μ_{FE} and the ON-resistance R_{ON} are not the only parameters to be considered in this technology. In fact, recessed-gate hybrid GaN MISHEMTs are often affected by instability phenomena of the threshold voltage V_{th}, when subjected to gate bias stresses.

These effects are associated to the charge trapping/de-trapping of defects located at the insulator/GaN interface and/or in the bulk of the gate insulator [58].

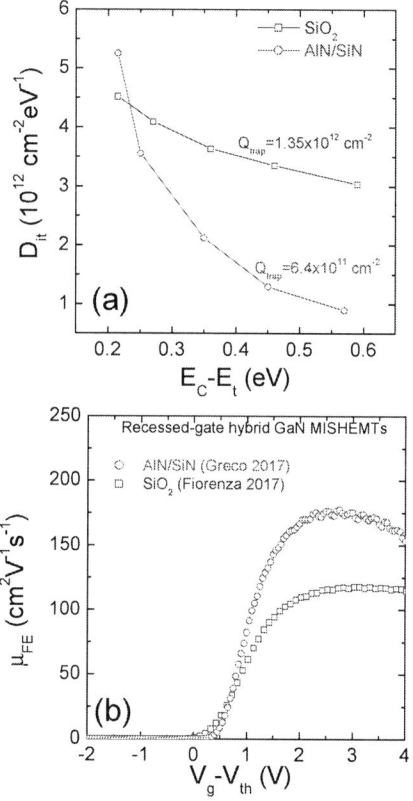

Fig. 6. (a) Interface state density D_{it} measured both in the SiO$_2$/GaN and AlN/SiN/GaN gates in hybrid MISHEMTs. (b) Channel mobility μ_{FE} as a function of the difference between the gate voltage and the threshold voltage (V_g-V_{th}) for recessed-gate hybrid GaN MISHEMTs using SiO$_2$, AlN/SiN and Al$_2$O$_3$ as gate insulators. The data are from Refs. [55,56].

Hence, a careful optimization of the properties of the interface and of the insulating materials is the route towards the achievement of stable MISHEMT devices.

4. Summary

In this paper, a short summary of some processing issues in SiC and GaN power devices technology was given. The main focus is put on transistors, i.e., 4H-SiC MOSFETs and GaN HEMTs. In particular, the importance of the channel mobility has been highlighted for both kind of devices. In 4H-SiC MOSFETs the most common trends to improve the channel mobility reported in literature are presented. Nitridations (NO or N_2O) remain the best processes to increase the channel mobility, without excessively compromising the device reliability.

In the case of GaN, the recessed hybrid MISHEMT is a currently debated solution to achieve a normally-off HEMT operation. For this technology, the choice of the dielectric and the control of its interface to GaN is fundamental to optimize the channel mobility and avoid a penalization of the R_{ON} and of the V_{th} stability..

Acknowledgments. The authors would like to thank the co-workers at CNR-IMM (F. Giannazzo, R. Lo Nigro, S. Di Franco, C. Bongiorno) for fruitful discussion and technical assistance. Colleagues of STMicroelectronics (F. Iucolano, A. Severino, S. Reina, A. Parisi, M. Saggio, S. Rascunà) are greatly acknowledged for support in device processing and characterization.

This work was partially supported by the ECSEL JU project WInSiC4AP (Wide Band Gap Innovative SiC for Advanced Power), Grant Agreement n. 737483.

References

[1] International Energy Agency (IEA) World Energy Outlook report for 2016 (WEO-2016)

[2] F. Roccaforte, P. Fiorenza, G. Greco, R. Lo Nigro, F. Giannazzo, F. Iucolano, M. Saggio, *"Emerging trends in wide band gap semiconductors (SiC and GaN) technology for power devices"* Microelectronic Engineering **187-188**, 66-77, 2018.

[3] F. Roccaforte, F. Giannazzo, F. Iucolano, J. Eriksson, M.H. Weng, V. Raineri, *"Surface and interface issues in wide band gap semiconductor electronics"* Appl. Surf. Sci. **256**, 5727-5735, 2010.

[4] S. Dimitrijev, *"SiC power MOSFETs: the current status and the potential for future developement"*, Proc. of 30th Int. Conf. on Microelectronics (MIEL2017), Nis, Serbia, 9-11 October 2017, pagg. 29-34.

[5] J. Baliga *"Silicon Carbide Power Devices"* World Scientific 2005

[6] M. Cabello, V. Soler, G. Rius, J. Montserrat, J. Rebollo, P. Godignon, Mater. Sci. Semicond. Proc. *"Advanced processing for mobility improvement in 4H-SiC MOSFETs: A review"* **78**, 22-31, 2018.

[7] V.V. Afanas'ev, F. Ciobanu, S. Dimitrijev, G. Pensl, A. Stesmans, *"Band alignment and defect states at SiC/oxide interfaces"* J. Phys.: Condens. Matter **16**, S1839–S1856, 2004.

[8] F. Ciobanu, G. Pensl, V.V. Afanas'ev, A. Schöner, *"Low Density of Interface States in n-Type 4H-SiC MOS Capacitors Achieved by Nitrogen Implantation"* Mater. Sci. Forum, **483-485**, 693-696, 2005.

[9] N.S. Saks, A. K. Agarwal, *"Hall mobility and free electron density at the SiC/SiO$_2$ interface in 4H–SiC"* Appl. Phys. Lett. **77**, 3281, 2000.

[10] E. Arnold, D. Alok, *"Effect of interface states on electron transport in 4H-SiC inversion layers"* IEEE Trans. Electron Devices **48**, 1870-1877, 2001.

[11] A. Frazzetto, F. Giannazzo, P. Fiorenza, V. Raineri, F. Roccaforte, *"Limiting mechanism of inversion channel mobility in Al-implanted lateral 4H-SiC metal-oxide semiconductor field-effect transistors"* Appl. Phys. Lett. **99**, 072117, 2011.

[12] H. Li, S. Dimitrijev, H. B. Harrison, D. Sweatman, *"Interfacial characteristics of N$_2$O and NO nitrided SiO$_2$ grown on SiC by rapid thermal processing"* Appl. Phys. Lett. **70**, 2028-2030, 1997.

[13] G. Y. Chung, C. C. Tin, J. R. Williams, K. McDonald, M. Di Ventra, S. T. Pantelides, L. C. Feldman, R. A. Weller, *"Effect of nitric oxide annealing on the interface trap densities near the band edges in the 4H polytype of silicon carbide"* Appl. Phys. Lett. **76**, 1713-1715, 2000.

[14] L.A. Lipkin, M.K. Das, J.W. Palmour, *"N$_2$O processing improves the 4H-SiC:SiO$_2$ interface"* Mat. Sci. Forum **389-393**, 985-988, 2002.

[15] C-Y. Lu, J.A. Cooper, T. Tsuji, G. Chung, J.R. Williams, K. McDonald, L.C. Feldman, *"Effect of process variations and ambient temperature on electron mobility at the SiO2/4H-SiC interface"* IEEE Trans. on Electron Dev. **50**, 1582-1588, 2003.

[16] D. Okamoto, H. Yano, K. Hirata, T. Hatayama, T. Fuyuki, *"Improved Inversion Channel Mobility in 4H-SiC MOSFETs on Si Face Utilizing Phosphorus-Doped Gate Oxide"* IEEE Electron Dev. Lett. **31** 710-712, 2010.

[17] L.K. Swanson, P. Fiorenza, F. Giannazzo, A. Frazzetto, F. Roccaforte, *"Correlating macroscopic and nanoscale electrical modifications of SiO2/4H-SiC interfaces upon post-oxidation-annealing in N2O and POCl3"* Appl. Phys. Lett. **101**, 193501, 2012.

[18] H. Yano, T. Araoka, T. Hatayama, T. Fuyuki, *"Improved Stability of 4H-SiC MOS Device Properties by Combination of NO and POCl3 Annealing"* Mat. Sci. Forum **740-742**, 727-732, 2013.

[19] Y. K. Sharma, A. C. Ahyi, T. Isaacs-Smith, A. Modic, M. Park, Y. Xu, E. L. Garfunkel, S. Dhar, L. C. Feldman, J. R. Williams, *"High-Mobility Stable 4H-SiC MOSFETs Using a Thin PSG Interfacial Passivation Layer"* IEEE Elect. Dev. Lett. **34**, 175-177, 2013.

[20] T. Umeda, K. Esaki,R. Kosugi,K. Fukuda,T. Ohshima,N. Morishita, J. Isoya, *"Behavior of nitrogen atoms in SiC-SiO2 interfaces studied by electrically detected magnetic resonance"* Appl. Phys. Lett. **99**, 142105, 2011.

[21] R. Kosugi, T. Umeda, Y. Sakuma, *"Fixed nitrogen atoms in the SiO2/SiC interface region and their direct relationship to interface trap density"* Appl. Phys. Lett. **99**, 182111, 2011.

[22] P. Fiorenza, F. Giannazzo , M. Vivona, A. La Magna , F. Roccaforte, *"SiO2/4H-SiC interface doping during post-deposition-annealing of the oxide in N2O or POCl3"* Appl. Phys. Lett. **103**, 153508, 2013.

[23] P. Fiorenza, L.K. Swanson, M. Vivona, F. Giannazzo, C. Bongiorno, A. Frazzetto, F. Roccaforte, *"Comparative study of gate oxide in 4H-SiC lateral MOSFETs subjected to post-deposition-annealing in N2O and POCl3"* Appl. Phys. A, **115**, 333-339, 2014.

[24] T. Akagi, H. Yano, T. Hatayama, T. Fuyuki, *"Effect of Interfacial Localization of Phosphorus on Electrical Properties and Reliability of 4H-SiC MOS Devices"* Mat. Sci. Forum **740-742**, 695-698, 2013.

[25] T. Sledziewski, A. Mikhaylov, S. Reshanov, A. Schoener, H.B. Weber, M. Krieger, *"Reduction of Density of 4H-SiC/SiO2 Interface Traps by Pre-Oxidation Phosphorus Implantation"* Mater. Sci. Forum **778-780**, 575-578, 2014.

[26] A. Mihaylov, T.Sledziewski, A. Afanasyev, V. Luchinin, S. Reshanov, A. Schoener, M. Krieger, *"Effect of Phosphorus Implantation Prior to Oxidation on Electrical Properties of Thermally Grown SiO2/4H-SiC MOS Structures"* Mater. Sci. Forum **806**, 133-138, 2014.

[27] A. Modic, G. Liu, A. C. Ahyi, Y. Zhou, P. Xu, M. C. Hamilton, J. R. Williams, L. C. Feldman, S. Dhar, *"High Channel Mobility 4H-SiC MOSFETs by Antimony Counter-Doping"* IEEE Electron Device Lett. **35**, 894-896, 2014.

[28] D. Okamoto, M. Sometani, S. Harada, R. Kosugi, Y. Yonezawa, H. Yano, *"Improved Channel Mobility in 4H-SiC MOSFETs by Boron Passivation"* IEEE Electron Device Lett. **35**, 1176-1178, 2014.

[29] D. Okamoto, M. Sometani, S. Harada, R. Kosugi, Y. Yonezawa, H. Yano, *"Effect of boron incorporation on slow interface traps in SiO2/4H-SiC structures"* Appl. Phys. A **123**, 133, 2017.

[30] M. Cabello, V. Soler, N. Mestres, J. Montserrat, J. Rebollo, J. Millan, P. Godignon, *"Improved 4H-SiC N-MOSFET Interface Passivation by Combining N2O Oxidation with Boron Diffusion"* Mater. Sci. Forum **897**, 352-355, 2017.

[31] M. Cabello, V. Soler, J. Montserrat, J. Rebollo, J.M. Rafi, P. Godignon, *"Impact of boron diffusion on oxynitrided gate oxides in 4H-SiC metal-oxide-semiconductor field-effect transistors"* Appl. Phys. Lett. **111**, 042104, 2017.

[32] D. J. Lichtenwalner, L. Cheng, S. Dhar, A. Agarwal, J. W. Palmour *"High mobility 4H-SiC (0001) transistors using alkali and alkaline earth interface layers"* Appl. Phys. Lett. **105**, 182107, 2014.

[33] D. J. Lichtenwalner, L. Cheng, S. Dhar, A.K. Argawal, S. Allen, J.W. Palour, *"High-Mobility SiC MOSFETs with Chemically Modified Interfaces"* Mater. Sci. Forum **821-823**, 749-752, 2015.

[34] D.J. Lichtenwalner, V. Pala. B. Hull, S. Allen. J.W. Palmour, *"High-Mobility SiC MOSFETs with Alkaline Earth Interface Passivation"* Mater. Sci. Forum **858**, 671-676, 2016.

[35] J.H. Dycus, W. Xu, D.J. Lichtenwalner, B. Hull, J.W. Palmour, J.M. LeBeau, *"Structure*

and chemistry of passivated SiC/SiO₂ interfaces" Appl. Phys. Lett. **108**, 201607, 2016.

[36] D. Lichtenwalner, J.H. Dycus, W. Xu, J.M. Lebeau, B. Hull, S. Hallen. J.W. Palmour, *"Electrical Properties and Interface Structure of SiC MOSFETs with Barium Interface Passivation"* Mater. Sci. Forum **897**, 163-166, 2017.

[37] K.J. Chen, C. Zhou, *"Enhancement‐mode AlGaN/GaN HEMT and MIS‐HEMT technology"* Phys. Status Solidi a, **208** 434-438, 2011.

[38] M. Su, C. Chen, S. Rajan, *"Prospects for the application of GaN power devices in hybrid electric vehicle drive systems"* Semicond. Sci. Technol., **28** 074012, 2013.

[39] M.J. Scott, L. Fu, X. Zhang, J. Li, C. Yao, M. Sievers, J. Wang, *"Merits of gallium nitride based power conversion"* Semicond. Sci. Technol., **28** 074013, 2013.

[40] Y. Uemoto, M. Hikita, H. Ueno, H. Matsuo, H. Ishida, M. Yanagihara, R. Ueda, T. Tanaka, D. Ueda, *"Gate injection transistor (GIT)—A normally-off AlGaN/GaN power transistor using conductivity modulation"* IEEE Transactions on Electron Device, **54** 3393-3399, 2007.

[41] G. Greco, F. Iucolano, F. Roccaforte, *"Review of technology for normally-off HEMTs with p-GaN gate"* Mater. Sci. Semicond. Process., **78** 96-106, 2018.

[42] T. Oka, T. Nozawa, *"AlGaN/GaN Recessed MIS-Gate HFET With High-Threshold-Voltage Normally-Off Operation for Power Electronics Applications"* IEEE Electron Device Lett., **29** (2008) 668-670, 2008.

[43] H. Kambayashi, Y. Satoh, S. Ootomo, T. Kokawa, T. Nomura, S. Kato, T. P. Chow, *"Over 100 A operation normally-off AlGaN/GaN hybrid MOS-HFET on Si substrate with high-breakdown voltage"* Solid State Electronics, **54** 660-664, 2010.

[44] F. Roccaforte, P. Fiorenza, G. Greco, M. Vivona, R. Lo Nigro, F. Giannazzo, A. Patti, M. Saggio, *"Recent advances on dielectrics technology for SiC and GaN power devices"* Appl. Surf. Sci., **301** 9-18, 2014.

[45] T. Hashizume, K. Nishiguchi, S. Kaneki, J. Kuzmik, Z. Yatabe *"State of the art on gate insulation and surface passivation for GaN-based power HEMTs"* Mater. Sci. Semicon. Processing **78**, 85-95, 2018

[46] K-S. Im, J-B.Ha, K-W. Kim, J-S.Lee, D-S. Kim, S-H. Hahm, J-H-. Lee, *"Normally Off GaN MOSFET Based on AlGaN/GaN Heterostructure With Extremely High 2DEG Density Grown on Silicon Substrate"* IEEE Electron Device Lett., **31** 192-194, 2010.

[47] K K-W. Im, S-D. Jung, D-S. Kim, H-S. Kang, K-S. Im, J-J. Oh, J-B. Ha, J-K. Shin and J.H. Lee, *"Effects of TMAH Treatment on Device Performance of Normally Off Al₂O₃/GaN MOSFET"* IEEE Electron Device Lett., **32** 1376-1378, 2011.

[48] H. Kambayashi, Y. Satoh, T. Kokawa, N. Ikeda, T. Nomura and S. Kato, *"High field-effect mobility normally-off AlGaN/GaN hybrid MOS-HFET on Si substrate by selective area growth technique"* Solid-State Electronics, **56** 163-167, 2011.

[49] Y. Wang, M. Wang, B. Xie, C.P. Wen, J. Wang, Y. Hao, W. Wu, K.J. Chen, B. Shen, *"High-performance normally-off Al₂O₃/GaN MOSFET using a wet etching-based gate recess technique"* IEEE Electron Device Lett., **34** 1370-1372, 2013.

[50] M. Wang, Y. Wang, C. Zhang, B. Xie, C.P. Wen, J. Wang, Y. Hao, W. Wu, K.J. Chen B. Shen, *"900 V/1.6 m·cm₂ normally off Al₂O₃/GaN MOSFET on silicon substrate,"* IEEE Transactions on Electron Devices, **61** (2014) 2035-2040, 2014.

[51] Y. Yao, Z. He, F. Yang, Z. Shen, J. Zhang, Y. Ni, J. Li, S. Wang, G. Zhou, J. Zhong, Z. Wu, B. Zhang, J. Ao, Y. Liu, *"Normally-off GaN recessed-gate MOSFET fabricated by selective area growth technique"* Appl. Phys. Express, **7** 016502, 2014.

[52] M. Hua, Z. Zhang, J. Wei, J. Lei, G. Tang, K. Fu, Y. Cai, B. Zhang, K.J. Chen,*" Integration of LPCVD-SiNx gate dielectric with recessed-gate E-mode GaN MIS-FETs: Toward high performance, high stability and long TDDB lifetime,",Proc. IEDM 2016, San Francisco USA, 3-7 December 2016, pagg. 260-263.

[53] Zhang Z., S. Qin, K. Fu, G. Yu, W. Li, X. Zhang, S. Sun, L. Song, S. Li, R. Hao, Y. Fan, Q. Sun, G. Pan, Y. Cai, B. Zhang, *"Fabrication of normally-off AlGaN/GaN metal insulator semiconductor high electron mobility transistors by photo-electrochemical gate recess etching in ionic liquid"* Appl. Phys. Express, **9** 084102, 2016.

[54] Q. Zhou, L. Liu, A. Zhang, B. Chen, Y. Jin, Y. Shi, Z. Wang, W. Chen B. Zhang, *"7.6 V threshold voltage high-performance normallyoff Al₂O₃/GaN MOSFET achieved by interface charge engineering,"* IEEE Electron Device Lett., **37** 165-168, 2016.

[55] P. Fiorenza, G. Greco, F. Iucolano, A. Patti, F. Roccaforte, *"Channel Mobility in GaN Hybrid MOS-HEMT Using SiO₂ as Gate*

Insulator" IEEE Transactions on Electron Devices, **64** 2893-2899, 2017.

[56] G. Greco, P. Fiorenza, F. Iucolano, A. Severino, F. Giannazzo, F. Roccaforte, *"Conduction Mechanisms at Interface of AlN/SiN Dielectric Stacks with AlGaN/GaN Heterostructures for Normally-off High Electron Mobility Transistors: Correlating Device Behavior with Nanoscale Interfaces Properties"* ACS Appl. Mater. Interfaces, **9** 35383−35390, 2017.

[57] H. Wang, J. Wang, J. Liu, M. Li, Y. He, M. Wang, M. Yu, W. Wu, Y. Zhou, G. Dai, *"Normally-off fully recess-gated GaN metal–insulator–semiconductor field-effect transistor using Al_2O_3/Si_3N_4 bilayer as gate dielectrics"* Appl. Phys. Express, **10** 106502, 2017.

[58] G. Meneghesso, M. Meneghini, C. De Santi, M. Ruzzarin, E. Zanoni, *"Positive and negative threshold voltage instabilities in GaN-based transistors"* Microelectronics Reliability, **80** 257–265, 2018.

DFT Calculations of Structure and Optical Properties in Wide Band-Gap Semiconductor Clusters for Dye-Sensitized Solar Cells

Corneliu I. Oprea,[1] Petre Panait,[1,2] Reda M. AbdelAal, Mihai A. Gîrțu[1]*

[1]Department of Physics and Electronics, Ovidius University of Constanța, Constanța, Romania
[2]Faculty of Physics, University of Bucharest, Bucharest, Romania
[3]Department of Chemical Engineering, Suez University, Suez, Egypt
*E-mail: mihai.girtu@univ-ovidius.ro

Abstract—We report results of a computational study of TiO$_2$ clusters to understand their structure and optical properties as well as the binding and charge transfer from organic dyes to such clusters. We perform density functional theory calculations of several coumarin-based and oligomethine cyanine-based dyes as well as complex systems consisting of the dye bound to a TiO$_2$ cluster. We provide the electronic structure of the dyes alone and adsorbed to the cluster, and discuss the matching with the solar spectrum. We display the energy level diagrams and the electron density of the key molecular orbitals and analyze the electron transfer from the dye to the oxide.

Keywords—TiO$_2$ clusters; electronic structure; density of states; dye-sensitized solar cells; photovoltaic conversion efficiency; organic dyes.

1. Introduction

Wide bandgap semiconductors, such as TiO$_2$, have stirred great and continuous interest in the past two decades due to their applications [1] in photovoltaic energy conversion [2,3], photocatalysis [4,5] and sensor electronics [6], antibacterial, anticorrosion, antifogging, self-cleaning coatings [7], drug delivery [8]. These new applications have added to the traditional uses such as producing a white color in paints, making substances more opaque, blocking UV rays in sunscreens etc. [9].

The physical and chemical properties of TiO$_2$ nanocrystals are influenced by their electronic structure, size, shape, organization, and surface properties [10,11]. To better understand the behavior of the TiO$_2$ nanostructured materials used in the multiple applications mentioned, numerous computational studies have tackled the modeling of clusters of various sizes [12,13]. They discuss the electronic properties of bulk

TiO$_2$ polymorphs (tetragonal), rutile (tetragonal), brookite (orthorhombic), and TiO2 (B) (monoclinic) [14], the structure and reactivity of anatase surfaces, and the modeling of bare and sensitized TiO$_2$ nanoparticles, nanosheets, and nanotubes.

Of the natural polymorphs of TiO$_2$, rutile, is the thermodynamically most stable bulk phase, whereas anatase is very common and stable in nanomaterials [13,15] and shows highest photocatalytic activity [16,17]. Anatase is most interesting phase of TiO$_2$ for photovoltaic and photocatalytic applications and, for this reason, we will focus in the following only on it. The crucial role in these applications is played by the electronic properties of oxide nanopraticles, as the energy level alignment between the conduction or valence band edges with the ground and excited states of the dye or with the redox level of the electrolyte determine whether a process can take place or not. Moreover, the states near the valence and conduction band edges have a major influence on the electrical conductivity and chemical reactivity [13]. To approach these topics and investigate the interaction of the dye with the TiO$_2$ surface it is natural to use the cluster approach [18]. In contrast, a periodic approach is to be preferred for the interaction of inherently periodic crystalline materials, for example, perovskites or other inorganic absorbers, on TiO$_2$ [18].

Although the TiO$_2$ nanoclusters have been extensively studied theoretically [12,13], the question regarding the proper cluster size to

978-1-5386-4483-6/18 $31.00 © 2018 IEEE

approach a particular problem or a certain complex system has remained controversial. For instance, it was argued that small cluster models reproduce the main features of the optical response, however, the $(TiO_2)_{15}$ cluster constitutes the minimal size to provide a complete picture in the case of the binding of a small molecule such as catechol [19]. However, for larger molecules, such as various coumarin-based dyes, even smaller $(TiO_2)_9$ cluster have been used [20].

In contrast, other authors, to study the anchoring and charge transfer for a small molecule such as pyridine chose a much larger $(TiO_2)_{46}$ cluster [21] from a set ranging from $n = 16$ to 68 [22]. The choice was made based not only on computational effort but also after concluding that the cluster shape has a strong influence on the quantum size effect. Later, the same group used for studying the binding of a larger iron complex, a cluster with $n = 92$ [23]. This choice was made after studying several $(TiO_2)_n$ clusters with $n = 32$ to 122 [24].

Early systematic computational studies of TiO_2 clusters have been performed by Persson et al. [25], whose strategy for n between 16 and 38, was to remove selected atoms from the (101) surfaces to keep the cluster neutral and stoichiometric. Independently, Jug and coworkers [26] had a different approach, in which they saturated all peripheral oxygen atoms of the clusters with hydrogen atoms and all less than fivefold coordinated titanium atoms with OH groups. The study of $(TiO_2)_n(H_2O)_m$ clusters, with $n = 33$–132 and $m = 17$–48, and small size adsorbed molecules indicated that the influence of the cluster size on the convergence of the adsorption energies and also of the number of relaxed surface atoms for the two considered levels of relaxation was insignificant [26].

Starting from the original concept of Persson et al. [21,25], De Angelis et al. used a cluster with $n = 38$ [27,28,29] to investigate the anchoring and the energy level alignment between the dye and the semiconducting cluster. Later on the cluster size was increased to $n = 82$ [30,31], the $(TiO_2)_{38}$ cluster being basically a part of the $(TiO_2)_{82}$ slab. Both clusters were shown to represent a good trade-off between accuracy and computational convenience and nicely reproduced the main electronic characteristics of TiO_2 nanoparticles [27,29], the larger one being used to check the accuracy of the results [32]. The larger cluster proved to be very useful when modeling dye aggregation by studying two dyes anchored on the substrate [33].

Larger clusters, of bipyramidal shape with n between 35 and 455, were studied [34], to find that, for smaller nanocrystals, compared to bulk anatase, a sizable structural relaxation was obtained, which involved a contraction of all the bond lengths, including those in the center of the nanoparticle. The changes in the surface structure of the nanoparticle were dominated by the outward relaxation of the O atoms, and the inward relaxation of the Ti atoms, which created a more rippled surface. The relaxation energy and the associated structural and electronic changes were found to decrease with increasing n [34].

A common feature of the optimized clusters mentioned above, including larger stoichiometric bipyramidal nanocrystals, is that they have at least two defective Ti=O groups at their surface [35]. The alternative stoichiometric cluster structures thus proposed a new set of $(TiO_2)_n$ ($n = 10$−16), did not contain terminal Ti=O defects but were found to form sphere- or rodlike compact structures, exhibiting some odd−even oscillations in the band gap, with no convergence in the considered n range.

Calculations on a $(TiO_2)_{29}$ cluster [36] showed four dangling oxygen atoms on the (001) surfaces. Surface hydration to lead to the most stable nanocrystal, in agreement with the experimental finding that the truncated bipyramidal morphology is typical of a moderately acidic environment.

A study of larger clusters, with n between 58 and 449 and different truncated-bipyramidal shape, chose to saturate the under-coordinated surface atoms by dissociated water molecules [37]. It showed that for nanoparticles larger than 2 nm, the band gap converges rapidly toward those of the extended (101) and (001) surfaces, implying that the quantum size effect may only be significant in very small TiO_2 anatase particles.

Later, De Angelis et al. examined clusters of $n = 367$ for which selected atoms from the (101) surfaces were removed to keep the cluster neutral and stoichiometric [38]. In parallel, an $n = 411$ cluster was built such that all the dangling oxygen atoms on the (001) surfaces were saturated by hydrogen atoms. Upon geometry optimization, the two relaxed models showed similar band gap and electronic density of states [38].

In the context of photovoltaic applications [39,40] we showed that, when dealing with small molecules an $n = 24$ cluster provides a reasonable compromise between accuracy and computational costs [41]. In contrast, when studying the photocatalytic activity of TiO_2 under UV and visible light [42], we examined the adsorption of various common antibiotics onto TiO_2 nanoclusters, we found that the penicillin molecule is strongly distorted when binding to the substrate [43]. Here, we report computational studies of different TiO_2 clusters, with $n = 14$ to 54, to better understand size effects on optimized geometries and charge transfers. For all these clusters, hydrogen atoms or –OH groups saturate the dangling bonds.

The goal is to understand their structure and optical properties as well as the binding and charge transfer from organic dyes to such clusters. We calculate the electronic structure and simulated UV-Vis spectra of the dyes alone and adsorbed to the cluster, and discuss the matching with the solar spectrum. Some of the dyes are well known, such as the coumarine based systems C343, NKX-2398 and NKX-2311 [44] whereas others are reported here for the first time, e.g. 5-carboxy-2-(3-(7-((4-(diphenylphospho) phenyl)ethynyl)-1,1,3-trimethyl-1H-benzo[e] indol-2(3H)-ylidene)prop-1-en-1-yl)-1,3,3-trimethyl-3H-indol-1-ium iodide (OMCD1).

We display the energy level diagrams and the electron density of the key molecular orbitals and analyze the electron transfer from the dye to the oxide.

2. Computational Details

The titania clusters of various sizes presented in the paper are modeling the (101) surface of anatase, and were initially cut from the experimental structure [45]. In order to ensure the charge neutrality in the presence of under-coordinated Ti atoms [35,36,37], four hydrogen atoms were used to solve the dangling bonds of the oxygen atoms bound to the two Ti atoms at the periphery of $Ti_{14}O_{30}H_4$ cluster, or to the three-fold coordinated Ti atoms in the corners of $Ti_{24}O_{50}H_4$, $Ti_{34}O_{70}H_4$, $Ti_{44}O_{90}H_4$, and $Ti_{54}O_{110}H_4$ clusters. These structures, along with the protonated species of coumarin-based dyes C343, NKX-2398, NKX-2311, isolated and adsorbed on the $Ti_{24}O_{50}H_4$ cluster, were energy minimized by density functional theory (DFT) [46] calculations using the generalized gradient approximation (GGA) BLYP exchange-correlation functional [47,48] and effective core potentials (ECP) for Ti atoms and double-ζ quality basis functions for all atoms via LANL2DZ [49].

For the density of electronic states, single-point calculations were performed using the hybrid B3LYP functional [50] with the same basis set and accounting for aqueous solvent effects via the conductor-like polarizable continuum model (C-PCM) [51]. The cavity used in the C-PCM calculation was built from spheres centered on heavy nuclei, based on the United Atom for Hartree-Fock procedure [52]. We used the similar method, where the water solvent was replaced by ethanol, for the geometry optimization of the cationic oligomethine cyanine dye OMCD1 in the presence of one negative iodine ion, and also adsorbed on the $Ti_{24}O_{50}H_4$ cluster. Calculations were performed with the GAUSSIAN09 quantum chemistry package [53], whereas the projection of the density of states on different system components was

978-1-5386-4483-6/18 $31.00 © 2018 IEEE

obtained with GaussSum [54].

3. Results and Discussion

To avoid the problem of the surface states in the gap, we performed geometry optimization of model clusters with a slight deviation from the TiO_2 stoichiometry, introducing H atoms or –OH groups to terminate the dangling bonds at the periphery ($Ti_{14}O_{30}H_4$, $Ti_{24}O_{50}H_4$, $Ti_{34}O_{70}H_4$, $Ti_{44}O_{90}H_4$, and $Ti_{54}O_{110}H_4$). This approach resulted in compact structures with 4-, 5-, and 6-fold coordinated Ti ions, together with 2- and 3-fold coordinated oxygen atoms [41,44]. Following the geometry optimization, the structure is slightly distorted (see Fig. 1) to minimize the surface stresses.

Table 1 reports the calculated and the experimental Ti–O distances, as well as the cluster length and width. The average distance is consistently smaller than the experimental value of 1.950 Å, valid for the bulk oxide. Also, the distribution of these distances widens significantly compared to the bulk, [55]. The deformation of the structure changes angles and distances such that the increase in the length of the cluster

from 12.04 Å in the bulk is opposite to the decrease in width,. The relative variation of the cluster distances compared to the values in the bulk, is less than 5%. We note that the values reported earlier [40,41,44] have slight deviations from those presented here, due to the different DFT functional and basis set used. The geometry relaxation leads to optical band gaps of 3.37 to 3.79 eV, larger than the experimental value of ~3.2 eV for anatase titania [39].

The optimized geometrical structures of all three coumarin-based dyes have been previously reported by other authors [56,57,58] and, therefore, here we only state that the structures are in agreement with the ones already presented. The structures of the dyes are presented in Fig. 2.

The anchoring modes of the dye to the TiO_2 surface are of crucial importance, the bonding type and the extent of electronic coupling between the dye-excited state and the semiconductor unoccupied states, directly influencing the electron injection, and, in this way, the short-circuit current and the overall photovoltaic conversion efficiency of the device [39,40]. Earlier studies have shown

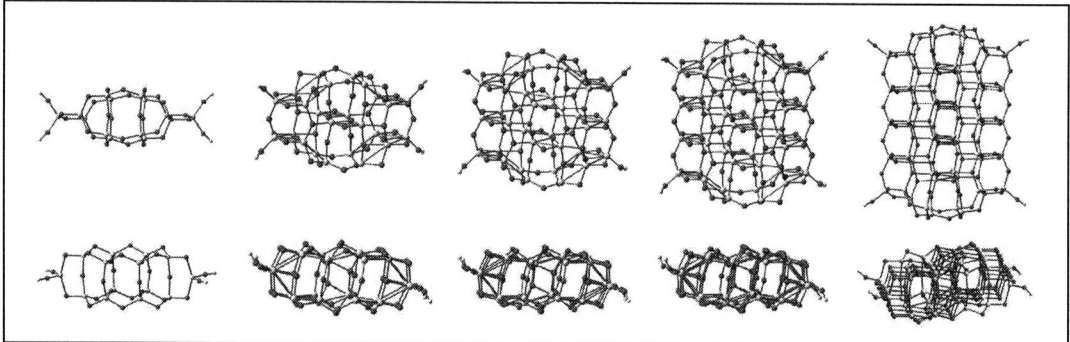

Fig. 1 Top and lateral views of the optimized structures of the $Ti_{14}O_{30}H_4$, $Ti_{24}O_{50}H_4$, $Ti_{34}O_{70}H_4$, $Ti_{44}O_{90}H_4$, and $Ti_{54}O_{110}H_4$ nanoclusters (from left to right, respectively), modeling the anatase titania (101) surface. The geometry optimization was performed by DFT at BLYP/LANL2DZ level.

Table 1. Average Ti–O distance and its standard deviation, as well as the cluster length and width, in Å, of the DFT/B3LYP/3-21G(d,p) optimized structures compared to the experimental value of the bulk TiO_2 [55].

Parameter	$Ti_{14}O_{30}H_4$	$Ti_{24}O_{50}H_4$	$Ti_{34}O_{70}H_4$	$Ti_{44}O_{90}H_4$	$Ti_{54}O_{110}H_4$	TiO_2 (bulk)
$r(Ti–O)$	1.834	1.876	1.889	1.896	1.901	1.950
$\sigma_{r(Ti–O)}$	0.074	0.100	0.100	0.103	0.105	0.022
Length	12.35	12.76	12.54	12.50	12.41	12.04
Width	3.74	7.39	10.91	14.34	18.05	–
Width (bulk)	3.80	7.59	11.37	15.14	18.92	–

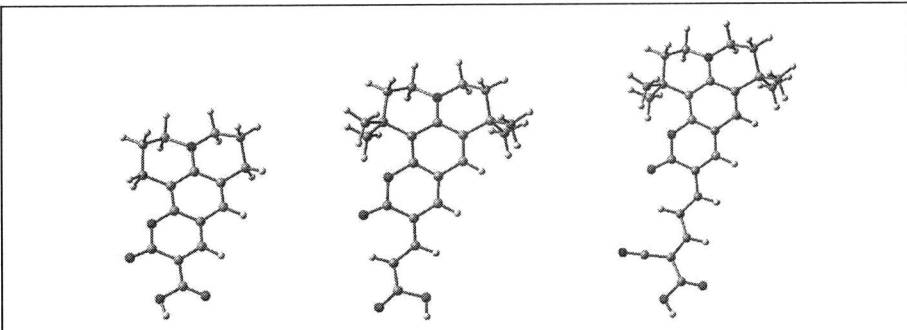

Fig. 2 Optimized structure of the dyes C343, NKX-2398 and NKX-2311 (from left to right) in their neutral, protonated form calculated by DFT at BLYP/LANL2DZ level. Colors: C - grey, O - red, N – blue, and H - white.

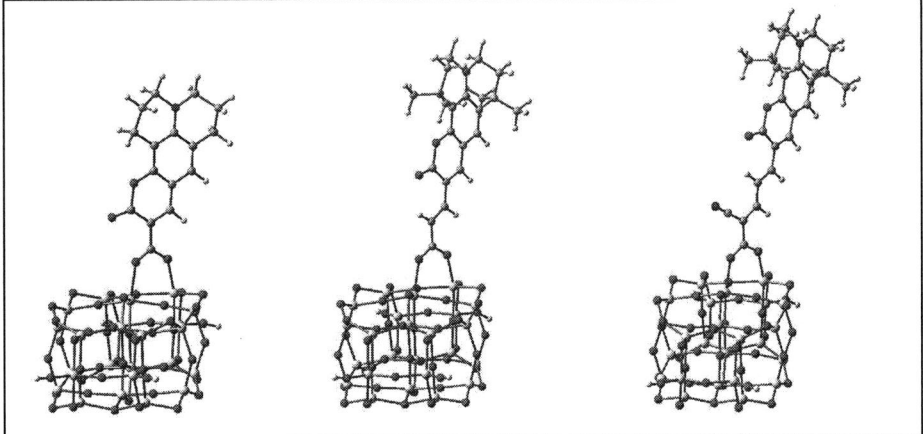

Fig.3 Optimized structure of the dye-oxide systems for C343, NKX-2398, and NKX-2311 on the $Ti_{24}O_{50}H_4$ cluster (from left to right), calculated by DFT at BLYP/LANL2DZ level. Colors: Ti - light grey, C - grey, O - red, N – blue, H - white.

that for the organic dyes bearing a carboxylic acid as the anchoring group, the preferred adsorption mode is bidentate bridging, with one proton transferred to a nearby surface oxygen [33,59,60].

Given the rigidity of all the dye backbone we chose to use the $n = 24$ cluster, as a good compromise between accuracy and computational time. Our calculations for the three coumarin-based dyes, C343, NKX-2398 and NKX-2311, showed that the preferred adsorption mode is indeed the bidentate bridging. Starting the optimization from the other types of binding (monodentate ester-like, bidentate chelating or through a hydrogen bond), finally lead to the same bidentate bridging configuration, as displayed in Fig. 3.

Along with the relatively well known coumarin-based dyes, we report here results obtained for a new dye, an oligomethine cyanine molecule, OMCD1 (see Fig. 4),

trying to explore whether it satisfies the criteria for good TiO_2 sensitizer in DSSCs [40]. The first condition that a sensitizer has to fulfill is the proper anchoring to the oxide. The OMCD1 molecule has a –COOH group at one end, leading to a bidentate binding configuration to the titania cluster, as displayed in Fig. 4. The carboxyl group is a very good anchor, as it insures not just the mechanical strength of the binding but also a good charge injection, since it has a π symmetry, which overlaps well with the d orbitals of the titanium atoms.

We note that the OMCD1 dye has to be a cationic species in order to ensure the π conjugation over the nitrogen atoms in the cyanine constituents. For the calculation, to insure that the overall neutrality is achieved, we place a negative iodine ion in the vicinity of the cyanine parts, at distances of 5 Å and 5.6 Å to the N atoms. This is a usual modeling and computational trick, which in the case of DSSCs comes natural as iodine is already present in the electrolyte.

A second criterion is the matching of the absorption spectrum of the dye with the solar irradiation spectrum [40]. Time-dependant DFT calculations show that the maximum of the absorption is at 537 nm, associated to the π-π^* electronic transition from the highest occupied molecular orbital (HOMO) to the lowest unoccupied molecular orbitals (LUMO), indicating that the spectral matching is good.

Another condition for the dye is the charge injection into the oxide, which is achieved if the molecule has an electron-rich donor part and an acceptor part with an anchoring carboxylic group, kept together by a π bridge. The key for charge injection is the alternant single/double bond character of the molecular backbone, which allows for electron delocalization. A push-pull effect occurs when the ground state of the dye has the charge localized on the donor part and the excited state, reached after light absorption, has the charge localized on the acceptor part, right next to the TiO_2.

OMCD1 has the conjugated backbone and the proper anchoring. The donor group is based on phenylacetylene, ending in an sp^3 hybridized P atom, the π bridge is based on benzindole, whereas the acceptor is based on indole, ending with a carboxyl group. Fig. 5 displays the key molecular orbitals of OMCD1, illustrating the push-pull effect towards the anchor, where the electronic density increases by 257%. Moreover, Fig. 6

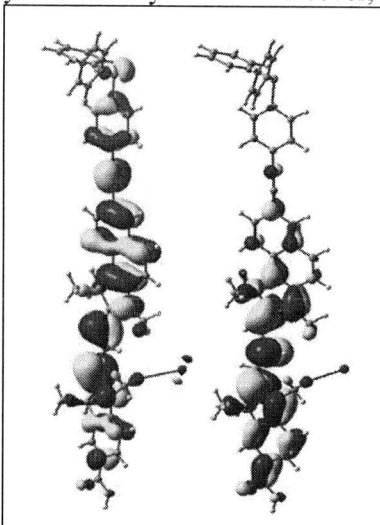

Fig. 5 Isodensity surfaces (0.03 e/bohr^3) of the key molecular orbitals of OMCD1 in ethanol: HOMO (left) and LUMO (right), showing the push-pull effect on the charge density. Calculatins performed by DFT at the LANL2DZ level.

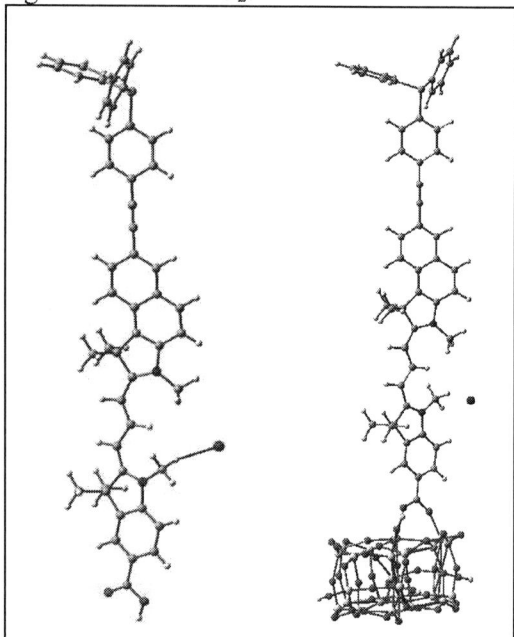

Fig. 4 Optimized structure of the OMCD1 dye (left) and the dye-oxide complex system OMCD1–$Ti_{24}O_{50}H_4$ (right) calculated at the DFT/B3LYP/LANL2DZ level. Color coding: Ti - light grey, C - grey, O - red, N - blue, P - yellow, I - purple, and H - white.

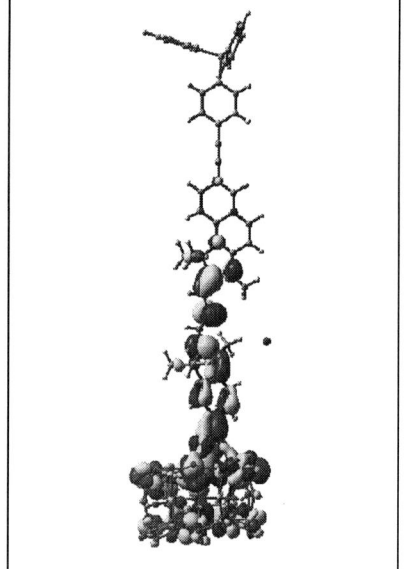

Fig. 6 Isodensity surfaces (0.03 e/bohr^3) of the LUMO+1 of OMCD1-$Ti_{24}O_{50}H4$ in ethanol, showing how the charge is pushed/pulled towards the oxide. Calculatins performed by DFT at the B3LYP/LANL2DZ level.

illustrates the push-pull effect when the oxide is also taken into account in the calculation. The localization of the electron density on LUMO+1 clearly suggests charge injection.

The next condition refers to the proper alignment of the ground/excited states of the dye with the conduction band of the oxide and the redox level of the electrolyte. This criterion can be easily analyzed from the density of states (DoS) of the dye-cluster systems (Fig. 7 and Fig. 8).

The DoS for $Ti_{24}O_{50}H_4$ shows high contribution from O p orbitals in the valence band and from Ti d orbitals in the conduction band. The additional H atoms or –OH groups have a minor contribution in states that are of minor importance. For all dye-oxide systems, the valence band has a mixed character with significant contributions from both the dye and the cluster. In contrast, the conduction band has dominating contributions from the titania cluster.

A key orbital is the HOMO of the dye, which is located in the gap between the oxide bands. Its position should be below the redox level of the electrolyte (experimental value of -5.04 eV for the I_3^-/I^- redox couple [39,61].),

which happens for all systems studied here. The reason is to allow for dye regeneration by electron transfer from the electrolyte anions. However, the HOMO should not be too low in the gap, such that the absorption spectrum is not pushed into the UV, as it is the case for C343.

The excited state of the dye is above the conduction band of the oxide for all systems, permitting the charge injection into the oxide. For NKX-2311 the excited states of the dye have a larger weight in the mixt states with the oxide, leading to stronger absorption and better matching with the solar spectrum.

For OMCD1-$Ti_{24}O_{50}H_4$, it can be observed that the excited state of the dye (at -3.17 eV, above the edge of the conduction band, situated at -3.49 eV), has mixed character with contribution of more than 59% from the oxide, as already seen qualitatively in Fig. 6. Partial delocalization of this electronic state over semiconductor states leads to the faster electron injection and lower rate of electron-hole recombination, due to the larger distance between surface and the acceptor part of the molecule.

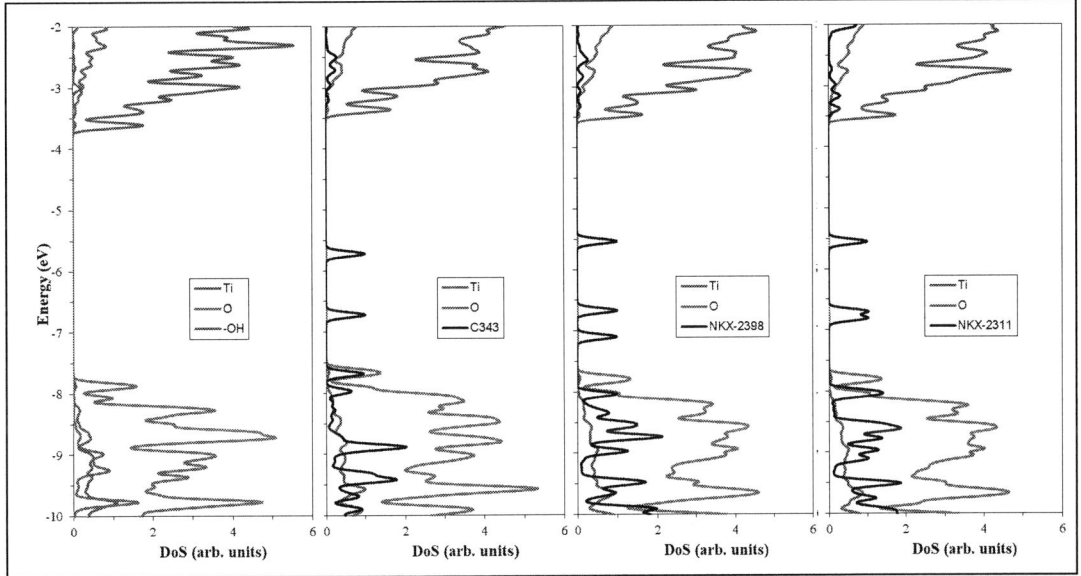

Fig. 7 Density of states of (from left to right): $Ti_{24}O_{50}H_4$, C343–$Ti_{24}O_{50}H_4$, NKX-2398–$Ti_{24}O_{50}H_4$, and NKX-2311–$Ti_{24}O_{50}H_4$, calculated in water at the DFT/B3LYP/LANL2DZ level. The contributions of the various atoms are: Ti - blue, O - red, dye molecules - black, -OH groups - gray. Energy levels were convoluted with Gaussian distributions with full width at half maximum 0.1 eV.

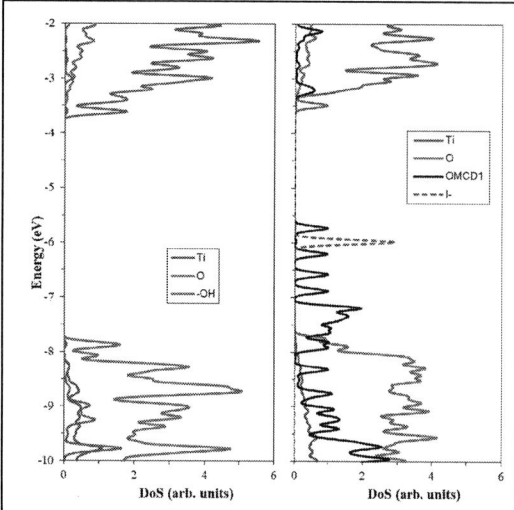

Fig. 8 Density of states of $Ti_{24}O_{50}H_4$ (left) and OMCD1–$Ti_{24}O_{50}H_4$ (right) calculated in water at the DFT/B3LYP/LANL2DZ level. The contributions of the various atoms are: Ti - blue, O – red, -OH groups – grey, dye molecules – black, iodine anion - green. Energy levels were convoluted with Gaussian distributions with full width at half maximum 0.1 eV

For OMCD1 present in the DoS are also the three occupied $5p$ orbitals of the negative iodine ion included in the modeling of the dye. They are localized just below the ground state, at -5.9 eV, thus not contributing to the regeneration process.

4. Conclusions

We reported DFT calculations of several dyes and dye-oxide systems to study the role of the cluster size and the suitability of the dyes as TiO_2 sensitizers for DSSCs. We examined various clusters and, due to the rigidity of the backbone of all dyes, we were able to use, in a compromise between accuracy and computation time, the $n = 24$ cluster.

Based on optimized geometries we discussed the anchoring of the dyes on the cluster, observing that the bidentate binding is energetically preferred. Based on the electronic spectrum we could discuss the matching of the absorption spectrum with the solar spectrum and the energy level alignment, which allows for both charge injection into the oxide and dye regeneration from the electrolyte.

We analyzed the key molecular orbitals and observed push-pull effects. We noted the superiority of the calculation for the entire dye-oxide system, which provides a mixed character of the molecular orbitals.

We justified that NKX-2311 has superior light harvesting properties to both NKX-2398 and (especially to) C343 and observed that OMCD1 is a candidate for DSSC dye.

Acknowledgments. M.A.G and C.I.O acknowledge the financial support received from SNSF and UEFISCDI under the Romanian-Swiss Research Programme, through the grant RSRP #IZERO-142144/1—PN-II-ID-RSRP-1/2012, and R.M.A.E. is grateful to the Science and Technology Development Fund, Egypt, for the STDF/25833 grant..

References

[1] X. Chen, A. Selloni, Introduction: Titanium Dioxide (TiO2) Nanomaterials, Chem. Rev., 114, 9281–9282, 2014.

[2] Y. Bai, I. Mora-Seró, F. De Angelis, J. Bisquert, P. Wang, Titanium Dioxide Nanomaterials for Photovoltaic Applications, Chem. Rev. 114, 10095−10130, 2014.

[3] M. Kapilashrami, Y. Zhang, Y.-S. Liu, A. Hagfeldt, J. Guo, Probing the Optical Property and Electronic Structure of TiO2 Nanomaterials for Renewable Energy Applications, Chem. Rev. 114, 9662−9707, 2014.

[4] J. Schneider, M. Matsuoka, M. Takeuchi, J. Zhang, Y. Horiuchi, M. Anpo, D.W. Bahnemann, Understanding TiO2 Photocatalysis: Mechanisms and Materials, Chem. Rev. 114, 9919−9986, 2014.

[5] Y. Ma, X. Wang, Y. Jia, X. Chen, H. Han, C. Li, Titanium Dioxide-Based Nanomaterials for Photocatalytic Fuel Generations, Chem. Rev. 114, 9987−10043, 2014.

[6] J. Bai, B. Zhou, Titanium Dioxide Nanomaterials for Sensor Applications, Chem. Rev. 114, 10095−10130, 2014.

[7] G. Liu, H.G. Yang, J. Pan, Y.Q. Yang, G.Q. Lu, H.-M. Cheng, Titanium Dioxide Crystals with Tailored Facets, Chem. Rev. 114, 9559−9612, 2014.

[8] T. Rajh, N.M. Dimitrijevic, M. Bissonnette, T. Koritarov, V. Konda, Titanium Dioxide in the Service of the Biomedical Revolution, Chem. Rev. 114, 10177−10216, 2014.

[9] A. Fujishima, K. Honda, Electrochemical Photolysis of Water at a Semiconductor Electrode, Nature 238, 37-38, 1972.

[10] M. Cargnello, T.R. Gordon, C.B. Murray, Solution-Phase Synthesis of Titanium Dioxide Nanoparticles and Nanocrystals, Chem. Rev. 114, 9319–9345, 2014.

[11] D. Fattakhova-Rohlfing, A. Zaleska, T. Bein, Three-Dimensional Titanium Dioxide Nanomaterials, Chem. Rev. 114, 9487–9558, 2014.

[12] C. Richard, A. Catlow, S.T. Bromley, S. Hamad, M. Mora-Fonz, A.A. Sokol, S.M. Woodley, Modelling nano-clusters and nucleation, Phys. Chem. Chem. Phys. 12, 786–811, 2010.

[13] F. De Angelis, C. Di Valentin, S. Fantacci, A. Vittadini, A. Selloni, Theoretical Studies on Anatase and Less Common TiO2 Phases: Bulk, Surfaces, and Nanomaterials, Chem. Rev. 114, 9708−9753, 2014.

[14] O. Carp, C.L. Huisman, A. Reller, Photoinduced reactivity of titanium dioxide, Prog. Solid State Chem. 32, 33-177, 2004.

[15] J.C. Conesa, The Relevance of Dispersion Interactions for the Stability of Oxide Phases, J. Phys. Chem. C 114, 22718–22726, 2010.

[16] L. Kavan, M. Grätzel, S.E. Gilbert, C. Klemenz, H.J. Scheel, Electrochemical and Photoelectrochemical Investigation of Single-Crystal Anatase, J. Am. Chem. Soc. 118, 6716, 1996.

[17] A. Selloni, Anatase shows its reactive side, Nature Mater., 7, 613-615, 2008.

[18] F. De Angelis, Modeling Materials and Processes in Hybrid/Organic Photovoltaics: From Dye-Sensitized to Perovskite Solar Cells, Acc. Chem. Res. 47, 3349−3360, 2014.

[19] R. Sanchez-de-Armas, M.A. San-Miguel, J. Oviedo, A. Marquez, J.F. Sanz, Electronic structure and optical spectra of catechol on TiO_2 nanoparticles from real time TD-DFT simulations, Phys. Chem. Chem. Phys. 13, 1506–1514, 2011.

[20] R. Sanchez-de-Armas, M.A. San-Miguel, J. Oviedo, A. Marquez, J.F. Sanz, Direct vs Indirect Mechanisms for Electron Injection in Dye-Sensitized Solar Cells, Phys. Chem. Chem. Phys. 14, 225–233, 2012.

[21] M.J. Lundqvist, M. Nilsing, S. Lunell, B. Akermark, P. Persson, Spacer and Anchor Effects on the Electronic Coupling in Ruthenium-bis-Terpyridine Dye-Sensitized TiO2 Nanocrystals Studied by DFT, J. Phys. Chem. B 110, 20513-20525, 2006.

[22] M.J. Lundqvist, M. Nilsing, P. Persson, S. Lunell, DFT Study of Bare and Dye-Sensitized TiO2 Clusters and Nanocrystals, Int. J. Quantum Chem. 106, 3214-3234, 2006.

[23] L.A. Fredin, K. Wärnmark, V. Sundström, P. Persson, Molecular and Interfacial Calculations of Iron(II) Light Harvesters, ChemSusChem 9, 667 – 675, 2016.

[24] M. Gałynska, P. Persson, Emerging Polymorphism in Nanostructured TiO2: Quantum Chemical Comparison of Anatase, Rutile, and Brookite Clusters, Int. J. Quantum Chem. 113, 2611–2620, 2013.

[25] P. Persson, J.C.M. Gebhardt S. Lunell, The Smallest Possible Nanocrystals of Semiionic Oxides, J. Phys. Chem. B 107, 3336-3339, 2003.

[26] T. Homann, T. Bredow, K. Jug, Adsorption of small molecules on the anatase (100) surface, Surf. Science 555, 135–144, 2004.

[27] F. De Angelis, A. Tilocca, A. Selloni, Time-Dependent DFT Study of [Fe(CN)6]4-Sensitization of TiO2 Nanoparticles, J. Am. Chem. Soc. 126, 15024-15025, 2004.

[28] A. Dualeh, F. De Angelis, S. Fantacci, T. Moehl, C. Yi, F. Kessler, E. Baranoff, M.K. Nazeeruddin, M. Grätzel, Influence of Donor Groups of Organic D-π-A Dyes on Open-Circuit Voltage in Solid-State Dye-Sensitized Solar Cells, J. Phys. Chem. C 116, 1572–1578, 2012.

[29] E. Ronca, M. Pastore, L. Belpassi, F. Tarantelli, F. De Angelis, Influence of the dye molecular structure on the TiO2 conduction band in dye-sensitized solar cells: disentangling charge transfer and electrostatic effects, Energy Environ. Sci. 6, 183–193, 2013.

[30] F. De Angelis, S. Fantacci, E. Mosconi, M.K. Nazeeruddin, M. Grätzel, First-Principles Modeling of the Adsorption Geometry and Electronic Structure of Ru(II) Dyes on Extended TiO2 Substrates for Dye-Sensitized Solar Cell Applications, J. Phys. Chem. C 114, 6054–6061, 2010.

[31] F. De Angelis, S. Fantacci, E. Mosconi, M.K. Nazeeruddin, M. Grätzel, Absorption Spectra and Excited State Energy Levels of the N719 Dye on TiO2 in Dye-Sensitized Solar Cell Models, J. Phys. Chem. C 115, 8825–8831, 2011.

[32] L. Lasser, E. Ronca, M. Pastore, F. De Angelis, J. Cornil, R. Lazzaroni, David Beljonne, Energy Level Alignment at Titanium Oxide−Dye Interfaces: Implications for Electron Injection and Light Harvesting, J. Phys. Chem. C 119, 9899−9909, 2015.

[33] M. Pastore, F. De Angelis, Intermolecular Interactions in Dye-Sensitized Solar Cells: A Computational Modeling Perspective, J. Phys. Chem. Lett. 4, 956−974, 2013.

[34] A.S. Barnard, S. Erdin, Y. Lin, P. Zapol, J. W. Halley, Modeling the structure and electronic properties of TiO2 nanoparticles, Phys. Rev. B 73, 205405, 2006.

[35] Z.-w. Qu G.-J. Kroes, Theoretical Study of Stable, Defect-Free (TiO2)n Nanoparticles with n = 10−16, J. Phys. Chem. C, 111, 16808–16817, 2007.

[36] A. Iacomino, G. Cantele, D. Ninno, I. Marri, S. Ossicini, Structural, electronic, and surface properties of anatase TiO2 nanocrystals from first principles, Phys. Rev. B 78, 075405, 2008.

[37] Y.-F. Li, Z.-P. Liu, Particle Size, Shape and Activity for Photocatalysis on Titania Anatase Nanoparticles in Aqueous Surroundings, J. Am. Chem. Soc., 133, 15743–15752, 2011.

[38] F. Nunzi, E. Mosconi, L. Storchi, E. Ronca, A. Selloni, M. Grätzel, F. De Angelis, Inherent electronic trap states in TiO2 nanocrystals: effect of saturation and sintering, Energy Environ. Sci., 6, 1221-1229, 2013.

[39] M. Grätzel, Photoelectrochemical cells, Nature 414, 338-344, 2001.

[40] A. Hagfeldt, G. Boschloo, L. Sun, L. Kloo, H. Pettersson, Dye-sensitized solar cells. Chem. Rev. 110, 6595–6663, 2010.

[41] C.I. Oprea, P. Panait, J. Lungu, D. Stamate, A. Dumbravă, F. Cimpoesu, M.A. Gîrţu, DFT study of binding and electron transfer from a metal-free dye with carboxyl, hydroxyl and sulfonic anchors to a titanium dioxide nanocluster, Int. J. Photoenergy, 893850, 2013.

[42] C.I. Oprea, P. Panait, M.A. Gîrţu, DFT study of binding and electron transfer from colorless aromatic pollutants to a TiO2 nanocluster: Application to photocatalytic degradation under visible light irradiation, Beilstein J. Nanotechnol. 5, 1016–1030, 2014.

[43] C.I. Oprea, L.C. Petcu, M.A. Gîrţu, DFT Study of Binding and Electron Transfer from Penicillin to a TiO2 Nanocluster: Applications to Photocatalytic Degradation, IEEE Proceedings 2015 E-Health and Bioengineering Conference (EHB) 7391481, 2016.

[44] C.I. Oprea, P. Panait, F. Cimpoesu, M. Ferbinteanu, M.A. Gîrţu, DFT Study of Coumarin-based Dyes Adsorbed on TiO2 Nanoclusters – Applications to Dye-Sensitized Solar Cells, Materials 6, 2372-2392, 2013.

[45] M. Horn, C.F. Schwerdtfeger, E.P. Meagher, Refinement of the structure of anatase at several temperatures, Z. Krist. 136, 273–281, 1972.

[46] W. Kohn, L.J. Sham, Self-consistent equations including exchange and correlation effects. Phys. Rev. 140, A1133–A1138, 1965.

[47] A.D. Becke, Density-functional exchange-energy approximation with correct asymptotic behavior. Phys. Rev. A38, 3098–3100, 1988.

[48] C. Lee, W. Yang, R.G. Parr, Development of the colle-salvetti correlation-energy formula into a functional of the electron density. Phys. Rev. B37, 785–789, 1988.

[49] P.J. Hay, W.R. Wadt, Ab initio effective core potentials for molecular calculations. Potentials for K to Au including the outermost core orbitals. J. Chem. Phys. 82, 299–311, 1985.

[50] A.D. Becke, Density-functional thermo-chemistry. III. The role of exact exchange, J. Chem. Phys. 98, 5648–5652, 1993.

[51] V. Barone, M. Cossi, Quantum calculation of molecular energies and energy gradients in solution by a conductor solvent model, J. Phys. Chem. A 102, 1995–2001, 1998.

[52] J. Tomasi, B. Mennucci, R. Cammi, Quantum mechanical continuum solvation models, Chem. Rev. 105, 2999–3093, 2005.

[53] GAUSSIAN 09, Revision A.02, M. J. Frisch, et al., Gaussian, Inc., Wallingford CT, 2016, citation available online at http://gaussian.com/g09citation/ (accessed on 22 June 2018).

[54] N.M. O'Boyle, A.L. Tenderholt, K.M. Langner, cclib: a library for package-independent computational chemistry algorithms, J. Comp. Chem. 29, 839-845, 2008.

[55] M. Lazzeri, A. Vittadini, A. Selloni, Structure and energetics of stoichiometric TiO2 anatase surfaces. Phys. Rev. B 63, 155409, 2011.

[56] K. Hara, T. Sato, R. Katoh, A. Furube, Y. Ohga, A. Shinpo, S. Suga, K. Sayama, H. Sugihara, H. Arakawa, Molecular design of coumarin dyes for efficient dye-sensitized solar cells. J. Phys. Chem. B 107, 597–606, 2003.

[57] Y. Kurashige, T. Nakajima, S. Kurashige, K. Hirao, Theoretical investigation of the excited states of coumarin dyes for dye-sensitized solar cells, J. Phys. Chem. A 111, 5544–5548, 2007.

[58] X. Zhang, J.-J. Zhang, Y.-Y. Xia, Molecular design of coumarin dyes with high efficiency in dye- sensitized solar cells. J. Photochem. Photobiol. A Chem. 194, 167–172, 2008.

[59] E. Mosconi, A. Selloni, F. De Angelis, Solvent Effects on the Adsorption Geometry and Electronic Structure of Dye-Sensitized TiO2: A First-Principles Investigation, J. Phys. Chem. C 116, 5932−5940, 2012.

[60] M. Pastore, F. De Angelis, Computational modelling of TiO2 surfaces sensitized by organic dyes with different anchoring groups: adsorption modes, electronic structure and implication for electron injection/recombination, Phys. Chem. Chem. Phys. 14, 920–928, 2012.

[61] F. De Angelis, S. Fantacci, A. Selloni, Alignment of the dye's molecular levels with the TiO2 band edges in dye-sensitized solar cells: A DFT-TDDFT study, Nanotechnology 19, 424002, 2008.

978-1-5386-4483-6/18 $31.00 © 2018 IEEE

Analytical modelling approach in study of the transient response of thermopile-based MEMS sensors applied for simultaneous detection of pressure and gas composition

D.V. Randjelović

Centre of Microelectronic Technologies, Institute of Chemistry, Technology and Metallurgy, University of Belgrade,
Njegoševa 12, 11000 Belgrade, Serbia
E-mail: danijela@nanosys.ihtm.bg.ac.rs

Abstract—This work demonstrates how binary gas mixture composition and pressure can be simultaneously determined based on the value of thermal time constant of thermopile-based MEMS sensor. Self-developed analytical model for transient analysis is applied initially to define optimal design of the thermal sensor in terms of the residual n-Si thickness and the number of thermocouples. Procedure for detection of gas mixture composition and pressure was implemented on thermal sensor placed in binary gas mixtures used as shielding gases in industrial processes of welding, cutting and melting of metals.

Keywords—gas sensor; analytical model; thermal time constant; thermopile.

1. Introduction

According to the latest reports [1], gas sensor market at the global level is expected to expand significantly by 2025. There is a growing demand for gas sensors in different fields like environmental monitoring, healthcare, industrial processing, industrial and occupational safety, anti-terrorism, etc.

Gas sensor based on multipurpose thermopile-based device is one of the topics of research at ICTM-CMT. This type of thermal sensor was developed earlier [2, 3] and the following applications of the device were verified so far: 1) flow sensor [4 5], 2) vacuum sensor [6], 3) helium gas sensor [7], 4) thermal converter [4], 5) intelligent vacuum sensor [8, 9].

One direction of research includes study of possibility of novel principle of operation, based on measurement of the thermal time constant. In [10] it was shown that this novel principle of operation could be used for vacuum detection, that is, for measurement of pressures below atmospheric. The aim of this work is to demonstrate how binary gas mixture composition and pressure can be simultaneously determined based on transient response of thermopile-based sensor.

One of the most widespread role of binary gas mixtures in industry is the shielding gas. Shileding gases are necessary in the processes of welding, cutting and melting of metals where they act as a barrier towards atmosphere. Binary gas mixtures obtained by adding specific percentage of carbon dioxide, oxygen, helium or hydrogen to argon, are applied as shielding gas mixtures in Metal Inert Gas (MIG) welding, Tungsten Inert Gas (TIG) welding and plasma welding and cutting processes [11]. Despite the fact that argon and helium belong to the class of noble gases, their cost justifies application in industry.

In this paper, transient response of thermal sensor in the atmosphere of the above mentioned binary gas mixtures for industrial use is investigated. After presenting the structure and relevant parameters of the studied sensor, analytical model for transient simulation is given. The main output of the modelling is the thermal time constant. Next, investigation of influence of geometrical parameters of the sensor as well as of the ambient gas composition on thermal time constant is performed. Afterwards, procedure for detection of binary gas mixture composition and pressure based on the value of thermal time constant is presented. Finally, simulation results and limitations of the proposed procedure are discussed and conclusions important for further research activities are deduced.

978-1-5386-4483-6/18 $31.00 © 2018 IEEE

2. Structure of the Thermal Sensor

Structure of the studied thermal sensor is illustrated in Fig. 1. Sensor is fabricated on standard n-type silicon wafer with 380 μm nominal thickness. In the central part of the chip, thermally and electrically isolating membrane is formed by bulk etching. The membrane consists of sputtered SiO_2 and residual n-Si layer of thickness d_{nSi}. In order to assure a multipurpose device, two independent thermopiles are placed to the left and right side of the p^+Si heater. Electrical resistance of the heater equals $R_H = 5.8$ kΩ. Each thermopile consists of $N/2$ thermocouples formed of p^+Si stripes obtained by boron diffusion and Al stripes obtained by sputtering. Hot thermopile junctions are situated in the vicinity of the heater, while the cold junctions are placed on the unetched part of the chip surrounding the membrane – the rim.

In this work structures with 60 and 120 thermocouples placed on the membrane area of the same size are studied. When $N = 60$ widths of p^+Si and Al stripes forming a thermocouple are $w_{p+Si} = 60$ μm and $w_{Al} = 40$ μm. For $N = 120$, $w_{p+Si} = 20$ μm, while $w_{Al} = 10$ μm.

Fig. 1 also shows the two rectangular zones defined for the purposes of analytical modelling. Length of the first zone is $l_0 = 180$ μm, while l_1 depends on thickness of the residual silicon and can easily be calculated for the assumed anisotropic etching conditions of Si (100) wafer. Two structures with different residual n-Si thickness are studied. For the structure with $d_{nSi} = 0$ μm, l_1 equals 790 μm, while in the case of $d_{nSi} = 3$ μm, $l_1 = 793$ μm.

Relevant properties and corresponding values of the materials from which sensor's elements are formed, are listed in Table 1.

Table 1. Material properties and corresponding values used in simulations

Parameter	Material			
	n-Si	*p^+Si*	*SiO_2*	*Al*
Thermal conductivity λ [W/(mK)]	150	75	1.2	218
Thickness d [μm]	3	0.3	1	0.7
Emissivity ε	0.5	0.5	0.2	0.8
Density ρ [kg/m3]	2330	2420	2220	2702
Thermal diffusivity a [10^{-6} m^2/s]	89.2	41.3	0.834	97.1
Specific heat capacity c [J/(kgK)]	712	750	745	903

3. Analytical Model for Transient Simulation

Thermopile based sensor shown in Fig. 1. has relatively complex structure, therefore, transient simulation is performed by studying the equivalent structure [12]. This equivalent structure is represented by equivalent parameters.

The applied two-zone model is explained

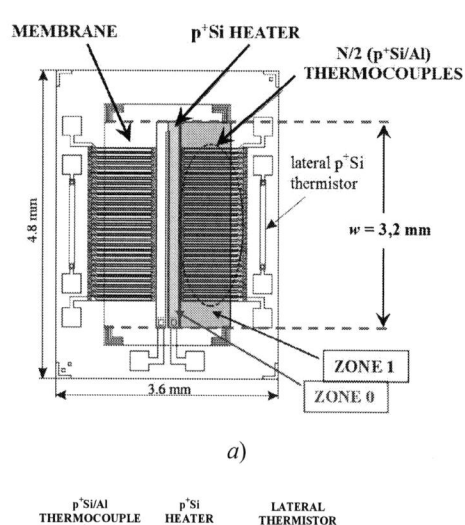

a)

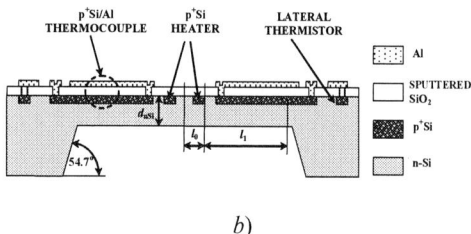

b)

Fig. 1 a) Top view of the sensor with marked main elements and the two zones used in analytical modelling, b) cross section of the sensor with depicted residual n-Si thickness, d_{nSi}, and the lengths of the two zones, l_0 and l_1.

in detail in [4]. The two zones are shown in Fig. 1. The first zone (zone 0) comprises the heater area, while the second one (zone 1) includes the rest of the membrane area along the thermocouples till its edge. The zones are delimited by the upper and lower edges of the heater which can be justified by comparison of simulation data obtained for the whole membrane area and the area marked in Fig. 1. Because of the symmetry, only one half of the structure is analysed.

In each zone we have elements fabricated of some of the materials listed in Table 1. In order to reflect their influence, coefficient of coverage, k_{mi} is calculated as a ratio of area covered by the specific material (m) to the total area of the zone i.

Equivalent thickness is defined as

$$d_{ei} = \sum_m k_{mi} d_{mi}, \qquad (1)$$

while other equivalent parameters are obtained using relation

$$\phi_{ei} = \frac{1}{d_{ei}} \sum_m k_{mi} d_{mi} \phi_{mi}. \qquad (2)$$

Now, instead of parameters given in Table 1, we have equivalent parameters in each zone as listed in Table 2.

Moreover, equivalent length of the whole analysed part of the structure is defined as

$$l_e = l_0 + l_1 \qquad (3)$$

By taking into account expressions derived

Table 2. Equivalent parameters of each zone.

Equivalent parameter	$i = 0$	$i = 1$
	Zone 0	Zone 1
Thermal conductivity λ_{ei} [W/(mK)]	15.96	30.15
Thickness d_{ei} [µm]	1.25	1.24
Density ρ_{ei} [kg/m3]	2260	2287.5
Thermal diffusivity a_{ei} [10^{-6} m²/s]	8.93	14.37
Specific heat capacity c_{ei} [J/(kgK)]	746	762

by transient analysis, which is explained thoroughly in [4], expression for thermal time constant of the studied sensor can be written as

$$\tau_{th}(x, p, T) = \frac{l_e^2}{(\pi/2)^2 a_e + B_e(x, p, T) l_e^2}. \qquad (4)$$

Equivalent thermal diffusivity of the whole area used in simulations, a_e, is calculated as

$$a_e = \frac{1}{l_e} \sum_{i=0}^{1} a_{ei} l_i = 13.36 \, 10^{-6} \, \frac{m^2}{s}. \qquad (5)$$

Parameter $B_e(x,p,T)$ can be written as

$$B_e(x, p, T) = \frac{1}{l_e} \sum_{i=0}^{1} \frac{D \lambda_{mix}(x, p, T) + E T^3}{c_{ei} \rho_{ei} d_{ei}} l_i, \qquad (6)$$

Where D and E are constants given by the following relations:

$$D = d_1^{-1} + d_2^{-1}, \qquad (7)$$

$$E = 4\sigma_B \left(\varepsilon_i^{upper} + \varepsilon_i^{lower} \right). \qquad (8)$$

Parameters d_1 and d_2 are the distances from the upper and the bottom side of the membrane to the housing cap and the housing base, respectively. Parameter d_1 is simply calculated by subtracting membrane thickness from the thickness of silicon wafer. It was assumed that $d_2 = 2$ mm. Emissivities of the upper and lower surfaces of each zone are denoted by ε_i^{upper} and ε_i^{lower}, while $\sigma_B = 5,67 \cdot 10^{-8}$ W/(m²K⁴) is Stefan-Boltzmann constant.

It can be concluded that thermal time constant depends on geometrical parameters of the sensor, like number of thermocouples and residual n-Si thickness. On the other hand, thermal time constant is also pressure and temperature dependent via the influence of thermal conductivity of the surrounding gas. Influence of gas composition is also present. Taking into account these facts, general implicit expression can be written

$$\tau_{th} = \tau_{th}(N, d_{n\text{-Si}}, \lambda_{mix}(x, p, T)). \qquad (9)$$

We will study the sensor surrounded by binary gas mixture with Ar as the carrier gas. Gas mixture is formed by adding a certain quantity, x, of a chosen gas to argon.

Expression for thermal conductivity of a binary gas mixture is obtained using procedure explained in [3]

$$\lambda_{mix}(x,p,T) = \frac{x\,\lambda_{GAS}(p,T)}{x+(1-x)\,F_{12}(T)} + \frac{(1-x)\,\lambda_{Ar}(p,T)}{(1-x)+x\,F_{21}(T)}. \quad (10)$$

Parameters F_{12} and F_{21} depend on the ratios of the molecular weights, M, and dynamic viscosities, μ, of argon and the added gas:

$$F_{12}(T) = \frac{1+\left(\mu_{GAS}(T)/\mu_{Ar}(T)\right)^{0.5}\left(M_{Ar}/M_{GAS}\right)^{0.25}}{\left(8\left(1+M_{Ar}/M_{GAS}\right)\right)^{0.5}}. \quad (11)$$

$$F_{21}(T) = \frac{1+\left(\mu_{Ar}(T)/\mu_{GAS}(T)\right)^{0.5}\left(M_{GAS}/M_{Ar}\right)^{0.25}}{\left(8\left(1+M_{GAS}/M_{Ar}\right)\right)^{0.5}}. \quad (12)$$

Temperature dependence of dynamic viscosity for each gas can be written as

$$\mu_{GAS}(T) = K_1(T/1000)^2 + K_2(T/1000) + K_3, \quad (13)$$

where coefficients K are constants determined by fitting data given in the NIST database [13].

On the other hand, thermal conductivity of each gas depends on temperature and pressure and it is given by [6, 10]:

$$\lambda_{GAS}(p,T) = \left(\left(\lambda_{hp}(T)\right)^{-1} + D\left(\gamma_{lp}(T)\,p\right)^{-1}\right)^{-1}. \quad (14)$$

Parameters $\lambda_{hp}(T)$ and $\gamma_{lp}(T)$ dominate in the high and low pressure region, respectively. Parameter $\lambda_{hp}(T)$ is commonly assumed as gas thermal conductivity in literature. Temperature dependence of λ_{hp} can be written as

$$\lambda_{hp}(T) = C_1(T/1000)^2 + C_2(T/1000) + C_3, \quad (15)$$

where constant coefficients C were determined by fitting data from the NIST database.

Parameter $\gamma_{lp}(T)$ is calculated using the following relation

$$\gamma_{lp}(T) = \frac{c}{3}\sqrt{\frac{8M}{\pi k_B T}}, \qquad (16)$$

where $k_B = 1{,}38\ 10^{-23}$ J/K is Bolzmann's constant, M is the molecular weight, while T is the ambient temperature.

4. Simulation results

Simulations were performed under assumption that the sensor is operating at constant temperature of 20 $^{\circ}$C. Sensor is placed in the atmosphere of binary gas mixtures formed of argon as a carrier gas and specific percentages of carbon dioxide, oxygen, helium or hydrogen.

A. Properties of the Chosen Gases

Molecular weights and specific heat capacities of the chosen gases are listed in Table 3. Molecular weights are expressed in atomic mass units (1u = 1.66 10^{-27} kg).

Coefficients necessary for calculation of thermal conductivities and dynamic viscosities of the gases using (13) and (15) are given in Table 4 and Table 5, respectively.

Table 3. Molecular weights and specific heat capacities of the chosen gases

Gas	Molecular Weight [u]	Specific Heat Capacity [J/kgK]
Argon (Ar)	39.95	520
Carbon dioxide (CO₂)	44.0	844
Hydrogen (H₂)	2.01	14320
Helium (He)	4.0	5190
Oxygen (O₂)	32.0	919

Table 4. Coefficients for calculation of thermal conductivities of the chosen gases

Gas	C_1	C_2	C_3
Argon (Ar)	-28.62	+67.35	+0.15
Carbon dioxide (CO_2)	+22.23	+65.97	-4.95
Hydrogen (H_2)	-137.63	+581.78	+23.36
Helium (He)	-131.41	+434.26	+37.31
Oxygen (O_2)	-4.11	+83.75	+1.85

Table 5. Coefficients for calculation of dynamic viscosities of the chosen gases

Gas	K_1	K_2	K_3
Argon (Ar)	-37.51	+86.53	+0.18
Carbon dioxide (CO_2)	-16.73	58.38	-0.99
Hydrogen (H_2)	-4.29	+22.20	+2.63
Helium (He)	-16.42	+55.10	+4.85
Oxygen (O_2)	-31.43	+74.36	+1.04

B. Influence of the Sensor's Design on the Thermal Time Constant

In order to analyse influence of the sensor's design on thermal time constant, simulations were performed for structures with different number of thermocouples and different thicknesses of residual n-Si layer. It was assumed that sensors are placed in binary gas mixture formed of 25% of helium added to argon.

Fig. 2 shows influence of residual n-Si thickness on pressure dependence of the thermal time constant for sensors with 60 thermocouples.

The first structure has residual n-Si layer of thickness $d_{nSi} = 3$ μm, while in the second one residual n-Si is completely removed, $d_{nSi} = 0$ μm. The change of thermal time constant over the given pressure range is 8 ms for sensor without residual n-Si, but only 0.44 ms when 3 μm of residual n-Si layer are present.

Next, the influence of the number of thermocouples was studied. Fig. 3 shows results obtained for sensors without residual

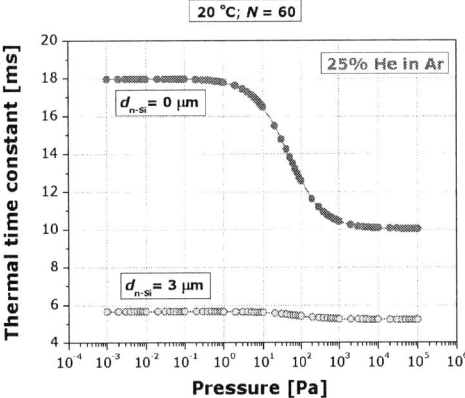

Fig. 2 Influence of residual n-Si thickness on pressure dependence of the thermal time constant for sensors with 60 thermocouples. Calculations were done for structures with $d_{nSi} = 3$ μm and $d_{nSi} = 0$ μm.

n-Si layer having $N = 60$ and $N = 120$ thermocouples. Structure with higher number of thermocouples exibits change of the thermal time constant of 15 ms, compared with previously mentioned change of 8 ms, for the sensor with 60 thermocouples.

Since these simulations imply that sensor without residual silicon and with higher number of thermocouples exibits more prominent change of the thermal time constant, the following simulations will be performed for structure with $d_{nSi} = 0$ μm and $N = 120$.

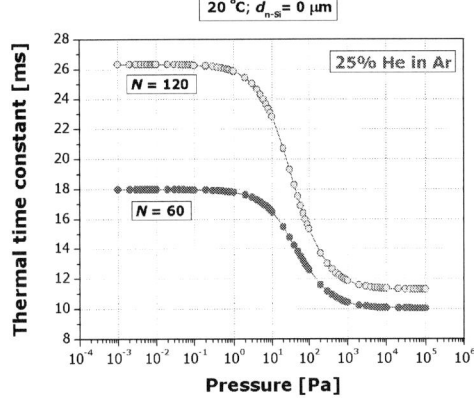

Fig. 3 Influence of the number of thermocouples on pressure dependence of the thermal time constant for sensors without residual n-Si layer. Calculations were done for structures with $N = 60$ and $N = 120$.

C. Influence of Ambient Gas Composition on Thermal Time Constant

In order to investigate how thermal time constant changes with ambient gas composition simulations were done for different gas mixtures under assumption that sensor is operating at atmospheric pressure. Each binary gas mixture was formed of argon as a carrier gas and a variable amount of carbon dioxide, oxygen, helium and hydrogen.

Results shown in Fig. 4 illustrate that by increasing the amount of the added gas, thermal time constant can increase or decrease. If the added gas has lower thermal conductivity than the carrier gas, thermal time constant will increase, like in the case of adding carbon dioxide. On the other hand, adding a gas with higher thermal conductivity compared with argon, will result in decrease of the value of thermal time constant.

As the extent to which difference in thermal conductivities of the added gas and the carrier gas is higher, slope of the curve increases resulting in higher total change of the thermal time constant.

D. Detection of Binary Gas Mixture Composition and Pressure Based on Measurement of the Thermal Time Constant

Procedure for the detection of presure and gas composition if only the measured value of the thermal time constant is known, will be presented for the sensor placed in the binary gas mixture of helium and argon with unknown fractions and at unknown pressure.

Let us assume that measured value of the thermal time constant is $\tau_0(p_0) = 16$ ms. As illustrated in Fig. 5, this value can correspond to a number of fractions, x, of helium in argon, but we will consider only two curves with He fractions x_1 and x_2. For each fraction, there is a corresponding value of the binary gas mixture pressure.

From previous simulations it can be concluded that for each specific composition of a binary gas mixture, there is a specific value of thermal time constant at the atmospheric pressure, p_{ATM}. The first step in the procedure is to measure thermal time constant of the unknown mixture at atmospheric pressure, $\tau(p_{ATM})$. For illustration, let's assume $\tau(p_{ATM}) = 11.3$ ms. Now we should consider $\tau(x)$ curve for He-Ar mixture at p_{ATM} which is shown in Fig. 6.

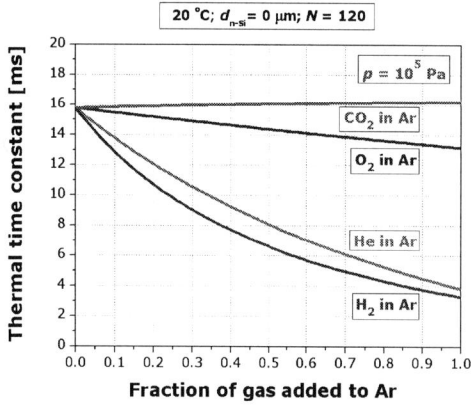

Fig. 4 Influence of fraction of carbon dioxide, oxygen, helium or hydrogen added ro argon on thermal time constant of the sensor at atmospheric pressure.

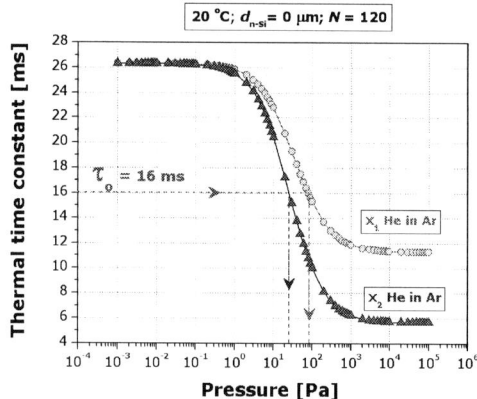

Fig. 5 Illustration of pressure dependence of the thermal time constant of thermal sensor ($d_{nSi} = 0$ μm, $N = 120$) at 20°C for only two unknown fractions, x_1 and x_2, of He in Ar. For the measured value $\tau_0 = 16$ ms, there exists two solutions (x_1,p_1) and (x_2,p_2).

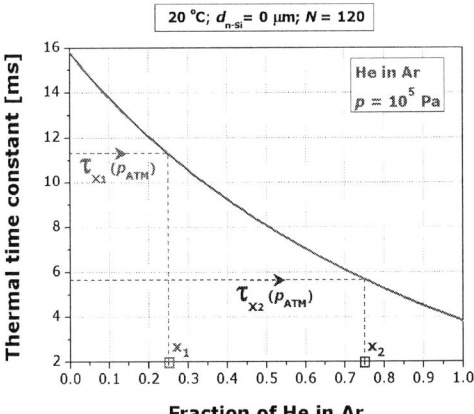
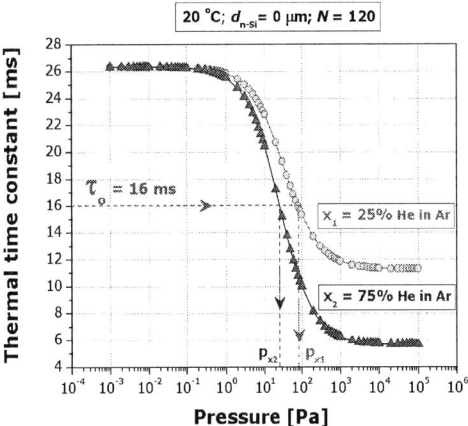

Fig. 6 Dependence of thermal time constant of the sensor on fraction of helium in He-Ar mixture at atmospheric pressure. If $\tau(p_{ATM})$ is measured then fraction x can be dermined

Fig. 7 Pressure dependence of thermal time constant for the determined fraction x leads to determination of the pressure corresponding to $\tau_0 = 16$ ms.

By solving equation

$$\tau(x) = \tau_0 = 16 \text{ ms}, \tag{17}$$

we obtain that fraction $x = x_1 = 25\%$.

Now that the helium fraction is determined, the next step is to draw the curve $\tau(p)$ for $x_1 = 25\%$ of helium in argon as shown in Fig. 7.

By solving equation

$$\tau(p)_{25\%He} = \tau_0 = 16 \text{ ms}, \tag{18}$$

we obtain that the helium-argon mixture pressure is $p = p_{x1} = 79.5$ Pa.

Using the same procedure, for the case of measured thermal time constant at atmospheric pressure, $\tau(p_{ATM}) = 5.7$ ms, the solution would be $x_2 = 75\%$ of helium in argon at the pressure of $p_{x2} = 25.6$ Pa, as illustrated also in the Figs. 5-7.

4. Conclusion

This work demonstrated that binary gas mixture composition and pressure can be simultaneously determined based on the measured value of thermal time constant of thermopile-based MEMS sensor.

Analytical model for transient simulation served to initially define optimal design of thermal sensor in terms of the residual n-Si thickness and the number of thermocouples. Afterwards, the model was applied to thermal sensor placed in the binary gas mixture of helium and argon with unknown gas fractions and at unknown pressure. Procedure for detection of binary gas mixture pressure and fraction of helium is presented in detail.

Due to the "plateaus" existing in the curves representing dependences of thermal time constant on pressure and binary gas composition, some limitations exist for application of the proposed procedure. For pressures below 0.1 Pa it is not possible to determine exactly the pressure and gas fractions. On the other hand, for pressures above 2 kPa, it is not possible to determine the pressure, but gas fractions can be determined if thermal time constant is measured also at any pressure belonging to the "plateau" range, for example, at atmospheric pressure.

Future research will be focused on further optimization of the sensor's design and exploration of limitations of application of the presented procedure. More complex study would include temperature effect which causes shift of the $\tau(p)$ and $\tau(x)$ curves thus implying difficulties in proper identification of the binary gas mixture constituents.

Acknowledgments. This work has been partially supported by the Serbian Ministry of Education, Science and Technological Development within the framework of the Project TR32008.

References

[1] https://www.grandviewresearch.com/industry-analysis/gas-sensors-market

[2] D. Randjelović, Ž. Lazić, M. Popović and M. Matić, *"Helium Sensing Using Multipurpose Thermopile-Based MEMS devices"*, Proc. 28th Int. Conf. on Microelectronics MIEL 2012, Niš, Serbia, May 13-16, pp. 147–150, 2012.

[3] D.V. Randjelović, O. Jakšić, P. Poljak *"Simulation of performance of thermopile based gas sensors applied for emission monitoring in thermal power plants"*, Proc. 40th Int. Semiconductor Conference CAS 2017, Sinaia, Romania, October 11-14, pp. 109-112, 2017.

[4] D. Randjelović et al., *"Multipurpose MEMS Thermal Sensor Based on Thermopiles"*, Sensor Actuat A-Phys, **141**, pp. 404–413, 2008.

[5] D. Randjelović et al., *"Analytical modelling of thermopile based flow sensor and verification with experimental results"*, Micro Engn, **86**, pp. 1293–1296, 2009.

[6] D. Randjelović, V. Jovanov, Ž. Lazić, Z. Djurić, M. Matić, *"Vacuum MEMS Sensor Based on Thermopiles – Simple Model and Experimental Results"*, Proc. 26th Int. Conf. on Microelectronics MIEL 2008, Niš, Serbia, May 11-14, Vol. 2, pp. 367-370, 2008.

[7] D. Randjelović, Ž. Lazić, M. Popović and M. Matić, *"Helium Sensing Using Multipurpose Thermopile-Based MEMS devices"*, Proc. 28th Int. Conf. on Microelectronics MIEL 2012, Niš, Serbia, May 13-16, pp. 147–150, 2012.

[8] D. Randjelović et al.,*"Intelligent thermopile-based vacuum sensor"*, Proc. Eng, **25**, pp. 575-578, 2011.

[9] D.V. Randjelović, M.P. Frantlović, B.L. Miljković, B.M. Popović, Z.S. Jakšić, *"Intelligent Thermal Vacuum Sensors Based on Multipurpose Thermopile MEMS Chips"*, Vacuum, Vol. 101, pp. 118-124, March 2014.

[10] D. V. Randjelović, A.G. Kozlov, O.M. Jakšić, *"Vacuum sensing based on the influence of gas pressure on thermal time constant"*, Proc. 29th Int. Conf. on Microelectronics MIEL 2014, Belgrade, Serbia, May 12-14, pp. 179-182, 2014.

[11] http://www.thermco.com

[12] A.G. Kozlov, *"Optimization of thin-film thermoelectric radiation sensor with comb thermoelectric transducer"*,Sensor Actuat, **75** pp. 139-150., 1999

[13] www.nist.gov

Second Harmonic Generation: a non-destructive characterization method for dielectric-semiconductor interfaces

I. Ionica*, D. Damianos*, A. Kaminski-Cachopo*, D. Blanc-Pelissier**, M. Lei***,
J. Changala***, A. Bouchard*, X. Mescot*, M. Gri*, G. Grosa* S. Cristoloveanu*, G. Vitrant*

* Univ. Grenoble Alpes, CNRS, Grenoble-INP, IMEP-LAHC, 38000 Grenoble, France
E-mails: Irina.Ionica@phelma.grenoble-inp.fr, dimitrios.damianos@minatec.grenoble-inp.fr,
anne.kaminski@minatec.inpg.fr, bouchard@minatec.inpg.fr, mescot@minatec.inpg.fr, gri@minatec.inpg.fr,
grosa@minatec.inpg.fr, sorin@minatec.inpg.fr guy.vitrant@minatec.grenoble-inp.fr
** INL - UMR 5270, INSA de Lyon, 7 avenue Jean Capelle, 69621 Villeurbanne, France
E-mail: daniele.blanc@insa-lyon.fr
*** FemtoMetrix, 1850 East Saint Andrew Place, Santa Ana, CA 92705, USA
E-mails: ming.lei@femtometrix.com, john.changala@femtometrix.com

Abstract—This paper reviews the application of second harmonic generation (SHG) to characterize dielectric-semiconductor interfaces used in microelectronics and photovoltaics. Based on non-linear optics, the method is non-destructive, so particularly advantageous for thin films. The theoretical background shows the possibility to access the electric field at interfaces and consequently to have a non-destructive measurement for interface state densities or fixed charges in oxides. Two more detailed examples of application of SHG characterization will be shown: field-effect passivation of silicon using thin film deposited alumina and interface analysis of silicon-on-insulator substrates.

Keywords—second harmonic generation, dielectric-semiconductor interfaces, interface electric field, fixed oxide charges, interface states.

1. Introduction

In many fields (microelectronics, photovoltaics, sensors etc.), dielectric-on-semiconductor stacks are key elements in active devices. For example, in metal-oxide-semiconductor field effect transistors (MOSFETS), this stack is found at the channel level; in solar cells or image sensors it is used as a passivation layer to improve the carrier collection. In all cases, the stack and most precisely the interfaces must be of high quality to guarantee the good performance of the device. The efforts at the fabrication level need to be accompanied by appropriate characterization techniques, giving access to relevant data. Apart from surface roughness, chemical composition and crystallographic defects, the measurements

must also provide parameters describing the electrically active defects such as interface state traps D_{it}, fixed oxide charges Q_{ox}, ionic mobile charges, etc. [1].

An ideal candidate for electrical characterization of the interface must be a sensitive, non-destructive technique, with mapping possibilities; furthermore it should not necessitate specific device fabrication and should allow direct extraction of the electrical parameters (e.g. D_{it}, Q_{ox}). The constraints are even stronger when the layers under study area few nm-thick.

Typical characterization techniques for measuring D_{it} and Q_{ox} are:

- electrical measurements (current or capacitance versus voltage), that allow direct D_{it} and Q_{ox} extraction; they need fabrication of specific test devices [1].
- corona oxide characterization of semiconductors (COCOS) performed at wafer level [2]; the technique requires strategies of discharging and does not allow charging/discharging studies.

Another parameter reflecting the electrical properties of an interface is the carrier lifetime [3]; its measurement can be directly performed after deposition of the thin oxide layer but quantitative extractions of D_{it} and Q_{ox} are challenging.

An interesting alternative for dielectric-semiconductor interface characterization is second harmonic generation (SHG), which is non-destructive (based on non-linear optics) and does not require fabrication of specific devices. In this paper we discuss the characterization ability of SHG starting with some theoretical aspects and a rapid state of the art of its application for interface characterization. Afterwards, we show two of our studies, one on single interface samples (alumina layers on silicon, Al_2O_3/Si) and the second on multiple interface samples (silicon-on-insulator, SOI substrates).

2. SHG – principle

A high-intensity laser with angular frequency ω (wavelength λ) and electric field amplitude $E(\omega)$, induces linear and nonlinear polarization of materials, which in general, contains terms of different order. The second harmonic is associated with the second order nonlinear polarization $\vec{P}^{(2)}$ [4]:

$$\vec{P}^{(2)}(2\omega) = \ddot{\chi}^{(2)} \cdot \vec{E}(\omega) \cdot \vec{E}(\omega) \quad (1)$$

where $\ddot{\chi}^{(2)}$ is the second order susceptibility tensor of the material.

For centrosymmetric materials, in the dipolar approximation, the crystal symmetry prohibits the bulk second order polarization. Consequently, materials such as silicon, amorphous SiO_2, alumina, HfO_2 and other high-k dielectrics do not exhibit bulk dipolar susceptibility. Nevertheless, if the symmetry is broken (for example at interfaces), the second order susceptibility tensor $\ddot{\chi}_{interface}^{(2)}$ is non-zero and the SH polarization becomes:

$$\vec{P}^{(2)}(2\omega) = \ddot{\chi}_{interface}^{(2)} \cdot \vec{E}(\omega) \cdot \vec{E}(\omega) \quad (2)$$

Moreover, a static electric field E_{DC} present at an interface is a source of symmetry breaking, so the second order polarization becomes:

$$\vec{P}^{(2)}(2\omega) = (\ddot{\chi}_{interface}^{(2)} + \ddot{\chi}^{(3)} \cdot E_{DC}) \cdot \vec{E}(\omega) \cdot \vec{E}(\omega) \quad (3)$$

with $\ddot{\chi}^{(3)}$ the third order susceptibility tensor.

The intensity emitted at the double frequency, $I^{2\omega}$ is given by the square modulus of the polarization:

$$I^{2\omega} \sim \left| \ddot{\chi}_{interface}^{(2)} + \ddot{\chi}^{(3)} \cdot \vec{E}_{DC} \right|^2 \cdot \left| I^{\omega} \right|^2 \quad (4)$$

An additional effect can appear in semiconductors: during laser illumination, electron-holes pairs are created and charged carriers can be injected and trapped in the dielectric material, thus modifying the electric field at the interface. In this case, E_{DC} can be decomposed into E_0, the field present initially due to fixed charges (Q_{ox}) and initially charged traps (including D_{it}) and $E(t)$, the time dependent electric field due to trapping/de-trapping effects at the interface and related to D_{it}:

$$E_{DC} = E_0 + E(t) \quad (5)$$

Consequently, the SHG intensity contains information on D_{it} and Q_{ox}.

3. SHG – a powerful characterization tool

The schematic experiment configuration is shown in Fig. 1. The main experimental parameters that can be used to fine-tune the SHG experiments are: the incident wavelength and power, the input and output polarization, the angle of incidence, the azimuthal angle of the sample. Each of them can provide access to a different kind of information on the sample.

For example, SHG versus wavelength (spectroscopic SHG) was used to identify inter-band resonances at silicon interfaces, as well as the different band contributions [5].

Depending on the incident power of the laser, multi-photon excitation processes can be activated and charging/discharging can occur in different types of traps in the sample. For example, for highly boron-doped silicon samples covered with SiO_2 [6], a low incident power allowed studying the trapping in boron-induced defects, while a higher incident power increased the probability of the 3-photons mechanism and allowed charging both boron-induced defects and oxygen traps on the top oxide surface.

Various input and output polarizations allowed selecting specific components of the susceptibility tensors and were used to determine susceptibility elements [7], [8], [9].

978-1-5386-4483-6/18 $31.00 © 2018 IEEE   36

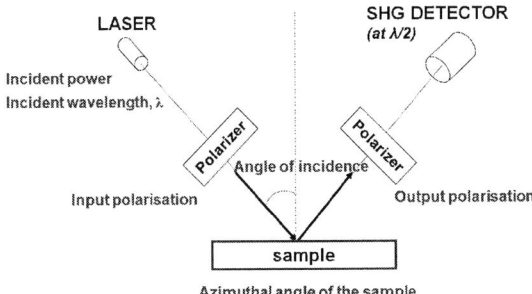

Fig. 1 Schematic of the experimental configuration for SHG measurement, with the key parameters

The SHG variation with the azimuthal angle of the sample (rotational angle around the normal to its surface) is known to depend on the substrate symmetry [10]. Moreover, the surface roughness is related to the SHG intensity versus azimuthal angle. Dadap et al. showed a good correlation between the surface roughness during the first stages of native oxidation and the SHG response to azimuthal angle variation [11]. The non-uniform distribution of interfacial defects that act as trapping centers was also evidenced using rotational angle SHG for ultra-thin high-k dielectric stacks [12].

As seen from equations (4) and (5), SHG variation with time can be used to study trapping/de-trapping phenomena and it was applied for a variety of dielectric materials: SiO_2 [6], [13] Al_2O_3[14], ZrO_2 etc.

The interface mostly studied by SHG was SiO_2/Si due to its numerous applications, especially in microelectronics. Reference [15] shows a wide review of these studies, including micro-roughness, strain effects, bonding and carrier injection.

4. Experimental SHG setup

We performed our measurements on the commercial tool Harmonic F1X from Femtometrix [16]. A pump laser of 780 nm wavelength emits femtosecond pulses with 80 MHz repetition rate, 95 fs pulse duration and an average power of 360 mW. The polarization of the incident beam is controlled by a half wave plate. In the following experiments only p-polarization

(laser electric field parallel to the plane of incidence) was used. The second harmonic light (at 390nm) generated in the sample is separated from the reflected fundamental light using proper filters. A rotating polarizer allows the selection of the SH polarization for the analysis (we only used p-polarization in the following experiments). The photons are detected by a photomultiplier tube coupled to a gated photon counter.

A reflectometer is integrated in the tool, to measure the layer thickness at the same position as the SHG. Wafers up to 300 mm diameter can be tested.

5. SHG for passivation quality evaluation: Al2O3/Si characterization with SHG

A. State of art

One of the main applications of alumina is to passivate image sensors or solar cells [17] in order to increase the collection rate of the charge carriers. Alumina can provide both types of passivation:

- chemical passivation, mainly due to the reduction of the electrically active interface states by chemical bonding Si-O-Al
- field effect passivation, due to negative charges attributed to interstitial oxygen and/or Al vacancies, that are activated by annealing.

SHG has already been applied for characterization of Al_2O_3/silicon samples. Spectroscopic and time-dependent SHG demonstrated the increase of the fixed negative charge in the Al_2O_3 layer after an annealing step [14]. Other studies (spectroscopic, azimuthal, time-dependent SHG) revealed that in fact the polarity of the charges contained in the SiO_2/alumina stack depends on the thickness of the interfacial intermediate SiO_2 layer (always present at the interface Al_2O_3/Si) [18].

With our following study, we took a step towards the quantification of SHG, by correlating it to carrier lifetime measurements.

978-1-5386-4483-6/18 $31.00 © 2018 IEEE

B. Samples and results

Samples were fabricated starting from float-zone p-type double-side polished Si(100) and Si(111) of resistivity 0.8Ωcm. The fabrication process detailed elsewhere [19] consists of a wet cleaning, followed by thermal atomic layer deposition of 15nm Al_2O_3. Half of each sample was annealed (400°C for 10 minutes) which activated hydrogen diffusion and the formation of negative charges in the layer. The surface passivation was first evaluated by spatially resolved effective minority carrier lifetime measurements performed by microwave photo-conductance decay (Semilab WC-2000 µW-PCD). Table 1 shows the average effective lifetime τ_{eff} values estimated from maps obtained on 4 samples, two Al_2O_3/Si(111) and two Al_2O_3/Si(100), before and after annealing. For both types of samples, annealing increases τ_{eff} thanks to reduction of the surface recombination, which is caused by the activation of the negative charges at the Al_2O_3/Si interface (field-effect passivation) and hydrogen diffusion (chemical passivation) [20]. Additionally, Si(111) is known to produce better quality surfaces than Si(100) [21] (confirmed by the higher lifetime values obtained when alumina is deposited on Si(111), in Table 1), suggesting a better passivation.

Table 1. Average values of minority carrier lifetime

τ_{eff} (µs)			
Al_2O_3/Si (100)		Al_2O_3/Si (111)	
As-deposited	Annealed	As-deposited	Annealed
57	140	81	250

Fig. 2 shows the time-dependent SHG signal obtained on these samples. The time variation of the signal is related to $E(t)$ in equation (5), due to trapping/detrapping phenomena, which are generally facilitated by higher trap densities. The faster time-variation for the Si(100) can be therefore related to a higher number of defects and a

lower passivation quality, as confirmed by the lifetime values. For both substrates, the increase of SHG intensity after annealing is consistent with a stronger static electric field at the interface, due to the negative charges activation at the interface during the thermal treatment [20].

The initial SHG value is related to the initial static electric field present at the interfaces, as shown by equation (5). Consequently, the initial SHG intensity is proportional to the square of the surface charge density as shown for SiO_2/Si interfaces [22] and for Al_2O_3/Si [14], [18]. The same dependence on the charge squared is also expected for the lifetime [23].

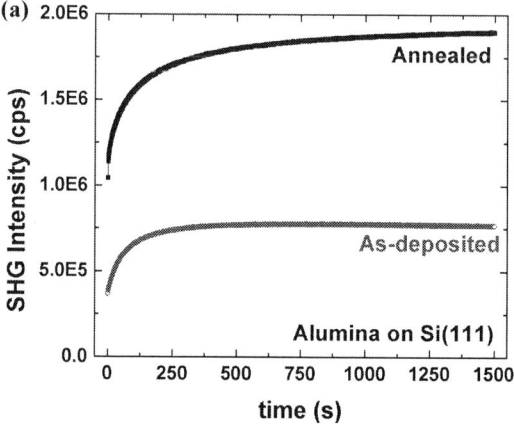

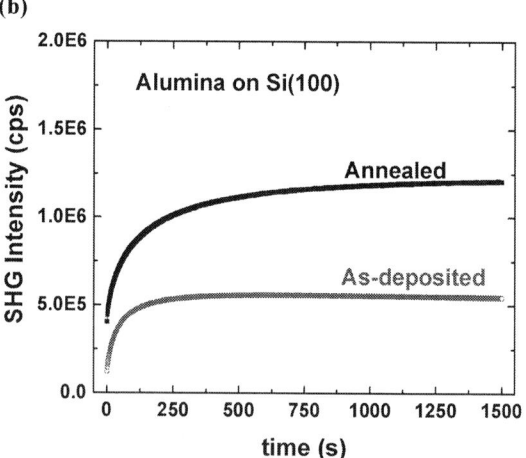

Fig. 2: Time-dependent SHG from the Al_2O_3 on Si(111) (a) and on Si(100) (b), before and after annealing. The same vertical scales were used in both graphs. The angle of incidence was 45°.

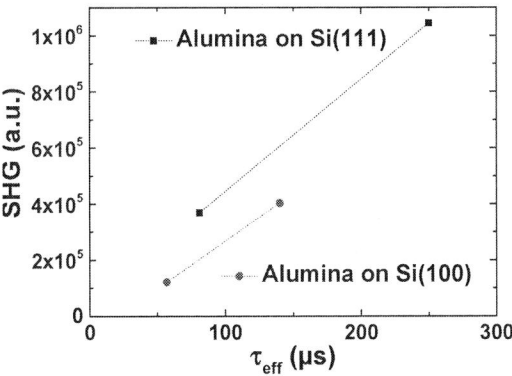

Fig. 3: Initial SHG versus effective carrier lifetime from the Al_2O_3 on Si(100) and on Si(111).

Fig. 3 shows the initial SHG intensity, extracted from measurements in Fig. 2 versus the corresponding effective carrier lifetime. Initial SHG is increasing with τ_{eff}. Note that all the 4 samples do not align on the same line, probably because of a different chemical passivation, which is most likely modifying the $\chi^{(2)}$ susceptibility tensor elements.

For an actual calibration, besides a larger number of samples, the correlation with D_{it} and Q_{ox} values extracted from capacitance-voltage characteristics on fabricated metal-oxide-semiconductor structures [24] would really allow to separately identify the impact of the chemical and the field effect passivation on the SHG.

6. SHG for SOI characterization

A. State of art

In microelectronics, high performance devices are fabricated on top quality silicon-on-insulator (SOI) substrates. The advanced substrates have very thin silicon film (~12nm) and buried oxide (~15-25nm) [25]. The characterization methods used to extract electrical properties of the SOI wafers such as interface trap densities necessitate either full fabrication of test devices (which is not cost-effective) or to place metallic probes on the film (which can be destructive for ultra-thin films) [26]. Within this context, the SHG, which is sensitive to interface electric fields and non-destructive, has a great potential. Previous studies showed that SHG

was able to evidence charging of the buried oxide due to radiation [27], metal contamination on the top of the silicon film [28] or even to reproduce pseudo-MOSFET characteristics [27].

In order to benefit from the obvious advantages of SHG for SOI characterization, we have to understand (and simulate) the SOI stacks as multilayer systems with multiple optical interferences. The optical propagation phenomena (at both fundamental I^ω and second harmonic $I^{2\omega}$) must be taken into account properly before comparing different geometry and/or quality SOI structures.

In our following studies we modeled the optical propagation phenomena and included the interface electric fields, in order to reproduce SHG experimental results from various SOI structures.

B. Experimental & modeled SHG on SOI

The absorption depth for the fundamental and the second harmonic wavelength is around 10μm and 70nm respectively, in the silicon layer. Consequently, when using SOI with silicon films thicker than 70nm, the SHG is mainly coming from the interface between the Si film and native oxide on the top. Therefore, the Si film thickness variations should not play any role. Nevertheless, interferences at the fundamental light can impact the I^ω distribution across the structure.

Fig. 4 shows the SHG intensity and the silicon film thickness, measured on the same spots, across two SOI wafers, one with a thick Si film (a) and the other with a thinner film (b). Despite the obvious correlation between SHG and film thickness for the two wafers, the tendency is opposite for the two cases. The origin of this different behavior is only related to the propagation phenomena which give different interference patterns for the fundamental in each of these wafers.

In order to confirm/predict the role of multiple reflections on the SHG signal, we developed a program for simulating optical propagation phenomena in multilayer

978-1-5386-4483-6/18 $31.00 © 2018 IEEE

structures. When light is incident on a stack, the optical phenomena to be modelled are: propagation and absorption inside each layer and transmission at each interface between two media. We used a matrix formalism [7] and calculated the optical electromagnetic field at 2ω exiting the structure, by properly accounting for the non-linear polarization at each interface, including the static field E_{DC} [29]. In order to compare with experiments, we simulated the SOI as a 5-layered structure, from top to bottom: air, native oxide, Si film, BOX, bulk silicon.

Fig. 5 shows the normalized SHG versus the film thickness. Experimental (stars) data were extracted from Fig 4, while the simulated data (dotted lines) were obtained with our simulator. The correlation is very good for thick SOI (Fig. 5a). Similar results are obtained for the thin SOI, although the variation of t_{Si} is very small (very homogenous wafer).

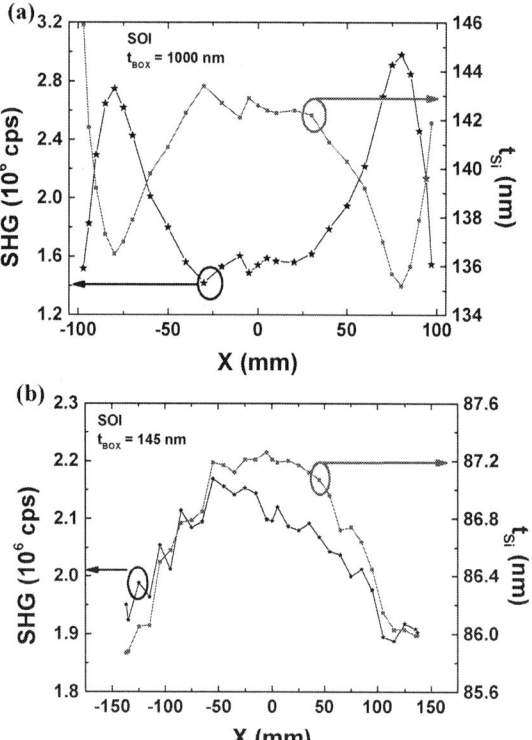

Fig. 4: SHG and Si film thickness measured on the same spots on thick (a) and thin (b) SOI wafers. The X-axis represents different measurement spots on each wafer, across its diameter. The angle of incidence was set at 45° and the input / output polarizations are P / P.

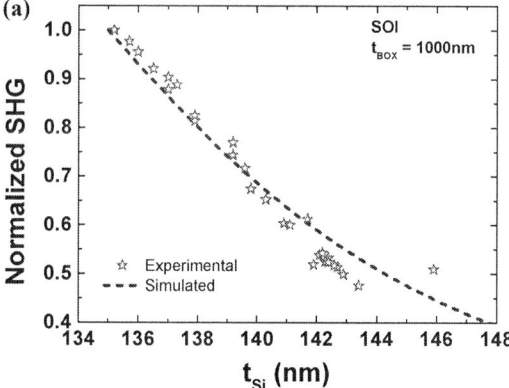

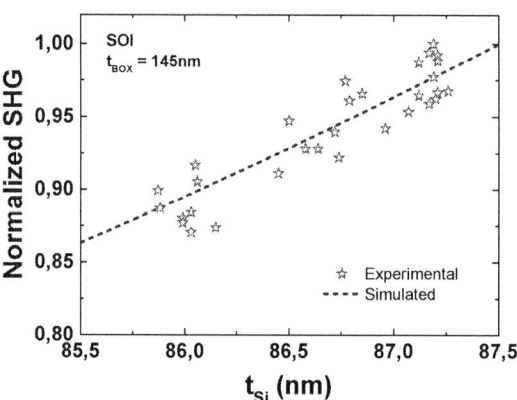

Fig. 5: Normalized SHG versus Si film thickness for thick (a) and thin (b) SOI: model (lines) and experiment (data points). The normalized experimental data points were calculated from Fig. 4a and 4b for both cases. The normalization was done by dividing each set (experimental and simulated) by its corresponding maximum value.

For a given film thickness, the optical path inside the structure can be modified by adjusting the angle of incidence (AOI) and in this case the intensity distribution of the fundamental beam across the structure will be also changed.

Fig. 6 shows normalized SHG versus AOI for SOI with 88nm Si film thickness and 145nm BOX thickness. The simulated curve with no electric field at the interface (solid line) fits quite well the experimental data. An electric field of 10^4 V/cm at the interfaces has a limited impact on the simulated curve (dashed line).

Fig. 7 shows normalized SHG versus AOI for SOI with 24nm Si film thickness and 25nm BOX thickness. The experimental

978-1-5386-4483-6/18 $31.00 © 2018 IEEE   40

SHG data and the simulation with no electric field are shifted by more than 20°. With a thin film, only the optical phenomena are not sufficient to explain the experimental results and the extra $\chi^{(3)}E_{DC}$ term must be taken into account. Two values of E_{DC} (10^4 V/cm and 10^5 V/cm) where simulated. The best fit is obtained for the largest electric field value (solid line). Indeed in thin SOI, the Si/SiO_2 interfaces are electrically coupled together and the electric field can be strong [30]. Therefore when thin Si films are used SHG measurements allow accessing the interfacial electric field partially due to the interface trap density.

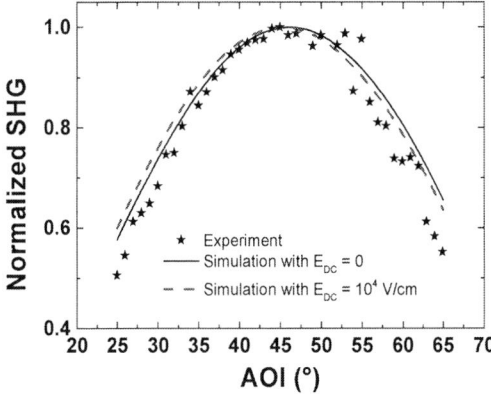

Fig. 6: Normalized SHG versus angle of incidence in SOI with 88nm Si film thickness and 145nm BOX thickness: model (lines) and experiment (data points).

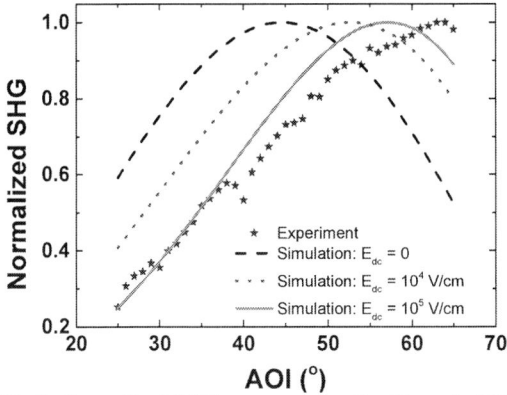

Fig. 7: Normalized SHG versus angle of incidence in SOI with 24nm Si film thickness and 25nm BOX thickness: model (lines) and experiment (data points). For the modeling, three different values of E_{DC} were tested: 0, 10^4 V/cm and 10^5 V/cm.

7. Conclusion

In this paper we present the use of second harmonic generation for characterization of dielectric-on-silicon stacks, with an emphasis on two particular applications: surface passivation and SOI characterization. The technique, based on non-linear optics, is non-destructive and well adapted for thin films testing. The SHG signal is partly related to the interface static field, therefore to the oxide charge and/or interface state densities. The analysis of the initial SHG value and the time-dependent SHG can lead to differentiation between the fixed charge and the interface traps contributions. Additionally, the technique is perfectly adapted to study charging and discharging dynamics in dielectric/silicon stacks. For a quantitative Q_{ox} and D_{it} characterization through stand-alone SHG, the optical phenomena that depend on the sample geometry need to be correctly de-correlated from the electrical properties of the samples.

Acknowledgments. Region Rhone-Alpes (ARC 6 program) is thanked for financial support. We would also like to thank SOITEC (O. Kononchuk, F. Allibert) for cooperation.

References

[1] D. K. Schroder, *Semiconductor Material and Device Chracterization*, 3rd Edition ed. New Jersey: John Wiley & sons, 2006.

[2] M. Wilson, J. Lagowski, L. Jastrzebski, A. Savtchouk, and V. Faifer, "COCOS (corona oxide characterization of semiconductor) non-contact metrology for gate dielectrics," *AIP Conference Proceedings,* vol. 550, pp. 220-225, 2001.

[3] D. K. Schroder, "Carrier lifetimes in silicon," *Electron Devices, IEEE Transactions on,* vol. 44, pp. 160-170, 1997.

[4] P. N. Butcher and D. Cotter, *The Elements of Nonlinear Optics*: Cambridge University Press, 1991.

[5] W. Daum, "Optical studies of Si/SiO 2 interfaces by second-harmonic generation spectroscopy of silicon interband transitions," *Applied Physics A,* vol. 87, pp. 451-460, 2007.

[6] H. Park, J. Qi, Y. Xu, K. Varga, S. M. Weiss, B. R. Rogers, *et al.*, "Boron induced charge traps near the interface of Si/SiO2 probed by second

978-1-5386-4483-6/18 $31.00 © 2018 IEEE   41

harmonic generation," *Physica Status Solidi (b),* vol. 247, pp. 1997-2001, 2010.

[7] K. Kotaro, T. Hideo, and F. Atsuo, "Symmetry and Second-Order Susceptibility of Hemicyanine Monolayer Studied by Surface Second-Harmonic Generation," *Japanese Journal of Applied Physics,* vol. 30, p. 1050, 1991.

[8] X. Li, J. Willits, S. T. Cundiff, I. M. P. Aarts, A. A. E. Stevens, and D. S. Dessau, "Circular dichroism in second harmonic generation from oxidized Si (001)," *Applied Physics Letters,* vol. 89, p. 022102, 2006.

[9] J. Sipe, D. Moss, and H. Van Driel, "Phenomenological theory of optical second-and third-harmonic generation from cubic centrosymmetric crystals," *Physical Review B,* vol. 35, p. 1129, 1987.

[10] H. W. K. Tom, T. F. Heinz, and Y. R. Shen, "Second-Harmonic Reflection from Silicon Surfaces and Its Relation to Structural Symmetry," *Physical Review Letters,* vol. 51, pp. 1983-1986, 1983.

[11] J. I. Dadap, B. Doris, Q. Deng, M. C. Downer, J. K. Lowell, and A. C. Diebold, "Randomly oriented Angstrom-scale microroughness at the Si(100)/SiO2 interface probed by optical second harmonic generation," *Applied Physics Letters,* vol. 64, pp. 2139-2141, 1994.

[12] V. Fomenko, E. P. Gusev, and E. Borguet, "Optical second harmonic generation studies of ultrathin high-k dielectric stacks," *Journal of Applied Physics,* vol. 97, p. 083711, 2005.

[13] B. Jun, Y. V. White, R. D. Schrimpf, D. M. Fleetwood, F. Brunier, N. Bresson, *et al.,* "Characterization of multiple Si/SiO2 interfaces in silicon-on-insulator materials via second-harmonic generation," *Applied Physics Letters,* vol. 85, pp. 3095-3097, 2004.

[14] J. J. H. Gielis, B. Hoex, M. C. M. van de Sanden, and W. M. M. Kessels, "Negative charge and charging dynamics in Al2O3 films on Si characterized by second-harmonic generation," *Journal of Applied Physics,* vol. 104, p. 073701, 2008.

[15] G. Lupke, "Characterization of semiconductor interfaces by second-harmonic generation," *Surface Science Reports,* vol. 35, pp. 75-161, 1999.

[16] http://femtometrix.com.

[17] *http://www.itrpv.net/.*

[18] N. M. Terlinden, G. Dingemans, V. Vandalon, R. H. E. C. Bosch, and W. M. M. Kessels, "Influence of the SiO2 interlayer thickness on the density and polarity of charges in Si/SiO2/Al2O3 stacks as studied by optical second-harmonic generation," *Journal of Applied Physics,* vol. 115, p. 033708, 2014.

[19] I. Ionica, D. Damianos, A. Kaminski, G. Vitrant, D. Blanc-Pélissier, J. Changala, *et al.,* "(Invited at

229th ECS Meeting) Non-Destructive Characterization of Dielectric - Semiconductor Interfaces by Second Harmonic Generation," *ECS Transactions,* vol. 72, pp. 139-151, 2016.

[20] B. Hoex, J. Schmidt, P. Pohl, M. C. M. van de Sanden, and W. M. M. Kessels, "Silicon surface passivation by atomic layer deposited Al2O3," *Journal of Applied Physics,* vol. 104, p. 044903, 2008.

[21] F. J. Grunthaner and P. J. Grunthaner, "Chemical and electronic structure of the SiO2/Si interface," *Materials Science Reports,* vol. 1, pp. 65-160, 1986.

[22] J. G. Mihaychuk, N. Shamir, and H. M. van Driel, "Multiphoton photoemission and electric-field-induced optical second-harmonic generation as probes of charge transfer across the Si/SiO2 interface," *Physical Review B,* vol. 59, pp. 2164-2173, 1999.

[23] G. Dingemans and W. M. M. Kessels, "Status and prospects of Al2O3-based surface passivation schemes for silicon solar cells," *Journal of Vacuum Science & Technology A,* vol. 30, p. 040802, 2012.

[24] D. Damianos and et.al., *submitted to Journal of Applied Physics.*

[25] www.soitec.com.

[26] S. Cristoloveanu, M. Bawedin, and I. Ionica, "A review of electrical characterization techniques for ultrathin FDSOI materials and devices," *Solid-State Electronics,* vol. 117, pp. 10-36, 2016.

[27] B. Jun, R. D. Schrimpf, D. M. Fleetwood, Y. V. White, R. Pasternak, S. N. Rashkeev, *et al.,* "Charge trapping in irradiated SOI wafers measured by second harmonic generation," *IEEE Transactions on Nuclear Science,* vol. 51, pp. 3231-3237, 2004.

[28] M. L. Alles, R. Pasternak, X. Lu, N. H. Tolk, R. D. Schrimpf, D. M. Fleetwood, *et al.,* "Second Harmonic Generation for Noninvasive Metrology of Silicon-on-Insulator Wafers," *Semiconductor Manufacturing, IEEE Transactions on,* vol. 20, pp. 107-113, 2007.

[29] D. Damianos, G. Vitrant, M. Lei, J. Changala, A. Kaminski-Cachopo, D. Blanc-Pelissier, *et al.,* "Second Harmonic Generation characterization of SOI wafers: Impact of layer thickness and interface electric field," *Solid-State Electronics,* vol. 143, pp. 90-96, 2018.

[30] G. Hamaide, F. Allibert, F. Andrieu, K. Romanjek, and S. Cristoloveanu, "Mobility in ultrathin SOI MOSFET and pseudo-MOSFET: Impact of the potential at both interfaces," *Solid-State Electronics,* vol. 57, pp. 83-86, 2011.

978-1-5386-4483-6/18 $31.00 © 2018 IEEE

Design Automation for Micro-Electro-Mechanical Systems

David Kriebel*, Henry Schmidt, Michael Schiebold, Markus Freitag,
Benjamin Arnold, Michael Naumann, Jan. E. Mehner**

University of Technology Chemnitz
*E-mail: david.kriebel@etit.tu-chemnitz.de
**E-mail: henry.schmidt@etit.tu-chemnitz.de

Abstract— This paper demonstrates and discusses a highly automated approach for the design of micro-electro-mechanical systems (MEMS) and system-level multi-domain reduced order model generation. The presented techniques in form of rigid body models (RBM) and modal superposition models (MSUP) in conjunction with component mode synthesis (CMS) enable fast and efficient model adaption and optimization of components in the different phases of the MEMS design process by providing sufficiently fast and accurate modeling solutions. Different aspects and requirements of individual methods are discussed and compared.

Keywords — *MEMS design, multi-field simulation, reduced order modeling, ROM, design automation, component mode synthesis, modal superposition*

1. Introduction

The preliminary determination and optimization of the function and form elements in a new MEMS design is a critical step in the layout process, as the foundation for the achievable performance of the product is already defined at this stage. Numerous approaches for reduced order modelling of components and system behavior based on lumped elements or numerical methods can be found in the literature. Achieving a time- and resource-efficient approach for the creation of the simulation models an automatable geometry and reduced order model generator is one of the essential requirements to enable an iterative automated design and optimization process during all stages of system development. In this paper a highly automated geometry and model generation strategy is presented. The first part presents a method for efficient geometry generation, featuring the automated consideration of manufacturing influences and procedurally generated design features. Starting from the initial geometry definition, reduced order models are automatically generated based on analytical and numerical modelling techniques.

2. Functional and Form Elements

In the conceptual phase of the design process basic functions and form elements of the MEMS are determined. This requires a first idea of the required elements and their mutual interactions. Commonly this initial design is defined by factors like the measurand, fabrication technology, available chip size or evaluation electronics to name just a few.

Starting from the initial design idea each component of the MEMS needs to be constructed and optimized till all requirements are fulfilled. A fast and easy modelling process is in this context essential to allow time-efficient design adaption. The basic geometry is effectively formed with a small library of parametric functional design elements. These elements are defined by a small number of different ASCII commands. The script-based approach allows the automated generation of parametric models and effective model regeneration (Fig 1). Typical structural design elements are seismic masses, anchors, beams and transducer cells like comb or bottom plate capacitors. These elements describe the connections of the mechanical, electrostatic and fluidic domains.

MEMS are typically fabricated by surface micromachining technologies. As a result, the mechanical structures are located in a layer of constant thickness which allows the description of the form elements in only a 2D plane.

978-1-5386-4483-6/18 $31.00 © 2018 IEEE

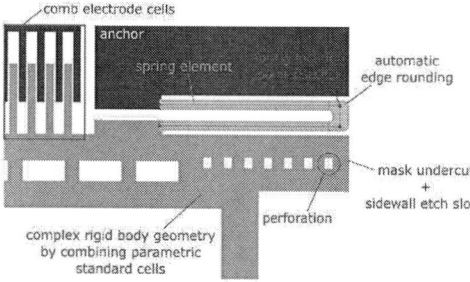

Fig. 1 Form elements for automatic ROM generation

Seismic masses are modelled by combining basic parametric primitives (rectangles, triangles, circle/-segments, polygons and ellipses) utilizing Boolean operations like addition and subtraction. Nearly all arbitrary shapes can be formed with this strategy. Anchors which are connected to the substrate surface are modelled with the same elements like seismic masses. Another commonly used structuring element are perforation patterns. Generally, perforations patterns are used because of technological requirements or for optimization of mass or damping properties. These perforation patterns are created by only one command which accesses a library of parametric perforation shapes.

The seismic masses and anchors are interconnected with suspension element beams or spring arrangements. Combined they form the mechanical domain as a spring-mass-system. The modelling effort is reduced to the creation of start and end points of a beam element. Specification of the beam width at both ends provides a sufficient description also for tapered beams. Transitions between connected trusses are calculated automatically. Rounded edges which are important in respect of stress distribution and device durability, are applied globally or to individual local beams.

Comb capacitors and top or bottom electrodes are defined by use of another parametric library with various predefined electrode shapes. Alternatively, user defined shapes are supported. Electrodes are defined by use of the associated rigid body number, location point, orientation angle, number of cells and the geometrical parameters. The electrodes are treated as a structural part of the related rigid body including the mass properties. Furthermore, cell like electrodes also create a capacitor and thus a connection to the electro-mechanical domain. Additionally, the relation to the fluid-mechanical domain is established through slide- and squeeze-film-damping. The models for the capacitance and damping are generated automatically using different approaches depending on the considered effects and requirements regarding precision of the results.

After describing the geometry with all of the functional elements (demonstration structure Fig. 2), the fabrications effects and tolerances are applied to the structure. This includes the layer thickness and etch effects like mask undercut and side wall slope. These global parameters are also accessible by an ASCII command to give the possibility for an iterative verification of the influences of all these effects and structural dimensions.

Additional functional elements like contact elements or different effects like package warping are also implemented but will not be further discussed in this paper.

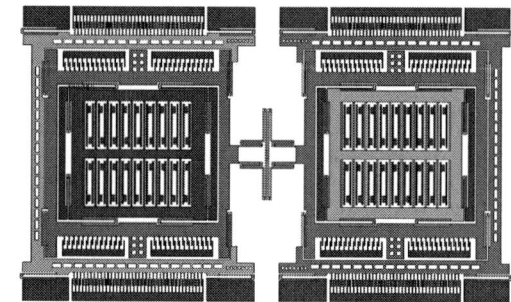

Fig. 2 Demonstrator structure (gyroscope)

3. Analytic Model Order Reduction

Analytical models are a common and rapid way to reduce the complexity of a simulation model. For most physical effects analytical descriptions are available which can also be extended to represent even advanced influences, like nonlinearities. The goal is to express the complex mechanical behavior of the MEMS in form of a second order differential equation with the use of a minimum amount of degrees of freedom.

$$\mathbf{M}\ddot{\mathbf{u}} + \mathbf{D}\dot{\mathbf{u}} + \mathbf{K}\mathbf{u} = F_{mech} + F_{el} \qquad (1)$$

In the mechanical domain the generated geometry can be separated into 2 types of elements. Beam elements are represented by a two node TIMOSHENKO beam [1], where each node has 6 degrees of freedom. By using the TIMOSHENKO beam theory, the stiffness and also mass matrix can be assembled based on analytical expressions. Non-rectangular cross sections are possible and enable an accurate description of the effects caused by symmetrical or nonsymmetrical etch slopes. High accuracy in terms of the stiffness of straight arbitrary suspension elements is already achieved by use of only one element per beam. For circular beams a denser discretization along the beam axis is required.

The well-understood TIMOSHENKO beam theory can be extended by the VLASOV torsion theory [2]. This analytical model describes precisely the increasing torsion stiffness of very short members. The analytical model was extracted from the 4th order differential equation of non-uniform torsion (2) where ϕ is the rotation of the beam cross section around the beam axis (x-axis), EC_w is the warping stiffness and GI_t is the torsion stiffness.

$$EC_w \frac{d^4\phi}{dx^4} - GI_t \frac{d^2\phi}{dx^2} = 0 \qquad (2)$$

Defining a 7th DOF per beam node, the warping DOF θ, the effect of restrained warping is considered by coupling the torsion DOF with the warping DOF.

$$\theta = \frac{d\phi}{dx} \qquad (3)$$

The seismic masses are modelled by reduction to a 6 DOF mass point. The mass properties of complex geometries are determined by a boundary representation model (BREP) [3]. This method works without discretization of the complex volumes but creates a non-deformable mass description. The mass points are connected to the beam elements by constraint equations:

$$u_{slave} = u_{master} + \theta \times d_{MS} \qquad (4)$$

$$M_{slave} = M_{master} + d_{MS} \times F_{trans} \qquad (5)$$

Equation (4) describes the translational displacement between a master and slave node, where θ is the rotational DOF of the master node and d_{MS} is the positioning vector from the center of mass to the beam connection point. Constraint equation (5) describes equivalent torque loads.

For most designs this assumption is acceptable and yields accurate results. In practice seismic masses can be flexible due to dense perforation patterns which are required in some manufacturing technologies (i.e. sacrificial layers).

Depending on the perforation pattern density seismic masses are alternatively modelled as a network of beam elements and small parts of rigid body regions. This allows the representation of flexible, densely perforated seismic masses by analytical methods. Both methods of seismic mass modelling always assume comb electrodes as rigid regions. These approaches are used for the assembly of the mass matrix $\mathbf{M}$ and stiffness matrix $\mathbf{K}$ of the system.

Out of the comb electrode library element types and the related geometrical parameters an analytical expression for the capacitance of the comb cells is automatically generated. This expression should be continuously differentiable twice to model the electrostatic force and electrostatic softening effect reliably. The capacitance is required to model charges and currents of the electrodes and to provide an opportunity to connect the MEMS model to electrical controller- or other electronic models. By user defined voltage or current ports quantities like the electrostatic force F_{el} and current I of a comb cell can be determined. To reduce the complexity of the capacitance equations at high numbers of cells, capacitance are only evaluated at a user defined number of gauss integration points called master nodes. Between these points the capacitance is interpolated. The resulting electrostatic force is then projected to the center of gravity of the related rigid body region.

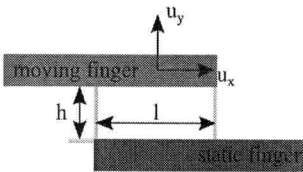

Fig. 3 Capacitance of homogenous electrical field

Capacitances can be generally described under different assumptions. The dominant part is typically the homogenous field between the comb electrode fingers (Fig 3). Using the homogenous field region alone already al-lows a quite accurate prediction in many cases [4]. Formula (6) calculates the capacitance in regards of plate movement, where t is the layer thickness, l is the overlapping finger length, h the gap between comb cell fingers and α is continuity factor for the capacitance function in u_z direction.

$$C\left(u_x, u_y, u_z\right) =$$

$$\epsilon\left(\frac{t\sqrt{\alpha^2 t^2 + 1}}{\sqrt{\alpha^2 t^2 + 1} - 1} - \frac{t\sqrt{\alpha^2 u_z^2 + 1}}{\sqrt{\alpha^2 t^2 + 1} - 1}\right)\frac{l + u_x}{h + u_y} \quad (6)$$

The second part of the capacitance model is the fringing field, which is modelled using an analytical expression to describe effects like levitation forces at unsymmetrical plate configurations. Different solution strategies already exist in the literature but will not be further discussed [5].

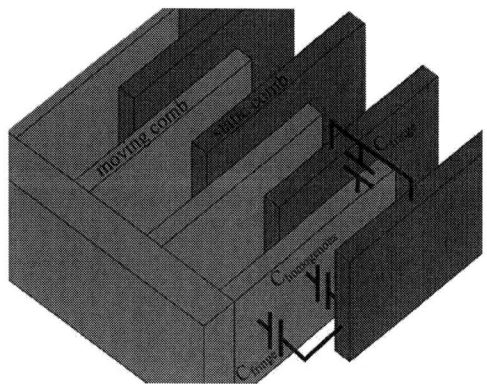

Fig. 4 Analytical comb cell capacitance models

It must be noted that the analytical capacitance models uses an averaged electrode gap. The etch slope is ignored in this approach. Numerical approaches which are discussed in a later chapter include these geometries and related effects.

The comb electrode library also allows to generate analytical models for slide- and squeeze-film damping automatically. The analytical models are related to just a few master nodes as the capacitance models. Simplified models for squeeze- and also wedge-film damping are built using the solution of the linearized REYNOLDS equation [6].

$$\frac{h^3}{12\eta}\nabla^2 p = \vec{v}_z + \frac{h}{p_0}\frac{\partial p}{\partial t} - \frac{\vec{v}_x}{2}\frac{\partial h}{\partial x} - \frac{\vec{v}_y}{2}\frac{\partial h}{\partial y} \quad (7)$$

The last part of the damping matrix forms the slide film damping. The analytical expressions for this damping phenomena was published in many papers. These two fluidic damping mechanism are used to assemble the damping matrix **D**.

4. Discussion Analytical ROMs

The main reason for using analytical models for preliminary designs is the speed of the assembly and solution process. Design parameters can be varied very fast and optimized till the desired behavior is achieved. Withal the coupling of the physical domains is taken into account and the most important effects can be evaluated. For example, frequency tuning optimization can be done without the need of time consuming transient simulation runs while still describing the effects of the coupling with the electrostatic domain. A design parameter or voltage can be swept over a user defined range. System matrices are rebuilt if the geometry is changing. In the next step the operation point is determined by using the nonlinear equations for the electrostatic force. The electrostatic and fluidic domain are linearized in the obtained operation point which leads to static system matrices. These matrices are used to perform a harmonic analysis. In this type of analysis all relevant coupling effects are observed and the solution process is much faster than numerical methods under assumption of some simplifications.

5. FE Based Model Order Reduction

By use of the RBM approach, it is possible to generate fast and efficient models which are yet able to represent complex interactions of the different physical domains at sufficient accuracy. At later stages of the MEMS design process, however, it becomes necessary to incorporate system mechanics that cannot be easily represented by rigid bodies and purely analytical approaches.

One critical influence observed in MEMS is flexure of the seismic masses caused by structurally weakened regions due to slenderness or perforation holes. This especially gains importance in the context of increasingly miniaturized systems and high frequency applications, as considerable impact on high order mode shapes and Eigen frequencies is observable.

Shell element models, due to their inherent areal homogenization approach, do not reliably cover effects outside of the element specifications; and while providing a computationally effective approach, need to be handled with care in the simulation of state-of-the-art MEMS devices. Locally applied beam models can also be a reasonable approach in specific cases but come with their own modelling challenges like geometrically overlap-ping beam portions that need to be managed accordingly. Up until now, full 3D FEM models are unmatched in terms of flexibility but lead to extensive computational load which renders coupled field simulations of the involved physical domains virtually un-feasible for feature-rich structures. Modal order reduction is a commonly used strategy to manage the computational load in such simulations of mechanical systems.

In the modal superposition approach (MSUP) the deformation state u of a structure is represented by a superposition of the lowest eigenvectors φ, weighted with the generalized modal degrees of freedom q:

$$u = \sum q_i \, \varphi_i \qquad (8)$$

In this context, eigenvectors can serve as the base functions for static, harmonic or transient simulation runs. The standard MSUP approach is extended by the CRAIG / BAMPTON component mode synthesis method [7] which allows the generation of reduced order component models that retain physical DOF at connection interfaces. Internal DOF u_i are represented by superimposed fixed-interface vibration modes Φ_i and static constraint modes Φ_{bi}.

$$u_i = \Phi_{bi} u_b + \Phi_i q \qquad (9)$$

A schematic representation is provided in figure 5. The physical interface DOF can be used for subsystem assembly, application of loads or boundary conditions, or as monitor points without an additional expansion step and are most importantly compatible to the modelling approaches in RBM. The treatment of MEMS as an assembly of interconnected subsystems enables in this way flexible system modelling that allows to choose the most-suitable modelling-approach for each of the inherent mechanics and facilitates a fluent transition between RBM and MSUP models during the design process. By utilizing symmetries or recurring parts in the simulated structures, the computational load for model generation can be significantly decreased.

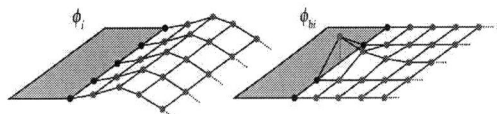

Fig. 5 Schematic representation of fixed interface vibrational (left) and static constraint modes (right).

The presented geometry generator offers automated access to suitable FEM models of the whole device or individual design elements in the ANSYS Multiphysics design environment, which are utilized to extract different

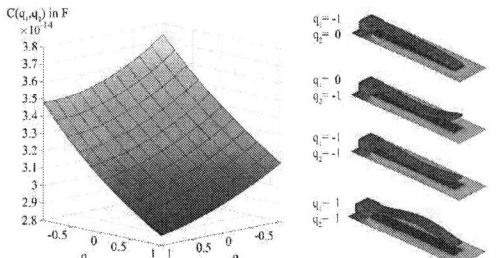

Fig. 6 Capacitance stroke function for the first two out-of-plane modes of a cantilever beam.

variations of reduced order models based on the modal superposition (MSUP) approach. The ultimate goal for model order reduction is to provide an automated procedure to extract fast dynamic transducer models which can be directly utilized in a system design environment.

MSUP models are linear in nature, so the implementation of non-linear mechanics is a challenge and needs to be determined for each cause of non-linear behavior individually. In the following, three MEMS-inherent causes of non-linearities are discussed.

The implementation of mechanical non-linearities in form of large deflection effects or stress stiffening has been demonstrated by DORWARTH et al. [8] for a basic one DOF system. As the stiffness matrix depends on the collective deformation state, stiffness matrix samples need to be pre-calculated along a representative trajectory and are used to determine a modal multi-variable response surface. In the context of MEMS, it is typically reasonable to assume linear flexible behavior for the seismic masses, as their deflections are still small compared to the dedicated bending beams.

Electrostatic fields show non-linear behavior whenever plate distance variations or fringing fields occur. Consequently, for the pure MSUP approach, capacitance stroke functions also need to be represented by a modal multivariable response surface as depicted exemplarily for the first two out-of-plane modes of a cantilever beam in figure 6. Capacitance function extraction requires in this case time consuming non-standardized data sampling of FEM capacitor models and advanced function fit approaches. It would be

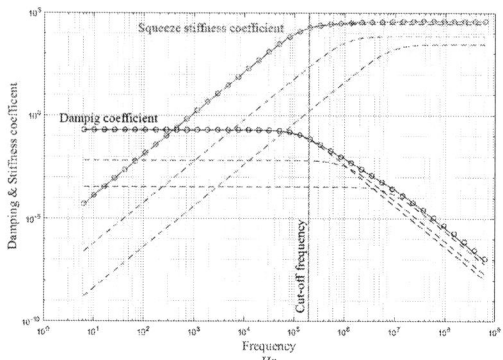

Fig. 7 Fluidic damping and stiffness coefficients in the frequency domain and approximation by a linear spring-damper network.

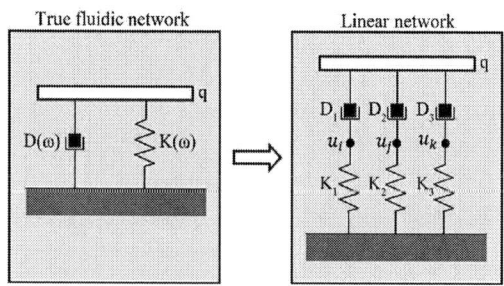

Fig. 8 Approaximation of frequency dependent stiffness and damping parameters by a linear spring damper network. [9]

further necessary to repeat the extraction procedure for each functional capacitance in the model individually, as the extracted modal response is bound to the location in the model. Ultimately, a universal and fully automated procedure is in this way hard to achieve for complex multi capacitor arrangements and would require restrictive model building. For this reason, it is typically more effective to use the physical DOF at interfaces and obtain capacitance stroke functions from 3D FEM capacitor models. If the capacitance values and their derivatives for the force response are determined based on physical displacements at interface points, the modeling is significantly simplified and allows a standardized and easily automatable data-sampling procedure.

Fluidic damping is one of the most challenging factors to consider accurately in the context of modal reduced order models, as the force response depends on the wall deflections h and velocities dh/dt as can be derived from REYNOLDS squeeze film equation (3). As a

978-1-5386-4483-6/18 $31.00 © 2018 IEEE   48

result, the resulting behavior is frequency dependent with a switching characteristic from damping- to stiffness-dominated response at the cut-off frequency as can be seen in figure (9). VEIJOLA [9] proposed a method to express the frequency dependence in form of linear spring-damper network by introduction of additional DOF $u_i, u_j, u_k, \ldots$ as depicted in figure 8.

$$D(\omega) = \sum_{i=1}^{n} \frac{D_i^{-1}}{D_i^{-2} + \omega^2 K_i^{-2}} \quad (10)$$

$$C(\omega) = \sum_{i=1}^{n} \frac{\omega^2 K_i^{-1}}{D_i^{-2} + \omega^2 K_i^{-2}} \quad (11)$$

The damping coefficients D_i and stiffness coefficients K_i are determined by a least square fit from data obtained from a harmonic response analysis. Figure 7 shows the resulting fit based on simulated data points. The dashed lines indicate how the individual spring damper pairs contribute to the overall behavior. A component wise implementation of modal damping has been demonstrated by Schmidt et al. [10] for a basic perforated plate device and its out-of-plane mode. The same procedure is also applicable for more complex devices and plate vibration modes.

6. Comparison of RBM and CMS

Table 1. Comparison of Accuracy between RBM and CMS in relation to a Full FEM Simulation.

	System Modes & Frequencies (kHz)						
	1	*2*	*3*	*4*	*5*	*6*	*7*
FEM	65.3	68.5	75.1	80.0	159.5	163.5	229.1
	Absolute & Relative Deviation to full FEM result (Hz / %)						
RBM	-1200 1.84 %	-1800 2.63 %	-2500 3.33 %	-3100 3.87 %			
CMS	45 0.07 %	18 0.03 %	176 0.24 %	172 0.22 %	-368 0.23 %	-356 0.22 %	714 0.31 %

The presented MEMS Model Builder can describe complex multi-domain models with mechanical system matrices of dramatically reduced size. Table 1 shows the comparison between a full 3D FEM model of the gyroscope, the RBM method with rigid edge rounding regions in spring connections, the

RBM method with an approximated stiffness of the spring connection regions (RBM flex con) and CMS. The CMS approach shows the best results, which are a very close approximation of the solution of the full FEM simulation, with a maximum relative error of 0.31% till mode 7 and below 1% until the 16th eigenmode. The eigenfrequencies observed with the RBM method are higher because of rigid seismic masses and rigid spring connections. If the transition region of the spring connections is also approximated with beam elements, the stiffness of the structure is in consequence lower and the results are close to the FEM/CMS solutions. Note that rigid seismic masses cannot describe flexible seismic mass modes. The modes five to seven could not be observed with the RBM method in comparable form, as already deformations of the mass bodies are involved.

7. Conclusion

In summary, with the reduction approaches of RBM and MSUP it is possible to enhance the design and optimization process of MEMS significantly. The high-speed solution process compared to full FEM simulations minimizes the time spent for data acquisition. With each of the presented techniques large scale statistical analyses are feasible. The method of RBM is especially suitable for geometry estimation and proof of concept evaluation in the early phases of the design process. MSUP with CMS extents these features in all phases of MEMS design because of the exact representation of flexible structures. The use of reduced order models in MATLAB/Simulink offers the possibility to link the MEMS model to electronics/controller models in higher abstraction layers. The behavior of the system can be simulated dynamically over a large number of time steps.

5. References

[1]   J. S. Przemieniecki, *Theory of matrix structural analysis*. New York, NY: McGraw-Hill, 1968.

[2]   V. Z. Vlasov and Y. Schechtman, *Thin-walled elastic beams*, 2nd ed. Jerusalem: Israel program for scientific translations, 1961.

[3]   S.-l. Lien and J. T. Kajiya, *"A symbolic method for calculating the integral properties of arbitrary nonconvex polyhedra,"* IEEE Comput. Grap. Appl., vol. 4, no. 10, pp. 35–42, 1984.

[4]   J. Mehner, *Entwurf in der Mikrosystemtechnik*. Dresden [u.a.]: Dresden Univ. Press, 2000.

[5]   H. Hammer, *"Analytical Model for Comb-Capacitance Fringe Fields,"* J. Microelectromech. Syst., vol. 19, no. 1, pp. 175–182, 2010.

[6]   B. J. Hamrock, *"Fundamentals of fluid film lubrication."* New York: McGraw-Hill, 1994.

[7]   R. R. Craig Jr., M. C. C. Bampton, *"Coupling of Substructures for Dynamic Analysis"*, AIAA Journal, vol. 6 nr.7, pp. 1313-1319, 1968.

[8]   M. Dorwarth et al., *"Nonlinear Model Order Reduction for high Q MEMS gyroscopes,"* 2014 IEEE 11th International Multi-Conference on Systems, Signals & Devices (SSD14), Barcelona, 2014, pp. 1-4.

[9]   T. Veijola et al., *"Equivalent circuit model of the squeezed gas film in a silicon accelerometer,"* Sensors and Actuators A48., , pp. 239–248, 1995.

[10]   H. Schmidt, A. Sorger and J. E. Mehner, *"Efficient reduced order modeling of fluid solid interactions for structurally complex perforated MEMS,"* 2017 19th International Conference on Solid-State Sensors, Actuators and Microsystems TRANSDUCERS 2017, Kaohsiung, 2017, pp. 2083-2086.

On Effective Graphene based Computing

N. Cucu Laurenciu, S.D. Cotofana
Computer Engineering Laboratory,
Delft University of Technology, The Netherlands.
{N.CucuLaurenciu, S.D.Cotofana}@tudelft.nl

Abstract—With CMOS feature size heading towards atomic dimensions, unjustifiable static power, reliability, and economic implications are exacerbating, prompting for research on new materials, devices, and/or computation paradigms. Within this context, Graphene Nanoribbons (GNRs), owing to graphene's excellent electronic properties, may serve as basic blocks for carbon-based nanoelectronics. In this paper, we present the two main avenues, i.e., graphene FET- and GNR- based, undertaken towards graphene based computing. The first approach is conservative and focuses on the realization of graphene FET transistor based switches as MOSFET replacements to maintain the state of the art logic Boolean algebra paradigm design methodology. The second one follows a different line of thinking and seeks GNR-based structures able to provide more complex behaviours by making better use of graphene's conduction properties. We first discuss Graphene Nanoribbon (GNR) based field Effect Transistors (GNRFETs) and Tunnelling GNR based Transistors (GNRTFETs) and their utilization as underlying elements for Boolean gate implementations. Subsequently, we present GNR-based structures that can directly compute Boolean functions, e.g., NAND, XOR, by means of one GNR only and a way to complementary arrange them in energy effective gates. To get inside into the potential of the two avenues we consider an inverter as discussion vehicle and evaluate the designs in terms of area and energy consumption. The GNR-based structure outperforms its counterparts by $15\times$ up to $104\times$ and $230\times$ smaller delay and 6 to 7 and 4 orders of magnitude smaller power than the GNRFET- and GNRTFET- based designs, respectively. Moreover, when compared with CMOS 7 nm Boolean gates GNR-based desgns exhibit up to $6\times$ smaller delay, and up to 2 orders of magnitude smaller active area, and total power consumption. Our analysis confirms that the alternative GNR-based design paradigm, which transcends the traditional switch based approach and takes better advantage of graphene intrinsicnproperties, is better suited for future carbon based nanoelectronics.

Index Terms—Graphene Nanoribbons, Conduction Maps, Boolean Gates, Graphene-based Boolean Gates, Carbon-Nanoelectronics, Energy Efficiency.

I. INTRODUCTION

In the past three decades, CMOS scaling has resulted in new technology generations every two to three years with doubled logic device density, lowered cost per operation, and increased chip performance. However, as CMOS feature size is approaching the atomic level, the faster switching speed comes at the expense of increased power density and leakage, decreased reliability and yield, increased production costs, and diminishing returns. In this landscape, and in line with the continuous impetus of device performance improvement, the development of new materials, structures, and computation paradigms are called for [1] [2]. One of the post-Si forerunners is graphene, which has enjoyed a surge of research during the

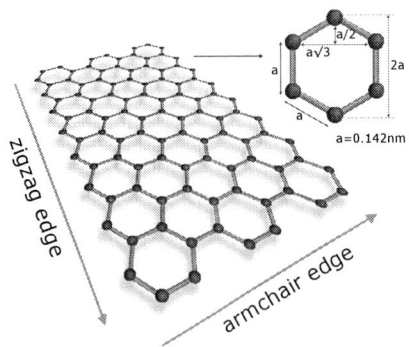

Fig. 1: Graphene Atomic Structure.

past decade, paving the way for a wide range of graphene-based applications, among which electronics, spintronics, photonics and optoelectronics, sensors, energy storage and conversion, flexible electronics, and biomedical applications [3].

Graphene is a 2-dimensional carbon atom monolayer lattice, as illustrated in Figure 1 for 2 types of edge terminations along the transport direction: zigzag and armchair. Virtue to the edge structures, graphene can present different electronic properties (i.e., the armchair terminated graphene can exhibit both metallic and semiconducting properties depending on the nanoribbon width, while the zigzag edge-patterned graphene is always metallic), offering appealing opportunities for the development of graphene-based electronic devices. Graphene has a wealth of unique, outstanding characteristics, which provide a strong drive to investigate its usage as a potent contender to Si-based technology and as a promising means towards carbon based nanoelectronics [4], [5], [6]. Notably, graphene exhibits: (i) atomic thinness and 2D structure (which allows for the direct excitation of charge carriers, and provides unique advantages that could be compatible with scalable fabrication processes), (ii) ballistic transport, with micron long mean free path and charge Fermi velocity $\nu_F \sim 10^6$ ms^{-1}, $10\times$ higher than in Si [7], (iv) ultrahigh intrinsic carrier mobility μ both at room temperature (over 2.5×10^5 cm^{2}V^{-1}s^{-1} [8]) and at low temperature (6×10^6 cm^{2}V^{-1}s^{-1} at 4 K [9]), outperforming existing materials with high mobility as InP (1.5×10^3cm^{2}V^{-1}s^{-1}), InAs (1.32×10^3cm^{2}V^{-1}s^{-1}), or strained Si (1.4×10^3cm^{2}V^{-1}s^{-1}) [10], (v) outstanding thermal properties (very high thermal conductivity $k \sim 3000$ to 5000 Wm^{-1}K^{-1} [11]), and ability to sustain very high current densities (10^6 greater than copper [12]), and (vi) it is the strongest material ever measured, with a Young modulus

978-1-5386-4483-6/18 $31.00 © 2018 IEEE

of 1 TPa and intrinsic tensile strength of 130 GPa, being able to withstand elastic deformations of $\sim 26\%$ without fracture [13]. Apart of those very attractive properties graphene is also transparent and by its very nature biocompatible, which makes it extremely fit for medical applications, e.g., implantable prosthetics.

Generally speaking, the principal impediments to graphene-based logic are twofold: design related, and manufacturing related [7], [14], [15], [16], [17]. From the manufacturing perspective, finding a cost-effective, scalable, and reliable manufacturing process, which enables mass-production with minimum defects density and with highly reproducible features, is the main desideratum. From the design point of view, the main caveat is graphene's absence of a bandgap, which impedes charge carriers depletion, results in high "off" state static power, and limits the achievable "on"-"off" current ratios ($I_{ON}/I_{OFF} < 10$ while $I_{ON}/I_{OFF} > 10^7$ is typical for current CMOS technology nodes). For digital Boolean logic applications, there are certain aspects that graphene-based structures need to comply with, foremost: (i) ability to control conductivity and yield distinguishable "on" and "off" states, while (a) not compromising any of the graphene intrinsic highly advantageous properties (e.g., high μ), and (b) providing acceptable I_{ON}/I_{OFF} ratio (in the range 10^4 to 10^7), (ii) finding the proper external electrical means (e.g., top gates, back gates) to control graphene behaviour and induce the desired logic functionality, (iii) ability to encode some desired logic transfer function onto the graphene electrical characteristics, and (iv) ensuring the conditions for digital circuits cascading (i.e., clean and compatible/matching electric levels, e.g., voltage, current, for the circuit inputs and outputs).

In this paper, we present the main approaches undertaken to comply with the aforementioned computing tenets and the approaches that pervade them for beyond CMOS computation. We first focus on device level efforts towards the realization of graphene based switches and by implication of graphene based logic circuits. The idea behind this approach is to maintain the state of the art logic design methodology, which constructs on the Boolean algebra paradigm, and just replace the MOS switches (transistors) with graphene based counterparts. To this end we describe Graphene Nanoribbon (GNR) based field Effect Transistors (FETs) and their underlying operation principle [18] and tunnelling GNR based FETs [19]. To put things into prospective we also summarize the potential performance of Boolean gates based on such graphene switches as reported in [20] and [21].

Subsequently, we follow a different line of thinking inspired by our previous investigation in [22], which provides strong evidence that GNRs can exhibit functionalities beyond the traditional switch. Thus, to take advantage of the full graphene potential, one can depart from the traditional switch based computation and envision novel GNR-based structures and computing avenues. In this regard, we present structures that can directly compute Boolean functions, e.g., NAND, XOR, by means of one GNR only and a way to arrange such structures instead of transistors in energy effective gates. Considering an inverter gate, the GNR-based structure outperforms the transistor-based ones as follows: its propagation delay is from $15\times$ up to $104\times$ and $230\times$ smaller and its total energy consumption is 6 to 7 and 4 orders of magnitude smaller than the one of GNRFET- and GNRTFET- based designs, respectively.

Finally, we briefly present some other graphene based structures which exploit some advantageous properties, such as weak intrinsic spin-orbit coupling and absence of hyperfine interactions, negative differential resistance, and ambipolar transport, in the context of spintronics, multiple valued logic, and in-field controllable dynamic and static logic.

This study suggests that: (i) bandgap opening is not an issue and can be energy effective dealt with by topological measures, in synergy with chemical and electrostatic, (ii) the alternative GNR-based design paradigm which transcends the traditional switch based approach takes better advantage of intrinsic graphene properties and outperform GFET based gate deigns in terms of area, delay, and energy consumption, and (iii) GNR-based Boolean gates can potentially outperform state of the art CMOS 7 nm counterparts by up to $6\times$ smaller delay, and up to 2 orders of magnitude smaller active area, and total power consumption.

The remaining of the paper is structured as follows: Section II presents an overview of graphene-based transistors. Specifically, FETs are discussed in Section II-A, and Tunnelling FETs, with both planar and out-of-plane tunnelling, are addressed in Section II-B. In Section III, we provide a brief encounter with GNR-based structures, which do not rely on the traditional switching mechanism as operation principle. Section IV concludes the paper with an outlook and opportunities for graphene-based computing.

II. GRAPHENE TRANSISTOR-BASED COMPUTING

According to ITRS, one of the requisites to continue devices scaling along the "More Moore" strategy, is the implementation of transistors that make use of high mobility channel materials [23]. In view of this, graphene comes as a natural channel material choice as its extremely high carrier mobility is surpassing by far currently utilized materials (e.g., Ge for pMOS transistors and III-V compound semiconductors - SiGe, InGaAs - for nMOS transistors). However, while a non-zero energy bandgap is not necessary for high speed analog circuits, for proper operation of digital logic it is a key property. Up to date, several approaches have been undertaken to induce a bandgap in graphene, noteworthy: (i) lateral confinement of a large sheet of graphene charge carriers in the form of narrow strips of graphene called Graphene Nanoribbons (GNRs), or in the form of Graphene Quantum Dots (GQDs), (ii) breaking the planar symmetry of the graphene crystal structure via chemical and/or structural modifications (e.g., substrate use, substitutional doping, chemical functionalization, straining), and (iii) applying a transverse electrical field to bilayer graphene [3]. For graphene-based transistors usage, a GNR quantum confinement and substrate induced bandgap opening approach is typically relied upon.

978-1-5386-4483-6/18 $31.00 © 2018 IEEE

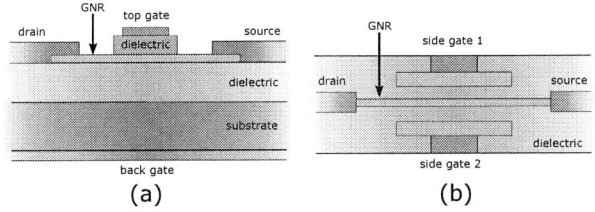

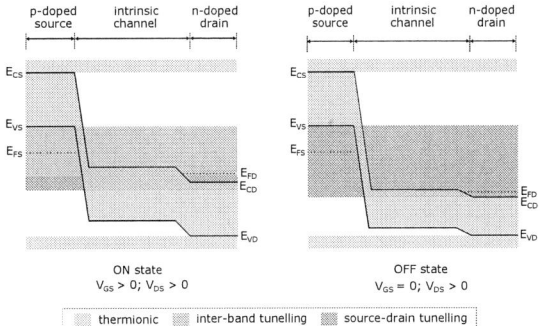

Fig. 2: GNRFET cross-section schematics: (a) GNRFET with top and/or back gate; and (b) FEBM GNRFET.

Fig. 3: GNRTFET Band Diagram.

Subsequently, we present graphene devices (transistors) meant to enable the road towards traditional switch-based logic design approaches.

A. GNR-based FETs

In GNRFET structures, as illustrated in Figure 2 (a), graphene serves as conduction channel, through which a current flow is induced by applying a bias voltage between the two graphene nanoribbon end-point contacts (source and drain). The top gate voltage, as local perturbation potentials, modulate the source-to-drain current, while the back gate shifts the Fermi level chemical potential away from the Dirac point, into the electron or hole conduction regime. Most past efforts typically employ a Si conducting substrate as back-gate and a thick layer of SiO_2 (~ 300 nm) as top gate dielectric. Recently, high-k dielectric materials (e.g., HfO_2, Al_2O_3) that reduce leakage gate tunneling currents are being increasingly utilized [24]. As for the source and drain electrodes, depending on wether they are metallic or semiconducting, two varieties of GNRFETs exist: Schottky Barrier (SB) type and Metal-Oxide-Semiconductor (MOS) type. For SB-GNRFETs, the source and drain contacts are metallic, resulting in formation of Schottky barriers at the metal-graphene junctions. MOS-GNRFETs on the other hand, have the source and drain contacts made of heavily doped graphene. While for MOS-GNRFETs the current flow is determined either by electrons or holes, depending on the dopant type of source/drain reservoirs, SB-GNRFETs exhibit ambipolar current conduction, which is not appropriate for CMOS-style logic.

While ambipolar devices-based logic designs have been investigated [25] such an approach is not particularly of interest as to obtain NMOS or PMOS transfer characteristics, SB-GNRFETs require extra work function engineering, which can result in unbalanced n-type and p-type characteristics, leading to robustness and performance loss. On the other hand MOS-GNRFETs exhibit a higher I_{ON}/I_{OFF} ratio and larger transconductance and cut-off frequency, however they are susceptible to doping variation (as it is difficult to control the exact doping level of source/drain reservoirs with several thousands atoms), and need to consider minimizing the ohmic contacts to graphene [26]. In practice the GNRFET channel usually consists of a dense array of parallel rectangularly shaped and equally spaced GNRs in order to increase the its drive strength.

Table I summarizes the power consumption and propagation delay figures reported in [20] for a set of Boolean gates constructed with SB-GNRFETs and MOS-GNRFETs ($V_{DD} = 0.5$ V), comparatively to Si-based CMOSFETs using HP 16 nm CMOS technology, with nominal $V_{DD} = 0.7$ V. These results indicate that relative to Si-based MOSFETs, SB-GNRFETs are better suited for high speed applications, while MOS-GNRFETs are more appropriate for low power applications.

To improve the GNRFET performance, several device architectures have been explored. One such structure is the GNR transistor with Field Effect Bandgap Modulation (FEBM) [27]. The rationale is to use the intrinsic bandgap for the "off" state, and a narrower bandgap enabled by the electrical field from two side gates - as illustrated in Figure 2 (b) - for the "on" state. Another structure that reduces the parasitic drain contact tunnelling current, and the "off" state current, uses an SB-GNRFET with an asymmetric top gate, which is situated closer to the source contact [28].

B. GNR-based Tunelling FETs

Other structures which have been recently investigated for their promising perspective in digital electronics, are GNR-based Tunnelling FETs (TFETs). Regular TFETs have either a single or double gate geometry (similarly to the GNRFET structure), and doped source and drain (via either chemical or electrostatic doping). Figure 3 illustrates a typical p-type GNRTFET energy band structure, noting that while for GNRFETs the transport is governed by both a thermionic emission current and a tunnelling current, for GNR-based TFETs the thermionic current component is negligible. The gate voltage shifts the energy bands, and has a big impact on the carriers tunnelling probabilities. Compared to GNRFETs, GNRTFETs benefit of superior gate control and higher I_{ON} current, and thus seems to be more attractive then GNRFETs for graphene-based computing. To get inside in GNRTFETs potential performance we present in Table II the evaluation results reported in [21] for a low-power inverter constructed with double-gated GNRTFETs with GNR channel widths of 10a, 13a, and 16a. One can observe in the Table that the GNRTFET avenue enables 8 to 9 orders of magnitude reduction of the static power when compared to the GNRFET counterpart.

978-1-5386-4483-6/18 $31.00 © 2018 IEEE

TABLE I: GNRFET-based Gates Propagation Delay and Power Consumption vs. CMOS 16 nm [20]

	Delay [ps]			Dynamic Power [W]			Leakage Power [W]		
	SB-GNRFET	MOS-GNRFET	CMOS	SB-GNRFET	MOS-GNRFET	CMOS	SB-GNRFET	MOS-GNRFET	CMOS
INV	4	28	15	$1.87 \cdot 10^{-5}$	$1.58 \cdot 10^{-6}$	$7.81 \cdot 10^{-6}$	$1.48 \cdot 10^{-6}$	$6.32 \cdot 10^{-11}$	$1.16 \cdot 10^{-8}$
NAND2	4	29	17	$5.83 \cdot 10^{-5}$	$1.13 \cdot 10^{-6}$	$6.85 \cdot 10^{-6}$	$1.89 \cdot 10^{-6}$	$1.11 \cdot 10^{-10}$	$1.35 \cdot 10^{-8}$
NOR2	4	29	22	$2.63 \cdot 10^{-5}$	$1.00 \cdot 10^{-6}$	$3.79 \cdot 10^{-6}$	$1.89 \cdot 10^{-6}$	$1.09 \cdot 10^{-10}$	$1.40 \cdot 10^{-8}$
XOR2	5	46	32	$4.10 \cdot 10^{-5}$	$1.22 \cdot 10^{-6}$	$9.47 \cdot 10^{-6}$	$8.84 \cdot 10^{-6}$	$4.87 \cdot 10^{-10}$	$7.09 \cdot 10^{-8}$

TABLE II: GNRTFET-based Inverter Propagation Delay, Static Power Consumption, and Energy [21].

nTFET / pTFET	Delay [ps]			Static Power [W]			Dynamic Energy [J]		
	10	13	16	10	13	16	10	13	16
10	$1.96 \cdot 10^{4}$	$1.11 \cdot 10^{4}$	$1.35 \cdot 10^{4}$	$1.49 \cdot 10^{-19}$	$7.39 \cdot 10^{-19}$	$4.93 \cdot 10^{-11}$	$2.14 \cdot 10^{-17}$	$2.52 \cdot 10^{-17}$	$2.96 \cdot 10^{-17}$
13	$1.11 \cdot 10^{4}$	$2.16 \cdot 10^{2}$	$1.41 \cdot 10^{2}$	$7.39 \cdot 10^{-19}$	$1.29 \cdot 10^{-18}$	$6.01 \cdot 10^{-11}$	$2.52 \cdot 10^{-17}$	$2.90 \cdot 10^{-17}$	$3.20 \cdot 10^{-17}$
16	$1.35 \cdot 10^{4}$	$1.41 \cdot 10^{2}$	$6.24 \cdot 10^{1}$	$4.93 \cdot 10^{-11}$	$6.01 \cdot 10^{-11}$	$1.20 \cdot 10^{-10}$	$2.96 \cdot 10^{-17}$	$3.20 \cdot 10^{-17}$	$3.65 \cdot 10^{-17}$

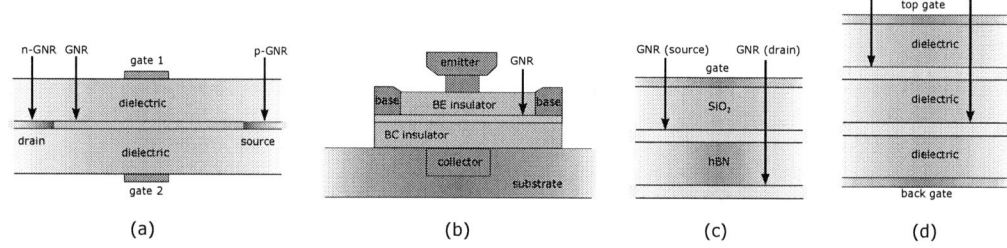

Fig. 4: GNRTFET cross-section schematics: (a) RTT; (b) GBT; (c) VTGNRFET; and (d) SymFET.

Vertical graphene-based structures (e.g., vertical tunnelling transistors, vertical Hot Electron Transistors (HET)), have been also proposed, which besides implications at the electronic transport level, enable integrated architectures with stacks of multiple transistors connected in series. Graphene Base Transistors (GBTs), as illustrated in Figure 4 (b), have a vertical structure composed of emitter, base, and collector just like a HET, with the base electrode made of graphene also [29] [30]. In the ON state, the emitter-base diode injects hot electrons which tunnel from emitter to collector. Operation in the THz frequency range and high current ratios are estimated to be obtained with GBTs.

A logical follow-up investigation of the tunnelling transistors refers to the Resonant Tunneling Transistors (RTT) [31]. Illustrated in Figure 4 (a), is a typical RTT structure, which enables barrier height modulation, and allows for resonant tunneling of the carriers. As RTTs can have several switching states (as a result of the negative differential resistance), they can also be potentially utilized for multiple valued logic.

For the previous transistor structures, the carrier transport was in the same plane as the graphene sheet. Changing the devices geometry, such that the tunnelling occurs between GNR layers (carrier transport vertical to the GNR), can

significantly increase the current. VTGRGETs, structurally illustrated in Figure 4 (c), are vertical tunnelling heterogeneous structures, which rely on effective voltage induced modulation of the GNR density of states and of the tunnel barrier height. Between the GNR made source and drain contacts a few layers (e.g., 3 to 7) of hexagonal boron nitride (hBN) [32], or molybdenum disulfide (MoS_2) [33] serve as tunnelling barrier. Tungsten disulfide (WS_2) can also be used as tunnelling barrier material, allowing one to switch between thermionic and tunnelling transport [34] and further increase the "on" current and by implication the I_{ON}/I_{OFF} current ratio.

Another vertical structure is the interlayer tunnelling transistor, SymFET [35], illustrated in Figure 4 (d). It has 2 layers of GNRs between which resonant tunnelling behaviour occurs, the resonant current peak being modulated by the applied gate bias and by the GNR chemical doping. The two GNR layers, are separated by a dielectric and flanked by a top and a bottom gate. An advantage of this structure is the current insensitivity to temperature.

While the previously introduced devices have different topologies, operation mechanisms, fabrication complexity, and performance they all target the realization of graphene based switches able to replace MOSFETs in the implementation of

978-1-5386-4483-6/18 $31.00 © 2018 IEEE

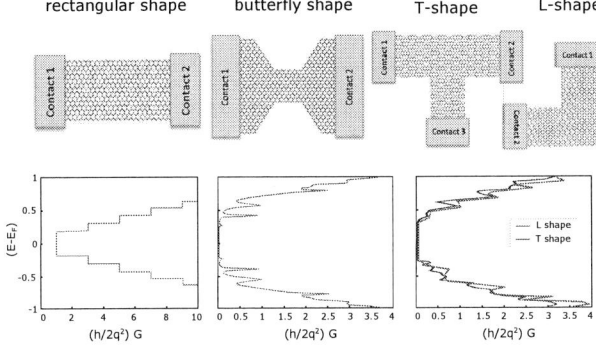

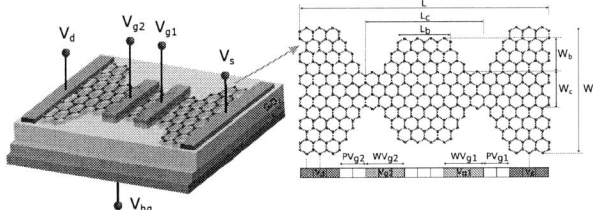

Fig. 5: Energy Bandgap vs GNR Geometry.

Fig. 6: Boolean Function Mirroring GNR-based Structure [22].

Boolean based graphene gates and circuits. In the next section we leave the traditional design avenue and investigate GNR potential to exhibit a more complex than switch behaviour and allow for the effective construction of basic blocks that my also go beyond traditional Boolean gates.

III. GRAPHENE NANORIBBON-BASED COMPUTING

In terms of novel devices and architectures, graphene's unique properties may enable operation modes which are fundamentally different than the traditional switching mechanism.

As mentioned in Section I, one of the main impediments of using graphene in logic design, is its lack of an energy bandgap. However, through GNR shape carving this problem can be overcome to some extent. In Figure 5, we exemplify 3 GNR shapes, which are subjected to a bias voltage via the 2 or 3 end-point contacts. The GNRs' associated conductance as a function of energy is depicted in the lower half of the figure. One can observe that for the standard rectangular shape the GNR is always conducting, while by carving the GNR geometry (e.g., into a butterfly shape, L-shape, or T-shape), a bandgap of approximately 0.5 eV can be induced, and the GNR conductance can be effectively switched off. GNR geometry shaping, together with the proper electrical external control means in order to modulate it conductance according to some desired logic function, provide the premises for a different perspective for logic design that is not based on transistors as basic building blocks. Specifically, the GNR can be patterned and biased in such a way that it can directly map a desired Boolean function onto its electrical characteristics [22]. Figure 6 presents such a GNR-based device architecture.

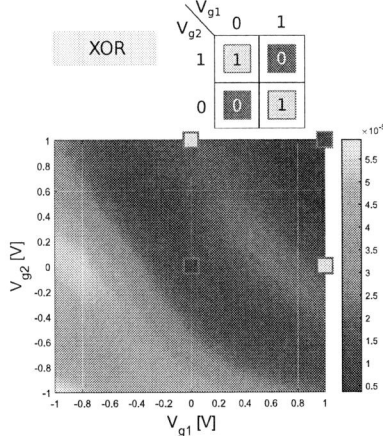

Fig. 7: 2-input XOR Gate Conductance Map.

The GNR-based basic building block is endowed with top gates, which modulate the current flow - through the GNR - induced by applying a bias voltage applied between the source and drain end point contacts. Underneath the graphene ribbon, there is a dielectric layer, the substrate, and a back gate. Different from GRNFETs, where the GNRs are rectangularly shaped, for this structure, a trapezoidal structure with zigzag edges is utilized. To obtain a certain, e.g., Boolean gate, behaviour GNR's geometry is shaped and the top gate contacts topology (distance between gate contacts and position relative to source/drain contacts) varied, until a conduction map which reflects the desired Boolean functionality is obtained. The Boolean function inputs are applied by means of top gate input voltages. For example, Figure 7 depicts the GNR structure conductance map (i.e., conductance G vs. top gate input controlling voltages, V_{g1} and V_{g1}), obtained for a GNR whose geometry was optimized such that it reflects the Boolean XOR operator functionality, for logic high and low voltage levels associated with 1 V and 0 V, respectively. The blue squares encode the XOR output logic "0", while the yellow squares represent the XOR output logic "1", in line with the afferent Karnaugh map. A similar procedure can be followed to obtain a GNR structure whose conductance maps a multi-input Boolean function, i.e., 3-input Boolean XOR, which is illustrated in Figure 8. An advantageous point for the aforementioned GNR structures, is that the voltage levels chosen for "0" logic and "1" logic, are not restrictive and the device can still properly operate when they are reduced into the order of hundreds or even tens of mV. In principle, every GNR structure which mirrors a certain Boolean function onto its conductance map, has its own V_{DD} limitation, which is highly dependent on the GNR geometry and contacts topology. As an example, it was found that 0.02 V is the lowest V_{DD} voltage value for which can still be obtained butterfly GNR structures able to mirror AND functionality, and which have an I_{ON}/I_{OFF} current ratio big enough to allow differentiation between logic low and logic high voltage levels [36].

978-1-5386-4483-6/18 $31.00 © 2018 IEEE   55

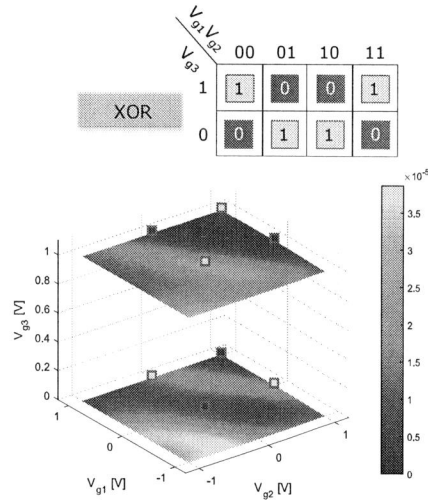

Fig. 8: 3-input XOR Gate Conductance Map.

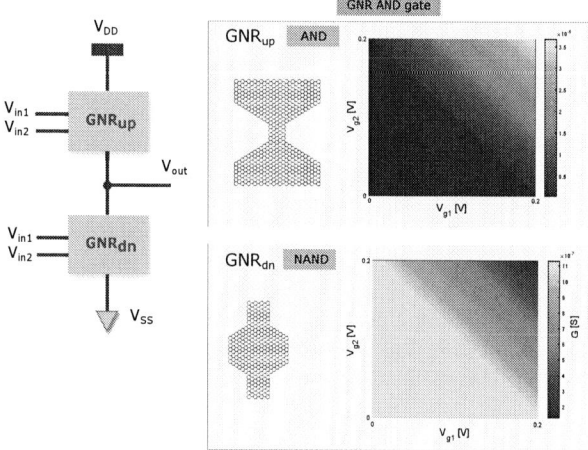

Fig. 9: GNR Boolean Gate [37].

In [37], the authors propose graphene-based Boolean gates, by arranging two such GNR structures as follows: a pull-up GNR structure, which has its drain terminal connected to V_{DD}, and a pull-down GNR, which has its source terminal connected to V_{SS}. The two GNR structures perform complementary functions: for instance for a graphene-based AND gate, the pull-up GNR mirrors the AND logic functionality onto it conductance, and the pull-down GNR maps the NAND Boolean functionality onto its conductance, as illustrated in Figure 9. For obtaining the GNR structures, which compose each GNR-based Boolean gate, the authors performed a design space exploration with respect to GNR's geometry and contacts topology. Figure 9 exemplifies an AND gate GNR structure and its conductance maps obtained as a result of the design space exploration. In Table III we summarize the propagation delay, active area (under the gate), and total power consumption reported in [37] for GNR-based gates operating at $V_{DD} = 0.2$ V and for CMOS 7 nm

TABLE III: GNR-based Gate Propagation Delay, Area, and Power Consumption vs. CMOS 7 nm [37]

	Delay [ps]		Active Area [nm^2]		Total Power [W]	
	GNR	CMOS	GNR	CMOS	GNR	CMOS
AND	1.38	9.618	$4.272 \cdot 10^1$	$1.452 \cdot 10^3$	$4.628 \cdot 10^{-9}$	$5.886 \cdot 10^{-7}$
NAND	2.15	7.556	$4.146 \cdot 10^1$	$9.680 \cdot 10^2$	$2.370 \cdot 10^{-9}$	$5.415 \cdot 10^{-7}$
XOR	7.48	9.168	$4.038 \cdot 10^1$	$2.420 \cdot 10^3$	$1.734 \cdot 10^{-9}$	$5.923 \cdot 10^{-7}$
BUFF	0.42	2.040	$3.283 \cdot 10^1$	$9.680 \cdot 10^2$	$0.937 \cdot 10^{-9}$	$4.704 \cdot 10^{-7}$
INV	0.27	1.110	$5.431 \cdot 10^1$	$4.840 \cdot 10^2$	$0.947 \cdot 10^{-9}$	$4.621 \cdot 10^{-7}$

($V_{DD} = 0.7$ V) counterparts. The Table indicates that the GNR-based approach substantially outperforms CMOS by up to $6\times$ smaller delay, and up to 2 orders of magnitude smaller active area, and total power consumption.

While a through comparison among graphene FET- and GNR- based logic gate implementations is not straight forward and also out the scope of this paper it is of interest to get a feeling about their potential and relative ranking. Let us consider as discussion vehicle the inverter, as cost and performance data have been reported for all its implementations, i.e., MOS-GNRFET, GNRTFET, and GNR-based, in Table I, Table II, and Table III, respectively. One can observe that the GNR-based inverter substantially outperforms its peers in terms of propagation delay, which is $15\times$, $104\times$, and $230\times$ smaller than the one of SB-GNRFET, MOS-GNRFET, and GNRTFET counterparts, respectively. Moreover, its power consumption is 4 and 3 orders of magnitude smaller than the one of SB-GNRFET and MOS-GNRFET, respectively. The GNRTFET inverter was specifically designed for extreme low power, thus at the expense of a very poor delay, its static power consumption is by 13 and 8 orders of magnitude smaller than the one of SB-GNRFET and MOS-GNRFET inverters, respectively. While no static power figures are available for the GNR-based inverter the 4 designs can be compared in terms of energy consumption, which is in the order of 10^{-17} J, 10^{-17} J, 10^{-17} J, and 10^{-21} J for the SB-GNRFET, MOS-GNRFET, GNRTFET, and GNR-based inverter, respectively. The area footprint is expected to be lower for the GNR-based gates as they comprise only two complementary GNR structures instead of several transistors.

These results clearly suggest that both the transistor-based and the GNR-based structures have the potential to outperform CMOS counterparts, with the GNR-based design style being the most promising one in terms of energy consumption.

By following the same avenue but a different line of reasoning in [38] the authors propose another GNR-based Boolean gate structure, arranged in a diapason like structure with 2 arms and 3 arms for 1-input and 2-input gates, respectively, as illustrated in Figure 10 for the particular case of a 2-input OR gate. The basic building block structure is this case is the L shaped GNR, which has a zig-zag side and an armchair side, which can give rise to an energy bandgap. The authors use -0.5 V for "0" logic, and 0.5 V for "1" logic. All 2-input

978-1-5386-4483-6/18 $31.00 © 2018 IEEE

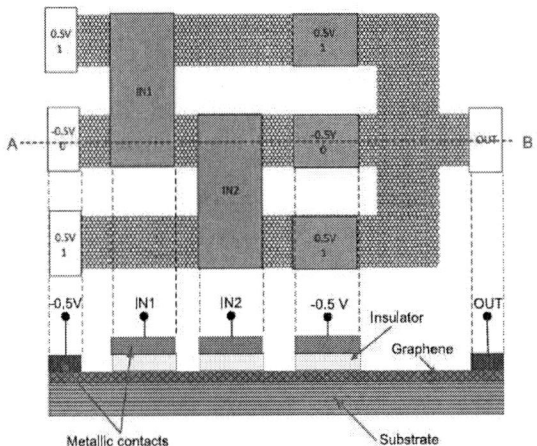

Fig. 10: Diapason GNR OR Gate [38].

gates (exemplified AND and OR) have an identical structure with 5 top gates and 4 electrodes, the only difference between the gates being the fixed applied bias voltages for 3 of the top gates, and for the left 3 electrodes, whose values can be either V_{DD} or V_{SS}. While being less effective in terms of delay and power the the gates introduced in [37] this GNR-based structure has some advantages which benefits fabrication, e.g., regularity and lack of a back gate, which can be proved useful for graphene-based biocompatible applications.

To conclude this section we would like to briefly highlight other computation approaches that can potentially benefit of graphene intrinsic properties. For instance, as graphene exhibits a negative differential resistance (peak-valley shaped I-V characteristic), one can envision graphene's potential for multi-valued non-binary logic [39]. For instance, in [40], the authors propose a one-digit radix-4 adder, composed of 2 GNRs, which exploits the quantized conductance and also uses the back gate as adder input. Graphene exhibits ambipolar transport, characterized by a superposition of electron and hole currents [41]. Thus, instead of suppressing this behavior as in the case of GNRFETs, one can also control it via an additional polarity top gate, with applications to in-field controllable dynamic logic, as well as static logic [42].

Besides electric charge, the fundamental electron property that was exploited for logic circuits, the electron spin and its associated magnetic moment can be used to control electrical conduction and create novel computing functionalities. Graphene's properties (e.g., negligible intrinsic spin-orbit coupling, absence of hyperfine interactions, long spin diffusion lengths) makes graphene an ideal candidate for spintronic devices expected to be faster, and to enable extremely low-power computing [43]. As an electron spin is inherently a quantum system that is in a superposition of states, it can serve as qubit for quantum information processing. To this end, graphene quantum dots have been proposed, as a host for spin qubits [44], as graphene holds the potential for long coherence time, as well as fast operating time.

IV. CONCLUSIONS

In this paper, we presented a comprehensive overview of state of the graphene-based computing. We have been interested in evaluating the potential impact graphene devices may have on circuit performance but also on circuit design style and underlying computation paradigm, thus we framed the discussion solely from a circuit design standpoint, without diving into any manufacturing and computer architecture related implications. We presented the mainstream switch-alike GNR-based transistors, namely GNRFETs and GNRT-FETs, followed by other transistor structures that improve their performance via a better modulation and control of the electronic transport. Then we focused on beyond switch based approaches and discussed GNR-based devices able to directly compute a Boolean function and on Boolean gates built with 2 such GNR structures with complementary behaviour. Both transistors- and GNR-based gate structures have been evaluated and compared with CMOS counterparts, in terms of area, delay, power consumption, and energy, to asses the potential viability of carbon based computation platforms. Simulation results indicated that the GNR-based inverter substantially outperforms its graphene based counter-candidates in terms of delay and energy consumption. Moreover, when compared with CMOS 7 nm Boolean gates GNR-based implementations exhibit a $6\times$ smaller propagation delay and a 2 orders of magnitude smaller total power consumption. Our analysis clearly indicated that graphene has great potential for the realization of beyond CMOS energy effective nanoscale circuits and that approaches that deviate from the traditional switch based design, in an attempt to take advantage of graphene's properties, are more successful and can catalyse the development of alternative computation avenues.

REFERENCES

[1] J. J. Liou, F. Schwierz, and H. Wong, *Nanometer CMOS*. Pan Standford Publishing, 2010.

[2] K. Rupp and S. Siegfried, "The economic limit to Moore's law." in *IEEE Transactions on Semiconductor Manufacturing*, vol. 24, no. 1, 2011, pp. 1–4. [Online]. Available: https://doi.org/0.1109/JPROC.2010.2040205

[3] A. Ferrari and et al., "Science and technology roadmap for graphene related 2D crystals, and hybrid systems." in *Nanoscale*, vol. 7, no. 11, 2015, pp. 4587–5062. [Online]. Available: https://doi.org/10.1039/C4NR01600A

[4] W. Choi and J. W. Lee, *Graphene syntehsis and applications*. CRC Press, 2012.

[5] J. M. Allen, V. C. Tung, and R. B. Kaner, "Honeycomb carbon: a review of graphene." in *Chemical Reviews*, vol. 110, no. 1, 2010, pp. 132–145. [Online]. Available: https://doi.org/10.1021/cr900070d

[6] K. Matsumoto, *Frontiers of graphene and carbon nanotubes - devices and applications.* Springer Japan, 2015. [Online]. Available: https://doi.org/10.1007/978-4-431-55372-4

[7] A. H. Castro Neto and et al., "The electronic properties of graphene." in *Reviews of Modern Physics*, vol. 81, no. 1, 2009, pp. 109–162. [Online]. Available: https://doi.org/10.1103/RevModPhys.81.109

[8] A. Mayorov and et al., "Micrometer-scale ballistic transport in encapsulated graphene at room temperature." in *Nano Letters*, vol. 11, no. 6, 2011, pp. 2396–2399. [Online]. Available: https://doi.org/10.1021/nl200758b

[9] J. Baringhaus and et al., "Exceptional ballistic transport in epitaxial graphene nanoribbons." in *Nature*, vol. 506, 2014, pp. 349–354. [Online]. Available: https://doi.org/10.1038/nature12952

978-1-5386-4483-6/18 $31.00 © 2018 IEEE

[10] K. Kim and et al., "A role for graphene in silicon-based semiconductor devices." in *Nature*, vol. 479, no. 7373, 2011, pp. 338–344. [Online]. Available: https://doi.org/10.1038/nature10680

[11] A. A. Balandin, "Thermal properties of graphene and nanostrcutured carbon materials." in *Nature Materials*, vol. 10, 2011, pp. 569–581. [Online]. Available: https://doi.org/10.1038/nmat3064

[12] J. Moser, A. Barreiro, and A. Bachtold, "Current-induced cleaning of graphene." in *Applied Physics Letters*, vol. 91, 2007, p. 163513. [Online]. Available: https://doi.org/10.1063/1.2789673

[13] V. M. Pereira and A. H. Castro Neto, "Tight-binding approach to uniaxial strain in graphene." in *Physical Review B*, vol. 80, 2009, p. 045401. [Online]. Available: https://doi.org/10.1103/PhysRevB.80.045401

[14] W. Ren and H. M. Cheng, "The global growth of graphene." in *Nature Nanotechnology*, vol. 9, 2014, pp. 726–730. [Online]. Available: https://doi.org/10.1038/nnano.2014.229

[15] Z. F. Wang and et al., "Emerging nanodevice paradigm: Graphene-based electronics for nanoscale computing." in *ACM Journal on Emerging Technologies in Computing Systems (JETC)*, vol. 5, no. 1, 2009, pp. 1–19. [Online]. Available: https://doi.org/10.1145/1482613.1482616

[16] M. R. Stan and et al., "Graphene devices, interconnect and circuits - challenges and opportunities." in *IEEE International Symposium on Circuits and Systems (ISCAS)*, 2009, pp. 69–72. [Online]. Available: https://doi.org/10.1109/ISCAS.2009.5117687

[17] M. J. Marmolejo and J. Velasco-Medina, "Review on graphene nanoribbon devices for logic applications." in *Microelectronics Journal*, vol. 48, 2016, pp. 18–38. [Online]. Available: https://doi.org/10.1016/j.mejo.2015.11.006

[18] L. Liao and et al., "Graphene field-effect transistors." in *Journal of Physics D: Applied Physics*, vol. 44, no. 31, 2011, p. 313001. [Online]. Available: https://doi.org/10.1088/0022-3727/44/31/313001

[19] D. Jena, "Tunneling transistors based on graphene and 2-D crystals." in *Proceedings of the IEEE*, vol. 101, no. 7, 2013, pp. 1585–1602. [Online]. Available: https://doi.org/10.1109/JPROC.2013.2253435

[20] Y. Y. Chen and et al., "Schottky-barrier-type Graphene Nano-Ribbon Field-Effect Transistors: A study on compact modeling, process variation, and circuit performance." in *IEEE/ACM International Symposium on Nanoscale Architectures*, 2013, pp. 82–88. [Online]. Available: https://doi.org/10.1109/NanoArch.2013.6623049

[21] X. Yang and et al., "Graphene tunneling FET and its applications in low-power circuit design." in *20th Symposium on Great Lakes Symposium on VLSI (GLSVLSI)*, 2010, pp. 263–268. [Online]. Available: https://doi.org/10.1145/1785481.1785544

[22] Y. Jiang, N. Cucu Laurenciu, and S. D. Cotofana, "On Carving Basic Boolean Functions on Graphene Nanoribbons Conduction Maps." in *IEEE International Symposium on Circuits and Systems*, 2018. [Online]. Available: https://doi.org/10.1109/ISCAS.2018.8351421

[23] D. Jena, "International Road for Semiconductors 2.0 - More Moore." 2015. [Online]. Available: https://www.semiconductors.org/main/2015_international_technology_roadmap_for_semiconductors_itrs/

[24] L. Liao and et al., "Single-layer graphene on Al2O3/Si substrate: better contrast and higher performance of graphene transistors." in *Nanotechnology*, vol. 21, no. 1, 2010, p. 015705. [Online]. Available: https://doi.org/10.1088/0957-4484/21/1/015705

[25] R. Sordan, F. Traversi, and V. Russo, "Logic gates with a single graphene transistor." in *Applied Physics Letters*, vol. 94, no. 7, 2009, p. 073305. [Online]. Available: https://doi.org/10.1063/1.3079663

[26] F. Giubileo and A. DI Bartolomeo, "The role of contact resistance in graphene field-effect devices." in *Progress in Surface Science*, vol. 92, no. 3, 2017, pp. 143–175. [Online]. Available: https://doi.org/10.1016/j.progsurf.2017.05.002

[27] L.-T. Tung and E. C. Kan, "Sharp Switching by Field-Effect Bandgap Modulation in All-Graphene Side-Gate Transistors." in *IEEE Journal of the Electron Devices Society*, vol. 3, no. 3, 2015, pp. 144–148. [Online]. Available: https://doi.org/10.1109/JEDS.2015.2397694

[28] M. Gholipour and et al., "Asymmetric gate Schottky-barrier graphene nanoribbon FETs for low-power design." in *IEEE Journal of the Electron Devices Society*, vol. 61, no. 12, 2014, pp. 4000–4006. [Online]. Available: https://doi.org/10.1109/TED.2014.2362774

[29] W. Mehr and et al., "Vertical graphene base transistor." in *IEEE Electron Device Letters*, vol. 33, no. 5, 2012, pp. 691–693. [Online]. Available: https://doi.org/10.1109/LED.2012.2189193

[30] S. Vaziri and et al., "A graphene-based hot electron transistor." in *Nano Letters*, vol. 13, no. 4, 2013, pp. 1435–1439. [Online]. Available: https://doi.org/10.1021/nl304305x

[31] H. Mohamadpour and A. Asgari, "Graphene nanoribbon tunneling field effect transistors." in *Physica E: Low-dimensional Systems and Nanostructures*, vol. 46, 2012, pp. 270–273. [Online]. Available: https://doi.org/10.1016/j.physe.2012.09.021

[32] N. Ghobadi and M. Pourfath, "A comparative study of tunneling FETs based on graphene and GNR heterostructures." in *IEEE Transactions on Electron Devices*, vol. 61, no. 1, 2014, pp. 186–192. [Online]. Available: https://doi.org/10.1109/TED.2013.2291788

[33] L. Britnell and et al., "Field-effect tunneling transistor based on vertical graphene heterostructures." in *Science*, vol. 335, no. 6071, 2012, pp. 947–950. [Online]. Available: https://doi.org/10.1126/science.1218461

[34] T. Georgiou and et al., "Vertical field-effect transistor based on graphene-WS$_2$ heterostructures for flexible and transparent electronics." in *Nature Nanotechnology*, vol. 8, no. 2, 2012, pp. 100–103. [Online]. Available: https://doi.org/10.1038/nnano.2012.224

[35] P. Zhao and et al., "SymFET: a proposed symmetric graphene tunneling field-effect transistor." in *IEEE Transactions on Electron Devices*, vol. 60, no. 3, 2013, pp. 951–957. [Online]. Available: https://doi.org/10.1109/TED.2013.2238238

[36] Y. Jiang, N. Cucu Laurenciu, and S. Cotofana, "Basic Boolean functions GNR conductance mapping." in *Technical Report, TU Delft*, 2018.

[37] Y. Jiang, N. Cucu Laurenciu, and S. D. Cotofana, "Complementary Arranged Graphene Nanoribbon-based Boolean Gates." in *IEEE International Symposium on Nanoscale Architectures*, 2018. [Online]. Available: https://doi.org/10.1109/ISCAS.2018.8351421

[38] S. Moysidis, I. G. Karafyllidis, and P. Dimitrakis, "Graphene logic gates." in *IEEE Transactions on Nanotechnology*, vol. 17, no. 4, 2018, pp. 852–859. [Online]. Available: https://doi.org/10.1109/TNANO.2018.2846793

[39] G. Liu and et al., "Graphene-based non-Boolean logic circuits." in *Journal of Applied Physics*, vol. 114, no. 15, 2013, p. 154310. [Online]. Available: https://doi.org/10.1063/1.4824828

[40] K. Rallis and et al., "Multi-valued logic circuits on graphene quantum point contact devices." in *IEEE/ACM International Symposium on Nanoscale Architectures (NANOARCH)*, 2018. [Online]. Available: https://doi.org/10.1145/3232195.3232214

[41] X. Yang and K. Mohanram, "Ambipolar electronics." in *Techical report, Rice University ECE Department, TREE1002*, 2010, pp. 1–5. [Online]. Available: http://hdl.handle.net/1911/27467

[42] A. E. Moutaouakil and et al., "Room temperature logic inverter on epitaxial graphene-on-silico ndevice." in *Japanese Journal of Applied Physics*, vol. 50, no. 7R, 2011, p. 070113. [Online]. Available: https://doi.org/10.1143/JJAP.50.070113

[43] X. Li and et al., "Large-area synthesis of high-quality and uniform graphene on coer foils." in *Science*, vol. 324, no. 5932, 2009, pp. 1312–1314. [Online]. Available: https://doi.org/10.1126/science.1171245

[44] P. Recher and B. Trauzettel, "Quantum dots and spin qubits in graphene." in *Nanotechnology*, vol. 21, no. 30, 2010, p. 302001. [Online]. Available: http://stacks.iop.org/0957-4484/21/i=30/a=302001

978-1-5386-4483-6/18 $31.00 © 2018 IEEE

Session N&N 1

NANOSCIENCE AND NANOENGINEERING 1

978-1-5386-4483-6/18 $31.00 © 2018 IEEE

Enhanced photoconductivity of SiGe-trilayer stack by retrenching annealing conditions

M.T. Sultan[1], J.T. Gudmundsson[2,3], A. Manolescu[1], M.L. Ciurea[4,5], C. Palade[4], A. V. Maraloiu[4], H.G. Svavarsson[1]

[1]Reykjavik University, School of Science and Engineering, IS-101 Reykjavik, Iceland
muhammad16@ru.is, manoles@ru.is, halldorsv@ru.is
[2]Department of Space and Plasma Physics, School of Electrical Engineering and Computer Science, KTH-Royal Institute of Technology, SE-100 44, Stockholm, Sweden
[3]Science Institute, University of Iceland, Dunhaga 3, IS-107 Reykjavik, Iceland
tumi@hi.is
[4]National Institute of Materials Physics, 077125 Magurele, Romania
[5]Academy of Romanian Scientists, 050094 Bucuresti, Romania
ciurea@infim.ro, catalin.palade@infim.ro, maraloiu@infim.ro

Abstract—We studied the effect of short term furnace annealing over the photoconductive properties of tristacked layer i.e. $TiO_2/(SiGe/TiO_2)_3$. The structure was prepared by depositing alternate layers of TiO_2 and SiGe films, using direct-current magnetron sputtering technique. A transmission electron microscopy and grazing incidence spectroscopy was used to analyze the morphology of the structure. Photoconductive properties were studied by measuring photocurrent spectra at different applied voltages and temperatures. Tristack layers were obtained with 5-10 nm SiGe nanocrystals (NCs) by annealing at 600 °C for 5 min. No sign of SiO_2 formation was found inside stacked layers. A maximum in the photocurrent spectra was observed at 994 nm at 300 K but it red-shifted gradually to 1045 nm with decrease in temperature to 100 K. This transition in peak maxima is attributed to SiGe NCs, due to lattice vibration and to contribution of non-radiative recombination at low temperatures.

Keywords—SiGe; TiO_2; nanocrystals; magnetron sputtering, photoconductivity, annealing.

1. Introduction

At present, there is a considerable interest in formation of SiGe quantum dots (QDs) due to the compatibility of Ge (Germanium) with Si (Silicon) and the possibility of altering its bandgap in the infrared region of the spectrum. Specifically, self-assembled SiGe dots have drawn interest because of its tuning ability by quantization with respect to optimal power and energy conversion efficiency and optical properties [1, 2]. In past decades, great effort has been made to increase the photosensitivity of TiO_2 films from visible to infrared spectra, by affecting the structure and crystallization by means of thermal annealing [1, 3, 4]. One of the critical challenges, in this relation, is to obtain structuring of NCs embedded in the oxide matrix at low-temperature to make them available for optoelectronic applications where low-processing temperature is required to preserve the functionality of other incorporated Si-based electronics. Although attempts have been made to obtain NCs by low-temperature annealing at 700 °C or lower for several minutes [5, 6], both the time and the temperature are still too high for appropriate processing of devices. Another important issue is to avoid the formation of the SiO_2 layer, which tends to occur after annealing in $TiO_2/SiGe$ system [7, 8] due to inter-diffusion of oxygen from TiO_2 into SiGe, which in turn blunt the interface and deteriorates the multilayer structure.

The aim of this study is to demonstrate the feasibility of using mild annealing process to create SiGe NCs within Si-sub/SiO_2-buffer/$(TiO_2/SiGe/TiO_2)_3$/Al system without the formation of a SiO_2 insulating layer; thus increasing the spectral intensity of TiO_2 for visible to near infrared regime.

978-1-5386-4483-6/18 $31.00 © 2018 IEEE

2. Experiment

Tri-stacked multilayer structure (TLs) of SiO_2-buffer/TiO_2-cap/$(GeSi/TiO_2)_3$ were deposited on Si (100) wafers by reactive direct current magnetron sputtering (dcMS) for the TiO_2 layers and co-sputtering Si and Ge by dcMS for the GeSi layer. The flow rate for Ar (q_{Ar}= 37 sccm) and O_2 (q_{O_2}= 1.2 sccm) were controlled by mass flow controller, and throttle valves were adjusted to stabilize growth pressure of 0.7 Pa during the TiO_2 deposition. A 30 nm thick GeSi film was deposited by dcMS in constant-power mode (Advanced Energy MDX500 power supply) using 25 W dc for Ge target and 80 W dc power for Si target, under 0.7 Pa Ar (5N) pressure. For obtaining GeSi NCs, furnace annealing was performed at 600°C in N_2 ambient for 5 min.

Photocurrent measurements (voltage dependences, spectral curves) were performed in dedicated setups for electrical and photoelectrical characterization. For photoelectrical measurements, coplanar aluminium contacts were thermally evaporated on top of the annealed structures. The current passing through the contacts at a constant bias of 1 V under illumination was measured. The structure and morphology of samples were studied by grazing incidence XRD (GIXRD, Philips X'pert diffractometer $Cu_{K\alpha}$), transmission electron microscopy (TEM, Jeol ARM 200F microscope).

3. Result and Discussion

Fig. 1 shows a GiXRD scan of the structure annealed at 600 °C for 5 min. The measured diffractogram have peak positioned at 27.60, 45.62 and 54.4 degree, which is between the tabulated maximum for Ge and Si [9].

Fig. 2(a) shows an HRTEM image of central SiGe layer evidencing the presence of spherical SiGe nanocrystals ranging from 5-10 nm in size. Fig. 2(b) shows the HRTEM image of bottom SiGe layer, where the SiGe NCs forms parallel columns perpendicularly to the GeSi layer.

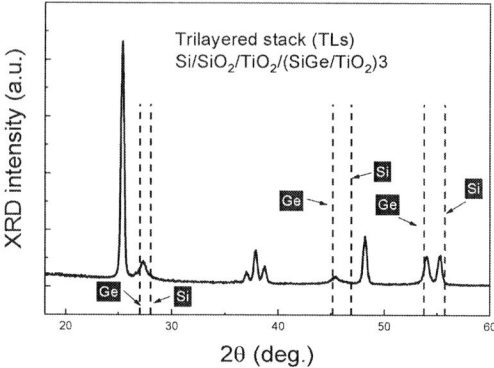

Fig. 1. GIXRD curve for TiO_2/GeSi/TiO_2/Si TL annealed at 600 °C for 5 min; vertical dashed lines correspond to standard tabulated positions for cubic Ge (2θ = 27.45; 45.59; 54.04 - ASTM 01-079-0001), cubic Si (28.45; 47.31; 56.13 - ASTM 01-070-5680) and anatase TiO_2 (26.53; 36.95; 37.80; 38.58; 48.05; 53.89; 55.06; 62.11- ASTM 00-021-1272)

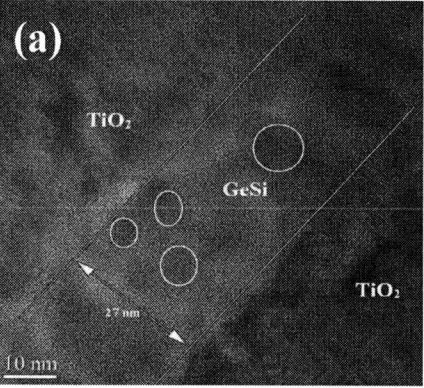

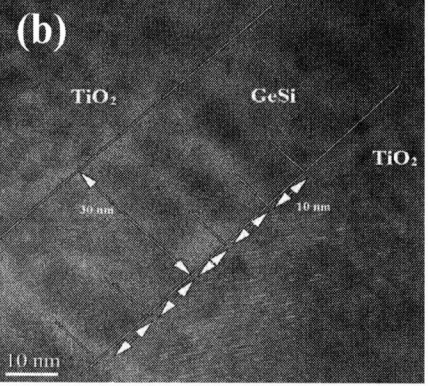

Fig. 2. HRTEM images showing the presence of NCs. (a) central and (b) bottom SiGe layer of TL annealed at 600 °C for 5 min.

Moreover, there is a periodicity of ~10 nm between columns; this can probably be due to the orientation of the TiO$_2$ crystal on which GeSi NCs columns stand, but this periodicity is not observed overall. In addition, it is noticed that the columns are composed of 5 − 6 nm GeSi regions/NCs that are separated from each other by amorphous regions with 5 − 6 nm thickness. Some studies [10, 11] have shown that the nucleation and growth of QDs in one layer are affected by the strain field produced by QDs in another layer. Such strain caused by the QDs, experienced by the surrounding matrix, will cause the adjacent layer QDs to align along the growth direction to minimize the strain field. However, as observed in Fig. 2, there is no sign of SiO$_2$ between the TiO$_2$ and SiGe layers after annealing.

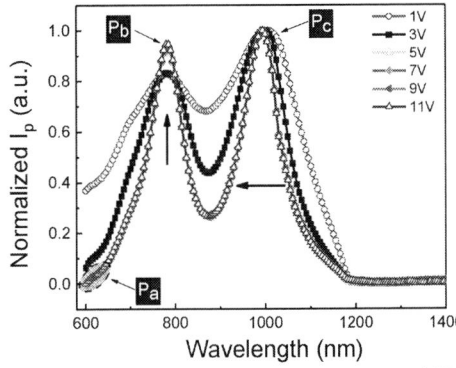

Fig. 3. Room temperature photocurrent spectra of TLs at different applied bias. The arrows shows the shift in peak height and position of P$_b$ and P$_c$ respectively.

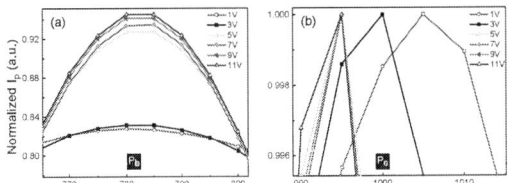

Fig. 4. Zoomed in view of peak P$_b$ and P$_c$ photocurrent spectra at different applied bias.

Fig. 3 shows the photocurrent spectra of the sample annealed at 600 °C for 5 min. A shoulder and two peaks at a wavelength of 620, 784 and 994 nm (P$_a$, P$_b$, and P$_c$) are clearly visible. We attribute the peak at higher

wavelength P$_c$ to SiGe nanocrystals. The small shoulder (on the right of P$_c$) at ~ 1100 nm wavelength could be explained by Si substrate influence through a capacitive coupling (surface photovoltage and gating effect) [12]. The photocurrent intensity increases with applied bias due to the creation of hole-depleted zone as a result of field effect and is being comprehensively discussed in [12].

The shoulder and peak (P$_a$, P$_b$) present at lower wavelength can be attributed to the formation of defects which results from stress relaxation in an oxide matrix and due to the presence of polycrystalline TiO$_2$ nanocrystals as evident in Fig. 2(b) [7, 8]. Fig. 4 shows a zoomed in view of peaks P$_b$ and P$_c$. The peak P$_c$ shows a blue shift (Fig. 4(a)) from 1006 to 994 nm while peak P$_b$ shows an increase (Fig. 4(b)) in intensity with increasing voltage, while the peak position is constant. The current vs voltage measured over TLs (Fig. 5) shows a symmetric and linear behavior in +11 to -11 V range and is due to high density of SiGe nanoparticles and polycrystalline TiO$_2$ and defects associated with it.

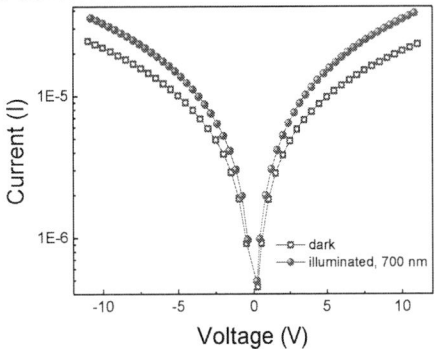

Fig. 5 Current- voltage characteristic measured at room temperature under dark and illuminated condition.

A normalized photocurrent spectrum of TLs, measured at constant bias of 11 V at different temperatures is shown in Fig. 6. One can clearly see that the peak P$_c$ shifts towards higher wavelength, from 994 to 1045 nm, as the temperature gets lower Concurrently, peak P$_b$ becomes a shoulder as the temperature is reduced from 300 to 100K.

978-1-5386-4483-6/18 $31.00 © 2018 IEEE

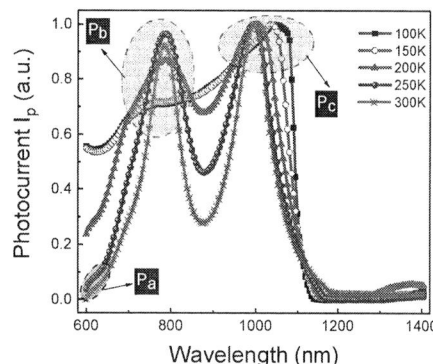

Fig. 6. Photocurrent spectra at 11 V bias at different temperatures.

4. Conclusion

Stacked layers films, composed of triple layers of SiGe/TiO$_2$ on top of the TiO$_2$ film were prepared by dcMS. The films underwent furnace annealing at 600 °C for 5 min to form nanocrystals in TiO$_2$ matrix. The as-deposited samples were amorphous while the annealed ones contained 5-10 nm SiGe NCs surrounded by polycrystalline TiO$_2$. No sign of SiO$_2$ after annealing was seen on the HRTEM images. The photoconductive properties are related to the morphology i.e. to the presence of NCs in the TLs and have the main maximum which shifts from 994 to 1045 nm for 300 to 100 K respectively. It is also attributed to the polycrystalline TiO$_2$ and to the formation of defects in structure (Pa and Pb). From the temperature measurement, we can expect that such peak is due to defect in structure resulting in variation in intensity rather than in a shift of peak position.

Acknowledgments. This work was supported by M-ERA.NET projects PhotoNanoP UEFISCDI Contract no. 33/2016 and GESNAPHOTO UEFISCDI Contract no. 58/2016, and the Technology Development Fund of the Iceland Centre for Research no. 159006-0612 and by Romanian Ministry of Research and Innovation through NIMP Core Program PN16-480102.

References

[1] C. P. Church, E. Muthuswamy, G. Zhai, S. M. Kauzlarich, S. A. Carter, "*Quantum dot Ge/TiO$_2$ heterojunction photoconductor fabrication and performance,*" Applied Physics Letters, 103 (22), 223506, November 2013.

[2] G .G. Pethuraja, R. E. Welser, A. K . Sood, C. Lee, N. J. Alexander, H. Efstathiadis, P. Haldar, J. L. Harvey, "*Effect of Ge Incorporation on Bandgap and Photosensitivity of Amorphous SiGe Thin Films,*" Materials Sciences and Applications, 03 (02), pp. 67-71, February 2012.

[3] A. F. Khan, M. Mehmood, M.; T. Ali, H. Fayaz, "*Structural and optical studies of nanostructured TiO2-Ge multi-layer thin films,*" Thin Solid Films, 536, pp. 220-228, 2013.

[4] X. Li, F. He, G. Liu, Y. Huang, C. Pan, C. Guo, "*Fabrication of Ge quantum dots doped TiO$_2$ films with high optical absorption properties via layer-by-layer ion-beam sputtering*", Materials Letters, 67 (1), pp. 369–372, January 2012.

[5] B. Zhang, S. Shrestha, P. Aliberti, M. A. Green, G. Conibeer, "*Synthesis and structural properties of Ge nanocrystals in multilayer superlattice structure*", Nanoscale Photonic and Cell Technologies for Photovoltaics II , Proc. SPIE 7.411 741103, August 2009.

[6] I. M. Ortiz, A. Rodríguez, J. Sangrador, T. Rodríguez, M. Avella, J. Jiménez, C. Ballesteros, "*Luminescent nanostructures based on Ge nanoparticles embedded in an oxide matrix*", Nanotechnology 16(5), pp. 197–201, March 2005.

[7] C. Palade, I. Dascalescu, A. Slav, A. M. Lepadatu, S. Lazanu, T. Stoica, V. Teodorescu, M. L. Ciurea, F. Comanescu, R. Muller, A. Dinescu, A. Enuica, "*Photosensitive GeSi/TiO2 multilayers in VIS-NIR*", 2017 International Semiconductor Conference (CAS), Sinaia, pp. 67-70, 2017.

[8] A. Slav, C. Palade, I. Stavarache, V. S. Teodorescu, M. L. Ciurea, R. Muller, A. Dinescu, M. T. Sultan, A. Manolescu, J. T. Gudmundsson, H. G. Svavarsson,"*Influence of preparation conditions on structure and photosensing properties of GeSi/TiO$_2$ multilayers*", 2017 International Semiconductor Conference (CAS), Sinaia, pp. 63-66, 2017.

[9] M.L. Ciurea, I. Stavarache, A.M. Lepadatu, I. Pasuk, V.S. Teodorescu, "*Electrical properties related to the structure of GeSi nanostructured films*", Phys. Status Solidi B, 251(7), pp. 1340–1346, July 2014.

[10] X. F.Yang, K. Fu, W. Lu, W. L. Xu, Y. Fu, "*Strain effect in determining the geometric shape of self-assembled quantum dot*", Journal of Physics D: Applied Physics, 42 (12), 125414, June 2009.

[11] P. Howe, E. C. L. Ru, E. Clarke, B. Abbey, R. Murray, T. S. Jones, "*Competition between strain-induced and temperature-controlled nucleation of InAs/GaAs quantum dots*", Journal of Applied Physics, 95 (6), pp. 2998–3004, March 2004.

[12] A. M. Lepadatu, A. Slav, C. Palade, I. Dascalescu, M. Enculescu, S. Iftimie, S. Lazanu, V. S. Teodorescu, M. L. Ciurea, T. Stoica, "*Dense Ge nanocrystals embedded in TiO2 with exponentially increased photoconduction by field effect*", Scientific Reports, 8 (1), 4898, March 2018.

From Pentacene Thin Film Transistor to Nanostructured Materials Synthesis for Green Organic-TFT

Cristian Ravariu[1], Dan Eduard Mihaiescu[2], Daniela Istrati[2], Maria Stanca[2]

[1] UPB-University "Politehnica" of Bucharest, Faculty of Electronics ETTI, Dept. of Electronic Devices Circuits and Architectures
Splaiul Independentei 313, Sect.6, 060042, Bucharest, Romania; **E-mail:** cristian.ravariu@upb.ro
[2] UPB-University "Politehnica" of Bucharest, Faculty of Applied Chemistry, Dept. of Organic Chemistry "C. Nenitescu", Splaiul Independentei 313, Sect.6, 060042, Bucharest, Romania; **E-mail:** dan.mihaiescu@upb.ro

Abstract—As first aim, a start Pentacene-Organic Thin Film Transistor - OTFT - is simulated to capture the static characteristics and to find the matching parameters with the experimental set-up. The current vectors validate the main conduction way and the OTFT functionality. In a second stage, the basic technology of an alternative polymer grafted on nanomaterial synthesis, is depicted. The Fe_3O_4 core-shell nano-particles are assembled by an external shell of para-aminobenzoic acid (PABA). The final scope will be OTFT construction by these green technologies. The first step: the Fe_3O_4/PABA thin films synthesis and characterization, is successfully performed.

Keywords—Organic-TFT, simulation, NCS/PABA

1. Introduction

Despite of a discouraging debut of the organic semiconductors with sub-10^{-5}cm^2/Vs carriers mobility - 25 years ago [1], the huge research effort from the last decade brings the Organic Thin Film Transistors (OTFT) among the competitive devices, [2]. Today, the mobility rises to 40cm^2/Vs, [3]. The main OTFT advantages are: accessible room temperature technology, easy grown on flexible substrates, low-cost deposition-techniques by inkjet-printing or spin coating, plastic foils bendable to manufacture foldable displays. One of the most successful organic semiconductor used for the OTFT fabrication is pentacene. Therefore, this paper start from the simulation of a pentacene-OTFT with the same configuration as an experimental one, [4]. The aim is to establish the simulation parameters, so that the simulated characteristics are matched by the experimental points, as a first contribution of this paper. The second contribution concerns

the synthesis and the micro-physical characterization of nano-core materials [5] with organic compound shells, designed for future implementations of OTFT, searching friendly environmental technologies.

Traditional organic semiconductors like pentacene are based on polynuclear aromatic hydrocarbons, being susceptible to high toxicity/carcinogenic precursors [6]. Hence, another optimization that must be accounted in the next future envisages non-toxic polymers grafted on nano-core-shell (NCS) nano-composites, appealing to green synthesis routes. In this category enters para-aminobenzoic acid (PABA) that naturally occurs as organic compound and respects the molecular conjugation as the main condition to ensure electronic conduction in OTFT,[7].

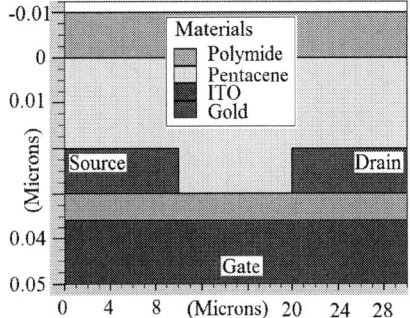

Fig. 1. The OTFT conceptual structure.

2. The Pentacene-OTFT simulations

The Atlas from Silvaco software is used for OTFT simulations. Most accessed models of the organic devices are used in order to activate the Poole-Frenkel mobility

978-1-5386-4483-6/18 $31.00 © 2018 IEEE

model and the Langevin recombination models, [8]. The gold source/drain contacts with well aligned work function of 4.9eV to the highest occupied molecular orbital level of Pentacene 4.8eV, all default Pentacene properties from the Silvaco library [9], ODEFECTS function to capture the Acceptor/Donor-like trap density, HA, HD (dual DOS in organic layers).

The simulated OTFT gets similar features as a fabricated Pentacene-OTFT, in the configuration Bottom-Contacts Bottom Gate (BCBG): pentacene film of 30nm thickness doped p-type of $7x10^{17}cm^{-3}$, channel length of 5µm, the same mobility parameters - μ_P=0.45cm^2/Vs and Poole-Frenkel parameters as beta.pfmob=$7.7x10^{-5}$ (V/cm)$^{0.5}$, deltae.pfmob=0.018eV, [4], fig. 1.

Few minor differences are applied: bottom insulator is polyimide of 6nm thickness instead Al$_2$O$_3$ of 5.7nm, ITO conductor instead Al bottom gate, as more compatible materials, channel depth on Oz is 1µm instead 100µm as the default Atlas Oz distance. So, 5µA/100µm = 0.05µA/µm, [4].

Figure 2 presents the simulated transfer characteristics, when the Pentacene-OTFT is biased at V$_S$=0V, V$_{DS}$=-3V or -5V and V$_G$ $\in$ (-5, 4)V. At linear scale, the I$_D$ current starts to increase from V$_G$>V$_T$ $\cong$ -1V and reaches 0.1µA/1µm at V$_{GS}$=-5V. The log scale reveals two firm ON (V$_G$<-2V, I$_D$~10^{-8}A) / OFF (V$_G$>2V, I$_D$~10^{-16}A) states, in agreement with the literature, [7].

point is visible as black/white contours on electrodes: V$_S$=0V, V$_D$=-5V, V$_G$ =3V. Along the entire channel, the holes concentration is decreased up to 10^{-4}cm^{-3} and lower, under V$_G$>0, so that only leakage drain-source current vectors are captured, fig. 3.

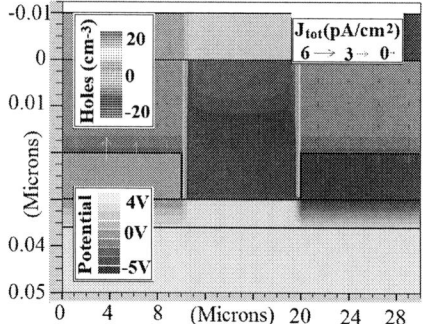

Fig. 3. A cross-section through OTFT biased to V$_G$=+4V.

3. Matching simulations with experiments

Because the DOS states are not explicited in [4], we started the output characteristics simulation from comparable DOS values as the doping, HA=HD=$7x10^{17}cm^{-3}$ at a temperature of TCA=TCD=300K, fig. 4. In this case, the I$_D$-V$_D$ curve is far away from the experimental picked points at V$_{GS}$=-3V, [4]. To match the simulated curve over the experimental points, the Acceptor/Donor-like trap density must be increased to $3x10^{20}cm^{-3}$. For higher density of $10^{21}cm^{-3}$, the current is alleviated. Both extreme cases are analyzed in fig. 5 that emphasizes a current density decreasing from 1000A/cm^2

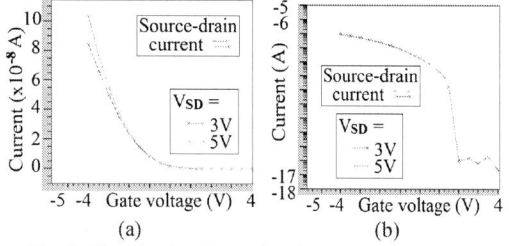

(a)   (b)

Fig. 2. The simulated transfer characteristics of OTFT at: (a) linear; (b) log scale.

Figure 3 explains the OFF state induction by a positive gate voltage. The OTFT bias

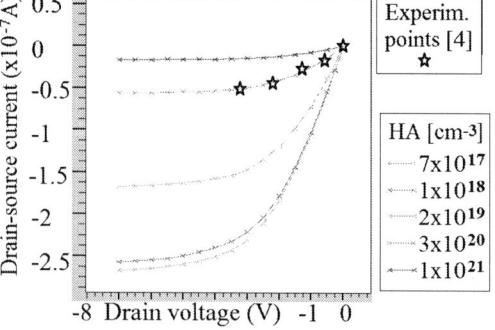

Fig. 4 The simulated output characteristics of OTFT at V$_{GS}$=-3V and experimental I$_D$-V$_D$ picked points [4].

at HA=$7x10^{17}cm^{-3}$ to 186A/cm^2 at

978-1-5386-4483-6/18 $31.00 © 2018 IEEE   66

$HA = 10^{21} cm^{-3}$, fig. 5 a, b.

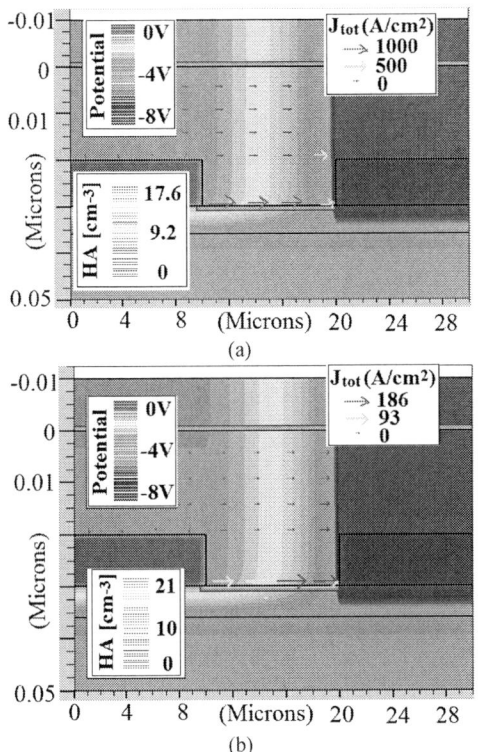

(a)

(b)

Fig. 5. The current density vectors in the OTFT biased to $V_{DS}=-8V$, $V_{GS}=-3V$ at $HA[cm^{-3}]$ of: (a) 7×10^{17}; (b) 10^{21}.

4. Experimental NCS materials

4.1. Synthesis

The basic nano-core-shell material contains a primary core-shell np's ferrite nano-particle (Fe_3O_4). The II-nd shell synthesis is achieved by co-precipitation method and a specific organic compounds as PABA - para-aminobenzoic acid. PABA respects the molecular conjugation, alternating single and double bonds between covalently bound carbon atoms and efficiently binds to metallic ions of the core. The final material Fe_3O_4/PABA is achieved by co-precipitation of the ions FeII and FeIII in NaOH solution, under molar ratio of Fe_3O_4:PABA = 1:7. After the subsequent shells capturing, the nanoparticle distribution modeling in a volume of liquid is made by the 3D modeling software Tomviz, fig. 6.

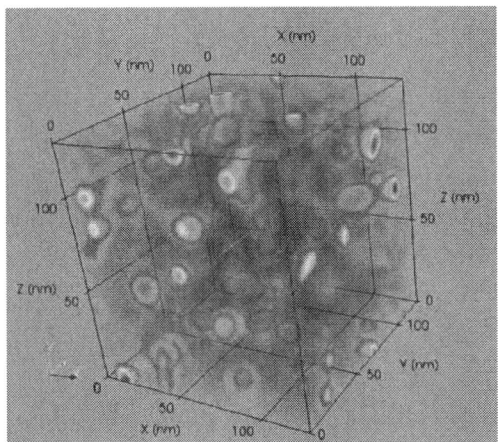

Fig. 6. Nanoparticles distribution in a aqueous phase.

The surface morphology is analyzed by SEM. A heterogenic structure, with granular clusters of uniform shape and size, separated by aleatory spaces, is recorded, fig. 7.

Fig. 7 The SEM image of the synthesized core-shell Fe_3O_4/PABA nano-material.

2.2. Experimental Characterization

The PABA shell presence is identified by FT-IR, finding the characteristics band of the free group $-NH_2$ at 3160.5 cm^{-1}.

The X-ray powder diffraction (XRD) analysis is achieved by PANalytical Empyrean equipment, using the $CuK\alpha$ (1.541874) radiation. The apparatus gets a programmable divergence slot on the incident direction and a programmable anti-diffusion slot, mounted on the PIXcel3D detector, in the diffraction direction. The scanning was possible by a Bragg Brentanno geometry, with 0.02 step and time per step of 100seconds in the range $2\theta = 20\text{-}70^0$, fig. 8.

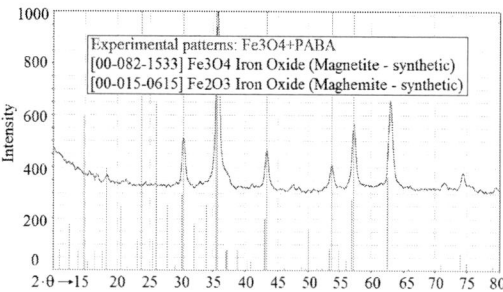

Fig. 8. The XRD spectrum of the core-shell synthesized materials of the same XRD spectrum.

The XRD spectrum still confirms the crystalline structure, spinel-type. The XRD highlights the specific peaks for the magnetite crystalline lattice: (220), (311), (400), (422), (511) and (440), as the literature confirms, too, [10].

5. Discussions

Already the ferrite NCS compounds are used in the Schottky diode construction, [11]. So, in the next future, we intend to deposit thin film of Fe_3O_4/PABA compounds on thin insulator to offer alternative green organic materials for OTFT, under a friendly environmental technology.

The main OTFT matching parameters are: dopings, contacts, interface charges, Odefects traps density, Materials parameters like band-gap, affinity, Nc-Nv energetic states, Mobility parameters like Poole-Frenkel beta.pfmob, deltae.pfmob or Hooping mobility parameters.

5. Conclusions

The matching parameters of the simulated OTFT with experiments were studied. Green electronic technologies for organic transistors, based on non-toxic green polymers attached to NCS nanocomposites

were identified. The synthesized Fe_3O_4/PABA nanomaterials are promising candidate for this purpose.

Acknowledgments. This work is supported by grants of the Romanian National Authority for Scientific Research and Innovation, CNCS/CCCDI UEFISCDI: PN-III-P4-ID-PCE-2016-0480 project number 4/2017 (TFTNANOEL).

References

[1] G. Horowitz, R. Hajlaoui, et al, "Temperature Dependence of the Field-Effect Mobility of Sexithiophene. Determination of the Density of Traps," *Journal de Physique III, EDP Sciences*, vol. 5, no. 4, pp.355-371, 1995.

[2] Y. Yuan, G. Giri, et al, "Ultra-high mobility transparent organic thin film transistors grown by an off-centre spin-coating method," *Nature Communications*. vol. 5, no. 3005, pp. 1-9, 2014.

[3] E. Bittle, J. Basham, et al, "Mobility overestimation due to gated contacts in organic field-effect transistors," *Nature Communications*, vol. 7, no. 10908, pp. 1-7, 2016.

[4] P. Mittal, B. Kumar, et al, "Channel length variation effect on performance parameters of organic field effect transistors," *Microelectronics Journal*, vol. 43, no. 12, pp. 985-994, 2012.

[5] A. Buteica, D. Mihaiescu, et al, "The anti-bacterial activity of magnetic nanofluid: Fe3O4 /oleic acid/cephalosporins core/shell/adsorption-shell proved on S. aureus and E. coli and possible applications as drug delivery systems," *Digest Journal of Nanomaterials and Biostructures*, vol. 5, no. 4, pp. 927-932, 2010.

[6] A. Solhaug, M. Refsnes, et al, "Polycyclic aromatic hydrocarbons induce both apoptotic and anti-apoptotic signals in Hepa1c1c7 cells," *Carcinogenesis*, vol. 25, no. 5, pp. 809-819, 2004.

[7] H. Klauk, "Organic thin-film transistors," *Chem. Soc. Rev.*, vol. 39, no.4, pp. 2643–2666, 2010.

[8] C. Ravariu, D. Dragomirescu, Different Work Regimes of an Organic Thin Film Transistor OTFT and Possible Applications in Bioelectronics, *American Journal of Bioscience and Bioengineering*, vol. 3, no. 3, pp. 7-13, 2015.

[9] ***, Atlas Manual 2012, available at: www.silvaco.com

[10] X. Zhu, S. Kalirai, et al., "What is the correct Fe L23 X-ray absorption spectrum of magnetite?," *Journal of Electron Spectroscopy and Related Phenomena*, vol. 199, no. 1, pp. 19–26, 2015.

[11] Ö. Metin, Ş. Aydoğan, et al, "A new route for the synthesis of graphene oxide–Fe3O4 (GO–Fe3O4) nanocomposites and their Schottky diode applications," *Journal of Alloys and Compounds*, vol. 585, no. 2, pp. 681-688, 2014.

ELECTRICAL PROPERTIES OF AS-DEPOSITED ALD HfO₂ FILMS RELATED TO SILICON SURFACE STATE

Cornel Cobianu[1,2], Florin Nastase[1,2], Niculae Dumbravescu[1,2], Octavian Buiu[1,2], Adrian Albu[1], Bogdan Serban[1,2], Mihai Danila[1], Cosmin Romanitan[1], Octavian Ionescu[1,2]

[1] National Institute for Research and Development in Microtechnologies-IMT Bucharest, Romania.
[2] Research Center for Integrated Systems, Nanotechnologies and Carbon-Based Nanomaterials (CENASIC)-IMT Bucharest, Romania.

*E-mail: cornel.cobianu@imt.ro; florin.nastase@imt.ro

Abstract-In this paper, we present an experimental study of the electrical properties of the as-deposited HfO₂ films obtained by atomic layer deposition (ALD) method from tetrakis dimethylamino hafnium and water vapors at 200°C as a function of the silicon substrate preparation, in terms of Si-H and Si-OH terminated surfaces. High frequency C-V characteristics have proven that relatively higher effective dielectric constant, lower fixed charge at the Si-HfO₂ interface and lower oxide trapped charge were obtained on MOS capacitors with HfO₂ dielectric performed on Si-OH terminated Si surface with respect Si-H terminated surface, proving a more robust Si-O-Hf interface with respect to Si-Hf-O interface.

Keywords-Atomic layer deposition, HfO₂, hydrogenated and hydroxylated Si surface, C-V measurement

INTRODUCTION

The atomic layer deposition (ALD) of HfO₂ ultrathin films has become a mature process for the preparation of the high-k gate dielectrics in the IC technology nodes below 45 nm [1]. This high value of the relative dielectric permittivity (k>15) has allowed the use of relatively "thicker" gate dielectrics, with much smaller leakage currents, but with an equivalent oxide thickness (EOT) below 2 nm. To preserve this high value of "k", solid state reactions of HfO₂ with silicon or gate of the MOS transistor should be minimized during device fabrication, as such reactions will create interface layers (IL) which have lower dielectric constants and thus an overall lower-k will be obtained. For the same reason, standard cleaning processes (RCA-1 and RCA-2) of the silicon surface [2] are followed by an HF dip 2% (HF/H₂O=2/100), which will remove the monolayer of SiO₂ and hydroxyl terminations (Si-OH) and will leave a hydrogenated Si surface (Si-H terminated) for a limited time before ALD deposition. However, the atomistic modeling of

reaction mechanism for the formation of the first HfO₂ monolayer between the ALD precursor like tetrakis (ethylmethylamino) hafnium and Si-H terminated Si surface has shown that such reaction is less thermodynamically favorable than the reaction between the same precursor and Si-OH terminated surface [3]. Therefore, higher activation barriers should be surmounted by the precursor reaction in the case of Si-H terminated surfaces and this may require higher deposition temperatures for this type of Si surface [3].

On the other hand, the presence of one or a few monolayers of SiO₂ due to RCA1 or RCA2 cleaning may help to the quality of the Si-HfO₂ interface. Recently, we have shown that silicon termination had an influence on the early breakdown distribution of as-deposited HfO₂ films under constant current stress [4]. It is the purpose of this paper to study the effect of Si-H and Si-OH silicon surface terminations on other electrical properties of the as-deposited HfO₂ films.

1. EXPERIMENTAL

HfO₂ films with thickness in the range of 9.63-10.61 nm were deposited by ALD method from Tetrakis(DimethylAmino)Hafnium (TDMAH) and water vapours at 200°C on p-type silicon (100) wafers of resistivity equal to (1-10) Ω*cm, by means of an "OpAL reactor" from Oxford Instruments. The Si-H terminated Si surface was obtained by cleaning the wafers in RCA 1 solution (1vol. H₂O₂-1vol. NH₄OH-5vol. DI H₂O) followed by dip in HF2%. The Si-OH terminated Si surface were obtained by cleaning wafers in RCA1-HF2%-RCA2 (1 vol. HCl-1vol H₂O₂-6 vol. DI H₂O). MOS capacitors with aluminium metallization and an area of $1*10^{-4}$ cm² were fabricated and annealed in nitrogen at 250°C so that as-deposited state of the HfO₂ thin films not to

978-1-5386-4483-6/18 $31.00 © 2018 IEEE

be altered. X-Ray-Reflectometry (XRR) by SmartLab XRD System" from Rigaku has been used for measuring the film thickness, the width of the lower and upper interface layer and the density of the HfO_2 films, while the Keithley 4200 Semiconductor Characterization System was used for C-V measurements.

2. RESULTS

In agreement with the results from Ref. [3], for the formation of the first monolayer of HfO_2 our precursor of TDMAH will react with Si-H terminated silicon, and Hf atoms will be directly bonded to the silicon atoms. This Si-Hf bonding will be present at least in the as-deposited state of HfO_2 film. In the case of Si-OH terminated the first half-reaction between TDMAH and Si-OH surface will end up with the formation of a Si-O-Hf bonding at the interface between silicon and HfO_2. This difference in the atomic arrangement at the Si-HfO_2 interface may influence the structural and electrical properties of the HfO_2 film, as shown below.

The XRR results of HfO_2 films deposited on both Si-H and Si-OH terminated Si wafer are presented in the Table 1 from below, where one can notice that the width of the Si-HfO_2 interface for Si-OH terminated surface is greater than that corresponding to Si-H terminated surface, in agreement with presence of oxygen atoms at the Si surface which will serve as bridge between silicon and Hf atoms for the case wafer cleaning in RCA1+HF2% +RCA2. Gravimetric density of HfO_2 film was also higher for the films deposited on Si-OH terminated surfaces. Such XRR data prove that even for the case of as-deposited HfO_2 films, we have very thin interlayers at the two interfaces, which may decrease the value of the relative permittivity of the HfO_2.

Table 1. XRR data on HfO_2 films related to cleaning process

Cleaning process	RCA1+HF2%	RCA1+HF2% +RCA2
Si-HfO_2 inter width (nm)	0.3	0.33
HfO_2 film thick (nm)	10.61±0.12	9.63±0.16
HfO_2-air int width (nm)	0.77±0.12	0.7±0.15
Film density (g/cm³)	8.09	8.67

From this reason, we operate in this paper with the concept of effective dielectric constant, k_{eff}.

For the electrical characterization of the interface between silicon and HfO_2 films, C-V plots of MOS capacitors were performed for both types of silicon terminations. In Fig. 1, we present a typical high frequency (1 MHz) C-V characteristic of a MOS capacitor with as-deposited ALD HfO_2 deposited on Si-H terminated silicon. The measurement has started at +2V applied on the Al gate and ended in the same point in order to determine the oxide trapped charge [5].

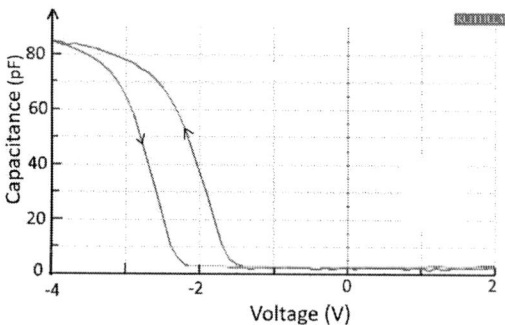

Fig. 1. High frequency (1 MHz) C-V plot of a MOS capacitor having as-deposited ALD HfO_2 film deposited on Si-H terminated silicon (p type) and Al gate, annealed in N_2 at 250°C.

In Fig. 2, we present the high frequency characteristic of a MOS capacitor with as-deposited ALD HfO_2 film deposited on Si-OH terminated silicon.

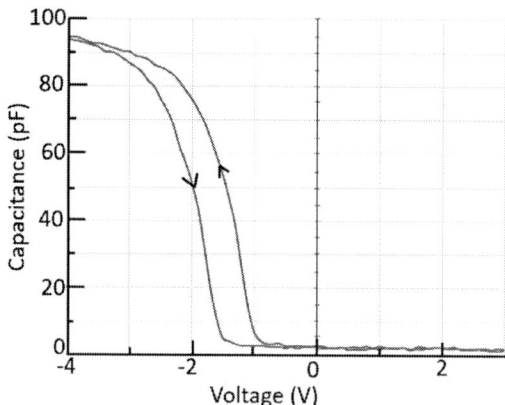

Fig. 2. High frequency (1 MHz) C-V plot of a MOS capacitor having as-deposited ALD HfO2 film deposited on Si-OH terminated silicon (p-type) and Al gate, annealed in N_2 at 250°C

From the above C-V characteristics, we have extracted the following MOS parameters: (i) the flatband voltage, (ii) the effective dielectric constant of as-deposited HfO$_2$ film (k$_{eff}$), (iii) the fixed charge from Si-HfO$_2$ interface, and (iv) the oxide trapped charge at the interface Si-HfO$_2$ due to dual sweep of the ramping voltage.

For the determination of the flatband voltage of MOS capacitor, we have used high frequency characteristic C-V and represented the dependence of 1/C^2 as a function voltage [6] as shown in Fig. 3.

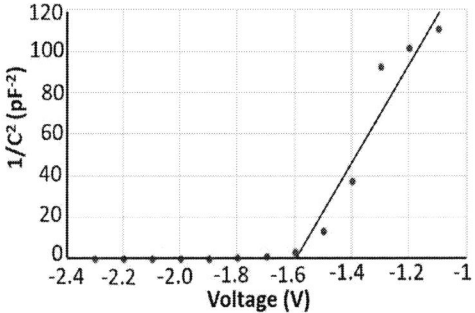

Fig. 3. Mott-Schottky plot for the extraction of the flatband voltage of MOS capacitor with as-deposited HfO$_2$ as dielectric

According to the Mott-Schottky equation (1):

$$1/C^2 = (2/(\varepsilon_o \, \varepsilon_{Si} \, A^2 \, e \, N_A)) * (V - V_{FB} - k_B \, T/e) \qquad (1)$$

the flatband voltage (V$_{FB}$) can be extracted from the intercept of the straight line on the V axis [6], and it was calculated on each sweep direction (V$_{FB1}$ associated to C-V plot with voltage sweep from inversion to accumulation and V$_{FB2}$ associated to sweep from accumulation to inversion), so that ΔVFB to be also calculated, as shown in Tables 2 and 3.

The effective dielectric constant (k$_{eff}$) was calculated from the high frequency MOS capacitance (C$_{ox}$) in accumulation, at V$_G$=-3.5 V by using simple formula from below:

$$C_{ox} = \varepsilon_o \, k_{eff} \, A/d \qquad (2)$$

where A is the area of the capacitor, "d" is the thickness of the dielectric, while ε_o is the dielectric permittivity of vacuum. This k$_{eff}$ value includes also the contribution of the interlayer from the Si-HfO$_2$ interface.

The fixed charge from Si-HfO$_2$ interface was calculated from the formula [5]:

$$N_f = (\Phi_{Al-Si} - V_{FB}) \, C_{ox}/(e \, A) \qquad [cm^{-2}] \qquad (3)$$

where Φ_{Al-Si} which is equal -0.88 V is the work function difference between Al gate and (100) boron doped p-type silicon with doping concentration N$_A$=10^{15} cm^{-3}, Cox is the capacitance of MOS device in accumulation.

The oxide trapped charge due to the dual voltage sweep in the high frequency C-V characteristic of MOS capacitor was calculated with the formula [5].

$$N_{ot} = -\Delta V_{FB} \, C_{ox}/(e \, A) \qquad [cm^{-2}] \qquad (4)$$

where $\Delta V_{FB} = V_{FB2} - V_{FB1}$. N$_{ot}$ it is caused by the net injected charge in the oxide during voltage sweep. This trapped charge remained near the Si-HfO$_2$ interface, where it has the highest influence on the C-V characteristic.

Based on this theoretical support, in Tables 2 and 3, we present these extracted parameters for a couple of MOS capacitors fabricated on Si-H and Si-OH terminated silicon wafers, respectively.

Table 2. Extracted parameters of MOS capacitors performed on Si-H terminated silicon

Chip #	1	2	3	4	5
Cox (pF)	83.46	80	80	79.47	79.7
V$_{FB1}$ (V)	-1.6	-1.3	-1.3	-1.3	-1.1
V$_{FB2}$ (V)	-2.5	-2.3	-2.3	-2.3	-1.2
ΔV$_{FB}$ (V)	-0.9	-1	-1	-1	-1.2
k$_{eff}$	10	9.6	9.6	9.52	9.55
N$_f$(cm^{-2}) x 10^{12}	4	2.1	2.1	2.09	1.1
N$_{ot}$ (cm^{-2}) x 10^{12}	4.69	5	5.25	4.96	5.97

Table 3. Extracted parameters of MOS capacitors performed on Si-OH terminated silicon.

Chip #	1	2	3	4	5
C$_{ox}$ (pF)	92.6	91.55	91.65	91.48	91.34
V$_{FB1}$ (V)	-1	-1.2	-1.1	-1.05	-1.27
V$_{FB2}$ (V)	-1.5	-1.7	-1.65	-1.55	-1.65
$\Delta$$_{VFB}$ (V)	-0.5	-0.5	-0.55	-0.5	-0.38
k$_{eff}$	10.1	9.96	9.97	9.95	9.94
N$_f$(cm^{-2}) x 10^{12}	0.7	1.83	1.26	0.9	2.23
N$_{ot}$(cm^{-2}) x10^{12}	2.9	2.9	3.15	2.85	2.16

3. DISCUSSION

Our previous research on the reliability of the ALD HfO$_2$ films deposited from TDMAH and water vapours at 200°C on Si-H and Si-OH terminated silicon surfaces has shown an increased quality of the HfO_2 films after the wafer cleaning in the sequence RCA1+HF2%+RCA2, in terms of a smaller number of early breakdown events and a relatively larger charge to breakdown capabilities [4].

Present experimental results are also suggesting better interface and bulk electrical properties of the as-deposited ALD HfO_2 films deposited on Si-OH terminated silicon surface as follows.

In the absence of higher thermal annealing of the (as-deposited) ALD HfO_2 films and maybe further deposition kinetics optimization, the effective dielectric constant k_{eff} is still enough low (around 10), but overall this is higher on the hydroxylated surfaces with respect to hydrogenated Si surface. This result may be unexpected if we remember that there are a few additional monolayers of SiO_2 on the Si surface at the end RCA2 cleaning, as revealed by XRR results at the lower interface of HfO_2 films. Such lower k_{eff} values for Si-H terminated Si surface could be also explained by the contribution of existing Si-Hf bonds on the first monolayer of HfO_2 film for the case of hydrogenated Si surface, which may decrease the dielectric constant at the interface. The dielectric constant of as-deposited ALD HfO_2 obtained at 225°C from TDMAH and water vapours on Si-H terminated wafers and annealed at 350°C in N_2/H_2 was equal to 12.3 [7], so our k_{eff} results are rather similar considering our lower temperature deposition and anneal 200°C and 250°C respectively) and absence of H_2 during final thermal treatment.

From a simple inspection of the Figs.1 and 2, one can see that the C-V characteristics are translated towards negative voltages, and this is an indication that at the end of deposition process a big amount of positive charge is remaining in the as-deposited film, near the interface with silicon [5]. Such shift of the C-V plot to negative voltage result has been also reported in the literature for the case of Si-H terminated surfaces [7]. In our case we have found it in both types of silicon terminations. As shown in Tables 2 and 3, the flatband voltages (V_{FB1} and V_{FB2}) for the dual voltage sweep are both smaller for the MOS devices performed on Si-OH terminated surfaces. From these tables it is obtained that the oxide trapped charge due to dual voltage sweep (Not) is two-fold smaller for MOS devices performed on Si-OH terminated Si surface, proving a more stable interface of Si-HfO_2 in the presence of a few monolayers of SiO_2.

4. CONCLUSIONS

This paper presents a comparative analysis of the MOS capacitors properties of as-deposited ALD HfO_2 films as a function of Si-H and Si-OH terminations of the Si surface. The study has shown that despite the presence of a few monolayers of SiO_2 for the HfO_2 films deposited on hydroxylated surfaces, as also revealed by XRR data, the effective dielectric constant of the as-deposited HfO_2 film was increased on average from about 9.5 (for HfO_2 on Si-H terminated surface) to 9.95 (for HfO_2 on Si-OH terminated surfaces) while the oxide trapped charge at the end of voltage dual sweep was two-fold smaller for MOS capacitors processed on Si-OH terminated silicon. This work proves for the first time the robustness of Si-O-Hf interface with respect to Si-Hf-O interface.

5. REFERENCES

[1] R. Clark, *Emerging Applications for High-k Materials in VLSI Technology*, Materials **2014**, 7, 2913-2944.

[2] W. Kern, Ed. Handbook of Semiconductor Cleaning Technology, Noyes publishing; Park Ridge, NJ, 1993, Ch.1.

[3] W. Chen, Q-Q Sun, M. Xu, S-J Ding, D.W. Zhang and L-K Wang, *Atomic Layer Deposition of Hafnium Oxide from tetrakis (ethyl methyl amino) hafnium and Water Precursors*, J. Phys. Chem. C 2007, 111, 6495-6499.

[4] C. Cobianu, F. Nastase, N. Dumbravescu, O. Buiu, B. Serban, M. Danila, R. Gavrila, O. Ionescu, C. Romanitan *Effect of Surface Cleaning on Reliability of ALD HfO_2 films Deposited from TDMAH*, WOCSDICE 2018 Conference, Bucharest, 14-26 May 2018.

[5] Dieter K. Schroeder, *Semiconductor Material Device Characterization*, Second Edition, John Wiley &Sons, Inc, pp. 360-364, ISBN 0-471-24139-3, 1998.

[6] K. Gelderman, L. Lee and S.W. Donne, *Flat-Band Potential of a Semiconductor: Using The Mott-Schottky Equation*, Journal of Chemical Education, vol. 84, No.4 April 2007.

[7] H. Garcia, H. Castan, S. Duenas and L. Bailon, *Electrical characterization of ALF hafnium oxide films from TDMAH and water/ozone: Effect of growth temperature, oxygen source and postdepostion annealing*, J. Vac. Sci. Technol., A31 (1) Jan/Feb 2013, pp 01A127-1-01A127-7.

Enhanced photocurrent in GeSi NCs / TiO₂ multilayers

C. Palade*, A. Slav*, O. Cojocaru*, V.S. Teodorescu*, S. Lazanu*, T. Stoica*,
M.T. Sultan***, H.G. Svavarsson***, M.L. Ciurea*,**,a

*National Institute of Materials Physics, 077125 Magurele, Romania
**Academy of Romanian Scientists, 050094 Bucuresti, Romania
***Reykjavik University, School of Science and Engineering, IS-101 Reykjavik, Iceland
[a]E-mail: ciurea@infim.ro

Abstract— GeSi NCs / TiO₂ multilayers with enhanced photocurrent properties were prepared and studied. Multilayers of TiO₂ /(GeSi/TiO2)x2 /Si-p were deposited by magnetron sputtering and annealed by RTA at 700 ºC for GeSi NCs formation. A post-annealing hydrogenation in plasma was performed on multilayers for healing of defects acting as traps and/or recombination centers and consequently producing the photocurrent enhancement. We studied the electrical and photoconductive properties of multilayers annealed by RTA and post-annealing hydrogenated. The current – temperature dependence reveals the conduction mechanisms in GeSi NCs / TiO₂ multilayers RTA annealed, i.e. thermal activation of carriers to extended states (0.31 eV activation energy), the electron tunneling mechanism to nearest neighbors ($T^{-1/2}$ behavior) and Mott variable range hopping ($T^{-1/4}$ dependence). The photocurrent spectra made on multilayers structures hydrogenated for 10, 20 and 30 min evidence the photocurrent increasing up to 50%, showing that the hydrogenation is a suitable treatment for enhancing photocurrent. All photocurrent spectra present a dominant maximum (920 nm) and two shoulders (~770 and ~1060 nm).

Keywords—GeSi nanocrystals; TiO₂; photosensing; hydrogenation.

1. Introduction

It is well known that GeSi NCs have the advantage of continuous tuning of bandgap energy from that of Ge NCs to the value of Si NCs as Ge and Si are completely miscible [1,2]. Beside this, the bandgap energy can be engineered by tailoring the NCs size for evidencing the quantum confinement effect in optical and electrical properties of GeSi NCs [3–8]. Therefore, the films and multilayers of GeSi NCs embedded in oxides (TiO₂, SiO₂) prepared by magnetron sputtering [9–13] and subsequent thermal annealing are promising materials for applications in optical sensors for selective spectral windows in visible and near infrared range [6, 14–16].

In order to avoid the defects in oxides, i.e. traps and recombination centers, that in turn diminish the photocurrent, it is necessary to control the oxygen density [17,18].

In this paper, we report on enhancing the photocurrent in multilayers GeSi NCs / TiO₂ by using optimal preparation parameters and hydrogenation treatment in plasma performed post-annealing.

2. Experimental

We prepared multilayers of **TiO₂ /(GeSi/TiO₂)x2 /Si-p** with a stack of two pairs of GeSi/TiO₂. Firstly, the multilayers (MLs) were deposited by magnetron sputtering on Si-*p* substrate followed by rapid thermal annealing (RTA) and hydrogenation processing under different conditions. For magnetron sputtering deposition we used Surrey NanoSystems 1000 Gamma equipment, for RTA Annealsys AS-Micro processor and the samples hydrogenation was made in plasma by using a mixture of H₂ and Ar (70:30 composition) in RF CESAR© 136 power generator. The MLs were deposited on heated substrate (300 – 500 ºC) by using alike conditions with those in [19], i.e. using TiO₂, Ge and Si targets, and powers of 45 W RF, 9 W DC and 40 W DC, respectively. As working gas, we used Ar 6N at a working pressure of 4 mTorr. The ML was sputtered on Si wafer previously oxidized in RTP processor for electrical isolation in respect of Si substrate. The thicknesses of layers in ML presented in this paper are 4 – 5 nm TiO₂, 10 - 11 nm GeSi. The GeSi layer is bordered by 2 nm Ge layer. For obtaining GeSi NCs, the as-deposited samples were annealed by RTA, in Ar atmosphere at 700 ºC.

The samples were contacted with thermally deposited Al. On the top side of ML we deposited

978-1-5386-4483-6/18 $31.00 © 2018 IEEE

Al contacts in coplanar geometry (5 mm gap between electrodes) and also on the back side of Si substrate.

For morphology and structure studies of MLs we used Jeol ARM 200F microscope, and for (photo)electrical measurements, a setup consisting of a Keithley 236 electrometer, a Stanford SR810 Lock-in Amplifier and chopper, and also a Janis optical cryostat.

3. Results and discussion

The structure and morphology of nanostructured ML (700 °C RTA) are presented in low magnification and high resolution TEM (HRTEM) images in Fig.1.

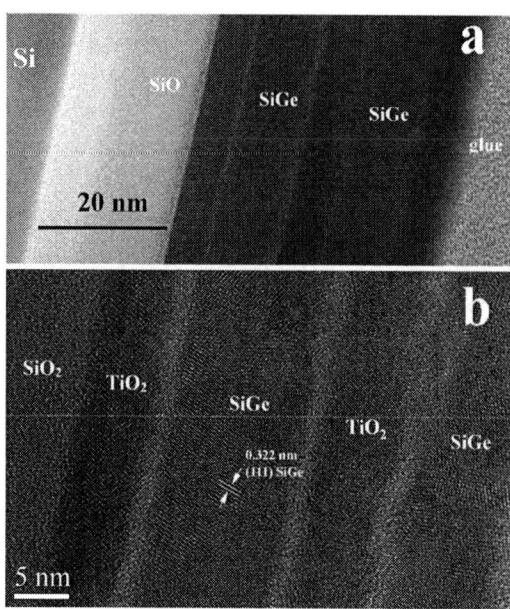

Fig. 1. XTEM images taken on 700 °C RTA MLs: a) low magnification and b) HRTEM – both SiGe and TiO₂ are crystallized.

In Fig. 1b, one can see that ML is crystallized, so that GeSi layer is formed of cubic NCs with diameters of 3 – 5 nm and TiO₂ layer is formed of anatase NCs. These layers have thicknesses of 10 nm TiO₂ and 6 nm GeSi as shown in Fig. 1a. More than that, at the interfaces between GeSi and TiO₂ layers (Fig. 1b), SiO₂ layers are formed during RTA. These layers are not present on as-deposited ML.

We performed electrical measurements in dark, namely current – voltage characteristic (*I-V*) and current – temperature dependence (*I-T*). In Fig. 2 the *I-V* curves taken in top-down geometry of contacts are shown. The rectifying

behaviour is mainly due to the back side contact together with SiO₂ buffer layer.

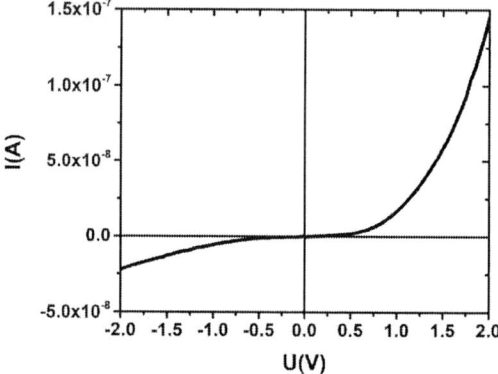

Fig. 2. *I-V* characteristic.

In Fig. 3, the *I-T* characteristic measured on the ML biased at 2V in the 100 – 300 K temperature range is presented.

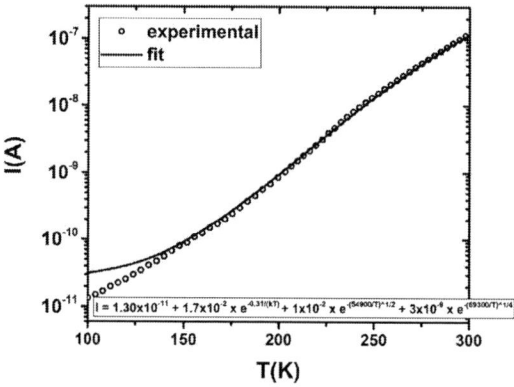

Fig. 3. *I-T* characteristic: experimental - doted curve, fit – continous curve.

The *I-T* experimental data was fitted using the equation,

$$I = C_1 \times e^{-\frac{E_a}{k_B T}} + C_2 \times e^{-\left(\frac{T_{0-tunn}}{T}\right)^{\frac{1}{2}}} + C_3 \times e^{-\left(\frac{T_{0-hopp}}{T}\right)^{\frac{1}{4}}} + C$$

where $C_1 = 1.7 \times 10^{-2}$ A, $C_2 = 1 \times 10^{-2}$ A, $C_3 = 1 \times 10^{-2}$ A, $C = 1.3 \times 10^{-11}$ A, $E_a = 0.31$ eV, $T_{0-tunn} = 54900$ K and $T_{0-hopp} = 69300$ K.

From the fitted *I-T* curve it results that at high temperatures the curve has an Arrhenius dependence meaning a thermal activation of carriers to extended states with $E_a = 0.31$ eV activation energy, while at lower temperatures a $T^{-1/2}$ dependence is observed, explained by tunneling mechanism between GeSi NCs to the nearest neighbors, and then around 130 K, Mott variable range hopping takes place, i.e. a $T^{-1/4}$ dependence. The characteristic temperatures corresponding to electron tunneling and hopping were determined by fit, i.e. $T_{0\text{-}tunn}$ = 54900 K and $T_{0\text{-}hopp}$ = 69300 K. C_1, C_2, C_3 and C were also determine by fit, C could match a temperature independent hopping mechanism.

The photoconductive properties of ML structures were studied after annealing of samples by RTA and after post-annealing hydrogenation. We hydrogenated ML structures with the aim to heal the defects acting as traps and/or recombination centers, that in turn will lead to improvement of the ML photosensitivity.

In Fig. 4, the photocurrent spectra for ML structures annealed at 700 °C by RTA together with the subsequent hydrogenated ones for different durations (10, 20 and 30 min) are shown. As one can see, with the increase of the hydrogenating time, the photocurrent intensity increases (for all hydrogenated samples, the photocurrent is higher than for the 700°C ML structure) as consequence of defects healing. The curve taken on 700 °C RTA ML sample presents two maxima at 920 and 1060 nm and a shoulder at about 770 nm whereas for all hydrogenated samples the photocurrent curves have a domniant maximum at 920 nm and two shoulders (770 nm and 1060 nm).

All photocurrent curves measured on hydrogenated structures present a dominant maximum (920 nm) and two shoulders (770 nm and 1060 nm)

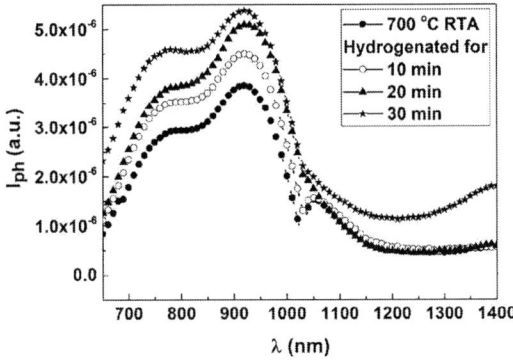

Fig. 4. Spectral distribution of the photocurrent on the as annealed and post-annealing hydrogenated ML that was measured at room temperature.

We attribute the dominant maximum positioned at 920 nm to GeSi NCs contribution and the shoulder at ~ 770 nm could be explained

by the Ge related defects [19, 20]. The small maximum at 1060 nm measured on 700 °C RTA sample only probably illustrates the influence of Si substrate by capacitive coupling [13].

4. Conclusions

We prepared ML of a stack of 2 pairs of GeSi/TiO$_2$ on Si-*p* substrate, **TiO$_2$ /(GeSi/TiO$_2$)x2 /Si-p**. For this we deposited ML structures by using RF magnetron sputtering and then we annealed them by RTA for GeSi NCs formation. Also post-annealing treatment of samples by hydrogenation in plasma was performed for improving the photocurrent intensity by healing defects acting as traps and/or recombination centers. We performed electrical (*I-V*, *I-T*) and photocurrent measurements (I_{ph}-λ). From electrical investigations it results that the *I-V* characteristic is rectifying mainly due to back side Al contact and *I-T* dependence reveals thermal activation of carriers to extended states (Arrhenius dependence) at high temperatures with an activation energy of E_a = 0.31 eV. The tunneling mechanism of electrons to the nearest neighbors at lower temperatures ($T^{-1/2}$ dependence) and Mott variable range hopping ($T^{-1/4}$ behaviour) around 130 K, were evidenced. The hydrogenation treatment in plasma produces the increase of photocurrent with up to 50% versus hydrogenation time, the aspect of photocurrent spectra (I_{ph}-λ) being conserved. All I_{ph}-λ curves present a dominant maximum located at 920 nm attributed to GeSi NCs contributions and two shoulders, one at ~770 nm attributed to Ge related defects, and the other one at ~1060 nm, probably showing the influence of Si substrate.

Acknowledgments. This work was supported by UEFISCDI through M-ERA.NET PhotoNanoP no. 33/2016, M-ERA.NET GESNAPHOTO no. 58/2016, PCE no. 122/2017 and by Romanian Ministry of Research and Innovation through NIMP Core Program 2018.

References

[1] E.G. Barbagiovanni, D.J. Lockwood, P.J. Simpson and L.V. Goncharova, *"Quantum confinement in Si and Ge nanostructures: theory and experiment"*, Appl. Phys. Rev., **1**(1), 011302, March 2014.

[2] I. Sychugov, J. Valenta and J. Linnros, *"Probing silicon quantum dots by single-dot techniques"*, Nanotechnology, **28**(7), 072002, February 2017.

[3] N.N. Ha, N.T. Giang, T.T.T. Thuy, N.N. Trung, N.D. Dung, S. Saeed and T.

Gregorkiewicz, "Single phase $Si_{1-x}Ge_x$ nanocrystals and the shifting of the E1 direct energy transition", Nanotechnology, 26(37), 375701, September 2015.

[4] I. Stavarache, A.M. Lepadatu, V.S. Teodorescu, A.C. Galca and M.L. Ciurea, "Annealing induced changes in the structure, optical and electrical properties of $GeTiO_2$ nanostructured films", Appl. Surf. Sci., 309, pp. 168–174, August 2014.

[5] C. Mehringer, C. Kloner, B. Butz, B. Winter, E. Spiecker and W. Peukert, "Germanium-silicon alloy and core-shell nanocrystals by gas phase synthesis", Nanoscale, 7(12), pp. 5186–5196, 2015.

[6] S. Cosentino, Pei Liu, Son T. Le, S. Lee, D. Paine, A. Zaslavsky, D. Pacifici, S. Mirabella, M. Miritello, I. Crupi and A. Terrasi "High-efficiency silicon-compatible photodetectors based on Ge quantum dots", Appl. Phys. Lett. 98, 221107, June 2011.

[7] A. M. Smith and S. Nie, "Semiconductor nanocrystals: Structure, properties, and band gap engineering", Acc Chem Res. 43, pp. 190–200 (2010).

[8] M. H. Kuo, W.T. Lai, T.M. Hsu, Y.C. Chen, C.W. Chang, W.H. Chang and P.W. Li, "Designer germanium quantum dot phototransistor for near infrared optical detection and amplification", Nanotechnology 26, 055203, February 2015.

[9] N.T. Giang, L.T. Cong, N.D. Dung, T. Van Quang and N.N. Ha, "Nanocrystal growth of single-phase $Si_{1-x}Ge_x$ alloys", J. Phys. Chem. Sol., 93, pp. 121–125, June 2016.

[10] J.N. Aqua, I. Berbezier, L. Favre, T. Frisch and A. Ronda, "Growth and self-organization of SiGe nanostructures", Phys. Rep., 552(2) pp. 59–189, January 2013.

[11] A.F. Khan, M. Mehmood, M. Aslam and S.I. Shah, "Nanostructured multilayer TiO_2 –Ge films with quantum confinement effects for photovoltaic applications", J. Colloid. Interf. Sci., 343(1), pp. 271–280, March 2010.

[12] H. Hussain, G. Tocci, T. Woolcot, X.Torrelles, C.l. Pang, d.S. Humphrey, c.M. Yim, D.C. Grinter, G. Cabailh, O. Bikondoa, R. Lindsay, J. Zegenhagen, A. Michaelides and G. Thornton, "Structure of a model TiO_2 photocatalytic interface", Nat. Mater. 16, pp. 461–466, 2017.

[13] J. Choi, S. Song, M. T. Hörantner, H. J. Snaith and T. Park, "Well-defined nanostructured, single crystalline TiO_2 electron transport layer for efficient planar perovskite solar cells", ACS Nano 10, pp. 6029–6036, May 2016.

[14] V.S. Teodorescu, M.l. Ciurea, V. Iancu, M-G. Blanchin, "Morphology of Si nanocrystallites embedded in SiO_2 matrix", J. Matter. Res. 23, pp. 2990-2995, November 2008.

[15] A.-M. Lepadatu, I. Stavarache, T.F. Stoica, M.L. Ciurea, "Study of Ge Nanorparticles embedded in an amorphous SiO_2 matrix with photoconductive properties", Dig. J. Nanomater. Biostruct. 6, pp. 67-73, January – March 2011.

[16] M.L. Ciurea, "Quantum confinement in nanocrystalline silicon", J. Optoelectron. Adv. M. 7, pp. 2341-2346, October 2005.

[17] Y. Li, J.K. Cooper, W. Liu, C.M. Sutter-Fella, M. Amani, J.W. Beeman, A. Javey, J.W. Ager, Y. Liu, F.M. Toma and I.D. Sharp, "Defective TiO_2 with high photoconductive gain for efficient and stable planar heterojunction perovskite solar cells", Nat. Commun. 7, 12446, August 2016.

[18] H. Wang, Q. Sun, Y. Yao, Y. Li, J. Wang and L. Chen "A micro sensor based on TiO_2 nanorod arrays for the detection of oxygen at room temperature", Ceram. Int. 42, pp. 8565–8571, May 2016.

[19] A. Slav, C. Palade, I. Stavarache, V.S. Teodorescu, M.L. Ciurea, R. Muller, A. Dinescu, M.T. Sultan, A. Manolescu, J.T. Gudmundsson, H.G. Svavarsson, "Influence of preparation conditions on structure and photosensing properties of $GeSi/TiO_2$ multilayers", IEEE CAS 2017 Proceedings; ISBN: 978-1-5090-3985-2; ISSN: 1545-827X.

[20] A.M. Lepadatu, A. Slav, C. Palade, I. Dascalescu, M. Enculescu, S. Iftimie, S. Lazanu, V.S. Teodorescu, M.L. Ciurea and T. Stoica, "Dense Ge nanocrystals embedded in TiO_2 with exponentially increased photoconduction by field effect", Sci. Rep. 8, 4898, March 2018.

Session N&N 2

NANOSCIENCE AND NANOENGINEERING 2

978-1-5386-4483-6/18 $31.00 © 2018 IEEE

TiO₂ – graphene oxide thin films obtained by Spray Pyrolysis Deposition

I. Tismanar*, L. Isac*, A. C. Obreja, O. Buiu**, A. Duta***
*Transilvania University of Brasov, Romania
E-mail: tismanarioana@yahoo.com; isac.luminita@unitbv.ro; a.duta@unitbv.ro
**National R&D Institute R&D for Microtechnologies, Bucharest, Romania
*E-mail: cosmin.obreja@imt.ro; octavian.buiu@gmail.com

Abstract - Graphene and graphene derivatives have a set of remarkable properties due to their 2D structure; to use these properties in various applications asks for increasing the thermal stability of the carbon-based components and this is why various composites are reported. The development of composites with metal oxide matrix and performant interfaces asks for using polar graphene derivatives as fillers, e,g, graphene oxide (GO). The development of thin multilayered composite films of TiO₂ – GO – TiO₂ using Spray Pyrolysis Deposition is discussed and the structural and surface properties are reported, considering further potential application of these composite layers as photocatalysts in advanced wastewater treatment or in self-cleaning surfaces.

Keywords — TiO₂-GO thin films; crystallinity; surface composition; surface morphology;

1. Introduction

Graphene and graphene-derivatives are a new type of 2D-carbon materials intensively investigated since their discovery by Novoselov et al [1]. Because of their unique properties as high transparency, mechanical strength, electrical and thermal conductivity [2], graphene materials are used in different applications: nanoelectronics, sensors, displays and energy conversion [3]. Recently, these types of carbon nanostructures were embbeded in oxidic compounds to improve their properties depending on the applications.

Because of the low compatibility between the ionic metal oxide matrix and the non-polar graphene filler, the intefaces in this type of composites are weak, thus to enhance the interfacial strength various graphene derivatives are used, as the graphene oxide (GO) or the reduced graphene oxide (rGO), that are chemically modified graphene through oxidation processes [4]. This type of composites were recently reported in photocatalytic application for organic polutants removal from wastewater and for self-cleaning surfaces [5, 6]. Mixing the oxide photocatalyst with the graphene derivatives can prevent the electron-hole recombination and can narrow the band gap of the composite at values lower then 2.43eV, allowing its VIS-activation [7].

Sol-gel or hydrothermal methods are frequently reported for obtaining the composites, mainly as powders. For depositing composite thin films, a simple and efficient method can be spray pyrolysis deposition (SPD); however, SPD is avoided because of the thermal sensitivity of the graphene compounds.

This paper reports on TiO₂-graphene oxide thin film composites obtained by SPD coupled with the drop casting method for GO deposition. These films will be further used in photocatalytic processes for wastewater treatement. The thin films crystallinity and their surface morphology and composition were investigated to outline the compatibility of the two compounds in the composite.

2. Method and samples

Regular glass substrates (1.5 cm x 1.5 cm) were used for deposition. The substrates were cleaned with water and detergent by ultra-sonication, followed by rinsing in ethanol and drying in air.

The GO dispersion was prepared using Hummers method, [8]. 2g of graphite (99,99% purity) and 46 mL of H_2SO_4 were

Acknowledgment: this work was supported by a grant of the Romanian Ministry of Research and Innovation, CCCDI –UEFISCDI, project number PN –III –P1 –1.2 –PCCDI –2017 –0619, contract no. 42 PCCDI / 2018 within PNCDI.

978-1-5386-4483-6/18 $31.00 © 2018 IEEE

mixed and stirred for 24 hours. Then, 1g of NaNO3 was added and the mix was placed in an ice bath at a temperature kept between 0 and 5°C. Further on, KMnO4 (6g) was slowly added to keep the temperature at 5°C. This mixture was stirred for 4 days. The KMnO4 excess was eliminated with 50 mL of H2O2 5% and 25 mL of HCl 5% and then the supernatant was filtered. Graphite oxide powder was washed several times with deionized water and the graphite oxide-water mix was placed in an ultrasonic bath to support the exfoliation of the GO sheets. The concentration of GO dispersion in water was 3 mg/mL.

The TiO2 layers were deposited using titanium chloride (TiCl4, Acros Organics, 99.9%) as precursor, mixed with ethanol (C2H5OH, Chemical Company, 99.3%) and deionized water in the 1:12.5:7.5 volume ratio. The precursor solution was deposited by SPD, using an ABB/IRB5400 robot and a heating plate. The deposition temperature was optimized at 350 °C after attempts have been made to lower down the temperature because of the thermal sensitivity of GO. The first layer of TiO2 was annealed at 450°C for 3 hours to increase the crystallinity degree of this base layer. The outer layer of TiO2 wasn't annealed to prevent the GO removal.

To incorporate GO in the TiO2 matrix the drop casting method was used: 0.5 mL of GO dispersion was dropped above the first TiO2 layer using a micropipette and the deposition was performed at 100 °C to support a fast water removal. Then, the temperature on the heater was raised up to 350 °C and the second thin layer of TiO2 was deposited by SPD over the GO layer.

Thus, three different samples were deposited: one containing only TiO2, the second one consisting of GO deposited on TiO2 (TiO2 – GO) and the third one with a multilayered TiO2 – GO - TiO2 structure.

The thin films crystallinity was investigated by X-ray Diffraction (XRD, Bruker D8 Discover, step size 0.024, scan speed 1.5 s/step, 2θ range from 5 to 70° with

Diffrac.EVA – 4.3.0.2 software analysis). Energy Dispersive X-ray spectrometer (EDX Thermo) was used to evaluate the elemental composition of the thin films while the surface morphology of the thin films was investigated using Scanning Electron Microscopy (SEM, Hitachi model S-3400 N type II). The thicknesses of the layers was estimated based on reflectance data obtained using a UV–VIS-NIR spectrophotometer (Perkin Elmer Lambda 950).

3. Results and Discussions

The diffraction patterns for the thin films are presented in Fig. 1.

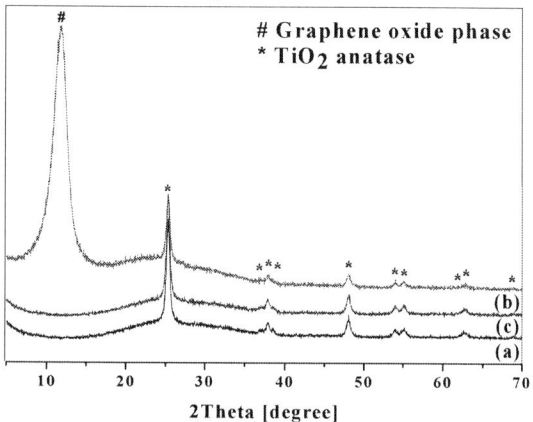

Fig. 1 XRD pattern of the thin films
a) TiO2; b) TiO2-GO; c) TiO2-GO-TiO2

The TiO2 sample shows characteristic peaks of anatase as main component. For the TiO2-GO sample, an additional peak can be observed at 11.7° corresponding to the GO phase, [9]. It can be also noticed that the peaks corresponding to anatase TiO2 can be observed in the TiO2 – GO and in the TiO2 - GO - TiO2 composite thin films, confirming the deposition and coverage of the GO layer; however, because the GO plates are very thin, their corresponding diffraction peak does no longer appear. The thickness measurements confirm that the GO layers are thin and the results presented in Table 1 allow to conclude that multiple overlapping layers of graphene oxide are deposited. Previous results, [10] showed that a single

978-1-5386-4483-6/18 $31.00 © 2018 IEEE

graphene oxide flake/layer has a thickness 1.2 nm, as result of the functional groups from both sides of the layer.

Table 1 Crystallite size and crystallinity degree in the thin films

Sample	Thickness [nm]
TiO_2	652,934
GO in TiO_2 - GO	12,547
TiO_2 – GO - TiO_2	802,639

Table 2 Crystallite size and crystallinity degree in the thin films

Sample	Crystallite size [nm]	Crystallinity degree [%]
TiO_2	16.95	40.2
TiO_2 in TiO_2 - GO	15.37	70.2
GO in TiO_2 - GO	4.38	
TiO_2 outer layer in TiO_2 – GO - TiO_2	16.48	31.9

The crystallite sizes were calculated using the Scherrer's equation and the crystallinity degree is presented in Table 2. The results show that the TiO_2 crystallinity degree is higher in the metal oxide thin film than in the composite layers, confirming that annealing plays a key role in increasing the thin films crystallinity.

The elemental composition of the films varies depending on the outer layer, as the EDX results in Table 3 show.

Table 3 Elemental average composition of the thin films

Thin film sample		Element [%]			
		O	Ti	Si	C
(a) TiO_2		62.11	29.03	8.85	-
(b) TiO_2-GO	Point 1	58.2	16.11	-	25.68
	Point 2	46.74	3.45	-	48.81
	Point 3	44.56	4.98	-	50.46
(c) TiO_2- GO-TiO_2		62.76	33.87	-	3.36

As the EDX results show, the TiO_2 thin film in sample (a) is thin, allowing to observe the specific signal of Si from the glass substrate. A similar effect is registered for the TiO_2-GO-TiO_2 thin film, in sample (c), that allows to evidence the carbon (GO) content, even when covered with TiO_2.

It is also to mention that the deposition of the second TiO_2 layer through SPD could partially remove the GO because of the carrier gas pressure. However, as the EDX results show, this process still leaves GO in the multi-layered structure.

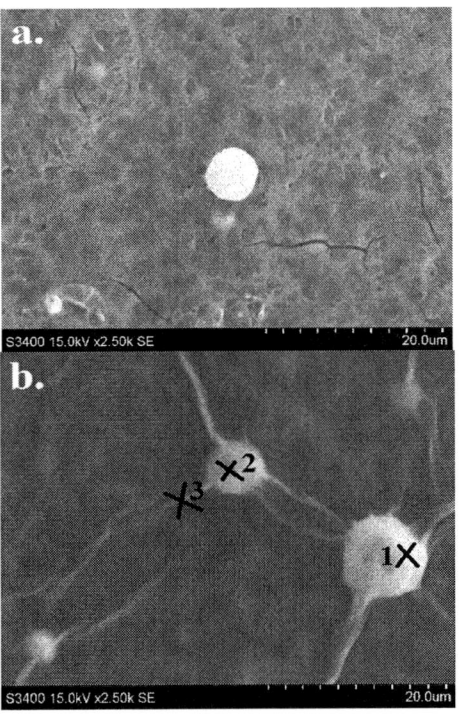

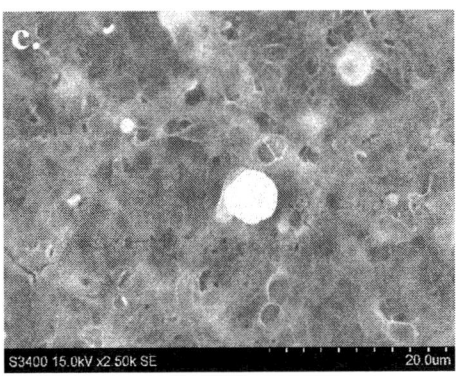

Fig. 2 SEM images of the thin films:
a) TiO_2; b) TiO_2-GO; c) TiO_2-GO-TiO_2

The SEM images in Fig. 2 show that the TiO_2 thin film (Fig. 2a) contains TiO_2 aggregates on the rather porous surface which may favor the adsorption of GO from the dispersion that will be further dropped above. The SEM image also shows fine cracks that can be the result of the lower deposition temperature.

The SEM image of the TiO$_2$-GO composite (Fig. 2b) shows that the graphene oxide covers each surface component. The elemental content of the selected points 1, 2 and 3 on the TiO$_2$-GO SEM image (Fig. 2b) and included in Table 2, allows to conclude that TiO$_2$ aggregates (Point 1) support the GO deposition and GO is denser deposited on the smaller anatase particles (Point 2). The image also outlines that GO is not uniformly distributed on the TiO$_2$ surface, where no obvious aggregates are positioned, but it is developed in thin structures resembling filaments (Point 3).

The SEM image of the TiO$_2$-GO-TiO$_2$ composite shows a surface similar to that of the TiO$_2$ films confirming the good and even deposition of the outer TiO$_2$ layer.

4. Conclusions

Thin films of TiO$_2$ and thin composite layers of TiO$_2$-GO and TiO$_2$-GO-TiO$_2$ were deposited using SPD coupled with drop casting.

The XRD results outline that the anatase TiO$_2$ polymorph is obtained using the lower temperature deposition conditions; in the compozite thin films, GO is clearly evidenced through its characterisitc peak that appears in the TiO$_2$-GO diffractogram. After the deposition of the second/outer TiO$_2$ layer the GO peak can no loger be observed suggesting the proper coverage of the GO layer with TiO$_2$. This conclusion is also supported by the EDX results, which outline a certain carbon content in the TiO$_2$-GO - TiO$_2$ sample.

The SEM images show that GO covers the TiO$_2$ surface forming distinct structures with filament shape.

This work offers new insight on the development of TiO$_2$-graphene oxide composite thin films using upscalable deposition techniques that support a good compatibility between the components in the composite.

References

[1] K. S. Novoselov, A. K. Geim, S. V. Morozov, D. Jiang, Y. Zhang, S. V. Dubonos, I. V. Grigorieva and A. A. Firsov, "*Electric Field Effect in Atomically Thin Carbon Films*", Science, **306**, pp. 666-669, October 2004

[2] Y. Liu and D. Zhang, "*Synergetic effect in the multifunctional composite film of graphene-TiO$_2$ with transparent conductive, photocatalytic and strain sensing properties*", Journal of Alloys and Compounds, **698**, pp. 60-67, November 2016

[3] M. J. Allen, V. C. Tung and R. B. Kaner, "*Honeycomb carbon: a review of graphene*", Chem. Rev., **110**, pp. 132-145, July 2009

[4] A. Bianco, H-M. Cheng, T. Enoki, Y. Gogotsi, R. H. Hurt, N. Koratkar, T. Kyotani, M. Monthioux, C. R. Park, J. M. D. Tascon and J. Zhang, "*All in the graphene family – A recommended nomenclature for two-dimensional carbon materials*", CARBON, **65**, pp. 1-6, December 2013

[5] X. Li, R. Shen, S. Ma, X. Chen and J. Xie, "*Graphene - based heterojunction photocatalysts*", Applied Surface Science, **430**, pp. 53-107, February 2018

[6] M. J. Nine, M. A. Cole, L. Johnson, D. N. H. Tran and D. Losic, "*Robust Superhydrophobic Graphene-Based Composite Coatings with Self-Cleaning and Corrosion Barrier Properties*", ACS Appl. Mater. Interfaces, **7 (51)**, pp. 28482–28493, 2015

[7] L. Zhang, Q. Zhang, H. Xie, Ji. Guo, H. Lyu, Y. Li, Z. Sun, H. Wang and Z. Guo, "*Electrospun titania nanofibers segregated by graphene oxide for improved visible light photocatalysis*", Applied Catalysis B: Environmental, **201**, pp. 470-478, February 2017

[8] A. C. Obreja, S. Iordanescu, R. Gavrila, A.Dinescu, F. Comanescu, A. Matei, M. Danila, M. Dragoman and H. Iovu, "*Flexible films based on graphene/polymer nanocomposite with improved electromagnetic interference shielding*", 2015 International Semiconductor Conference (CAS), Sinaia, 2015, pp. 49-52, IEEE Xplore, doi: 10.1109/SMICND.2015.7355156.

[9] L. Stobinski, B. Lesiak, A. Malolepszy, M. Mazurkiewicz, B. Mierzwa, J. Zemek, P. Jiricek, I. Bieloshapka, *Graphene oxide and reduced graphene oxide studied by the XRD, TEM and electron spectroscopy methods*, Electron Spectros. Relat. Phenomena., **195**, pp. 145–154, 2014 http://dx.doi.org/10.1016/j.elspec.2014.07.003

[10] A.C. Obreja, D. Cristea, R. Gavrila, V.Schiopu, A. Dinescu, M. Danila, F. Comanescu, *Isocyanate functionalized graphene/P3HT based nanocomposites*, Applied Surface Scienec, **276**, pp. 458-467, 2013

Effect of the Deposition Conditions on Titanium Oxide Thin Films Properties

M. Pustan*, C. Birleanu*, A. Trif*, S. Garabagiu, D. Marconi**, L. Barbu-Tudoran****

*Technical University of Cluj-Napoca, Faculty of Building Machine, 103-105, Bd. Muncii, 400641, Cluj-Napoca, Romania
E-mail: **Marius.Pustan@omt.utcluj.ro**
**National Institute for Research and Development of Isotopic and Molecular Technologies, 67-103 Donat Street, 400293, Cluj-Napoca, Romania
****E-mail**: **sorina.garabagiu@itim-cj.ro**

Abstract—This paper presents the fabrication and characterization of titanium oxide thin films deposited by Pulsed Laser Deposition in different experimental conditions. The scope of this work is to investigate the effect of the oxygen pressure in the deposition chamber on the material properties. Thin films characterizations include the mechanical and tribological properties such as the modulus of elasticity, hardness and the adhesion force. The mechanical and tribological properties of the materials are experimentally determined by using the atomic force microscopy technique. The effect of the oxygen pressure on the film thickness is analyzed. As the pressure in the deposition process decreases, the thickness of the thin films increases, respectively. The surfaces roughness increases as the deposition pressure decreases that leads to a decrease of adhesion forces. Hardness and modulus of elasticity increases as the deposition pressure decreases. This study shows that the mechanical and tribological properties of the investigated thin films strongly depend on the grain size and the films density, which are influenced by the deposition conditions (the oxygen pressure in the deposition chamber).

Keywords—thin films; pulsed laser deposition; modulus of elasticity, hardness, adhesion force.

1. Introduction

Titanium dioxide (TiO_2) is a material successfully implemented in thermoelectric applications including thermoelectrical generator, electrodes or thermal energy sensor [1, 2]. This material is characterized by low absorption coefficient, high dielectric constant and good biocompatibility [1]. Due to their excellent biocompatibility, the other application for TiO_2 is for human implants.

The grain of the TiO_2 material depends on the deposition parameters including the medium pressure and temperature [3]. By increasing the temperature of treatment, the grain size of the specimens increases [3]. The higher particle sizes can be obtained at low

oxygen pressure in the deposition process, as it is presented in this paper. The grain size has effect on the thickness of deposited thin films. The thickness of the deposited thin films is experimentally measured by using an atomic force microscope (AFM).

The other deposition parameters such as the temperature of substrate, the deposition angle or by exposing the growing film to a beam of accelerated ions change the material mechanical properties [4].

2. Samples Deposition

Titanium dioxide thin films were fabricated by Pulsed Laser Deposition (PLD) technique from a TiO_2 (rutile) target. PLD system is equipped with a laser excimer (KrF, λ=248nm, pulse duration is 20ns), the target to substrate distance is set to 55mm. N-doped Si (100) substrates, 5mm×10mm, have been cleaned in acetone, ethanol and water, in ultrasonic baths, and then thoroughly rinsed with distilled water prior deposition. The SiO_2 layer (3nm) that naturally forms onto the surface of Si substrates has not been removed. During the deposition process, the temperature of the substrate was maintained at 700°C. Laser fluence was set to $2J/cm^2$ with a frequency of 5Hz. Deposition time was set of 1 hour. The oxygen pressure in the deposition chamber has been varied between 1×10^{-6}mbar (vacuum) and 3×10^{-3}mbar, in order to optimize the structural, morphological, mechanical and tribological properties of the investigated TiO_2 thin films.

978-1-5386-4483-6/18 $31.00 © 2018 IEEE

3. Experimental Characterizations

The scope of experimental tests is to determine the effect of deposition pressure on mechanical and tribological properties of investigated TiO$_2$ thin films. The tests are performed using an atomic force microscope XE70 with a nanoindentation module manufactured by Park Systems Co. The testing temperature in cleanroom was kept at 20°C and the relative humidity at 40%. The indentation of the thin films was performed using a diamond AFM tip.

A. Thickness Measurement

The first step in the experimental characterization was orientated to measure the film thickness as a function of the oxygen pressure in the main chamber during the deposition process. The AFM tip is positioned at the edge of the deposited thin films. In this situation, the scanning area includes the investigated thin film as well as the substrate (**Fig.1a**).

The scan size is 5µm × 45.5µm as presented in **Fig.1b**. The cross-section of the scanning area provides information about the film thickness, as the difference between the substrate and the bottom surface of films (difference between cursors from **Fig.2**). The measurements were performed in different location of the edge of thin films and the results were in good agreement that confirms the uniformity of the deposited process.

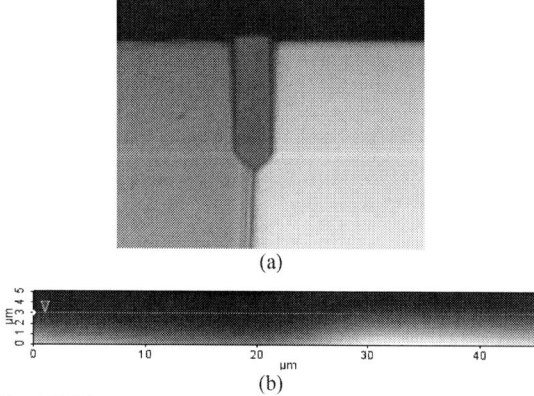

Fig. 1 Thickness measurement of TiO$_2$ thin films: (a) AFM tip is positioned at the edge of the film; (b) the selected scanning area (5µm × 45.5µm).

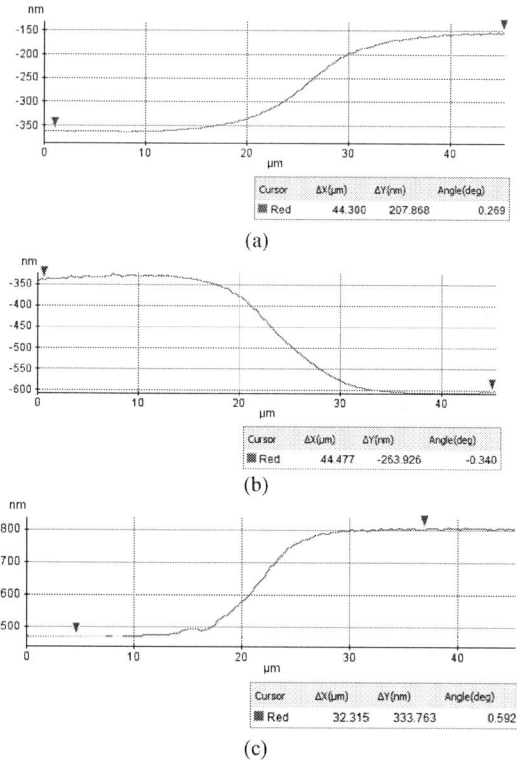

Fig. 2 Thickness of TiO$_2$ thin films with the deposition oxygen pressure of: (a) 10^{-3}mbar, (b) 10^{-4}mbar, (c) in vacuum.

As the oxygen pressure is decreases from 10^{-3}mbar to 10^{-4} mbar, and then in vacuum, the thickness of the investigated TiO$_2$ thin films increases, respectively. The thickness of the investigated films is provided in **Fig.2** as the difference between cursors (the ΔY values). The thin film thickness is 207.868nm for TiO$_2$ deposited at 10^{-3}mbar oxygen pressure (**Fig.2a**), it increases at 263.926nm (**Fig.2b**) if the pressure in the deposition process decreases to 10^{-4}mbar and 333.763nm (**Fig.2c**) for the TiO$_2$ thin film deposition in vacuum (10^{-6}mbar).

B. Surface Characterization

The roughness parameters were measured using the tapping mode of the AFM. The scanning area was selected at 5µm × 5µm for all samples.

Fig. 3 presents the 3D images and the roughness parameters for the TiO$_2$ films with different deposition oxygen pressure. As the oxygen pressure decreases from 10^{-3}mbar to vacuum, the R$_a$ roughness increases from 3.068nm to 4.525nm.

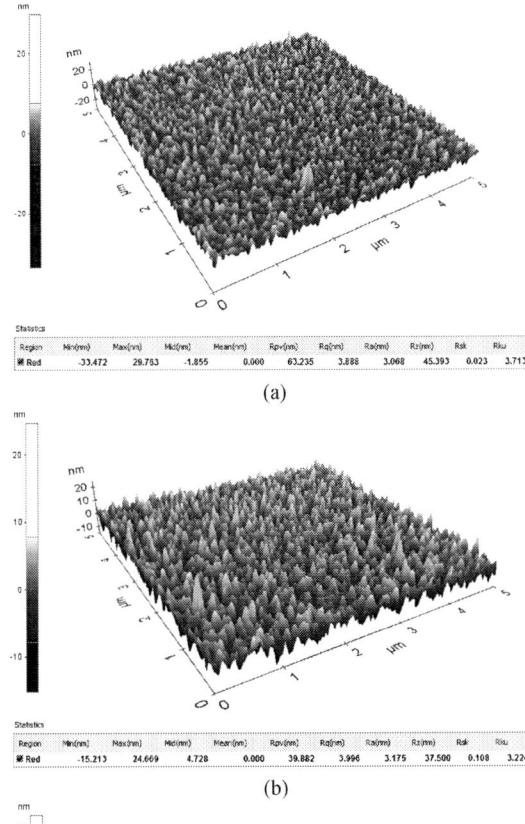

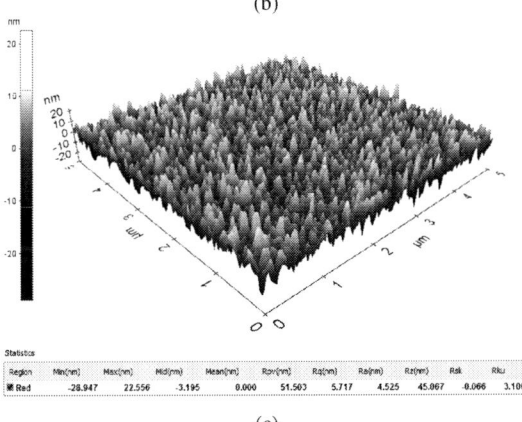

(a)

(b)

(c)

Fig. 3 Roughness parameters of TiO$_2$ films with the deposition oxygen pressure of: (a) 10^{-3}mbar, (b) 10^{-4}mbar, (c) in vacuum.

C. Mechanical Properties

The mechanical properties under interest are the modulus of elasticity and hardness. The measurements were done using the indentation module of AFM. The experimental values of hardness and modulus of elasticity were determined from the load-displacement curves using the Oliver and Pharr method. Since 1992, the

analysis method proposed by Oliver and Pharr has been established as the standard procedure for determining the hardness and elastic modulus from the indentation load-displacement data for bulk materials [5].

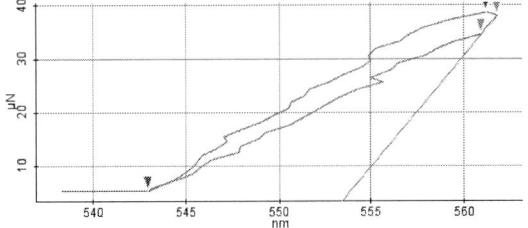

Cursors						
Cursor	ΔX(nm)	ΔY(μN)	Left X(nm)	Left Y(μN)	Right X(nm)	Right Y(μN)
Force	18.257	33.491	542.934	5.137	561.191	38.628
Slope	0.884	3.626	560.945	34.498	561.829	38.124

Slope Cursor Index 1 :	26
Slope Cursor Index 2 :	23
Contact Depth :	12.26nm
Tip Shape :	Berkovich
Poisson's ratio of the tip :	0.07
Poisson's ratio of the sample :	0.28
Hardness :	9.1GPa
Young's modulus :	208.58GPa

Fig. 4 Nanoindentation of the TiO$_2$ thin film obtained in vacuum.

Fig. 4 presents the nanoindentation curve used to determine the modulus of elasticity and hardness of TiO$_2$ film deposited in vacuum (10^{-6}mbar). An indentation force of 40μN was selected to obtain an indentation depth less than 10% from the films thickness to avoid the substrate effect. The same experiment was used for all TiO$_2$ samples and the hardness and modulus of elasticity increases if the oxygen pressure decreases, as presented in section 4 of paper.

D. Tribological Characterization

The oxygen pressure during deposition changes the roughness parameters of TiO$_2$ films with influence on the adhesion force. The other test was done to investigate the effect of the deposition oxygen pressure on the adhesion effect. The tests were performed using the spectroscopy-in-point of AFM. The adhesion between AFM tip (Si$_3$N$_4$) and TiO$_2$ films was determined from the unloading part of AFM curve (**Fig.5**). The peak given by the blue cursors and the ΔY- values from the retrace line represents the adhesion force.

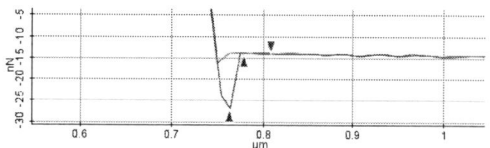

Cursor	ΔX(nm)	ΔY(nN)	Left X(nm)	Left Y(nN)	Right X(nm)	Right Y(nN)
▣ Trace	0.000	0.000	807.853	-14.167	807.853	-14.167
▣ Retrace	15.126	13.030	763.355	-27.018	778.481	-13.988

(a)

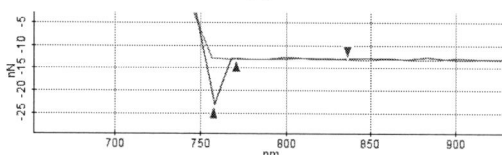

Cursor	ΔX(nm)	ΔY(nN)	Left X(nm)	Left Y(nN)	Right X(nm)	Right Y(nN)
▣ Trace	0.000	0.000	836.687	-13.218	836.687	-13.218
▣ Retrace	13.244	10.253	757.442	-23.244	770.686	-12.991

(b)

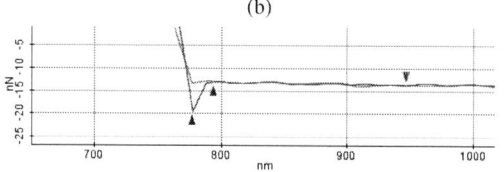

Cursor	ΔX(nm)	ΔY(nN)	Left X(nm)	Left Y(nN)	Right X(nm)	Right Y(nN)
▣ Trace	0.000	0.000	946.803	-13.611	946.803	-13.611
▣ Retrace	16.155	6.414	777.670	-19.697	793.825	-13.283

(c)

Fig. 5 Adhesion force between AFM tip (Si₃N₄) and TiO₂ films with the deposition oxygen pressure of: (a) 10^{-3}mbar, (b) 10^{-4}mbar, (c) in vacuum.

As the deposition pressure decreases, the roughness increases and the adhesion force decreases from 13.030nN to 6.414nN.

4. Results and Discussions

The effect of the oxygen pressure during deposition process changes the mechanical and tribological properties of TiO₂ thin films.

Table 1. Experimental results for TiO₂ thin films.

Deposition Pressure	Investigated Parameters*				
	h [nm]	R_a [nm]	E [GPa]	H [GPa]	F_a [nN]
10^{-3}mbar O₂	207.868	3.06	188.8	8.13	13.03
10^{-4}mbar O₂	263.926	3.17	190.52	8.45	10.25
10^{-6}mbar	333.763	4.52	208.58	9.1	6.41

* h- thickness, R_a- average roughness, E- modulus of elasticity, H- hardness, F_a- adhesion force.

Table 1 presents the experimental results obtained for the investigated TiO₂ films. As

the deposition pressure decreases, the thickness of the TiO₂ films and the roughness increase. This effect is based on obtaining higher particle sizes at low pressures. The adhesion force decreases as the surface roughness increases. The hardness and modulus of elasticity increase as the deposition pressure decreases, respectively.

5. Conclusions

The scope of this paper was to investigate the effect of deposition pressure on the modulus of elasticity, hardness, surface parameters and adhesion force. The obtained results lead to improve the fabrication process of TiO₂ films to obtain materials with good properties for thermoelectric applications. The deposition parameters change the material properties with effect on the wear resistance and the lifetime. As the deposition pressure decreases, the mechanical properties of TiO₂ films are improved and the adhesion effect is reduced.

Acknowledgments. This work was supported by a grant founded by the Romanian Space Agency, STAR Program C3, project no.193/2017.

References

[1] A. Pura, K. Rubenis, D. Stepanovs, L. Berzina-Cimdina and J. Ozolins, "*Semiconducting properties of nonstoichiometric TiO₂₋ₓ ceramics*", Process. Appl. Ceram., **6** (2), pp. 91–95, May 2012.

[2] S.B. Riffat, X. Ma, "*Thermoelectrics: a review of present and potential applications*" Appl. Therm. Eng., **23**, pp. 913-935, 2003.

[3] M.L. Vera, M.A. Alterach, M.R. Rosenberger, D.G. Lamas, C.E. Schvezov and A.E. Ares, "*Characterization of TiO₂ nanofilms obtained by sol-gel and anodic oxidation*", Nanomater. Nanotechno., **4**, pp. 1-11, April 2014.

[4] Y. Gaillard, V.J. Rico, E. Jimenez-Pique and A.R. Gonzales-Elipe, "*Nanoindentation of TiO₂ thin films with different microstructures*", J. Phys. D Appl. Phys., **42**, pp.1-9, 2009.

[5] D.S. Grierson, E.E. Flater and R.W. Carpick, "*Accounting for the JKR–DMT transition in adhesion and friction measurements with atomic force microscopy*", J. Adhes. Sci. Technol., **19**, pp. 291-311, 2005.

MOS DOSIMETER BASED ON Ge NANOCRYSTALS IN HfO$_2$

C. Palade, A. Slav, A.M. Lepadatu, I. Stavarache, I. Dascalescu,
O. Cojocaru, T. Stoica, M.L. Ciurea, S. Lazanu

National Institute of Materials Physics, 077125 Magurele, Romania
E-mail: lazanu@infim.ro

Abstract—Trilayer MOS capacitors gate HfO$_2$ / *floating gate of Ge nanocrystals in HfO$_2$ / tunnel HfO$_2$ / Si substrate were prepared in the aim to be used for the detection of ionizing radiation. Magnetron sputtering and rapid thermal annealing were used for their fabrication. Capacitance-voltage measurements showed that Ge nanocrystals are the most important charge storage centres in our structure. The possibility to use these trilayer MOS capacitors as dosimeters was investigated, and the sensitivity to α particle irradiation was extracted.*

Keywords—dosimeter, Ge nanocrystals, HfO$_2$, MOS capacitor.

1. Introduction

The interest for the detection and the measurement of ionizing radiation is related to a large variety of applications, *e.g.* medicine, and in particular radiotherapy [1], manned space missions [2], nuclear waste management [3], nuclear power stations [4], military sector [5] and research [6]. Specific requirements are imposed to dosimeters in different application fields, and due to this fact different types of dosimeters were developed based on the effects of ionizing radiation: ionization (in gas filled detectors, p-n junctions, MOS transistors); scintillation, thermoluminescence, chemical reactions.

Two types of MOSFETs are used in commercial dosimeters: radiation sensing FET (RADFET) and floating gate (FG) MOSFET.

RADFET was first designed for space applications [7]; the principle of operation is the alteration of MOSFET characteristics due to the space charge stored in the oxide following the irradiation with high energy particles or photons. It must be biased during operation to ensure the separation of electrons and holes created by exposure to radiation, this being its major drawback.

FGMOSFETs have a different working principle: the FG is charged before irradiation, thus creating a local electric field which ensures the sensitivity of the device. They were proposed to be used as dosimeters in 1991 [8]. The charge extracted from the FG during irradiation is proportional to the absorbed dose. These devices are of interest for both space and medical applications. For space applications, the advantages of FGMOSFETs are: low mass and dimension, possibility to be integrated with other sensors, and permanent storage of information, independent on the dose rate. Related to the utilisation in oncology, they have the advantage of working unbiased, of low sensitive volume and of the possibility to be used *in vivo*.

FGMOSFETs were in fact developed for non-volatile storage as memories. In this application, in the 1990s, the continuous polycrystalline FG layer was replaced with a FG made up from isolated nanocrystals (NCs), and thus the information is kept even if a leakage path is created.

In the present paper we report for the first time on the possibility to use FG capacitor with Ge NCs embedded in HfO$_2$ as a sensing device for ionizing radiation. There are few papers in the literature regarding dosimeters with FGMOS capacitors structure, with the FG made up of NCs: Si NC in SiO$_2$ [9] and Ge NCs in SiO$_2$ [10].

We prepared trilayer structures of the type *gate oxide / NCs in oxide / tunnel oxide/ Si substrate,* the NCs being Si or Ge ones, and the oxide SiO$_2$ or HfO$_2$ [11-14], and investigated the possibility to build dosimeters starting from them.

978-1-5386-4483-6/18 $31.00 © 2018 IEEE

Our group have studied in depth the system of Ge NCs embedded in SiO_2 from the point of view of the correlation between preparation conditions, morphology and structure and electric properties [15-19]. Trilayer capacitors benefited from the replacement of SiO_2 with HfO_2, which enabled obtaining a higher memory window.

We obtained the best dosimeter performances on the trilayer structure *gate HfO₂/floating gate of Ge NCs in HfO₂/tunnel HfO₂/Si substrate*, and we report the results in the present paper.

2. Experimental

The trilayer structures were deposited by magnetron sputtering (*Gamma 1000 C* from Surrey NanoSystems) on p-type Si wafers of (100) orientation and 7-14 Ωcm resistivity, into a single run. The wafers were previously cleaned in the clean room, using RCA standard procedure, and then mounted in the magnetron sputtering deposition chamber. Layer deposition takes place in 6N Ar atmosphere, at 4 mTorr working pressure. The intermediate layer is obtained by Ge and HfO_2 cosputtering, in the volume ratio 65:35. Rapid thermal annealing (*As-Micro* from Annealsys) is subsequently performed in N_2 atmosphere at 600-700 °C for nanostructuring. Al contacts are deposited by thermal evaporation on the front and backside of the structure.

The morphology of the resulting structure were measured by high resolution transmission electron microscopy (HRTEM) using JEM-ARM200F Electron Microscope and by X-ray photoelectron spectroscopy (XPS) in an Ultra High Vacuum System for XPS/UPS/AES (SPEC-S Phoibos 150 MCD electron energy analyzer, monochromatic x-ray radiation Al Kα 1486.74 eV).

Capacitance measurements (capacitance – voltage, *C-V* and capacitance – time, *C-t* characteristics) were performed in dark at room temperature (Agilent E4980A LCR meter). A [241]Am alpha source of 340 kBq activity from Lehr und Didaktiksysteme LD Didactic GmbH was used for irradiation.

3. Results and discussion

A. *Morphology and structure*

HRTEM analysis reveals that HfO_2 is crystallized in the whole trilayer depth. In the intermediate layer [13], Ge NCs of 2-3 nm diameter and $4–5 \times 10^{11}$ NC/cm^2 are located at the cross of boundaries of HfO_2 NCs. In this layer, Ge NCs are separated by HfO_2 NCs that have ~7 nm size.

XPS measurements were taken at the free surface of the trilayer and in the depth of the sample by Ar^+ sputtering. They show that at the free surface Ge is fully oxidized, and at the position of the intermediate layer the majority of Ge is in metallic form, with only a small contribution (less than 11 %) of oxidized Ge.

B. *C-V measurements*

The *C-V* characteristic measured at 1 MHz is presented in Fig. 1.

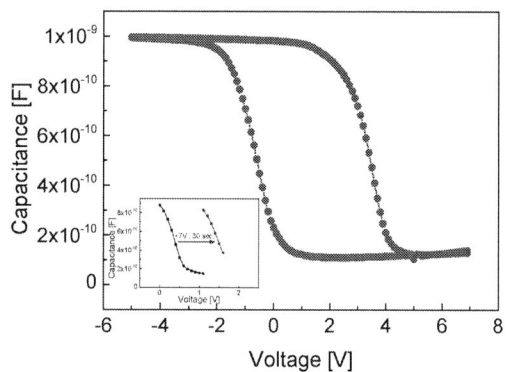

Fig. 1 C-V curve measured at 1 MHz on the trilayer structure

It shows a counter-clockwise hysteresis loop with a memory window of about 2.5 V. The memory window was measured with 30 sec. charging time at accumulation and inversion. In the inset, we present the displacement of the *C-V* curve after charging the NCs for 30 sec at 7V ($\Delta V_{FB} = 1$ V). The *C-t* curve, measured at 0 V, after charging the NC as before, shows a slow decrease of the capacitance at the beginning of the measurement, followed by an even lower descendent slope. Thus, a decrease of about 16% the initial value of the capacitance was found after 6000 sec.

C. α particle irradiation

The energy of the α particle at the surface of the [241]Am source is 5.638 MeV and decreases due to attenuation in air up to 3.80 MeV at the surface of the capacitor. It passes through the Al front contact and through the trilayer structure, and is stopped in the Si substrate, as results from the simualtion of the penetration of 500 α particles, using the Monte Carlo programme SRIM [20].

In Figs. 2 and 3 we present the positions of the stopped ions in the target, and target ionization, respectively.

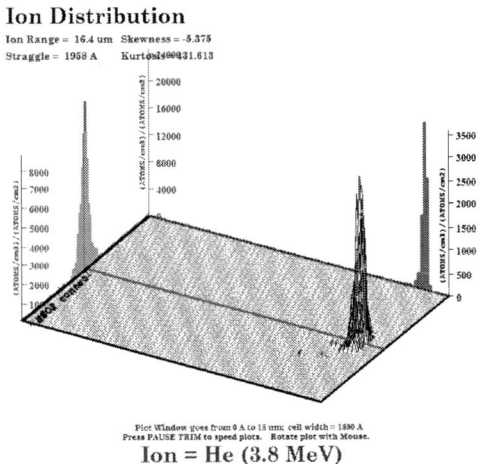

Fig. 2 Positions of stopped α particles

The range of α particles in our structure is 16.40 ± 0.19 μm, and the divergence of the beam in the sample is very low, both in longitudinal and transversal directions.

As shown in Fig. 3, the rate of energy loss by ionization is higher in the deposited layers than in the substrate. On the other hand, α particles lose energy mainly by ionization.

The trilayer capacitor was then tested for the sensitivity to α particle irradiation, at doses up to 100 Gy.

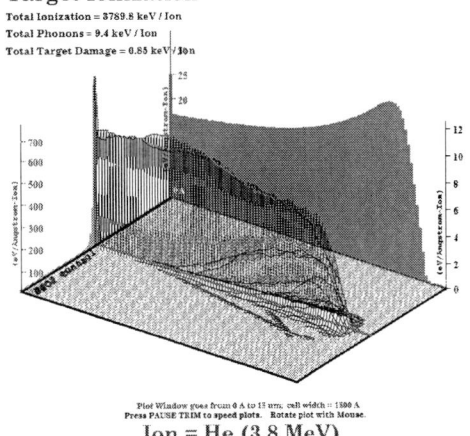

Fig. 3 Space distribution of the energy lost by ionization

In this aim, Ge NCs were charged with electrons, by keeping the MOS capacitor at inversion for 30 sec. This way, a local electric field was created. By exposure to radiation, electrical carriers are generated in the oxide, and they are moving in the electric field, producing NC discharging. In its turn, NC discharging produces the modification of the flat-band voltage of the MOS capacitor, ΔV_{FB}, which becomes a measure of the absorbed dose.

We found a linear dependence of ΔV_{FB} versus dose up to about 60 Gy, with a slope of 0.8 mV/Gy, followed by a tendency to saturation at higher doses. This is probably due to the decrease of the electric field in the oxide, in its turn determined by the discharge of Ge NCs. This sensitivity is situated between the values reported for structures with Ge NCs in SiO_2 [10], and Si NCs in SiO_2 respectively [9].

4. Conclusions

We have shown for the first time that trilayer capacitors having Ge NCs as floating gate and HfO_2 as tunnel and control oxide show good dosimeters qualities, and can be used as for α particle irradiation in the dose range 0-100 Gy. The structures were prepared by magnetron sputtering deposition and thermal annealing.

Corroborated structural and electrical characteristics indicate that Ge NCs formed during annealing are the most important storage centres in these structures.

The sensitivity to the dose was found to be intermediate between the values reported for similar structures with Ge NCs and Si NCs respectively, embedded in SiO_2, reported in the literature.

Acknowledgments

This work was supported by the Executive Agency for Higher Education, Research, Development and Innovation UEFISCDI (contracts PED 42/2017, PED 89/2017 and PED 203/2017) and by Romanian Ministry of Research and Innovation (NIMP Core Program 2018).

References

[1] B. Paliwal, D. Tewatia, "*Advances in radiation therapy dosimetry*", J. Med. Phys. 34 (2009) 108-116

[2] A. Beheshti et al., "NASA GeneLab Project: Bridging Space Radiation Omics with Ground Studies", Rad. Res.189 (2018) 553-559.

[3] B. Pang, H. Sauri Suarez, F. Becker, "*Individual dosimetry in disposal repository of heat-generating nuclear waste*", Radiat Prot Dosimetry **170** (2016) 387-392.

[4] A. Metcalfe et al., "*Diamond based detectors for high temperature, high radiation environments*", J. Instrumentation 12 (2017) C01066.

[5] A. Romanyukha, F. Trompier, L.A. Benevides, "*KEVLAR (R) as a potential accident radiation dosimeter for first responders, law enforcement and military personnel*", Health Phys. 111 (2016) 127-133.

[6] C. Baldock, "*Trends in Radiation Dosimetry: preliminary overview of active growth areas, research trends and hot topics from 2011-2015*", J. Phys: Conf. Series **777** (2017) 012030.

[7] A. G. Holmes-Siedle, "*A review of the use of metal-oxide–silicon devices as integrating dosimeters*", Nucl. Instr. Meth. Phys. Res. **121** (1974) 169-179.

[8] J. Kassabov, N. Nedev, N. Smirnov, "*Radiation dosimeter based on floating gate*

MOS transistor", Rad. Eff. And Defects in Solids **116** (1991) 155-158.

[9] D. Nesheva et al., "*Application of Metal-Oxide-Semiconductor structures containing silicon nanocrystals in radiation dosimetry*", Open Phys. **13 (2015) 63–71.**

[10] A. Aktag, E. Yilmaz, N.A.P. Mogaddam, G. Aygun, A. Cantas, R. Turan, "*Ge nanocrystals embedded in SiO_2 in MOS based radiation sensors*", Nucl. Instr. Meth Phys Res B **268** (2010) 3417-3420.

[11] A. Slav et al., "*How morphology determines the charge storage properties of Ge nanocrystals in HfO_2*", Scripta Mater.**113** (2015) 135-138.

[12] D. Vasilache et al., "*Non-volatile memory devices based on Ge nanocrystals*", Phys. St. Sol. A **213** (2016) 255-259

[13] A.M Lepadatu et al., "*Single layer of Ge quantum dots in HfO_2 for floating gate memory capacitors*", Nanotechnology **28** (2017) 175707.

[14] C. Palade et al., "*Material parameters from frequency dispersion simulation of floating gate memory with Ge nanocrystals in HfO_2*", Appl. Surf. Sci. **428** (2018) 698–702.

[15] I. Stavarache, A.M. Lepadatu, A.V. Maraloiu, V.S. Teodorescu, M.L.. Ciurea, "*Structure and electrical transport in films of Ge nanoparticles embedded in SiO_2 matrix*", J. Nanopart. Res. 14 (2012) 930.

[16] A.M. Lepadatu, T. Stoica, I. Stavarache, V.S. Teodorescu, D. Buca and M.L. Ciurea, "*Dense Ge nanocrystal layers embedded in oxide obtained by controlling the diffusion–crystallization process*", J. Nanopart. Res.**15** (2013) 1981.

[17] I. Stavarache, A.-M. Lepadatu, T. Stoica, M. L. Ciurea, "*Annealing temperature effect on structure and electrical properties of films formed of Ge nanoparticles in SiO_2*", Appl. Surf. Sci. **285 B** (2013) 175-179.

[18] M.L. Ciurea and A.M. Lepadatu, "*Tuning the properties of Ge and Si nanocrystals based structures by tailoring the preparation conditions Review*", Dig. J. Nanomater. Bios., **10** (2015) 59.

[19] I. Stavarache, V.A Maraloiu, P. Prepelita, G. Iordache, "*Nanostructured germanium deposited on heated substrates with enhanced photoelectric properties*", Beilstein J. Nanotechn. 7 (2016) 1492-1500.

[20] SRIM, Stopping and Range of Ions in Matter, www.srim.org.

Session MW

MICROWAVE AND MILLIMETER WAVE CIRCUITS AND SYSTEMS

978-1-5386-4483-6/18 $31.00 © 2018 IEEE

Investigation of Liquid Metal and FDM 3D Printed Microwave Devices

*K. Y. Chan, X. Li, R. Ramer

School of Electrical Engineering and Telecommunications, The University of New South Wales, Sydney, Australia
*kyc@unsw.edu.au

Abstract— The paper proposed a new fabrication technique for microwave devices using liquid metal conductor and 3D printed dielectric containers. 3D printed conductive dielectric materials provide electrical connections to the liquid metal at the input and output ports. These input/ output ports and the dielectric container were analyzed to determine their impact on the RF performance. Simulation results of a WR62 rectangular waveguide transmission line, waveguide resonator, a three-pole iris filter and a horn antenna were presented. The RF performance achieved satisfactory RF performance validating the proposed fabrication concept.

Keywords— microwave, waveguide, liquid metal, 3D printing.

1. Introduction

In the last decade, wireless and mobile communication technologies have experienced exponential growth, with almost every object around us, ranging from personal computers and cars to industrial automated tracking systems, relying on wireless technologies. Typically, high-performance RF and passive microwave devices are predominantly fabricated from bulk metal pieces, requiring expensive manufacturing techniques utilizing computer numerical control (CNC) machines and significant investments in equipment, thus are unsuitable for large mass production [1]-[5]. For traditional microwave and millimetre-wave applications, selective laser sintering (SLS), selective laser melting (SLM) and binder jetting (BJ) metal 3D printing could be viewed as the most suitable technologies, as metal waveguide components can be manufactured directly with high dimensional accuracy [6]-[8]. However, these metal printings suffer from high capital investment and require costly specialized materials in the form of alloy powders. [9]-[11]. Many RF components such as waveguide filters and antennas have been reported using this 3D printing technique [12]-[17]. However, as SLA 3D printing is unable to print metal directly, the printed components require further metallization. The quality of the metallization

affects the RF performance significantly, in particular, the reliability, losses and power handling. Many different metallization techniques associated with the SLA printing were reported [18], [19]. Some require complicated procedure involving industrial-grade electroplating with dangerous chemicals. Other, employed multilayer depositions such as painting the printed parts with silver paint before electroplating [20]. In this paper, a new method of fabricating 3D microwave devices using liquid metal and 3D printing is proposed. Our proposed technology allows 3D microwave devices manufactured in a single piece without using dangerous chemicals. Moreover, it could enable easy reconfiguration by relocating the liquid metal and introducing mechanical reconfiguration instead of the traditional, widely used RF MEMS-based approaches [21]-[49]. Our technique uses fused filament fabrication (FDM) 3D printed conductive dielectric flanges at the input and output with non-conductive dielectric enclosure for liquid metal, to create 3D microwave devices.

2. Investigation of Conductive Flanges

In our study, FDM conductive filaments purchased from Blackmagic3D and Proto-Pasta with a volumetric resistivity ranging from 0.6 to over 100 ohm-cm were investigated. These conductive filaments are mainly produced out of mixtures of polylactic acid (PLA) or acrylonitrile butadiene styrene (ABS) and graphene nanocomposites. Fig. 1(a) shows the simulated attenuation vs conductivity of a WR62 conductive dielectric rectangular waveguide flange, with different conductivities at three frequencies 10, 14, and 18 GHz. The worst simulated attenuation is 0.75 dB/mm at 18 GHz when the conductivity is 1 S/m. The simulation of 3D printed conductive materials for coaxial-based devices was also investigated. For

978-1-5386-4483-6/18 $31.00 © 2018 IEEE

SMA coaxial connector, simulated attenuations in dB/mm is shown in Fig. 1(b) at 1, 6 and 10 GHz. As can be noticed, the worst attenuation is 1.5 dB/mm, when the conductivity is 1 S/m. It could also be noticed that with the increase in conductivity, the attenuation diminishes significantly. As the best resistivity could be as low as 0.6 ohm-cm or conductivity 167 S/m, the attenuation due to the flanges can be as low as 0.2 dB/mm for both SMA coaxial and WR62 waveguide. To-date, the thinnest possible layer thickness is 20 μm from commercially available FDM printer with theoretically, attenuation as low as 0.004 dB per flange for a length of 20 μm. Therefore, for 2-port devices, the attenuation due to the flanges can be as low as 0.008 dB for 20 μm at each port.

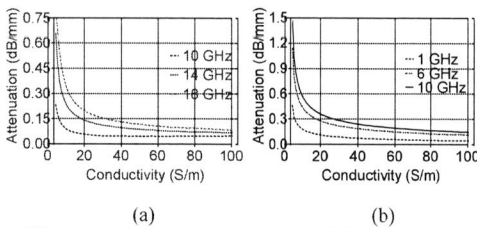

Fig. 1. Attenuation vs. conductivity at different frequencies: (a) waveguide, and (b) SMA coaxial cable.

3. Impact of The Dielectric Enclosures

In this section, the attenuations in a waveguide transmission line and a resonator will be discussed.

A. Rectangular Waveguide Section

Fig. 2 shows a proposed rectangular waveguide; the 3D model with its perspective view (Fig. 2(a)) and front-view (Fig. 2(b)) and its RF simulations (Fig. 2(c)) are shown. This waveguide has WR62 dimensions with A is 15.8 mm, and B is 7.9 mm and is to be operated in Ku-band (12-18 GHz). As can be seen in Fig. 2 (b), the waveguide liquid metal conductor is bounded by the 3D dielectric enclosure. A WR62 rectangular waveguide was simulated. In the simulations, we varied the inner dielectric enclosure wall thickness (t) together with its dielectric loss tangent, to determine its impact on the RF performance. Fig 2(c) shows the attenuation (dB/mm) vs wall thickness for different loss tangent values. Depending on the dielectric loss

tangent value and the wall thickness, the attenuation contribution from the inner wall can vary significantly. It is evidenced that thinner dielectric walls will result in smaller attenuation and better RF performance regardless of the dielectric tanδ.

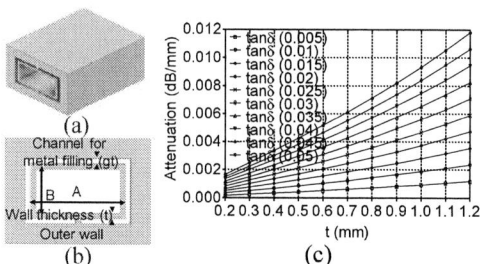

Fig. 2. Proposed 3D rectangular waveguide structure: (a) perspective view; (b) front-view with labeled dimensions; (c) simulated attenuation (dB/mm) vs inner wall thickness (t) for various loss tangent at 13GHz.

B. Microwave Resonator

Using a similar design as the waveguide transmission line, a WR62 waveguide-based cavity resonator was simulated to determine the achievable Q-factor. The resonator was designed irises having sloped walls of 45 degrees to allow 3D printing without supports. A minimal iris width (W) of 2 mm was deliberately selected with the intention to simplify the calculation, from the S-parameters. The cavity length (L) was designed such that the resonance is at 13 GHz. In the simulations, the resonant cavity with various dielectric wall thicknesses and loss tangent were investigated to determine the highest achievable Q-factor; gallium (Ga) was used as the liquid conductor. Fig. 3(a) and (b) show the perspective view and the top-view the resonator. Fig. 3(c) shows the simulation results; it can be noticed that the maximum achievable Q-factor using Ga as metal is over 2300 in Ku-band. Depending on the loss tangent of the 3D printed materials and the thickness of the dielectric wall, the Q-factor varies linearly.

4. Waveguide Filter and Antenna

To further verify the potential RF performance of the proposed technology, a Chebyshev three-pole filter with a bandwidth of 500 MHz centered at 13GHz and a horn antenna were designed and simulated in Ku-band. Similar to the resonator, the irises of the filter were designed with a slope of 45 degrees. A width of 0.5 mm was used for the inner wall with 1 mm metal filling channel to form the filter. Conductive flanges were included at the I/O ports. Fig. 4 shows the simulation results of the two devices. The simulated insertion loss was 0.35 dB which has an estimated Q-factor of over 1000, similar to the previous resonator simulation. The horn antenna also has minor gain degradation, from 8.6 dB to 8.4 dB, when compared with the same horn antenna simulation using perfect conductor (PEC). This result once again confirmed the simulation results for the rectangular waveguide.

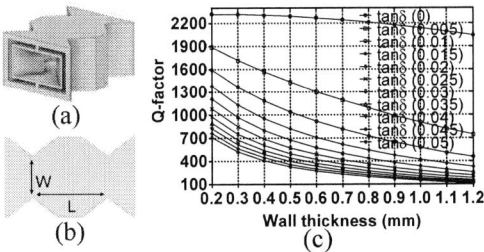

Fig. 3. The simulated waveguide resonator: (a) perspective view; (b) top-view showing the resonator length (L) and the coupling iris width (W) and (c) the simulated Q-factor with different loss tangent and dielectric wall thicknesses.

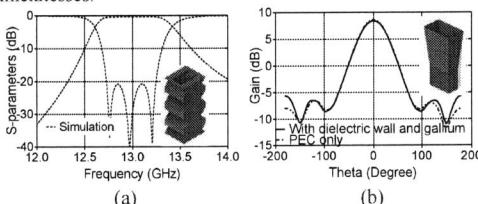

Fig. 4 Simulation of (a) a three-pole bandpass filter and (b) a horn antenna (dielectric walls, Ga liquid metal, and fabrication dimensions were used).

5. Conclusions

This paper presented a novel fabrication method of 3D type microwave devices. Simulations confirmed the possibility of using liquid metal conductors enclosed by FDM 3D printed dielectrics. Different dielectric loss tangent values were analysed to determine the impact to the waveguide attenuation and the Q-factor of a resonant cavity at 13 GHz. The use of 3D printed conductive dielectric to provide external electrical connections to the liquid metals

was also presented. It was shown that even when the dielectric flanges have very low conductivities, adequate RF performance can still be obtained.

References

[1] K. Y. Chan et al., "A Switchable Iris Bandpass Filter Using RF MEMS Switchable Planar Resonators," IEEE Microw. Wireless Compon. Lett., vol. 27, no. 1, pp. 34-36, 2017.

[2] K. Y. Chan et al., "Design of Waveguide Switches Using Switchable Planar Bandstop Filters," IEEE Microw. Wireless Compon. Lett., vol. 26, no. 10, pp. 798-800, 2016.

[3] K. Y. Chan et al., "Switchable Waveguide Iris Filter using Planar Dipoles," in IEEE Microwave Symposium Digest (IMS), 2013 IEEE MTT-S International, 2013, pp. 1-4.

[4] L. Gong et al., "A Third Order Bandpass Iris Filter with Reconfigurable Dielectric Irises," Microwave and Optical Technology Letters, vol. 60, no. 5, pp. 1287-1290, 2018.

[5] L. Gong et al., "A Four-State Iris Waveguide Bandpass Filter with Switchable Irises," in IEEE IMS, 2017, pp. 260-263.

[6] E. Decrossas et al., "Evaluation of 3D Printing Technology for Corrugated Horn Antenna Manufacturing," in IEEE Int. Symp. on Electromagn. Compat. (EMC), Ottawa, Canada, 2016, pp. 251-255.

[7] E. A. Rojas-Nastrucci et al., "Ka-Band Characterization of Binder Jetting for 3-D Printing of Metallic Rectangular Waveguide Circuits and Antennas," IEEE Trans. Microw. Theory Techn., vol. 65, no. 9, pp. 3099-3108, Sep. 2017.

[8] F. Calignano et al., "Overview on Additive Manufacturing Technologies," Proc. IEEE, vol. 105, no. 4, pp. 593-612, Apr. 2017.

[9] M. D'Auria et al., "3-D Printed Metal-Pipe Rectangular Waveguides," IEEE Trans. Compon. Packag. Technol., vol. 5, no. 9, pp. 1339-1349, Sep. 2015.

[10] B. Zhang et al., "Metallic 3-D Printed Rectangular Waveguides for Millimeter-Wave Applications," IEEE Trans. Compon. Packag. Technol., vol. 6, pp. 796-804, May. 2016.

[11] O. A. Peverini et al., "Selective Laser Melting Manufacturing of Microwave Waveguide Devices," Proc. IEEE, vol. 105, no. 4, pp. 620-631, Mar. 2017.

[12] C. Guo et al., "A 3-D Printed Lightweight X-Band Waveguide Filter Based on Spherical Resonators," IEEE Microw. Compon. Lett., vol. 25, no. 7, pp. 442-444, Jul. 2015.

[13] B. Rohrdantz, C. Rave, and A. F. Jacob, "3D-Printed Low-Cost, Low-Loss Microwave Components up to 40 GHz," in IEEE MTT-S Int. Micr. Symp. (IMS), San Francisco, CA, USA, 2016, pp. 1-3.

[14] C. Guo et al., "A Lightweight 3-D Printed X-Band Bandpass Filter Based on Spherical Dual-Mode Resonators," IEEE Microw. Compon. Lett., vol. 26, no. 8, pp. 568-570, Aug. 2016.

[15] E. G. Geterud et al., "Lightweight Waveguide and Antenna Components using Plating on Plastics," in 7th Eur. Conf. on Antennas and Propag. (EuCAP), Gothenburg, Sweden, 2013, pp. 1812-1815.

978-1-5386-4483-6/18 $31.00 © 2018 IEEE

[16] M. Ahmadloo et al., "A Novel Integrated Dielectric-and-Conductive Ink 3D Printing Technique for Fabrication of Microwave Devices," in IEEE MTT-S Int. Micr. Symp. (IMS), Seattle, WA, USA, 2013, pp. 1-3.

[17] R. Zhu et al., "Rapid Prototyping Lightweight Millimeter Wave Antenna and Waveguide with Copper Plating," 40th Int. Conf. Infrared, Millimeter, and Terahertz waves (IRMMW-THz), pp. 1-2, 2015.

[18] S. Mufti et al., "3D Electrically Small Dome Antenna," in IEEE Antennas and Propag. Conf., Loughborough, UK, 2014, pp. 653-656.

[19] N. Arnal et al., "3D Multi-Layer Additive Manufacturing of a 2.45 GHz RF Front End," in IEEE MTT-S Int. Micr. Symp. (IMS), Phoenix, AZ, USA, 2015, pp. 1-4.

[20] W. Su et al., "3D Printed Reconfigurable Helical Antenna Based on Microfluidics and Liquid Metal Alloy," in IEEE Int. Sym. Antennas and Propag. (APSURSI), Fajardo, Puerto Rico, 2016, pp. 469-470.

[21] K. Y. Chan et al., "Monolithic MEMS T-type Switch for Redundancy Switch Matrix Applications," in IEEE Microwave Conference, 2008. EuMC 2008. 38th European, 2008, pp. 1513-1516.

[22] K. Y. Chan et al., "A novel RF MEMS Switch with Novel Mechanical Structure Modeling," Journal of Micromechanics and Microengineering, vol. 20, p. 015031, 2009.

[23] L. Gong et al., "RHCP Pattern-Reconfigurable Spiral Antenna Biased with Two DC Signals," Microwave and Optical Technology Letters, vol. 56, no. 7, pp. 1636-1640, 2014.

[24] L. Gong et al., "A Reconfigurable Spiral Antenna with Wide Beam Coverage," in IEEE Antennas and Propagation Society International Symposium (APSURSI), 2013, pp. 206-207.

[25] L. Gong et al., "Beam Steering Spiral Antenna Reconfigured by PIN Diodes," International Journal of Microw. and Wireless Technol., vol. 6, no. 06, pp. 619-627, 2014.

[26] Y. Yang et al., "MEMS-Loaded Millimeter Wave Frequency Reconfigurable Quasi-Yagi Dipole Antenna," in IEEE Asia-Pacific Microwave Conference Proceedings (APMC), 2011, pp. 1318-1321.

[27] G. I. Kiani et al., "MEMS Enabled Frequency Selective Surface for 60 GHz Applications," in IEEE APSURSI, 2011, pp. 2268-69.

[28] K. Y. Chan et al., "Novel Miniaturized RF MEMS Staircase Switch Matrix," IEEE Microwave and Wireless Components Letters, vol. 22, no. 3, pp. 117-119, 2012.

[29] H. U. Rahman et al., "Investigation of Residual Stress Effects and Modeling of Spring Constant for RF MEMS Switches," in IEEE Mediterrannean Microwave Symposium (MMS), 2009, pp. 1-4.

[30] K. Y. Chan et al., "RF-MEMS Switches with New Beam Geometries: Improvement ff Yield and Lowering of Actuation Voltage," in Device and Process Technologies for Microelectronics, MEMS, Photonics, and Nanotechnology IV, 2008, vol. 6800, p. 680026.

[31] M. G. Banciu et al., "Microstrip Filter Design Using FDTD and Neural Networks", Microwave and Optical Technology Letters 34 (3), 219-224.

[32] K. Y. Chan et al., "RF MEMS Millimeter-Wave Switchable Bandpass Filter," in IEEE International Wireless Symposium (IWS), 2013, pp. 1-4.

[33] K. Y. Chan et al., "RF MEMS Switch with Low Stress Sensitivity and Low Actuation Voltage," in IEEE Antennas and Propagation Society International Symposium, APSURSI'09., 2009, pp. 1-4.

[34] Y. Yang et al., "Experimental Proof for Pattern Reconfigurability of 60-GHz Quasi-Yagi Antenna," Microwave and Optical Technology Letters, vol. 57, no. 1, pp. 84-88, 2015.

[35] L. Gong et al., "Phase Correction of the Electric Field for a Dielectric Loaded Substrate Integrated Waveguide H-Plane Horn Antenna," Microwave and Optical Technology Letters, vol. 59, no. 3, pp. 584-588, 2017.

[36] K. Y. Chan et al., "Miniaturized RF MEMS Switch Cells for Crossbar Switch Matrices," in IEEE Asia-Pacific Microwave Conference Proceedings (APMC), 2010, pp. 1829-1832.

[37] H. Rahman et al., "Fabrication of RF NEMS Series Switch Using Surface Micromachining," in IASTED Intl. Conf. on Nanotechnology and Applications, 2010.

[38] L. Gong et al., "A Split-Ring Structures Loaded SIW Sectorial Horn Antenna," in IEEE-APS Topical Conference on Antennas and Propagation in Wireless Communications (APWC), 2015, pp. 349-350.

[39] K. Y. Chan et al., "60 GHz to E-Band Switchable Bandpass Filter," IEEE Microwave and Wireless Components Letters, vol. 24, no. 8, pp. 545-547, 2014.

[40] Y. Yang et al., "60GHz Pattern Reconfigurable Quasi-Yagi Antenna—Proof Through Computational Design," in Antenna Technology:" IEEE iWAT, 2014, pp. 53-56.

[41] L. Gong et al., "RF MEMS for Reconfigurable RF Front-End: Research in Australia," Advanced Materials Research, vol. 901, pp. 105-110, 2014.

[42] K. Y. Chan et al., "Low-Cost E-Band Lange Coupler with Vialess Load," Electronics Letters, vol. 51, no. 11, pp. 839-841, 2015.

[43] K. Y. Chan et al., "A Novel RF MEMS Switch with Novel Mechanical Structure Modeling," Journal of Micromechanics and Microengineering, vol. 20, p. 015031, 2009.

[44] K. Y. Chan et al., "Monolithic MEMS T-Type Switch for Redundancy Switch Matrix Applications," IEEE European Microwave Conference (EuMC), 2008, pp. 1513-1516.

[45] K. Y. Chan et al., "RF MEMS Millimeter-wave Reconfigurable Bandpass Filters," International Journal of Microwave and Wireless Technologies, 2014.

[46] K. Y. Chan et al., "Novel Beam Design for Compact RF MEMS Series Switches," IEEE APMC, 2007, pp. 1-4.

[47] L. Gong et al., "A Beam Steering Single-Arm Rectangular Spiral Antenna with Large Azimuth Space Coverage," IEEE WAMICON, 2013, pp. 1-4.

[48] X. Li et al., "Fabrication of Through via Holes in Ultra-Thin Fused Silica Wafers for Microwave and Millimeter-Wave Applications," Micromachines, vol. 9, no. 3, p. 138, 2018.

[49] L. Gong et al., "Substrate integrated waveguide H-plane horn antenna with improved front-to-back ratio and reduced sidelobe level," IEEE Antennas and Wireless Propagation Letters, vol. 15, pp. 1835-1838, 2016.

Wafer Level Packaging of GaN/Si SAW Band Pass Filters with Operating Frequencies above 5 GHz

Alina-Cristina Bunea*, Dan Neculoiu*,**, Adrian Dinescu*

*National Institute of R&D in Microtechnologies Bucharest, Romania
**"Politehnica" University of Bucharest, Romania
alina.bunea@imt.ro, dan.neculoiu@imt.ro

Abstract—This paper proposes a quasi-wafer level packaging approach of surface acoustic wave band pass filters (SAW-BPF) with operating frequencies above 5 GHz. A Poly(methyl methacrylate) (PMMA) cap is designed, fabricated and glued using an epoxy resin directly on the GaN/Si chip. First a coplanar waveguide transmission line is packaged and measured up to 65 GHz. Results show additional insertion losses of only 0.1 dB up to millimeter wave frequencies. A 5.6 GHz SAW-BPF is then packaged using the same approach. Measurement results show excellent properties of the packaged device for a temperature range between -150...+150°C.

Keywords—band pass filters; coplanar waveguide; gallium nitride; packaging; surface acoustic wave.

1. Introduction

Wafer Level Packaging (WLP) is quite a recent technique mainly employed for RF-MEMS encapsulation. Since bulk acoustic wave and surface acoustic wave (SAW) devices suffer from some of the same issues as RF-MEMS, WLP has become an attractive solution to protect the piezoelectric resonators from environmental factors [1].

The Surface Mount Device (SMD) packaging is a useful solution for frequencies below 3 GHz. In this approach, the SAW chip pads are connected to the package case through wire bonding and a top metal plate provides a completely sealed capsule. But it is a large size, high cost solution and at frequencies higher than 3 GHz the wires can affect the SAW filter performances [2].

One commercial solution used for SAW-based devices are chip-size packages (CSP), where the SAW device is bonded to a ceramic carrier and laminated with a polymer layer [3]. Cavities and through substrate vias are common requirements, as well as a metallized lid for hermetic packaging.

In a common WLP approach one or more layers (for ex. SU-8, TMMF, BCB) are used to form cavities over the SAW areas [4]. These approaches involve quite a lot of front-end processes (additional photolithography steps, sputtering, localized etching, etc.) and can cause issues during dicing.

There are two approaches to predict the microwave behavior of the packages: (i) electromagnetic modeling of the 3D complex configuration of the packaged device; (ii) measurement of the key parameters of prototype samples.

In this paper, a Poly(methyl methacrylate) (PMMA) cap is designed, fabricated and glued directly onto the GaN/Si chip. This quasi-WLP approach simplifies the packaging by eliminating wire bonding and integration in a bulky ceramic package.

For the proof of concept, a coplanar waveguide transmission line (CPW-TL) is designed and processed on a GaN/Si substrate (Fig. 1) and the package performance is evaluated up to 65 GHz.

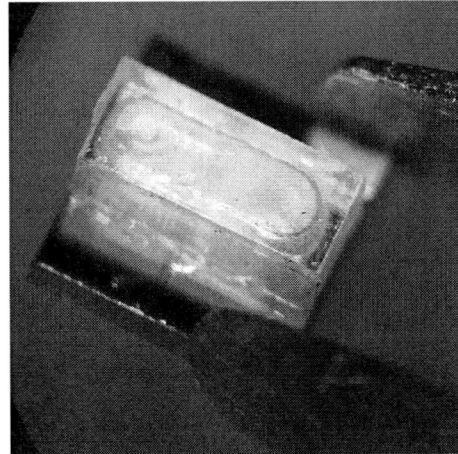

Fig. 1. Photo of the packaged GaN/Si chip

978-1-5386-4483-6/18 $31.00 © 2018 IEEE

A 5.6 GHz SAW Band Pass Filter (BPF) is packaged using this approach. The active area is of 0.9 x 2 mm², with a chip size of 2.8 x 2.8 mm². Measurement results for the packaged and un-packaged SAW-BPF are compared for temperatures between −150°C…+150°C.

2. 3D electromagnetic modeling

PMMA has several advantages: it is very cheap and widely available, easy to process and fares well with temperature (melting point of 160°C) and has a relative permittivity between 3.06-3.92 [5].

The 3D electromagnetic model of the PMMA cap, placed on top of a 2.8 mm CPW-TL (gap-signal-gap of 50-100-50 μm), is shown in Fig. 2 (a). The epoxy glue was also included in the model and a detail of the interface between the CPW-TL, the epoxy glue and the PMMA cap is shown in Fig. 2 (b). The model was developed in CST Microwave Studio. The exterior size of the cap is 2.7x1.4 mm², with a wall thickness of 0.2 mm and a total height of about 1 mm. The cavity was designed for processing with a 1 mm diameter cutting tool.

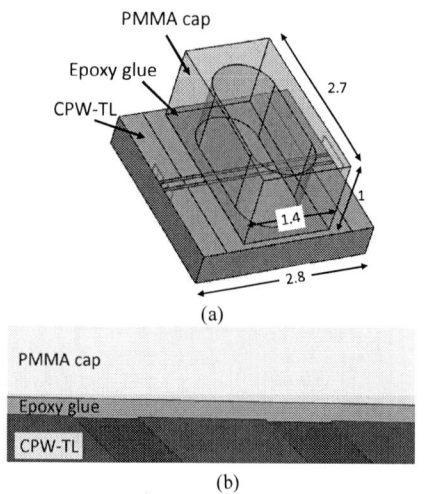

(a)

(b)

Fig. 2. (a) 3D electromagnetic model of PMMA cap mounted on a CPW-TL (dimensions given in millimters); (b) detail of the interface between the CPW-TL, the epoxy glue and the PMMA cap

The PMMA version under the trade name "Plexiglas" was used in the model. In the CST material library, Plexiglas has a permittivity of 3.6 and a resistivity of 5 kOhm·cm. The epoxy glue was simulated as a 5 μm thick layer covering

the metal traces of the CPW (Fig. 2 (b)). An estimate relative permittivity of 3.5 was considered in the model [6-7]. The electric field distribution at the CPW-TL – PMMA cap interface is shown in Fig. 3. The extension of the electromagnetic wave in the cap results in the localized increase of the line effective permittivity.

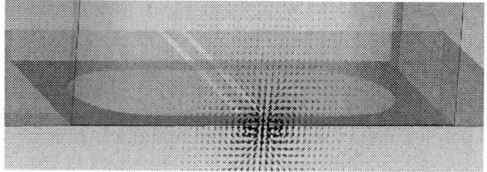

Fig. 3. Electric field distribution at the CPW-TL-PMMA cap interface

The effect of the thicknesses of the PMMA cap wall on the insertion losses of the CPW-TL is shown in Fig. 4. As expected, the PMMA wall thickness has only a slight influence on the insertion losses for values between 0.1-0.8 mm. The increase in losses and the ripple of the frequency response is due to the fact that under the PMMA walls the CPW-TL characteristic impedance decreases to a simulated value of 46.5 Ohm (outside the walls, the characteristic impedance is 51 Ohm). Considering the mechanical stability of the cap a processed wall thickness of 200 μm was used.

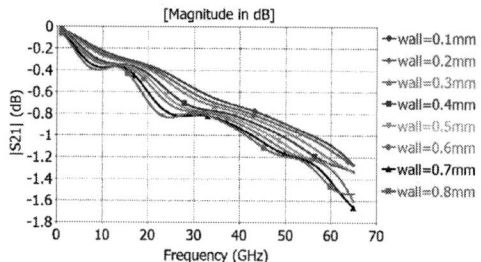

Fig. 4. Insertion losses as a function of the thickness of the PMMA cap wall

3. Fabrication and characterization of a CPW-TL with PMMA cap

A common type of PMMA (Plexiglas) was aquired and machined to the dimensions presented in the previous section. A resin epoxy glue (reaction product: bisphenol-A-(epichlorhydrin), hardener: 4-tert-Butylphenol; 1,3-Cyclohexanedimethanamine) was used to integrate the cap.

A 2.8 mm long CPW-TL was processed on a GaN/Si wafer acquired commercially from

EpiGaN in Belgium. The GaN layer was 1 μm thick and it was grown on a 0.5 μm thick (AlGa)N buffer. The silicon wafer had a thickness of 675 μm and a resistivity of ~5000 Ohm·cm. The chip was diced using a diamond saw.

The epoxy resin reaction product and hardener were mixed in a 1:1 ratio. The Plexiglas cap was pressed on the epoxy glue, to facilitate uniform layer formation, and then placed over the CPW-TL chip. The resin hardens in about 1 hour and was left at room temperature (23°C) overnight.

The test structure was characterized on wafer using a Vector Network Analyzer. A comparison of measurement (solid traces) and simulation (dotted traces) results for the Plexiglas cap integrated on the GaN/Si chip with the 2.8 mm long CPW-TL with (red traces) and without (black traces) the cap is shown in Fig. 5 (a) and (b), for the insertion losses and the reflection losses, respectively.

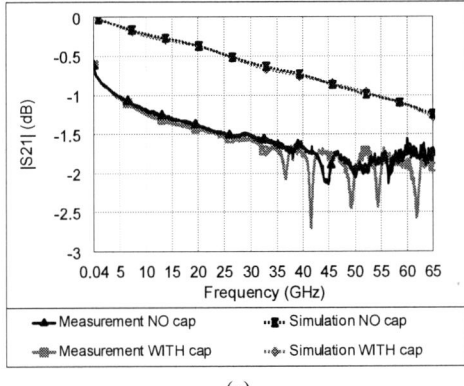

(a)

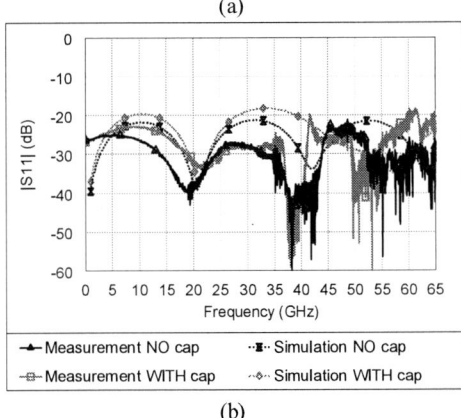

(b)

Fig. 5. Comparison of measurement and simulation results for the Plexiglas cap integrated on GaN/Si chip with CPW-TL: (a) S21 parameter; (b) S11 parameter

The agreement is fair, with measurement results showing insertion losses lower than 1.5 dB up to 25 GHz and lower than 3 dB up to 65 GHz for the standalone and for the packaged CPW-TL. The influence of the Plexiglas cap (fixed with resin glue) starts to show at frequencies over 35 GHz, where some resonances add up to 1 dB loss. Up to 30 GHz, the measured difference between the standalone 2.8 mm long CPW-TL and the CPW-TL integrated with the Plexiglas cap are lower than 0.1 dB. The measured reflection losses are better than – 20 dB up to 65 GHz.

4. Integration of a SAW-BPF with a PMMA cap

A SAW-BPF filter chip processed on a wafer obtained from the European provider EpiGaN, was integrated with the proposed Plexiglas cap and resin glue. The filter chips were characterized before and after gluing the Plexiglas cap and the results are summarized in Table 1. The effect of the package on the main filter parameters is negligible.

Table 1. Comparison of main parameters for packaged and un-packaged SAW BPF

Filter Parameter	Initial chip	Plexiglas package
Insertion losses	15.9 dB	16.44 dB
Return loss (S11)	10.5 dB	10.1 dB
Return loss (S22)	10.9 dB	10.7 dB
Out of band rejection	24 dB	23 dB
-3dB bandwidth	16.5 MHz	17.7 MHz
Central frequency	5.6035 GHz	5.6052 GHz

The packaged SAW-BPF was tested with temperature, with S parameter measurements performed at temperatures between −150°C and +150°C. The measurements were performed over the duration of about 5 hours using a cryostat equipped with a custom on-wafer probing station (as described in [8]). First the packaged chip was cooled down to −150°C and then gradually heated up to +150°C.

A comparison between the resonance frequencies of the standalone SAW-BPF chip and the packaged chip show very little influence of the package, with a total frequency shift only slightly larger (0.7 MHz) for the packaged device. The differences can be attributed to the measurement uncertainties (exact placement of probe tips, different calibration etc.). These results are highly encouraging suggesting that both the Plexiglas and the epoxy resin glue did not have adverse effects on the SAW resonators, such as contamination of the nanometric IDTs or mass loading.

Table 2. Comparison of central frequency for packaged and un-packaged SAW BPF at different temperatures

Temperature	SAW-BPF chip	Plexiglas package
- 150°C	5.6138 GHz	5.6137 GHz
+20°C	5.6035 GHz	5.6052 GHz
+ 150°C	5.5883 GHz	5.5875 GHz
Total difference	25.5 MHz	26.2 MHz

5. Conclusions

A PMMA cap was proposed as a low-cost packaging solution for a 5.6 GHz SAW-BPF. A CPW-TL processed on a GaN/Si substrate was used for the initial testing of the package.

The package was fabricated using common Plexiglas and was integrated with an epoxy resin on the CPW-TL chip. The package added losses lower than 0.1 dB up to millimeter wave frequencies.

Using this approach a 5.6 GHz SAW-BPF was also packaged on wafer. The total active area of the filter was 0.9 x 2 mm^2, with a chip size of 2.8 x 2.8 mm^2. On wafer measurements showed a very small influence of the package. The packaged device was measured between −150°C…+150°C and showed excellent frequency stability. No contamination or mass loading of the IDTs were observed.

Unlike traditional packages with thru substrate vias, the proposed package doesn't introduce any parasitics and keeps the chip area small. The packaged chip can easily be wire bonded or probed directly on-wafer. The extremely low losses added by the package up to millimeter wave frequencies as well as the low cost and easy manufacturing make this a viable solution for SAW-based band pass filter encapsulation.

Acknowledgements. This work was supported by the Romanian Ministry of Research and Innovation, under the project no. PN-III-P2-

2.1-PED-2016-0976, as well as by the European Space Agency (Contract No. 40000115202/15/NL/CBi). The authors would like to thank Mr. Mircea Pasteanu for the manufacturing and mounting of the Plexiglas caps.

References

[1]. Li Xiao and H. Li, "Chapter 7: Chip Size Packaging (CSP) for RF MEMS Devices", in *RF and Microwave Packaging II*, K. Kuang, R. Sturdivant (eds.), Springer International Publishing AG, 2017, pp. 83-97.

[2]. M. P. Goetz and C. E. Jones, "Modular integration of RF SAW filters", *IEEE Ultrasonics Symposium,* 2004, pp. 1090-1093 Vol.2.

[3]. F. M. Pitschi, C. Bauer, R. D. Koch and K. C. Wagner, "Approaches to wafer level packaging for SAW components", *2013 IEEE MTT-S International Microwave Symposium Digest (MTT)*, Seattle, WA, 2013, pp. 1-3.

[4]. J. H. Kuypers, S. Tanaka and M. Esashi, "Imprinted laminate wafer-level packaging for SAW ID-tags and SAW delay line sensors", in *IEEE Transactions on Ultrasonics, Ferroelectrics, and Frequency Control*, vol. 58, no. 2, pp. 406-413, February 2011.

[5]. https://plastics.ulprospector.com/generics/3/c/t/acrylic-properties-processing

[6]. L. Zong, S. Zhou, R. Sun, L.C. Kempel and M.C. Hawley, "Dielectric properties of crosslinking epoxy resins at 2.45 GHz", in *Conference Proceedings of the AIChE Annual Meeting*, pp. 1455, 2004

[7]. J. H. Wang, G. Z. Liang, H. X. Yan, S. B. He, "Mechanical and dielectric properties of epoxy/dicyclopentadiene bisphenol cyanate ester/glass fabric composites", in *Express Polym Lett,* 2008; vol. 2, No. 2, pp. 118–125.

[8]. Alina-Cristina Bunea, Dan Neculoiu, Adrian Dinescu, "GaN/Si monolithic SAW lumped element resonator for C- and X- band applications", *IEEE Asia Pacific Microwave Conference (APMC)*, pp. 1010-1013, 2017.

Metal-Insulator Transition in Monolayer MoS$_2$ for Tunable and Reconfigurable Devices

Martino Aldrigo*, Mircea Dragoman* and Diego Masotti**

*IMT-Bucharest, 077190 Voluntari (Ilfov), Romania
E-mail: {martino.aldrigo;mircea.dragoman}@imt.ro
**D.E.I.-University of Bologna, 40136 Bologna, Italy
*E-mail: diego.masotti@unibo.it

Abstract—In this paper, we show the electromagnetic design of a small patch antenna based on a molybdenum disulphide (MoS$_2$) monolayer, with an area of only 22mm^2, that exhibits high radiation efficiency and large tunability in microwaves at 10GHz thanks to a metal-insulator transition (MIT) induced by electrostatic gating. Furthermore, the MIT in MoS$_2$ is used to reconfigure a tunable carbon nanotube-based filter, conferring it different functionalities: low-pass, high-pass and band-pass around 2GHz, while its carbon nanotube varactors allow tuning the cutoff frequency or central frequency.

Keywords—Microstrip antennas, tunable filters, molybdenum compounds, carbon nanotubes.

1. Introduction

The applications of phase-transition properties, such as metal-insulator transition (MIT), in microwave devices (e.g. antennas and switches) is known since a long time due to vanadium dioxide (VO$_2$) – the best known MIT material studied in this respect since half-century [1-3]. VO$_2$ is still an actively researched material for microwave and photonic devices, although other MIT materials are being discovered. Atomically-thin transition metal dichalcogenides (TMDs) are the latest MIT materials under scrutiny, like molybdenum ditelluride (MoTe$_2$, MIT induced at room temperature by strain [4] or electrostatic gating [5]) and rhenium disulphide (ReS$_2$, MIT caused by temperature and electrostatic gating [6]). MoS$_2$ monolayer is the most studied atomically-thin material after graphene [7]. Molybdenum disulphide (MoS$_2$) monolayer is a direct bandgap semiconductor, with a bandgap of about 1.9eV. The bandgap of MoS$_2$ decreases as the number of monolayers increases, its bulk counterpart being an indirect semiconductor with a bandgap of about 1.2eV. MoS$_2$ monolayers have numerous applications in electronics and photonics [8]. Among the most important physical properties of MoS$_2$ are the two types of metal-insulator transitions induced by various parameters. The first type of phase transition is an irreversible or structural phase transition, accompanied by a structural transformation between semiconducting (2H) and metallic (1T) phases. This is possible via in situ scanning transmission electron microscopy, with re-arrangement of either sulfur or molybdenum atoms [9]. The second type of phase transition is a reversible one, encountered, for instance, in few-layer MoS$_2$ under an applied pressure. When the applied pressure is beyond 19Pa, the few layers of MoS$_2$ become metallic, a subsequent removing of the pressure rendering the few layers of MoS$_2$ semiconducting [10]. Another example of a MoS$_2$-specific reversible phase transition is induced via electrostatic gating in a field-effect device, such as a transistor [11]. In this case, the MIT in MoS$_2$ monolayers or few layers is evidenced by a crossover in the drain current versus gate voltage dependences at various temperatures. Moreover, at a certain temperature, close to room temperature, there is a jump of 4-5 orders of magnitude in the transistor trans-conductance, which indicates a jump from an insulating to a metallic state. The MIT effect in 2D semiconductor monolayers was not further extended to tunable/reconfigurable devices. In this paper, we show that MIT in 2D semiconductors has applications in electrically tunable devices at high frequencies, a topic which is gaining an increasing interest in microwaves, especially in view of the upcoming 5G and internet-of-things (IoT) telecommunication systems.

978-1-5386-4483-6/18 $31.00 © 2018 IEEE

2. MOS₂-based X-Band Patch Antenna

The first proposed device is a tunable microwave antenna, depicted in Fig. 1 together with the biasing technique (with a cross-section on the right). It is in fact a field-effect antenna, since the MIT is induced by an electrostatic gating, a few-layer graphene sheet playing the role of the gate. In Fig. 1, we have $L_S=W_S=10$mm, $L_P=4$mm, $W_P=5.5$mm and the substrate is made of high-resistivity silicon (HRSi) and silicon dioxide (SiO₂). Semiconducting antennas are not widespread in microwaves and millimeter waves, but integrated semiconducting antennas were reported at 44GHz [12]. Semiconducting antennas are used mostly in the optical spectrum, since metals exhibit high losses. For example, germanium antennas are used in the mid-infrared spectrum [13] and silicon antennas in the visible spectrum [14]. In the case of MoS₂, microwave measurements on samples of various thicknesses have shown low losses in this spectral range [15] (in deep contrast with graphene), so that a higher radiation efficiency is expected. Indeed, looking at the MoS₂ antenna configuration in Fig. 1, we see that graphene, which must not disturb the MoS₂ antenna through parasitic radiation, acts as a gate electrode, having the role to induce the metal-insulator transition in MoS₂ at a certain gate voltage. Initial simulations (carried out by using CST Microwave Studio®) have shown that when MoS₂ is in the metal state (at +5V bias), the antenna has a radiation efficiency of 54.6% at 10.12GHz, while the graphene electrode alone, with the same dimensions as the MoS₂ patch, has a radiation efficiency between 1%–2%. This way, the graphene electrode is not disturbing the MoS₂ radiation, whereas a very serious perturbation would have been produced if we had used a metal electrode as gate. We determined the electromagnetic radiation properties of the MoS₂ antenna considering the conductance values extracted from [11] for the monolayer MoS₂ and scaled to our antenna dimensions. The MoS₂ monolayer was modelled accordingly as a 6.5Å thick

dielectric or lossy metal, depending on its state. As regards the contact resistance between 2D materials and metals, it was taken into account by the electromagnetic simulator by solving Maxwell's equations using the Finite Integration Technique (FIT). The transient (time domain) solver is based on hexahedral meshes and it is noticeably efficient for most high frequency applications such as connectors, transmission lines, filters, antennas etc. At a gate voltage of -5V we have a conductivity $\sigma=1.26\times10^{-2}$S/m, which corresponds to an insulating state (i.e. MoS₂ antenna not radiating). Further, by increasing the gate voltage at -2V, we get a conductivity $\sigma=5.23\times10^3$S/m, at +0.84V we have $\sigma=6.28\times10^4$S/m, and finally at +5V we obtain $\sigma=1.88\times10^5$S/m. The conductivity retakes its original values as the gate voltage decreases down to -5V. This way, within a biasing span of only 7V, the conductivity value varies with seven orders of magnitude, indicating a reversible MIT behaviour. In the case of MoS₂ thin-films, we have observed also experimentally a MIT behaviour associated to a four-orders-of-magnitude change in conductivity [16], but this on-off ratio is much higher in the case of monolayer MoS₂.

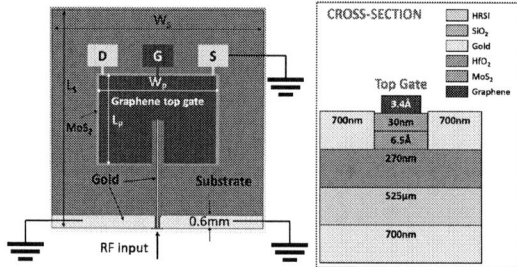

Fig. 1 Electromagnetic design of the monolayer MoS₂ patch antenna: (left) top view with electrode configuration for DC biasing. The letters D-G-S refer to drain-gate-source, respectively; (right) cross-section.

When the monolayer MoS₂ experiences a MIT transition, the antenna is switching from a poor radiation state, due mainly to its metallic connection, to a highly efficient radiation state. This behavior is illustrated in Fig. 2, which shows the simulated input resistance of the MoS₂ patch. From this figure it follows that increasing the voltage applied to the monolayer MoS₂, the matching

978-1-5386-4483-6/18 $31.00 © 2018 IEEE

to 50Ω in the X-band improves, until it reaches -30.71dB at 10.12GHz. It is apparent a 560-MHz frequency shift from 9.56GHz at -2V to 10.12GHz at +5V. At -5V, the patch cannot be matched to 50Ω and the radiation mechanism is basically due to the sole graphene top gate.

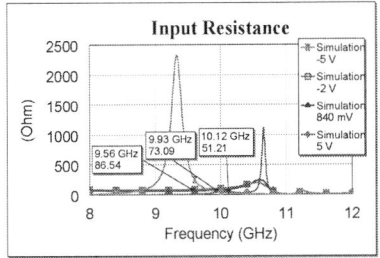

Fig. 2 Simulated frequency-dependent input resistance of the monolayer MoS₂ patch antenna.

Switching the DC voltage from -5V to +5V allows controlling the radiation properties as well. In Fig. 3 we show the antenna gain at 10GHz for different bias values, on the xz (ϕ=0°) plane: when the antenna is biased at -2V, the gain is -15.87dB and the radiation efficiency is 1.2%, whereas at +5V the gain is +2.41dB and the radiation efficiency is 54.6%, hence a difference of 18.28dB. These data confirms that the electromagnetic radiation can be turned on and off, the MIT acting as an ultrafast switch, with a switching time of few picoseconds. The characterisation of MoS₂ monolayers and the fabrication process of the antenna are currently under study.

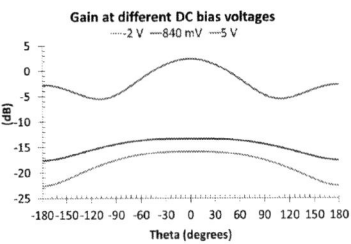

Fig. 3 Simulated gain at 10GHz of the monolayer MoS₂ patch antenna at different DC bias voltages.

3. MoS₂-based Microwave Filter

A second device revealing the importance of MIT in monolayer MoS₂ for high-frequency devices is the tunable and reconfigurable filter represented in Fig. 4. The filter is formed by meander inductors

and carbon nanotube (CNT) array-based varactors actuated by an applied DC voltage. The meander inductors have a measured value of about 1.4nH up to 5GHz. The CNT varactors modeling, fabrication and measurements are described in [17] and the references therein. The SPDT (W_{SPDT}=2.25mm and L_{SPDT}=4.2mm) consists of three MoS₂ monolayers arranged in the form of the Greek letter Π (the choice of 3 ports was done to provide a basic proof-of-concept). Each MoS₂ monolayer acts as a switch, where the on state is the conduction state and the off state corresponds to the insulator state. Denoting with 1, 2, and 3 the MoS₂ switches in front of the ports with the same numbers, we can implement a band-pass filter when switches 1 and 2 are on and switch 3 is off. When switches 2 and 3 are on and switch 1 is off, we have a high-pass filter. Finally, when switches 1 and 3 are on and switch 2 is off, we have a low-pass filter. This way, MoS₂ monolayers allow reconfiguring the filter to any desired architecture, while CNT-based varactors confer DC-controlled tunability. All the biasing circuits could be realized by creating vias in the HRSi/SiO₂ and designing the DC circuitry on the back of the substrate.

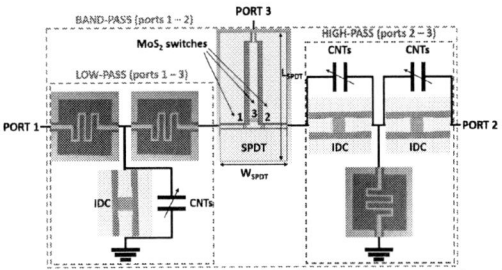

Fig. 4 Tunable and reconfigurable filter configuration based on MIT in monolayer MoS₂.

Fig. 5 shows filter's tunability, the characteristics being simulated for different values of the capacitance. From Fig. 5 it can be seen that the band-pass is 40MHz–2.43GHz, and $|S_{11}|$=-26.81dB at 1.74GHz. A very good isolation is observed at port 2 due to the MIT in the insulation state of MoS₂ switch #2. Considering that the measured capacitance of the CNT matrix is about 0.36pF when no bias is applied, by varying

978-1-5386-4483-6/18 $31.00 © 2018 IEEE   103

the applied DC bias voltage up to +6.5V, the CNT matrix capacitance should reach about 2.15pF [17], thus obtaining a tunable return loss with a frequency bandwidth Δf=2.94GHz. Similar results (not shown) are obtained for high-pass and bandpass filters in a 2GHz bandwidth, with an isolation well below -40dB up to 10GHz. In the case of high-pass filter, the bandpass is 1.8–7.1GHz; for the band-pass filter we have a central frequency of 2GHz and a 3-dB bandwidth of 1.3GHz. Losses at the central frequency are around 5dB.

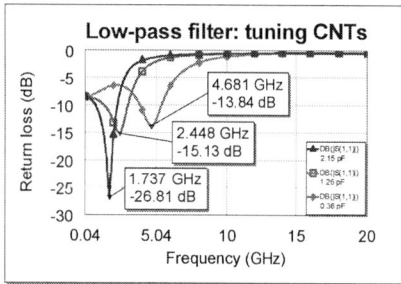

Fig. 5 Tunability of the low-pass filter.

Conclusions. In this paper, we have demonstrated that the MIT effect in monolayer MoS$_2$ is of utmost importance for high-frequency reconfigurable and tunable devices in modern communication systems. Thus, 2D materials are expected to play a major role in future 5G/IoT applications.

Acknowledgments. The author Martino Aldrigo thanks the financial support from the Romanian Ministry of Research and Innovation, via the project PN-III-P1-1.1-PD-2016-0535, contract no. 58/2018.

References

[1] S.D. Ha, Y. Zhou, A.E. Duwel, D.W. White and S. Ramanathan, *"Quick switch"*, IEEE Microwave Magazine 32, pp. 33–44, 2014.

[2] M. Dragoman, A. Cismaru, H. Hartnagel and R. Plana, *"Reversible metal-semiconductor transitions for microwave switching applications"*, Appl. Phys. Lett. 88, 073503 (2006).

[3] M.M. Fadlelmula, E.C. Sürmeli, M. Ramezani and T. Serkan Kasırga, *"Effects of thickness on the metal-insulator transition in free-standing vanadium dioxide nanocrystals"*, Nano Lett. 17, pp. 1762–1767, 2017.

[4] S. Song, H. Keum, S. Cho, D. Perello, Y. Kim and Y.H. Lee, *"Room temperature semiconductor-metal transition of MoTe2 thin films engineered by strain"*, Nano Lett. 16, pp. 188–193, 2016.

[5] J. Heo, et al., *"Reconfigurable van der Waals heterostructured devices with metal-insulator transition"*, Nano Lett. 16, pp. 6746–6754, 2016.

[6] N.R. Pradhan, et al., *"Metal to insulator quantum-phase transition in few-layered ReS2"*, Nano Lett. 15, pp. 8377–8384, 2015.

[7] M. Dragoman, D. Dragoman and I. Tiginyanu, *"Atomically thin semiconducting layers and nanomembranes: a review"*, Semiconductor Science and Technology 32, 033001 (2017).

[8] M. Dragoman and D. Dragoman, *"2D Nanoelectronics. Physics and Devices of Atomically Thin Materials"*, Springer (2017).

[9] T.-C. Lin, D.O. Dumcenco, Y.-S. Huang and K. Suenaga, *"Atomic mechanism of the semiconducting-to-metallic phase transition in single-layered MoS2"*, Nature Nanotechnology 9, pp. 391–396, 2014.

[10] A.P. Nayak, et al., *"Pressure-induced semiconducting to metallic transition in multilayered molybdenum disulphide"*, Nature Commun. 5, 3731 (2014).

[11] B. Radisavljevic and A. Kis, *"Mobility engineering and a metal-insulator transition in monolayer MoS2"*, Nature Materials 12, pp. 815–820, 2013.

[12] B.M. Brown, F.C. Jain and R. Bansal, *"A 44 GHz monolithic semiconductor antennas"*, Int. J. Infrared and Millimeter Waves 11, pp. 937–945, 1990.

[13] L. Baldassarre, et al., *"Midinfrared plasmon-enhanced spectroscopy with germanium antennas on silicon substrates"*, Nano Lett. 15, pp. 7225–7231, 2015.

[14] R. Regmi, et. al., *"All-dielectric silicon nanogap antennas to enhance the fluorescence of single molecules"*, Nano Lett. 16, pp. 5143–5151, 2016.

[15] M. Dragoman, A. Cismaru, M. Aldrigo, A. Radoi, A. Dinescu, and D. Dragoman, *"MoS2 thin films as electrically tunable materials for microwave applications"*, Appl. Phys. Lett. 107, 243109 (2015).

[16] M. Dragoman, A. Cismaru, M. Aldrigo, A. Radoi, and D. Dragoman, *"Switching microwaves via semiconductor-isolator reversible transition in a thin-film of MoS2"*, J. Appl. Phys. 118, 045710 (2015).

[17] M. Aldrigo, M. Dragoman, A.-C. Bunea, D. Neculoiu, S. Xavier and A. Ziaei, *"CNT-Based Microwave Filter for C and X-Band Applications"*, 47th European Microwave Conference (EuMC), pp. 308–311, 2017.

Design Aspects and Experimental Results on Broadband Monopole Dielectric Resonator Antenna

Stefan Simion and Sergiu Iordanescu*

Dept. of Electronics and Communications Engineering, MTA – Bucharest, Romania
E-mail: stefan.simion@yahoo.com
*National Institute for Research and Development in Microtehnologies, IMT – Bucharest, Romania
E-mail: sergiu.iordanescu@imt.ro

Abstract— A configuration of broadband dielectric resonator (DR) based monopole antenna covering the whole X–frequency band is analyzed numerically, as well as experimentally. The analysis is performed for two cases: with and without PTFE covering the part of the monopole antenna which is extended above the DR. It is shown that the presence of the PTFE may improve the return loss of the antenna. Experimental results for the return loss and antenna directivity are presented, too.

Keywords—monopole antenna; dielectric resonator; broadband antenna.

1. Introduction

Antennas consisting of quarter wavelength monopole on a large enough ground metal plane is a common solution for narrowband applications, when omnidirectional directivity as well as low elevation angle are required [1,2]. In order to extend the frequency bandwidth of this type of antennas for more than one octave, a few configurations based on dielectric resonators (DRs) have been proposed and analyzed [3–5].

In this paper, using commercial ring DR and SMA connector, a configuration of broadband DR based monopole antenna is analyzed and experimental results are presented. Compared to the antenna presented in [3,4], the part of the monopole antenna extended above the DR is covered by a PTFE cylinder, and its influence on the antenna return loss and frequency bandwidth is analyzed.

2. Description and Design Antenna

The configuration of the antenna analyzed in this paper is presented in Fig. 1. It consists of a monopole antenna loaded by a DR. The antenna is fed through a SMA connector. The central conductor of the SMA connector is passing through a metal disk (ground plane) and then it is extended into the upper half space up to the antenna height, h_a, equal to the quarter

wavelength, computed at the resonance frequency of the monopole antenna. The part of the monopole antenna extended above the DR is covered with a PTFE cylinder.

DR antenna has been designed in order to cover the whole X-band. Large frequency bandwidth of the antenna may be obtained if the $TM_{01\delta}$ resonance frequency of the DR and the resonance frequency of the monopole antenna are chosen such as to optimize the return loss, which must be better than a minimum imposed value (let say, the return loss greater than 10 dB), into the whole working frequency bandwidth of the antenna.

In this paper, a commercial ring DR with the outer and inner diameters equal to 6 mm and 1.9 mm, respectively, the height equal to 4.25 mm, the dielectric constant equal to 28 and the unloaded quality factor larger than 6000, has been used. For this ring DR, the resonance frequency on the $TM_{01\delta}$ obtained by using analytical means [6] is equal to 11.58 GHz, being in very good agreement with the simulated results obtained by using Ansoft's high frequency structure simulator (HFSS). The frequency bandwidth of the antenna may be optimized by varying the antenna height.

By simulation, it was observed that for any values of the antenna height, the return loss values of the antenna may be increased over a large frequency bandwidth, if the diameter of the monopole antenna is smaller than the central

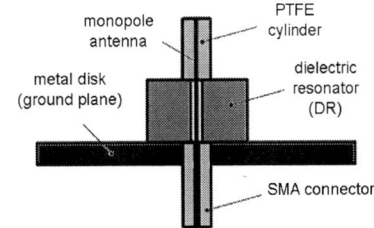

Fig. 1 Configuration of the dielectric resonator antenna analyzed in this paper.

conductor of the SMA connector, which is equal to 1.3 mm. In particular, the best results have been obtained if the diameter of the monopole antenna is equal to 1 mm; consequently, the diameter of the SMA – above the ground metal plate – was reduced from 1.3 mm to 1 mm. All results presented in the following have been obtained for the diameter of the monopole antenna equals to 1 mm.

The ground plane radius of the antenna has been chosen to 20 mm, and the influence of this geometrical parameter is analyzed in the last part of this section.

The height of the monopole antenna has been varied in order to find the highest values for the return loss, maximizing the frequency bandwidth. The simulation results obtained with HFSS, without and with PTFE cylinder, are presented in Fig. 2a,b, for h_a equals to 8.5, 9 and 9.5 mm. It is noticed that the return loss values are a little bit higher for the case when the upper part of the monopole antenna is covered by PTFE. Also, it is observed that, in both situations (without and with PTFE cylinder), as the height of the monopole antenna increases, the frequency bandwidth increases, but, unfortunately, the return loss in the middle of the frequency bandwidth decreases.

To maximize the return loss, as well as the frequency bandwidth, the antenna has been fabricated for $h_a = 9$ mm, assuming the upper part of the monopole antenna as being covered by PTFE cylinder. Therefore, the following simulations have been performed for $h_a = 9$ mm, only.

For a direct comparison between the two cases, with and without PTFE cylinder, the results for $h_a = 9$ mm are shown in Fig. 3. It is observed that for the case with PTFE cylinder, the frequency bandwidth is ~1GHz shifted to the lower frequencies, while the return loss in the middle of the frequency bandwidth is about 1.5 dB better compare to the case without PTFE cylinder. If necessary, the shift of the frequency bandwidth may be compensated by changing the geometrical dimensions of the DR. In our case, in spite of this frequency shift, the frequency bandwidth of the antenna still covers the whole X-band.

The influence of the ground plane radius, R_g, on the antenna return loss is presented in Fig. 4. It is noticed that for $R_g \gtrsim 20$ mm, no significant

improvements in terms of return loss are obtained. In order to minimize the size for the fabricated antenna, $R_g = 20$ mm has been chosen in this paper.

3. Simulation versus Experimental Results

The antenna with PTFE cylinder has been fabricated for $R_g = 20$ mm (the ground plane is

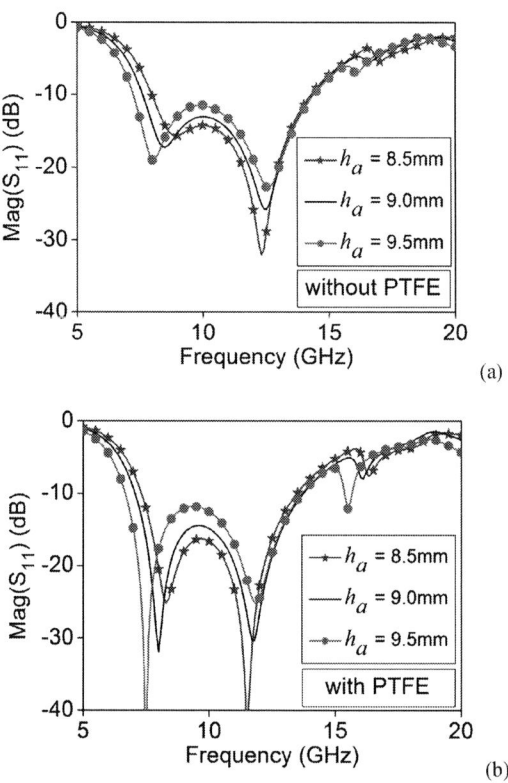

(a)

(b)

Fig. 2 Magnitude of S_{11} versus the frequency for different values of the monopole height, h_a, in the case of the antenna without (a) and with (b) PTFE cylinder (simulation results).

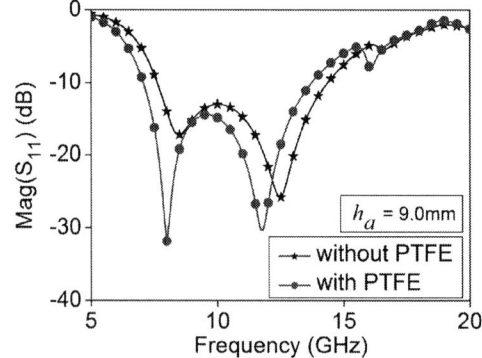

Fig. 3 Magnitude of S_{11} versus the frequency, with and without PTFE cylinder, when $h_a = 9$mm (simulation results).

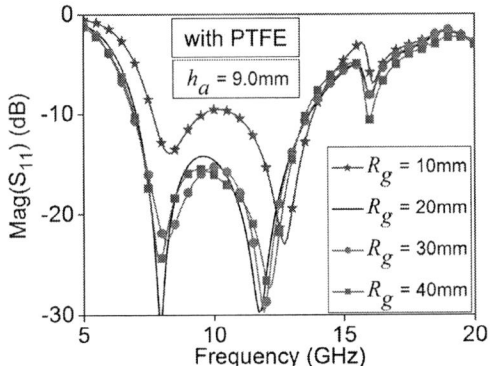

Fig. 4 Magnitude of S_{11} versus the frequency for different values of the metal disk (ground plane) radius, R_g, when h_a is equal to 9 mm (simulation results).

an aluminum disk of 3 mm thickness), $h_a = 9$ mm, the diameter of the monopole antenna is 1 mm, while the dielectric resonator has the parameters already mentioned in the previous section. A SMA female jack connector with 18 mm long extension of the central conductor, from MACOM company, has been used to feed the antenna. To protect the antenna against the whether condition, a rubber based dielectric material of ~200μm was used to coat the whole antenna (in this experiment, this dielectric material has a black color). A photograph of the fabricated antenna is presented in Fig. 5.

In the following, the experimental results for the return loss have been obtained using the Anritsu MS46122A Vector Network Analyzer. For the experimental results of antenna directivity, Agilent E8257D signal generator feeding the reference horn antenna and the Anritsu MS2668C spectrum analyzed connected to the antenna under test, have been used.

The return loss has been measured before and after the rubber based coat was deposited on the antenna. These experimental results are given in Fig. 6, showing that the rubber coat has no significant influence on the return loss. In the following, the all experimental results are given for the antenna covered by the rubber coat.

In Fig. 7, a comparison between the return losses obtained by simulation and measurement is presented. An excellent agreement between the expected and experimental results is obtained. The experimental frequency bandwidth for the return loss greater than 10 dB extends from ~7 GHz to ~13 GHz.

The antenna directivity versus the elevation angle has been simulated and measured for three

frequencies chosen into the antenna frequency bandwidth (vertical polarization). The results are presented in Fig. 8, when the frequency from the signal generator is equal to 8, 10 and 12 GHz. A good agreement between the expected and experimental results is noticed.

The maximum value for the simulated gain is 3 dB at the working frequency of 10 GHz, for elevation angles equal to ±45 deg., very closed to the experimental value. The simulated gain is greater than 2.5 dB, for the elevation angles

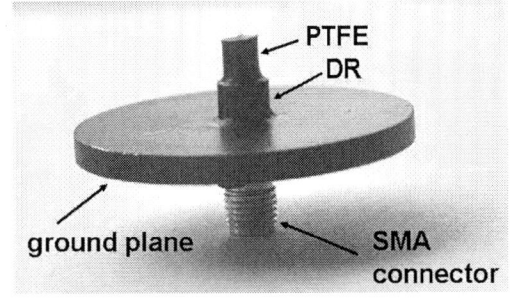

Fig. 5 Photograph of the fabricated antenna.

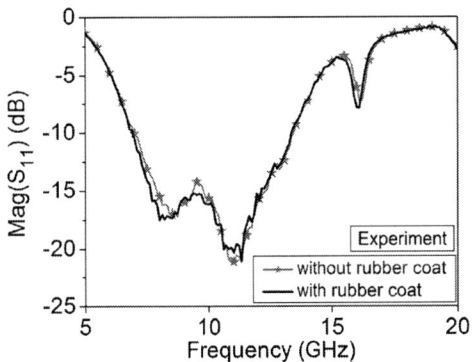

Fig. 6 Experimental magnitude of S_{11} versus the frequency, for the antenna with and without rubber coat of ~200μm.

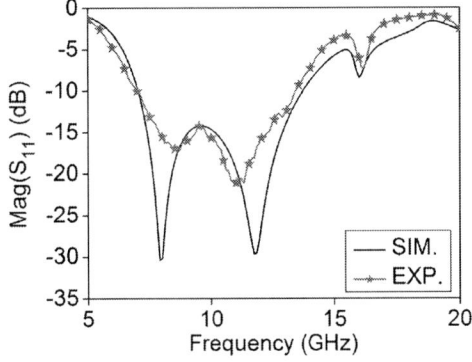

Fig. 7 Comparison between simulation and experiment, for the magnitude of S_{11} versus the frequency.

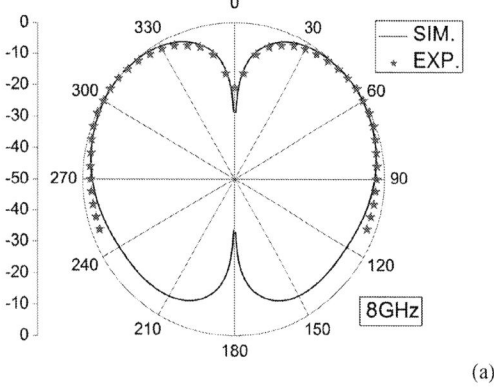

(a)

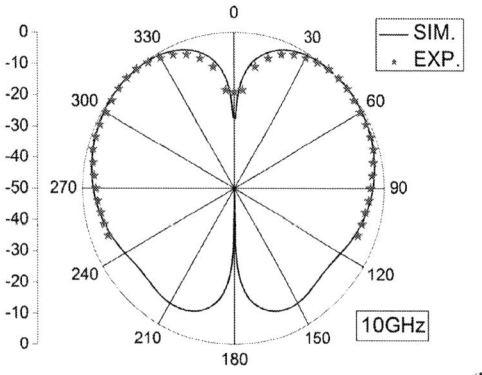

(b)

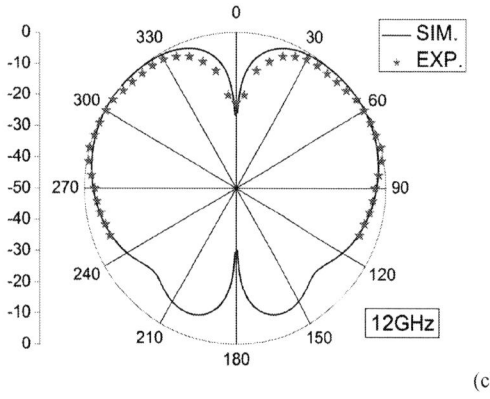

(c)

Fig. 8 Comparison between simulation (SIM.) and experimental (EXP.) results obtained for the antenna directivity versus the elevation angle, when the frequency is equal to 8 GHz (a), 10 GHz (b) and 12 GHz (c).

ranging from 30 to 60 deg. and from -30 to -60 deg., for any frequencies between 8 GHz and 12 GHz.

Conclusions

The return loss of the DR based monopole antenna may be improved, if the part of the monopole antenna which is extended above the

ring DR is covered by a dielectric cylinder (i.e. PTFE, in this paper). In the presence of this PTFE cylinder, the frequency bandwidth is slightly shifted to the lower frequencies. The experimental results for the return loss and antenna directivity are in good agreement with the expected ones obtained by simulation. For the fabrication of the antenna, commercial components (SMA connector and ring DR) have been used, only. Based on the same design rules, broadband antennas may be realized by using ring DRs having other dielectric constants values, by optimizing the DR's dimensions and the height of the monopole antenna, for imposed working frequency bandwidth.

References

[1] C. A. Balanis, *"Antenna theory, analysis and design – third edition"*, John Wiley & Sons, Inc., 2005.

[2] F. Pang, J. Yin, S. Li, J. Cui, Y. Zheng, C. Hu, L. Ma, X. Wang and Q. Chi, *"A low elevation angle conical beam antenna for CAPS-based vehicle monitoring system"*, Progress in Electromagnetics Research Letters, **74**, pp. 17–22, 2018.

[3] M. Lapierre, Y. M. M. Antar, I. Ittipiboon and A. Petosa, *"Ultra wideband monopole/ dielectric resonator antenna"*, IEEE Microwave and Wireless Components Letters, **15**(1), pp. 7–9, January 2005.

[4] D. Guha, Y. M. M. Antar, A. Ittipiboon, A. Petosa and D. Lee, *"Improved design guidelines for the ultra wideband monopole-dielectric resonator antenna"*, IEEE Antennas and Wireless Propagation Letters, **5**, pp.373–376, 2006.

[5] S. Keyrouz and D. Caratelli, *"Dielectric resonator antennas: basic concepts, design guidelines, and recent developments at millimeter-wave frequencies"*, International Journal of Antennas and Propagation, 2016.

[6] R. K. Mongia and P. Bhartia, *"Dielectric resonator antennas – A review and general design relations for resonant frequency and bandwidth"*, International Journal of Microwave and Millimeter-Wave Computer-Aided Engineering, **4**(3), pp. 230–247, 1994.

978-1-5386-4483-6/18 $31.00 © 2018 IEEE

Permittivity Characterization Using a Double-Sided Parallel-Strip Line Resonator

Dusan A. Nesic* and Ivana Radnovic**

*Centre of Microelectronic Technologies, Institute of Chemistry, Technology and Metallurgy, University of Belgrade,
Njegoseva 12, Belgrade, Serbia
nesicad@nanosys.ihtm.bg.ac.rs
**Institute IMTEL Komunikacije a.d, Bulevar Mihajla Pupina 165b, New Belgrade, Serbia
ivana@insimtel.com

Abstract - The paper introduces the new type of a microwave permittivity sensor with an open stub realized as a double-sided parallel-strip line without substrate. It can be totally immersed into the measured material and obtains high sensitivity of the resonant frequency nearly proportional to ratio of square roots of the dielectric constants of the measured materials.

Index terms - Microwave sensor, Microstrip, Double-sided parallel-strip line, Permittivity measurement

1. Introduction

Microwave sensors are being more and more used as sensing components in many applications [1]. They are sensitive, able to survive overdrives and their signal can be directly transmitted over a distance [2]. One type of microwave sensors is a resonant sensor. Great advantage of this type of a sensor is its principle of operation that is based on the resonance frequency and is generally immune to the environmental noise. Besides, the use of the planar technology enables an easy, fast and inexpensive fabrication. Advantages of the planar microwave fabrication process finds wide application in planar structures such as microstrip, CPW and strip line [1,3]. Microwave microstrip resonator is a good choice for sensors [4-9].

The location of the Material Under Test (MUT) is usually above the microstrip line [4,9], under the pattern etched in the microstrip ground plane [5,6] or above the coupling area of the coupled microstrip structures [7,8]. However, there is one main problem - the fact that the sensitivity depends on the extent of the field penetration inside the MUT [3]. In all three mentioned positions of the MUT only a part of the field lines are inside the MUT because the field lines in microstrip are predominantly concentrated within the substrate area (ε_{r-eff} - effective dielectric constant).

It is obvious that locating the MUT inside the substrate results in a higher sensitivity [3]. Still, one can insert the MUT (i.e. fluid) through the substrate [10, 11]. That solution is difficult especially in case of thin substrates and is good only for microfluids. Another solution can be double-sided parallel-strip line printed on dielectric pipes for testing fluids [12]. It is good for pipes but not for immersing stub into a fluid. Also resonance is on low frequencies and open stubs are too long (around 25 cm). Some analogy to coaxial open stub is given in [4]. Its resonance is also on low frequencies and open stubs are too long (around 33 cm), not practical for number of applications. Microstrip sensor for immersing into a fluid is presented in [5]. It has disadvantages of construction and protection problems during measurements. One solution to problems in [5] is in substrate integrated waveguide (SIW) technology [13]. Disadvantage of [13] is SIW technology with many vias.

In this paper a new type of a modified microstrip as a $\lambda/4$ - open stub resonant sensor is introduced. It is good for immersing into a fluid and has a short open stub less than 20 mm. The whole structure is in the form of a double-sided parallel-strip line [14,15]. Double-sided parallel-strip line technology is chosen in order to obtain such sensing structure. The pair of two symmetrical metal strips without a substrate represents the sensing part of the stub. Effective dielectric constant is then near equal to the dielectric constant of the MUT ($\varepsilon_{r-eff} \cong \varepsilon_{r-mut}$).

2. Design and Fabrication

The structure is designed in printed planar technology as a double-sided parallel-strip line T-junction with an open stub without a substrate. The photos of the both sides of the fabricated structure are in Fig. 1. Rigid metal strips, 20 mm long, 4.5 mm wide and 0.3 mm thick, are bonded on the 4.75 mm long stubs (A in Fig.1) on the both sides of the substrate. Free parts of the rigid metal strips are forming a 15.25 mm long part of the open stub without any substrate (B in Fig.1).

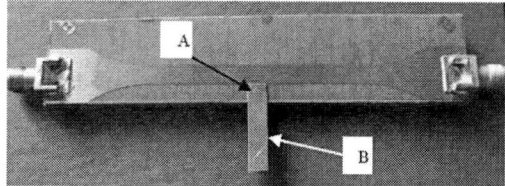

a) Bottom side.

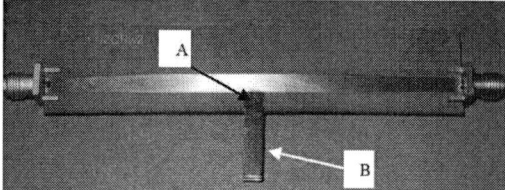

b) Top side.

Fig. 1. The sensor with SMA connectors and bonded strips on both sides: A – Part of the metal strip bonded on the substrate; B - Part of the metal strip without substrate.

The main part of the proposed structure is realized on CuClad 217 substrate (ε_r = 2.17 and h = 1.143 mm) as a double-sided parallel-strip line T-junction. Layouts of the bottom and the top parts of the structure are presented in Fig.2 denoted by gray and black color, respectively. The structure consists of a 4.5 mm wide 50 Ω-double-sided parallel-strip line with a double-sided parallel-strip line open stub in the middle, 4.75 mm long and 4.5 mm wide, Fig.2. The part of the stub printed on the dielectric substrate serves for bonding the rigid metal strips (A in Fig.1) on both sides. There is also a transition (BAL-UN) [15] to unsymmetrical (conventional) 3.5 mm width 50 Ω-microstrip line at its both ports.

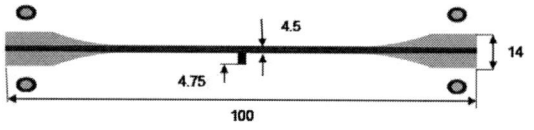

Fig. 2. Bottom (gray) and top metallization (black) of the proposed double-sided parallel-strip line T-junction with BAL-UN transitions to the conventional microstrip.

Only the part B, Fig. 1, can be immersed into the MUT. Ideal sensitivity, ratio of square roots of the dielectric constants, can be nearly achieved.

3. Simulation

Simulations were carried out using 3D WIPL-D

Microwave Pro program package [16]. Simulated $\varepsilon_{r\text{-mut}}$ related to the resonant frequencies is presented in Fig. 3. and Fig. 4. Simulation results are obtained for two specific ranges of $\varepsilon_{r\text{-mut}}$: from 1.5 to 3 (oils) and from 20 to 80 (water-ethanol mixture).

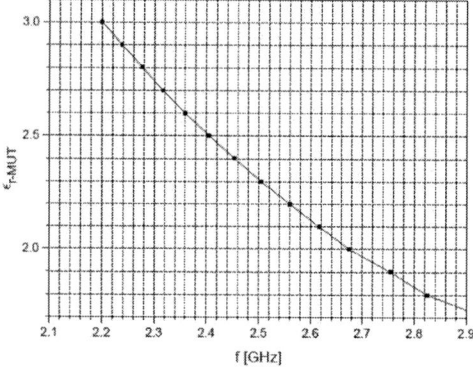

Fig. 3. The first range $\varepsilon_{r\text{-mut}}$ simulation vs. the resonant frequencies.

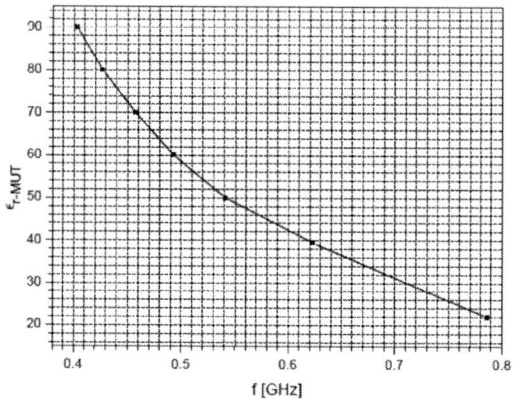

Fig. 4. The second range $\varepsilon_{r\text{-mut}}$ simulation vs. the resonant frequencies.

4. Measurement

The measurements are performed in the steady state and in the temperature around 300 K in order to obtain stable results. Humidity is not included. Measurement setup with the sensing open stub and the container with the MUT are presented in Fig.5. The container, Fig. 5, inserts itself a negligible frequency shift. Transmission is measured using the Agilent Technologies Network Analyzer N5227A.

Several materials were tested: air, gasoline (medical), paraffin oil and sunflower oil. Next materials were water and ethanol. The measured S_{21} parameters in both cases are presented in Fig. 6, 7 and 8, respectively.

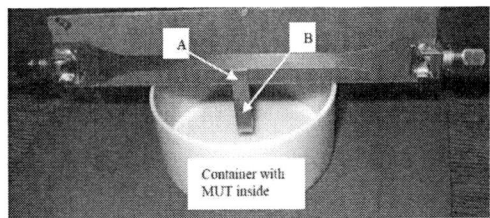

Fig. 5. Mesurement setup with the container: A-Part on the substrate; B-Part without the substrate to be immersed into the MUT.

According to the diagrams presented in Fig. 3. and Fig. 4. results are shown in Table 1. All results reasonably match values from the available references [17-21].

Frequency shift between two measured materials is nearly proportional to the ratio of square roots of their relative dielectric constants ε_r for the both ranges. The ratio between the air and the water resonant frequencies is around 8.3 and the ratio between square roots of the water and the air dielectric constants is around 8.5. For gasoline it is 1.36 and 1.38 respectively.

1. CONCLUSION

The new type of a microwave resonant sensor is realized as a T-junction with an open stub as a sensing part. The sensing part of the stub represents a pair of two metal strips in the form of a double-sided parallel-strip line without a substrate. The absence of the substrate enables each stub strip to be totally surrounded by the MUT.

The proposed sensor is fabricated in planar technology with no dimension tolerance problem: narrowest line width is 3.5 mm much wider than common tolerances of about 0.030 mm. There are no technological processes such as vias, air-bridges, defected ground structures (DGS) or many vias as in substrate integrated waveguide (SIW). The only additional process is bonding of the rigid metal strips on the line on the substrate.

The sensing stub can be simply immersed into the MUT without additional preparing or auxiliary structures like cavity. The sensor is tested in two dielectric constant ranges: oils (1.5-3) and ethanol-water mixture (20-80). It also uses two frequency ranges: around 2.5 GHz and below 1 GHz respectively. In both cases frequency shift between two measured materials is nearly proportional to ratio of the square roots of their relative dielectric

constants $\varepsilon_{r\text{-MUT}}$. All results reasonably match values from the available references. The sharp stopband always exists. The difference between materials can also be in the second resonance.

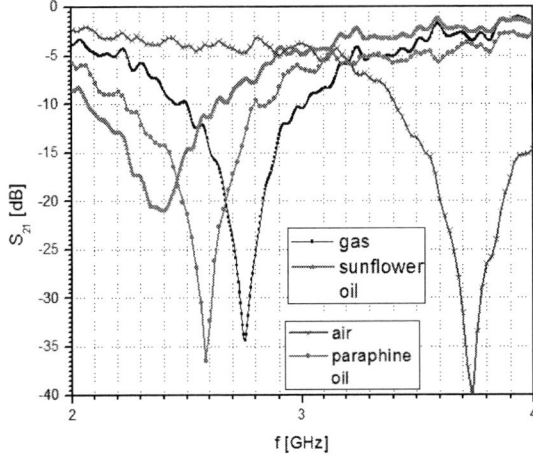

Fig. 6. Measured S_{21} coefficient of several different MUT.

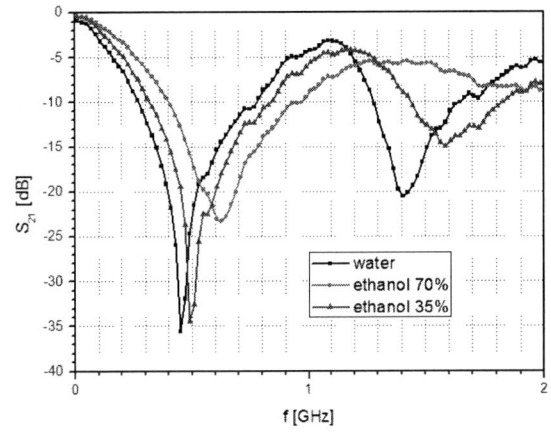

Fig. 7. Measured S_{21} coefficient of water and Ethanol.

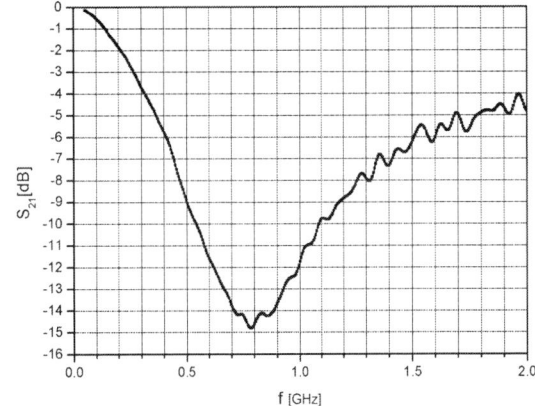

Fig.8. Measured S_{21} coefficient of Ethanol 96%.

Table 1.

MUT	Measured ε_r (f_R)	Ref. ε_r		(error)
Air	1 (3.74 GHz)	1		(<1 %)
Gasoline	1.90 (2.755 GHz)	1.92	[18]	(1 %)
Paraffin oil	2.16 (2.584 GHz)	2.2	[19]	(1.8 %)
Sunflower oil	2.5 (2.4 GHz)	2.56	[20]	(2.3 %)
Water	73 (0.449 GHz)	76	[21]	(4 %)
Ethanol 35%	61 (0.49 GHz)	58.9	[17]	(3.6 %)
Ethanol 70%	37 (0.629 GHz)	39.5	[17]	(6.3 %)
Ethanol 96%	22 (0.787 GHz)	22	[17]	(<1 %)

It can be used in installations. For example if 1 mm of the open stub remains above the measured fluid the frequency shifts are not more than 1 %.

ACKNOWLEDGMENT

The authors thank M. Pesic, N. Tasic, Lj. Radovic, N. Popovic and P. Manojlovic, Ins. IMTEL and professor M. Potrebic, Univ. Belgrade, School of EE. Funded by Serbian Ministry of Education and Science, project TR 32008.

References

[1] S. Dey, J.K. Saha, and N.C. Karmakar, "Smart Sensing", *IEEE Microwave Magazine*, November 2015, pp. 26-39

[2] J. Polivka, "An Overview of Microwave Sensor Technology", April 2007, *High Frequency Electronic*, pp.32-42

[3] K. Saeed, M. F. Shafique, M. B. Byrne and I. C. Hunter (2012). Planar Microwave Sensors for Complex Permittivity Characterization of Materials and Their Applications, *Applied Measurement Systems*, Prof. Zahurul Haq (Ed.), InTech, https://www.intechopen.com/books/applied-measurement-systems

[4] A. Hoog, M.J.J. Mayer, H. Miedema, W. Olthuis, F.B.J. Leferink and A. van den Berg, "Modeling and simulations of the amplitude–frequency response of transmission line type resonators filled with lossy dielectric fluids", *Sensors and Actuators A*, 216 (2014) 147–157

[5] C. Liu and Y. Pu, "A Microstrip Resonator With Slotted Ground Plane for Complex Permittivity Measurements of Liquid", *IEEE Microwave and Wireless Components Letters*, vol. 18, no. 4, 2008, pp. 257-259

[6] C.-S. Lee and C.-L. Yang, "Complementary Split-Ring Resonators for Measuring, Dielectric Constants and Loss Tangents", *IEEE Microwave and Wireless Components Letters*, Vol. 24, No. 8, 2014, pp. 563-565

[7] A. A. Abduljabar, D. J. Rowe, A. Porch, and D. A. Barrow, "Novel Microwave Microfluidic Sensor Using a Microstrip Split-Ring Resonator", *IEEE Transactions on Microwave Theory and Techniques*, Vol. 62, No. 3, 2014, pp. 679-688

[8] M. T. Jilani, W. P. Wen, L. Y. Cheong, M. Z. U. Rehman, and M. T. Khan, "Determination of Size-Independent Effective Permittivity of an Overlay Material Using Microstrip Ring Resonator", *Microwave and Optical Technology Letters*, Vol. 58, No. 1, 2016, pp. 4-9

[9] Lescopa, F. Galléeb, S. Riouala, "Development of a radio frequency resonator for monitoring water diffusion in organic coatings", *Sensors and Actuators A*, 247 (2016) pp. 30–36

[10] L. Le Cloirec, A. Benlarbi-Delaï and B. Bocque, "New concept of RF functions by microfluidic coupling", *Microwave and Optical Technology Letters*, Vol. 48, Iss.10, 2006, pp. 1912–1916

[11] D.L. Diedhiou, R. Sauleau, and A.V. Boriskin, "Microfluidically Tunable Microstrip Filters", *IEEE Transactions on Microwave Theory and Techniques*, vol. 63, no. 7, 2015, pp. 2245-2252

[12] M. A. Karimi, M. Arsalan and A. Shamim, "Low cost and pipe conformable microwave-based water-cut sensor", *IEEE Sensors Journal*, vol. 16, iss. 21, 2016, pp. 7636 – 7645

[13] C. Liu and F. Tong, An SIW Resonator Sensor for Liquid Permittivity Measurements at C Band, *IEEE Microwave Wireless Components Letters*, Vol. 25, No. 11, 2015, pp. 751-753

[14] S.-G. Kim and K. Chang, "Ultrawide-Band Transitions And New Microwave Components Using Double-Sided Parallel-Strip Lines", *IEEE Transactions on Microwave Theory and Techniques*, Vol. 52, No. 9, September 2004, p. 2148

[15] J.-X. Chen, C.-H. K. Chin and Q. Xue, "Double-Sided Parallel-Strip Line With an Inserted Conductor Plane and its Applications", *IEEE Transactions on Microwave Theory and Techniques*, Vol. 55, No. 9, 2007 p. 1899

[16] 3D WIPL-D Microwave Pro program package

[17] A. Megrichel, A. Belhadj and A. Mgaidi, "Microwave Dielectric Properties of Binary Solvent Water-Alcohol, Alcohol-Alcohol Mixtures at Temperatures Between -35°C and +35°C and Dielectric Relaxation Studies", *Mediterranean Journal of Chemistry* 2012, 1(4), 200-209

[18] A. K. Venna, Nasiinuddin , R. K. Garg, and A. S. Oinar, "Suspended Microstrip Patch Resonator Sensor for Determination of Complex Dielectric Constant of Liquid and Paste", 2003 Antennas and Propagation Society International Symposium, vol. 4, pp. 651- 654, 2003.

[19] https://www.engineeringtoolbox.com/relative-permittivity-d_1660.html

[20] J. Vrba and D. Vrba, Temperature and Frequency Dependent Empirical Models of Dielectric Properties of Sunflower and Olive Oil, Radioengineering, Vol. 22, No. 4, 2013 pp. 1281-1287

[21] Martin Chaplin, Water and Microwaves, http://www1.lsbu.ac.uk/water/microwave_water.htm

Session M&M

MICROSENSORS AND MICROSYSTEMS

978-1-5386-4483-6/18 $31.00 © 2018 IEEE

Low-temperature Packaging Methods as a Key Enablers for Microsystems Assembly and Integration

S. Stoukatch*, F. Dupont*,**, M.Kraft***

*Microsys lab, Department of Electrical Engineering and Computer Science, Liege University, Liege Scientific Park, Rue du Bois Saint Jean 15-17, B-4102, Seraing, Belgium
serguei.stoukatch@uliege.be (corresponding author)
**fff.dupont@uliege.be
***He is currently with Department of Electrical Engineering (ESAT) – MICAS, Microelectronics and Sensors, University of Leuven (KUL), Leuven, Belgium
***michael.kraft@kuleuven.be

Abstract—The paper reports on assembly and integration of MS (microsystems) into fully functional system. We show that among varieties of assembly techniques and methods commonly used for IC, some can be successfully used also for the assembly of microsystems. MS are specifically sensitive to thermal exposure that can occur during the assembly and integration process.

Keywords—Low-temperature assembly methods; MS first-level packaging; MS integration.

1. Introduction

MS (microsystems) or micro-electro-mechanical systems (MEMS) have become common in many applications; it describes miniaturized systems that, unlike as IC (integrated circuits), comprises non-electronic components. There are large varieties of MS [1], they can be classified by functionality, for example sensing, actuation, bidirectional transductions, etc., or by application: humidity, pressure, inertial chemical and biosensing, RF switches and resonators, optical, imaging sensors and power MEMS, etc. Many materials and a wide variety of technologies [2] are used to fabricate MS such as silicon, poly-silicon, piezoresistive and piezoelectric, glass, metals, plastics. As a result of that, all MS are very different and they only can be classified in groups based on specific criteria, and unlike as IC, require a specific packaging solution. Generally speaking, packaging technology and methods have several challenges, the most crucial one are miniaturization, cost reduction, performance, thermal management and reliability. Meanwhile, MS packaging dictates specific requirements, among them the most crucial is that the packaging processes must not damage and/or deteriorate integrity and compromise performance of the MS. The most demanding is a first-level MS packaging. There are several physical effects that commonly have to be considered during the packaging process: mechanical impact, thermal and chemical exposure and ultra-sound impact. They potentially can compromise MS integrity, and, in the ideal case, must be excluded fully or at least minimized and then used with special caution.

One of the most important factors that potentially can affect MS is exposure to thermal treatment, that includes factors such as temperature and duration of thermal treatment. In this paper, we will discuss effect of thermal exposure on mechanical integrity of the MS die.

2. Microsystems Die

The microsystem discussed here was processed on 8" silicon wafer using conventional silicon processing techniques, it was thinned to 300 μm thickness and was separated into individual sensor dies of lateral dimensions 8.2 x 8.4 mm² and sequentially released. The MS is a gyroscope, designed for rate of turn detection for applications in motion detection. The top and bottom views of an individual sensor die are depicted in figure 1a and 1b.

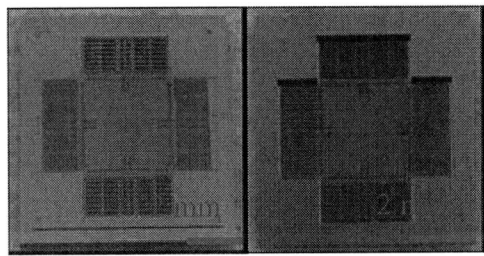

Figure 1: MS die top view (1a, left) and bottom die view (1b, right).

The individual die was robust enough to survive transportation and we did not observe any

978-1-5386-4483-6/18 $31.00 © 2018 IEEE

damage upon arrival. The assembly processing and integration was performed at Microsys Lab, Liege University ISO7 (class 10,000) clean room.

A schematic cross-sectional view of the sensor die is depicted in figure 2.

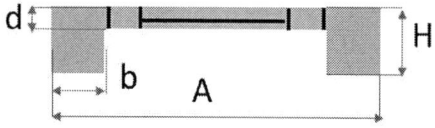

Figure 2: Cross-sectional view of the sensor die.
Legend to figure 2:
H is a total thickness of the die, H=300 µm
d is a thickness of a membrane, d=50 µm
A is lateral dimension of the sensor die, 8.2x8.4 mm2.
B is a width of the frame, B=1 mm.

The MS die sensitive area comprises an array of capacitive comb structures; one cell is depicted in figure 3.

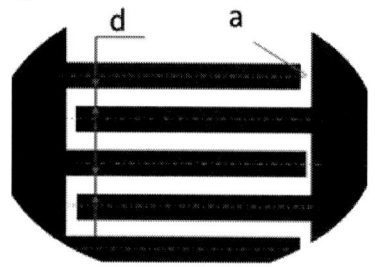

Figure 3: Schematic of the sensing comb element structure, top view.
Legend to figure 2:
d is a distance between two neighboring beams of the comb structure, d=10 µm
a is beam deflection from initial position, 0 means that there is no deflection.

In the nominal position, there is no beam deflection from the initial position (a=0 µm), and the distance (d) between two neighboring beams is 10 µm. Once electrostatic actuating is applied to the MS, the beams will deflect from the initial position; the distance between two neighboring beams will decrease; its value depending on the actuating voltage. The distance change can be detected capacitively, sequentially converted to an electrical signal and related to a value of the displacement. After the MS calibration, it is possible to qualitatively characterize the displacement.

The issue we have discovered during the assembly process development is that during the processing that the MS undergoes, the beams

deflect from its initial position without actuation voltage. Based on that, we concentrated our efforts on development and implementing an assembly process that result in no or little beam deflection from the initial position.

3. Assembly methods consideration

As it was described and discussed in [3] MS packaging and integration is, unlike IC CMOS packaging, a rather complex and delicate process. Specifically, MS gyroscopes [2] are very sensitive to thermal-mechanical stresses that are commonly taking place during the device assembly. The main source of stress are exposure to an elevated temperature (above 150°C) and stress caused by the die attach adhesive. Typically, the die attach adhesive is an epoxy based material that is relatively stiff (E Modulus >5 GPa) [4], and has a coefficient of thermal expansion (CTE) (CTE $\geq$ 10 1/K, [4]) that is larger than for silicon (CTE=3 1/K, [4]). The adhesive upon curing usually shrinks and due to the CTE mismatch between the adhesive and the silicon, mechanical stress develops in the MS silicon body. The mechanical stress will translate into a displacement of suspended mechanical parts from their nominal position.

Meanwhile the standard adhesive that used for die mounting requires for curing at least temperature of 150°C applied for 1 hour. The die mounting process is followed by wire bonding, this is to provide an electrical interconnection between the die and the package. Typical gold and copper wire bonding techniques require also an elevated temperature of 150°C and above during the bonding process.

The existing and commonly known packaging processes must be examined if they cause any effect on MS die. The required process must result in no damage to the MS, and thus suitable processes must be selected.

4. Assembly Process Flow

The first step of the assembly process is mounting the MS die into a corresponding package. To build a prototype we have selected LCC04438 package (from Spectrum Semiconductors); the package inner cavity is 12 x 12 mm². As interconnect material we have chosen an epoxy based, silver filled conductive die attach adhesive Ablestik 84-1, from Henkel [5]. A standard cure schedule is 1 h, at 150°C, and an alternative one is 2 h, at 125°C. The most important physical properties are listed in table 1.

978-1-5386-4483-6/18 $31.00 © 2018 IEEE

Table 1: Physical properties of Ablestik 84-1 [5].

Parameters	Ablestik 84-1 LMIT
Service temperature range, °C	-50..+150
Weight loss at 300°C, %	0.16
Viscosity at 25° C, cps	22000
E Modulus, GPa	8
Tensile strength , psi/MPa	1200/8.23
Hardness, Shore D	80
Tg (glass transition T) , °C	103
CTE, 1/K	50/150 (below and above Tg)
Thermal conductivity, W/m*K	2.4
Volume Resistivity, ohm-cm	0.0005

As we noticed during testing neither a standard no an alternative curing schedule (2 h, at 125°C) is suitable for the MS, both of them result in detectable beam displacement from the initial position. Figures 4 and 5 illustrate the beam displacement from the initial position as highlighted by a black contour.

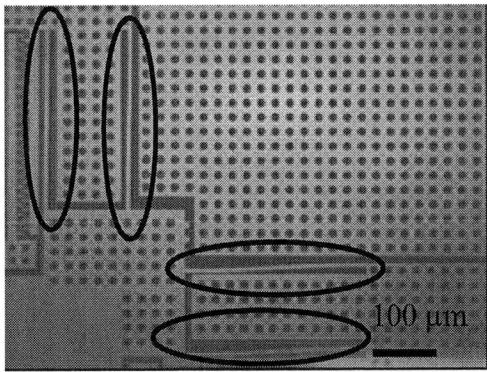

Figure 4: MS die top view (area sensitive to thermal exposure). Detectable beam displacement (vertical and horizontal).

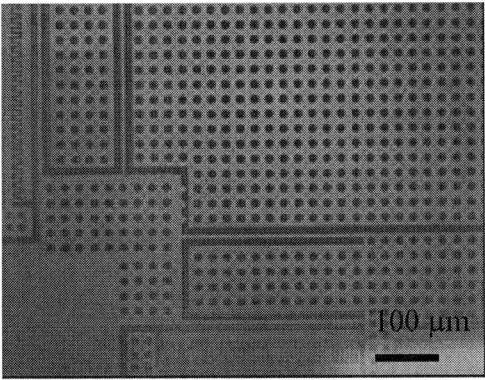

Figure 5: MS die top view (area sensitive to thermal exposure). No detectable beam displacement.

On each MS die there are thousands of beams as a part of comb sensing structures. Practically, it is not feasible to observe all of them. Therefore, we identified experimentally four structures on the MS die as indicators for an impact of the thermal treatment.

We performed a test to check if curing is possible at temperatures lower than 125 °C and if it does not result in the beam displacement. For that, we dispensed droplets of Ablestik 84-1 of 50 μm diameter on Si wafer and subjected it to curing in a convection heating chamber FED 53, manufacturer: Binder (temperature variation: ±2°C at 150°C). To minimize the thermal shock on the samples, we applied a slow heat up with gradient of 5°C/min and slow cool down of 1.5°C/min. Both curing schedules, Test 1 and Test 2, resulted in no detectable beam displacement. Although for Test 1, the epoxy was not fully cured.

Table 2: Curing schedule for die attach epoxy Ablestik 84-1.

	T, °C	t, h	Effect on MS	Result
Standard*	150	1	Yes	NOK
Alternative*	125	2	Yes	NOK
Test 1	80	6/8	No	NOK
Test 2	100	6	No	OK

Legend to table 1:
T,°C: temperature of thermal exposure
t, h: duration of the thermal exposure
NOK: not acceptable result
OK: acceptable result
* vendor's data's [5].

Because of the moving part on the MS die, the adhesive may only be applied on a specified area where there is no moving part (no-go area for the adhesive is marked in red, see figure 6a).

Figure 6: MS die bottom view: marked no-go area for adhesive (6a, left) and adhesive droplet deposited on the die (6b, right).

We dispensed four epoxy droplets of 50 μm

diameter on the designated area of the MS die (figure 6b), then we flipped and attached the die in the cavity of the package. Finally, the whole assembly was cured at 100 °C, for 6 h in a convection heating chamber. After epoxy curing the assembly was carefully inspected and images on the critical parts were investigated.

As explained above, the conventional Au and Cu wire bonding techniques may not be applied for the thermal sensitive MS die, therefore we selected Al wire bonding that requires room temperature processing. For that we used TPT HB16 semi-automatic wire bonder; the Al wire had a 25 µm diameter. The wire-bonded MS die is shown on the figure 7.

Figure 7: MS die mounted and wire-bonded to the package (top view).

The final inspection of the assembly demonstrated no detectable damage to the MS die caused by processing steps occurred during the assembly.

Figure 8: MS die top view observation (area sensitive to thermal exposure). No detectable beam displacement.

5. Conclusion

In the paper we have demonstrated that the thermal sensitive MS die can be assembled using carefully selected and adapted conventional assembly techniques without compromising MS die mechanical integrity. For the adhesive curing, we used lower than a conventional curing temperature, for the wire bonding we utilized room temperature Al wire bonding process. As a criteria for MS die mechanical integrity we introduced methods where we observed only four sensitive areas on the die. Such areas are more sensitive to thermal impact as an actual comb structures and are easy to observe and to quantify.

Acknowledgments

The research has been conducted in the framework of the Microsysteme_ULg Microsys project, funded by the Walloon Region of Belgium.

References

[1] Lau J, Lee C, Premachandran C, Aibin Y. Advanced MEMS Packaging. McGraw-Hill Education, New York, USA, 2009. 576 p.

[2] Gilleo, K., MEMS/MOEMS Packaging, McGraw-Hill, NY, 2005. 240 p.

[3] Stoukatch S., "Low Temperature Microassembly Methods and Integration Techniques for Biomedical Applications.", in: P.Salvo, M.Hernandez-Silveira (Eds.), Wireless Medical Systems and Algorithms, CRC Press, 2016, pp. 21-42.

[4] Licari, J. J. and Swanson, D. W. "Adhesives Technology for Electronic Applications: Materials, Processing, Reliability", William Andrew Publishing, 2011, p.512.

[5] Data Sheets & Certifications, from www.henkel-adhesives.com, (accessed 16.06.2018).

Sensing Applications Based on Cavity Perturbation Method - A Proof of Concept

Valentin Buiculescu[(*)], **Roxana Rebigan**[(*)]

[(*)] IMT Bucharest, 126A Erou Iancu Nicolae Street, 077190 Bucharest, ROMANIA

E-mail: valentin.buiculescu@imt.ro

Abstract—A new category of sensing devices based on perturbation of a resonant circuit or cavity is presented in this paper. The sensor uses a substrate integrated waveguide (SIW) resonator perturbed by the column length of a liquid-in-glass thermometer. Total 1.05°MHz/°C sensitivity is measured with contributions of 0.6 MHz/°C from a liquid-in-glass thermometer with 0.2 mm diameter of the ethanol column and 0.45 MHz/°C from the SIW resonator. Simulations based on liquid columns with diameters of 0.5 mm and 1 mm show that sensitivities at least one order of magnitude higher than values currently available from state-of-the-art SAW sensors can be achieved. A solar energy harvesting solution is also analyzed for increasing the reading distance in fully wireless sensing applications.

Keywords—Temperature transducer, wireless sensing, solar cell, light energy harvesting, substrate integrated waveguide (SIW), SIW resonator.

1. Introduction

The sensors based on resonance frequency change according to a specific environmental condition are used in different fields such as physics, chemistry, biology, and medicine. Currently, many sensors are based on surface acoustic wave (SAW) devices, due to their advantages: high quality factor, small size, operation without power supplies [1]. Since the SAW sensor's sensitivity increases with their operation frequency, continuous efforts are being made to take advantage of this property [2]. However, the attenuation of the acoustic wave along the propagation path increases with frequency, hence the SAW devices' applications are limited to the lower GHz range.

In this paper, a novel sensing principle is proposed: the depth of penetration of a liquid column inside a SIW resonant circuit, which is proportional to the ambient temperature, is converted into a corresponding variation of the resonance frequency of that circuit. This approach, derived from cavity perturbation method [3], is suitable for accurate determination of certain environmental parameters. In order to prove this concept, a liquid-in-glass thermometer is used as two-element temperature sensing device (TSD):

1. the liquid filled bulb of the liquid-in-glass thermometer, as *temperature sensor*, and
2. the temperature dependent column length of the liquid (*the actuator*) which perturbs a specific substrate integrated waveguide (SIW) structure, by changing its resonance frequency.

Despite the limited dynamic range of this sensor, its high sensitivity reveals a great application potential. In addition, the operating principle can be easily extended to other instruments using liquid columns of variable length, or even mechanical devices, for reading the measured values.

2. Novel RF Temperature Sensor

An admittance inverter-based resonator consists of a transmission line (TL) with the characteristic impedance Z_0 and length L_{res}, and equal reactances jX connected at each TL end (Fig. 1), as described in [4].

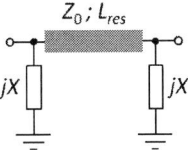

Fig. 1 Admittance inverter based resonator.

Metallic rods or diaphragms are the most used reactive elements in waveguide circuits [5]. Inductive diaphragms (metal plated via rows) at both ends of a L_{res} long SIW section were selected for the proposed resonator structure shown in Fig. 2.

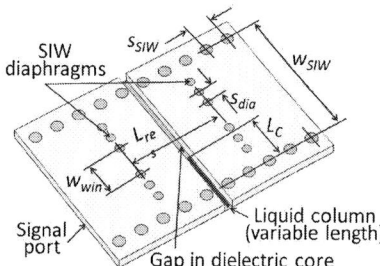

Fig. 2 SIW resonator with removed top ground plane.

The coupling windows within diaphragms have the width w_{win} dimensioned for critical coupling factor. The glass tube of an ethanol based thermometer is inserted in a gap located between diaphragms, across the SIW body. The SIW resonator was simulated with CST Microwave Studio® [6], considering 4.3 relative permittivity and 1.6 mm thickness of the FR-4 dielectric support. The metal plated via holes have a diameter of 2 mm for side walls and 1.5 mm for both diaphragms. Other dimensions defined in Fig. 2 are presented in Table 1.

Table 1. Main resonator dimensions.

No.	Layout element	Value
1.	w_{SIW}	24 mm
2.	h_{SIW}	1.6 mm
3.	s_{dia}	2.5 mm
4.	s_{SIW}	4.0 mm
5.	L_{res}	15 mm
6.	w_{win}	6 mm

The characteristics of a lossless resonant circuit were simulated for an ethanol column with 0.2 mm, 0.5 mm, and 1 mm diameters. The column length L_C used for temperature sensing covers 0-24 mm range (Fig. 3).

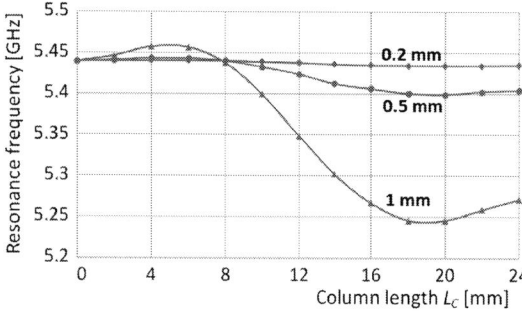

Fig. 3 Simulated characteristics for ethanol columns of different diameters.

Linear sensor response is observed over a limited L_C range, hence its dynamic range is restricted to less than 15°C temperature shift, using thermometer's scale factor of 1°C/mm.

The sensitivity s of the sensor is defined by the ratio between Δf, the resonance frequency deviation, and the corresponding temperature change ΔT:

$$s = \Delta f/\Delta T \quad [\text{MHz/°C}] \qquad (1)$$

The *relative* sensitivity S is obtained by dividing s to the measurement frequency f_M:

$$S = s/f_M \quad [\text{ppm/°C}] \qquad (2)$$

Table 2 shows the temperature sensitivity calculated for the results presented in Fig.3.

Table 2. SIW resonator's sensitivity.

Column type	Column diameter	Sensitivity s
Ethanol	0.2 mm	0.53 MHz/°C
	0.5 mm	5.1 MHz/°C
	1 mm	24.2 MHz/°C

Both the scale factor of the thermometer and the dynamic range of the sensor can be simultaneously changed by proper choice of both the reservoir volume and liquid column diameter. For example, ten-times scale factor reduction of the thermometer, to 0.1°C/mm, using a liquid column diameter of 1 mm, provides ten times dynamic range increase. Resulting sensitivity of 2.42 MHz is several times better than the best performing state of the art SAW sensors [7].

4. Sensor Reading and Energy Harvesting

The sensor presented above is included in the category of single or multiple resonator tags. Short distance wireless reading of these devices is possible with orthogonal polarization antennas at the resonator's ports, in order to minimize the interference between the interrogation and response signals [8].

Reading range extension can be obtained from low power oscillators with the operation frequency controlled by the SIW resonator. However, the oscillator requires a DC power level for biasing the active device(s) that is not available from RF tag readers. A solar energy driven wireless harvesting solution is

978-1-5386-4483-6/18 $31.00 © 2018 IEEE

therefore proposed: the output voltage of a solar cell panel is applied to a low input voltage (LIV) DC/DC converter to reach a stable power supply voltage for all circuits associated with the sensor tag (Fig. 4).

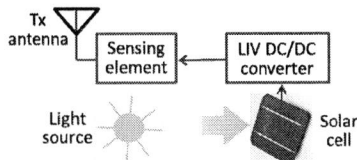

Fig. 4 Wireless sensing with solar energy harvesting.

Since the characteristics of this power supply structure depend on power required by all active circuit attached to the sensor and the solar cells' parameters, it will be analyzed as an independent block.

Efficient solar cells are available as multi-junction devices on $A_{III}B_V$ materials. A single junction solar cell manufactured on mono-crystalline silicon features 20-30 mA/cm^2 current density and 400-500 mV open circuit voltage at 1 sun standard illumination conditions (SIC) [9]. A solar panel with N_{cell} elements and V_{cell} DC voltage for each element was considered as primary energy source for the proposed sensor. The input current $I_{LIV,in}$ absorbed by the LIV converter is therefore evaluated in order to verify its compatibility with the available output current of a given solar cell model:

$$I_{LIV,in} = P_{tag}/(\eta \cdot N_{cell} \cdot V_{cell}) \qquad (3)$$

where P_{tag} is the power consumption of the tag, i.e. the product between the LIV output voltage and the current absorbed by the load resistance, and η is the efficiency of the LIV converter. The general operating conditions assumed for the DC/DC converter are:

- 3.3 V stabilized voltage at the LIV output;
- either 1 mA or 5 mA LIV load current;
- $V_{cell} = 0.42$ V;
- $\eta = 60\%$, independent on the input voltage or load current of the DC/DC converter.

Using (3), the simulation results shown in Fig. 5 provide a few values of $I_{LIV,in}$ calculated according to the number of solar cells used within a panel for energy harvesting of a single resonator tag. The proper sizing of this

panel becomes therefore possible according to the actual solar cell characteristics.

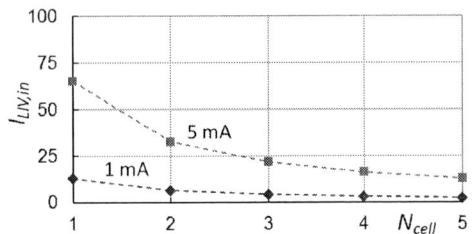

Fig. 5 LIV input current for 1 mA and 5 mA load currents.

4. Experimental Results

A printed circuit board was manufactured according to the dimensions used for electromagnetic analysis and defined in Section 2. The gap used to insert the glass tube of an ethanol-based thermometer was milled across the SIW body. A thin Cu foil soldered above this gap ensures the continuity of the SIW wall. The SIW-coplanar line transitions used at the signal ports ensure 3.48–9.28 GHz bandwidth with return loss better than 10 dB [10]. The photo of the SIW resonator after complete assembly is shown on top of Fig. 6.

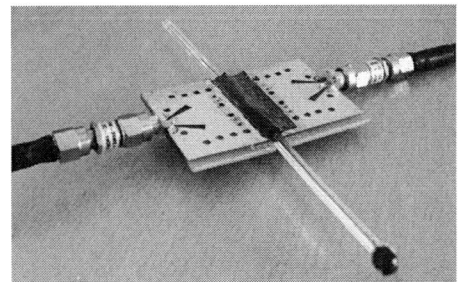

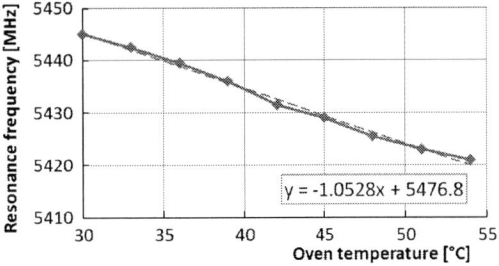

Fig. 6 Photograph of the assembled SIW resonator (top) and SIW resonator's response vs. ambient temperature (bottom).

The response of the resonant sensor put in a temperature controlled oven was measured with 37397D vector network analyzer model from Anritsu, over 24°C temperature range.

About 1.05 MHz/°C (or 194 ppm/°C) overall sensitivity was achieved from the trendline slope shown on bottom graph in Fig. 6, using 0.2 mm diameter ethanol column. The SIW resonator, measured separately, has a contribution of 0.45 MHz/°C to this figure, due to FR-4 thermal expansion, hence the difference of about 0.6 MHz/°C, obtained exclusively from the ethanol column, is very close to the simulated sensitivity of 0.53 MHz/°C.

Solar cells fabricated on p-type (100 cut) silicon wafers with $1 - 10\,\Omega\cdot\text{cm}$ resistivity, feature a single p-n junction with 1 μm depth and 1 cm^2 active area. The cells' response has been investigated under standard illumination conditions using the solar simulator ORIEL LSH-7320 ABA LED with the spectral range similar to sun radiation and the power of 1 sun (100 mW/cm^2): short circuit current $I_{sc} = 26\,\text{mA}$, open voltage $V_{oc} = 0.45$ V, and fill factor $FF = P_{max}/(I_{sc}\cdot V_{oc}) = 70\,\%$.

The DC2042A demo board with LTC3105 and LTC3108 LIV DC/DC converters from Analog Devices was tested as a solar cell harvesting solution. The LTC3105 section provides 3.3 V output voltage in a 750 Ω load using a solar panel with six cells and 0.8 sun minimum illumination. The selected value of the load impedance simulates approximately the bias requirements of an oscillator with low power consumption.

5. Conclusions

In this paper we present a new type of sensing devices based on the perturbation of a resonant circuit or cavity. The temperature sensor based on resonance frequency change uses a SIW resonant circuit perturbed by the column of the liquid-in-glass thermometer. Sensitivity close to 1.05°MHz/°C is experimentally measured using a thermometer with 0.2 mm diameter of the ethanol column. The simulations with 0.5 mm and 1 mm diameters of the liquid column show that sensitivities are at least one order of magnitude higher than currently available values from state-of-the-art SAW sensors.

A low power oscillator powered by a solar cell panel and controlled by a SIW resonator according to an environmental parameter will provide a full wireless sensing device in a future integrated solution.

Acknowledgment. This work was supported by a grant of the Romanian National Authority for Scientific Research and Innovation, CNCS/ CCCDI-UEFISCDI, project number PN-III-P2-2.1-PED-2016-0957, within PNCDI III.

References

[1] A. Pohl, "*A review of wireless SAW sensors*", IEEE Trans. Ultrason., Ferroelectr., Freq. Control, vol. 47, no. 2, pp 317-332, Mar. 2000

[2] A. Müller, et al, "*GaN/Si based single SAW resonator temperature sensor operating in the GHz frequency range*", Sensors and Actuators A: Physical, vol. 209, pp. 115-123, Mar. 2014

[3] S. Roberts, A. Von Hippel, "*A new method for measuring dielectric constant and loss in the range of centimeter waves*", J. Appl. Phys. 17, 610 (1946), pp. 610-616

[4] G.L. Matthaei, L. Young, E.M.T. Jones, "*Microwave filters, impedance-matching networks, and coupling structures*", Artech House, 1964

[5] N. Marcuvitz, "*Waveguide handbook*", Radiation Laboratory Series, vol. 10, 1951

[6] CST Microwave Studio® software suite

[7] A. Müller, et al, "*Sezawa propagation mode in GaN on Si surface acoustic wave type temperature sensor structures operating at GHz frequencies*", IEEE Electron Dev. Let., vol. 36, no. 12, pp. 1299-1302, Dec. 2015

[8] S. Preradovic, N.C. Karmakar, "*Chipless RFID: bar code of the future*", IEEE Microw. Mag., vol. 11, no. 7, pp. 87-97, Dec. 2010

[9] I. Mathews, D. O'Mahony, B. Corbett, A.P. Morrison, "*Theoretical performance of multijunction solar cells combining III-V and Si materials*", Optics Express, vol. 20, no. S5 / A754, 2012

[10] V. Buiculescu, M. Aldrigo, A. Ştefănescu, "*SIW choke-based technique for accurate dielectric measurements in the 3.5 – 5 GHz band*", Proc. European Microwave Conference EuMW 2017, pp. 312-315, 10-12 Oct. 2017, Nüremberg, Germany

978-1-5386-4483-6/18 $31.00 © 2018 IEEE

Continuous-wave Mm-wave Waveguide-based Probe for Skin Tissue Characterisation

*K. Y. Chan, X. Li, Y. Fu, R. Ramer

School of Electrical Engineering and Telecommunications, The University of New South Wales, Sydney, Australia

*kyc@unsw.edu.au

Abstract— This paper presents the study and simulation results of a millimetre-wave based device for cancerous skin tissue detection. A probe that could be implemented using inexpensive silicon planar fabrication is proposed. It permits easy system-on-chip integration with other silicon devices to achieve an entire measuring tool for easy deployment. We used the available open literature basal cell carcinoma (BCC) data considerations, for initial development and simulation validation. This study showed that the reflection coefficients vs frequency could capture useful information indicating the possible BCC presence at millimetre-wave frequencies by using both magnitude and phase of the reflection coefficients. It was found that a dual-band approach, 100 to 150 GHz and 200 to 250 GHz, has the ability to highlight deviations from the normal skin.

Keywords— millimetre-wave, on-wafer, probe, waveguide.

1. Introduction

During the past two decades, groups from around the globe conducted research on millimetre-waves and terahertz imaging. Most of their approaches used Terahertz Pulsed Imaging (TPI) with optical mixing to generate upper mm-wave and THz signals and performed analysis with finite difference time domain method (FDTD) [1-3]. The main drawbacks of optical mixing systems are their large size, high cost and their inherent stationary aspect. This is so as they are primarily based on optical sources, mirrors, and lenses. To overcome these problems, more recent research focuses toward near-field systems that use circuits in the millimetre-wave frequency range. Sensors operating below 40 GHz have been proposed and demonstrated; however, although they provide well-integrated solutions, their sensitivity is not sufficient due to the relatively large wavelengths [4-6]. Other systems, operating at higher frequencies around 90 GHz, have shown promising results [7, 8] but are very bulky and not suitable for integration in a readout device with the small form factor. We propose a novel device that would permit designing of a system that could quantify the signal reflected from the exposed skin area. The proposed technique is based on shining the skin tissue with millimetre-waves. Different reflections and absorptions occur for different skin tissues. The analysis of the reflected wave will allow the evaluation of the complex dielectric properties and the real and imaginary extraction.

2. Method and Technique

In the paper, we propose a system consisting of three parts: a signal generator / detector (the vector network analyser (VNA)), the proposed device itself, called 'the probe', and the sample under test (skin area). The key innovation of this work consists in designing a probe that offers a small footprint for high lateral resolution sensing with sufficient sensitivity for high contrast between the normal and diseased skin, and also in the EM modelling of the BCC using existing data from the literature. The system is designed based on the principle of continuous wave near-field electromagnetic wave reflectometry. The generated mm-wave signal illuminates the skin sample and travels through the probe; a part of the signal penetrates the sample with the remainder reflected back through the probe and reaching the detector. This reflected signal gives unique information (in terms of its amplitude and phase) at different frequencies when compared to the transmitted signal from the generator. The deviances of the magnitude and phase of different skin samples from the normal healthy skin are determined in this paper. In the suggested system, the probe is designed so that it permits limited penetration depth within typical skin thickness (e.g. less than 3mm) [9, 10]. The normal-dry skin and the most common type of skin cancers BCC were considered in this study. For simplicity, a flat and homogeneous skin was assumed. There have been three steps in the present study. First, we calculated the variation of relative electrical permittivity and total effective conductivity functions vs frequencies. Secondly, the near-field reflectometry probe was designed. Finally, the combined probe-skin samples (treated as terminated/loaded probes) were studied and followed by their HFSS simulations. The results accomplished in these steps are outlined below.

A. Step one: Normal Skin and BCC Electrical Parameters Calculation

With continuous-wave EM simulations, commercially available software packages (e.g. HFSS, CST, etc.) are unable to use complex frequency dependent electric permittivity functions. Instead, they require the relative permittivity ε_r and the electric conductivity σ. However, human skin dielectric properties are typically modelled as complex frequency-dependent permittivity $\varepsilon^*(\omega)$, based on Debye theory. Pickwell et al [11] used double Debye theory and modelled

978-1-5386-4483-6/18 $31.00 © 2018 IEEE

frequency-dependent permittivity $\varepsilon^*(\omega)$, from 0.1 to 4 THz for both normal skin and BCC, where $\varepsilon^*(\omega)$ is given by $\varepsilon^*(\omega)=\varepsilon_\infty+(\varepsilon_s-\varepsilon_2)/(1+j\omega\tau_1)+(\varepsilon_2-\varepsilon_\infty)/(1+j\omega\tau_2)$ (1) where ε_s, ε_∞, ε_2, τ_1, and τ_2 are the five double Debye parameters. In our study, we perform EM modelling using Pickwell's double Debye simulation parameters [11]. Here, (1) was separated into the real (ε') and imaginary (ε'') parts as $\varepsilon^*=\varepsilon'-j\varepsilon''$. The real part ε' represents the relative permittivity ε_r, and the imaginary part ε'' was further expressed as total effective conductivity (equivalent conductivity) σ_e of the material (skin) as $\varepsilon''=\sigma_e/(2\pi f \varepsilon_0)$ where $\varepsilon_0=8.85\times10^{-12}$ F/m is the permittivity of free space and f is the frequency in Hz. It should be noted that this total effective conductivity is not just the static (ionic) conductivity; it includes all the dielectric losses such as the standard static (ionic) conductivity and the associated loss (imaginary) terms from double Debye theory. These calculations have been performed for each frequency point, from 95 to 300 GHz, and have been plotted in Fig. 1.

B. Second Step: Probe Considerations

A typical coaxial type probe could be used as the sensor. Although some coaxial probes are commercially available, they are not suitable for this application as they suffer from having relatively large footprints [12]. Greater than 3 mm in diameter, the typical commercial probes, which will also sense surrounding normal skin due to the fringing field, are not able to provide sufficient properties discrimination between the BCC and the surrounding normal skin [13].

Fig. 2 illustrates the proposed rectangular waveguide probe that supports the transverse electric TE10 mode for the sensing signal, and that is easy to fabricate [14-37]. The rectangular waveguide probe sits on the skin area of interest at one end and is connected to the VNA (generator and detector) through a coplanar waveguide (CPW) transition at the other end. Front and back views of the skin sample in contact with the probe are given in Fig. 2(a) and (b). A higher operational frequency, > 90 GHz, is selected based on the lateral resolution and penetration considerations. In rectangular waveguide design, higher frequencies offer smaller waveguide cross-section due to waveguide cut-off frequency and also provide shallower electromagnetic wave penetration in the skin. The lateral resolution and electromagnetic penetration must be adequately addressed; this is because BCC can be very shallow with a thickness in sub-millimetre range and can have a lateral dimension of the order of mm². Glass, with a dielectric constant of 5.5,

was selected as the waveguide filling material due to its wide commercial availability for microfabrication. The dielectric constant is similar to typical health skin of 5.6 [38] resulting in better impedance matching at the interface probe-skin; this permits the mm-wave signal to penetrate into the skin and allows reflection to take place only when a mismatch occurs. As BCC exhibits departure from the dielectric constant and conductivity values of the normal skin, a mismatch will be encountered, and this can be quantified from the measured reflection coefficient.

With the selection of sensing frequencies above 95 GHz and glass as probe filling material, a minimal footprint, as small as 0.7 mm (W) × 0.35 mm (H) can be achieved. Also, as the sensing frequency is above 95 GHz and operation is in the near-field electromagnetic radiation condition, the system is interference immune.

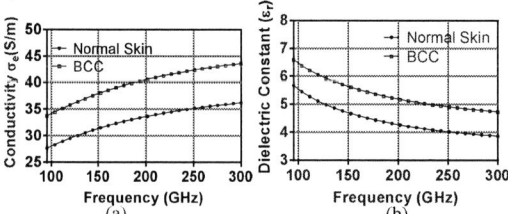

Fig. 1 (a) Extracted frequency-dependent relative permittivity ε_r and (b) total effective conductivity (equivalent conductivity) σ_e (S/m) for normal-dry skin and BCC for frequencies ranging from 95GHz to 300GHz.

C. Step three: Probe Loaded with Skin Samples Study

An entire combination of the probe loaded with different skin samples was simulated in HFSS. The electromagnetic signal was excited from the generator, has travelled through the probe, was reflected from the skin and has returned back to the detector. Two possible occurring scenarios were considered in order to emulate different BCC conditions. In 'Scenario 1', a shallow BCC layer of various thicknesses is located on the top of the normal skin, as shown in Fig. 3(a). In 'Scenario 2', a 500 μm thick BCC layer is embedded under the normal skin at various depths, as shown in Fig. 3(d).

3. Results and Discussions

The HFSS simulated reflected signals ($|S_{11}|$, in dB) and phase ($\angle S_{11}$ in degrees) were analysed, for both scenario 1 and 2. These results were compared with the $|S_{11}|$ and $\angle S_{11}$ of the healthy skin. The results were presented as differences between the magnitudes and phases of the reflection coefficients where the BCC values were considered first and then the normal skin, i.e. $\Delta|S_{11}| = |S_{11}|_{BCC} - |S_{11}|_{normal\ skin}$ and $\Delta\angle S_{11} = \angle S_{11\ BCC} - \angle S_{11\ normal\ skin}$. The study results for scenario 1 and 2 are illustrated in Fig. 4 and 5. For scenario 1, Fig. 4(a) show the differences in magnitude of the reflection coefficient $\Delta|S_{11}|$, while Fig. 4(b) illustrate the phase differences $\Delta\angle S_{11}$ of BCC of different thicknesses (10 μm, 50 μm, 100 μm, 500 μm, 1000 μm, and 3000 μm) and normal skin, over the entire frequency range.

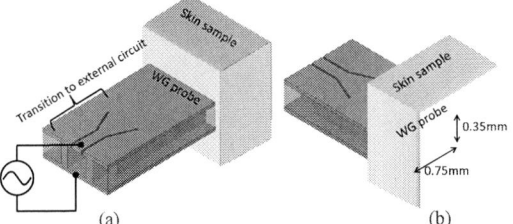

(a)   (b)

Fig. 2 Probe sitting on the skin sample of interest. (a) front-view of the probe connected to the VNA via a CPW transition, and (b) back-view showing the rectangular waveguide behind the skin sample.

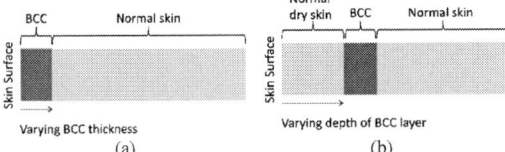

(a)   (b)

Fig. 3 (a) Scenario 1: BCC layer (of different thicknesses) on the surface of normal skin, (b) Scenario 2: (d) BCC layer embedded beneath normal skin at different depths.

According to these results, even a very shallow BCC layer resulted in a distinguishable difference in magnitude response compared to normal skin. This is even more pronounced for the difference in phase response, where other than the 10 μm thin BCC layer, a significant phase deviation from the normal skin phase is observed, in particular above 200 GHz. Fig. 5(a) shows the simulation results for the second scenario, where a 500 μm thick BCC layer is embedded in a normal skin layer at progressive depths 10 μm, 50 μm, 100 μm, 500 μm, and 1000 μm, and 3000 μm. Notable differences in magnitude $|S_{11}|$ (Fig. 5(a)) and phase $\angle S_{11}$ (Fig. 5(b)) values could be captured for all BCC layers embedded at depths from 10 μm to 100 μm, within these variations. In fact, the shallower the BCC layer is embedded under normal skin tissue, the more differences to the $\Delta|S_{11}|$ and $\Delta\angle S_{11}$ from those for the normal skin. It should be noted that according to the simulations, as the BCC layer is embedded beyond 500 μm under the normal skin, the responses from both magnitude and phase become similar to normal skin, across the entire 95 to 300 GHz frequency range. This is, in fact, a desirable feature for the proposed probe. According to researchers [39], the device should only detect unusual tissues that are embedded in the epidermis layer which has a typical thickness of fewer than 100 μm across body sites. The difference in magnitude, for

scenario 1 cases, shows small values with less than 2 dB values above 200 GHz; for scenario 2, the difference in magnitude is less than 3 dB above 200 GHz as shown in Fig. 4 and 5. This is not the case for the differences in phase, $\Delta\angle S_{11}$ data, where significant values were obtained, rendering the phase as the preferred parameter for our analysis. Therefore, a dual-band operation (95 – 150 GHz and 200 – 250 GHz) is ideal, where magnitude and phase are the preferred parameters respectively.

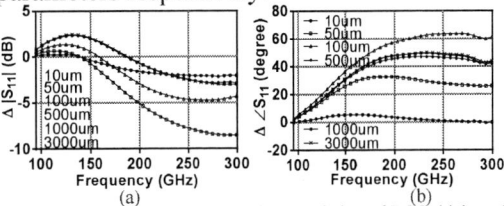

(a)   (b)

Fig. 4 Simulation results of scenario 1 (variation of BCC thickness) (a) differences in magnitude ($\Delta|S_{11}|$) and (b) phase ($\Delta\angle S_{11}$).

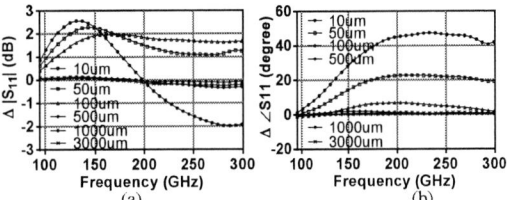

(a)   (b)

Fig. 5 Simulation results of scenario 2 (variation of the depth of BCC) (a) differences in magnitude ($\Delta|S_{11}|$) and (b) phase ($\Delta\angle S_{11}$).

5. Conclusions

Studies and simulation results of a millimetre-wave rectangular glass filled waveguide probe using continuous wave were carried out for the detection of skin conditions. In this study, normal skin and BCC skin electromagnetic models were created and simulated with a high-frequency electromagnetic simulator ANSYS HFSS. The simulation results validated the technique for satisfactory sensitivity detection of BCC while providing a small footprint and adequate sensing depth. Depending on the frequency range, either amplitude or phase could be the preferred parameter for the maximum contrast between BCC and normal skin.

References

[1] I. McAuley *et al.*, "Millimetre-wave and Terahertz Imaging Systems with Medical Applications." (in English), *Conference Digest of the 2006 Joint 31st International Conference on Infrared and Millimeter Waves and 14th International Conference on Terahertz Electronics,* pp. 371-371, 2006.

[2] V. P. Wallace *et al.*, "Terahertz pulsed spectroscopy of human Basal cell carcinoma," *Appl Spectrosc,* vol. 60, no. 10, pp. 1127-33, Oct 2006.

[3] R. M. Woodward *et al.*, "Terahertz pulsed imaging of skin cancer in the time and frequency domain," *J Biol Phys,* vol. 29, no. 2-3, pp. 257-9, Jun 2003.

978-1-5386-4483-6/18 $31.00 © 2018 IEEE

[4] K. Byoungjoong *et al.*, "Novel low-cost planar probes with broadside apertures for nondestructive dielectric measurement of biological materials at microwave frequencies," (in English), *IEEE Transactions on Microwave Theory and Techniques,* vol. 53, no. 1, pp. 134-143, Jan 2005.

[5] J. M. Kim *et al.*, "Planar type micromachined probe with low uncertainty at low frequencies," (in English), *Sensors and Actuators a-Physical,* vol. 139, no. 1-2, pp. 111-117, Sep 12 2007.

[6] J. M. Kim *et al.*, "In vitro and in vivo measurement for biological applications using micromachined probe," (in English), *IEEE Transactions on Microwave Theory and Techniques,* vol. 53, no. 11, pp. 3415-3421, Nov 2005.

[7] S. Danylyuk *et al.*, "Broadband microwave-to-terahertz near-field imaging," *IEEE/MTT-S International Microwave Symposium, 2007,* Honolulu, HI, USA, 2007, pp. 1383-1386.

[8] F. Topfer *et al.*, "Dermatological verification of micromachined millimeter-wave skin-cancer probe," *IEEE MTT-S Int. Microw. Symposium (Ims),* 2014

[9] A. Laurent *et al.*, "Echographic measurement of skin thickness in adults by high frequency ultrasound to assess the appropriate microneedle length for intradermal delivery of vaccines," *Vaccine,* vol. 25, pp. 6423-30, Aug 21 2007.

[10] S. A. M. Shuster *et al.*, "The influence of age and sex on skin thickness, skin collagen and density," *British Journal of Dermatology,* vol. 93, no. 6, pp. 639-643, 1975.

[11] E. Pickwell *et al.*, "Simulating the response of terahertz radiation to basal cell carcinoma using ex vivo spectroscopy measurements," *J Biomed Opt,* vol. 10, no. 6, p. 064021, Nov-Dec 2005.

[12] P. M. Meaney *et al.*, "Open-Ended Coaxial Dielectric Probe Effective Penetration Depth Determination," *IEEE Trans Microw Theory Tech,* vol. 64, no. 3, pp. 915-923, Mar 2016.

[13] D. M. Pozar, *Microwave Engineering, 4th Edition.* Wiley, 2012.

[14] L. Gong *et al.*, "Substrate integrated waveguide H-plane horn antenna with improved front-to-back ratio and reduced sidelobe level," *IEEE Antennas and Wireless Propagation Letters,* vol. 15, pp. 1835-1838, 2016.

[15] L. Gong *et al.*, "A beam steering single-arm rectangular spiral antenna with large azimuth space coverage," *IEEE Wireless and Microwave Technology Conference (WAMICON), 2013 IEEE 14th Annual,* 2013, pp. 1-4.

[16] K. Chan *et al.*, "A novel RF MEMS switch with novel mechanical structure modeling," *Journal of Micromechanics and Microengineering,* vol. 20, no. 1, p. 015031, 2009.

[17] L. Gong *et al.*, "RHCP pattern reconfigurable spiral antenna biased with two DC signals," *Microwave and Optical Technology Letters,* vol. 56, no. 7, pp. 1636-1640, 2014.

[18] L. Gong *et al.*, "A Reconfigurable Spiral Antenna with Wide Beam Coverage," *IEEE Antennas and Propagation Society International Symposium (APSURSI), 2013 IEEE,* 2013, pp. 206-207.

[19] L. Gong *et al.*, "Beam steering spiral antenna reconfigured by PIN diodes," *International Journal of Microwave and Wireless Technologies,* vol. 6, no. 06, pp. 619-627, 2014.

[20] Y. Yang *et al.*, "MEMS-loaded millimeter wave frequency reconfigurable quasi-Yagi dipole antenna," *IEEE Microwave Conference Proceedings (APMC), 2011 Asia-Pacific,* 2011, pp. 1318-1321.

[21] G. I. Kiani *et al.*, "MEMS enabled frequency selective surface for 60 GHz applications," *IEEE Antennas and Propagation (APSURSI),*

2011 IEEE International Symposium on, 2011, pp. 2268-2269.

[22] K. Y. Chan *et al.*, "Novel miniaturized RF MEMS staircase switch matrix," *IEEE Microwave and Wireless Components Letters,* vol. 22, no. 3, pp. 117-119, 2012.

[23] H. U. Rahman *et al.*, "Investigation of residual stress effects and modeling of spring constant for RF MEMS switches," *IEEE Microwave Symposium (MMS), 2009 Mediterranean,* 2009, pp. 1-4.

[24] K. Y. Chan *et al.*, "RF-MEMS switches with new beam geometries: improvement of yield and lowering of actuation voltage," *International Society for Optics and Photonics Device and Process Technologies for Microelectronics, MEMS, Photonics, and Nanotechnology IV,* 2008, vol. 6800, p. 680026.

[25] E. Siew *et al.*, "RF MEMS-integrated frequency reconfigurable quasi-Yagi folded dipole antenna," *IEEE Microwave Conference Proceedings (APMC), 2011 Asia-Pacific,* 2011, pp. 558-561.

[26] K. Y. E. Chan *et al.*, "RF MEMS millimeter-wave switchable bandpass filter," *IEEE Wireless Symposium (IWS),* 2013, pp. 1-4.

[27] K. Y. Chan *et al.*, "RF MEMS Switch with low stress sensitivity and low actuation voltage," *IEEE Antennas and Propagation Society International Symposium, 2009. APSURSI'09. IEEE,* 2009, pp. 1-4.

[28] Y. Yang *et al.*, "Experimental proof for pattern reconfigurability of 60 GHz quasi Yagi antenna," *Microwave and Optical Technology Letters,* vol. 57, no. 1, pp. 84-88, 2015.

[29] L. Gong *et al.*, "Phase correction of the electric field for a dielectric loaded substrate integrated waveguide H plane horn antenna," *Microwave and Optical Technology Letters,* vol. 59, no. 3, pp. 584-588, 2017.

[30] K. Y. Chan *et al.*, "Miniaturized RF MEMS switch cells for crossbar switch matrices," *IEEE Microwave Conference Proceedings (APMC), 2010 Asia-Pacific,* 2010, pp. 1829-1832.

[31] H. Rahman *et al.*, "Fabrication of RF NEMS Series Switch Using Surface Micromachining," *IASTED Intl. Conf. on Nanotechnology and Applications,* 2010.

[32] L. Gong *et al.*, "A split-ring structures loaded SIW sectorial horn antenna," *IEEE Antennas and Propagation in Wireless Communications (APWC),* 2015, pp. 349-350.

[33] K. Y. Chan *et al.*, "60 GHz to E-Band Switchable Bandpass Filter," *IEEE Microwave and Wireless Components Letters,* vol. 24, no. 8, pp. 545-547, 2014.

[34] K. Y. Chan *et al.*, "A Switchable Iris Bandpass Filter Using RF MEMS Switchable Planar Resonators," *IEEE Microwave and Wireless Components Letters,* vol. 27, pp. 34-36, 2017.

[35] Y. Yang *et al.*, "60GHz pattern reconfigurable quasi-Yagi antenna—Proof through computational design," *IEEE iWAW, 2014 International Workshop on,* 2014, pp. 53-56.

[36] L. Gong *et al.*, "RF MEMS for Reconfigurable RF Front-End: Research in Australia," *Advanced Materials Research,* pp. 105-110, 2014.

[37] K. Y. Chan *et al.*, "Low-cost E-band Lange coupler with vialess load," *Electronics Letters,* vol. 51, no. 11, pp. 839-841, 2015.

[38] P. A. Hasgall *et al.* (2015, Mar). *IT'IS Database for thermal and electromagnetic parameters of biological tissues.* Available: www.itis.ethz.ch/database

[39] J. Sandby-Moller *et al.*, "Epidermal thickness at different body sites: relationship to age, gender, pigmentation, blood content, skin type and smoking habits," *Acta Derm Venereol,* vol. 83, no. 6, pp. 410-3, 2003.

Session M

MODELLING

978-1-5386-4483-6/18 $31.00 © 2018 IEEE

Multi-scale finite element modeling of CNT-polymer-composites

Michael Schiebold & Jan Mehner
Chemnitz University of Technology
Chemnitz, Germany
michael.schiebold@etit.tu-chemnitz.de

Abstract—A hierarchical multi-scale approach is used to model a composite consisting of carbon nanotubes and a polymer which can be used as pressure sensor matrix to prevent people from decubitus ulcer. Starting with the modeling of a carbon nanotube and the calculation of its equivalent cylinder properties. Subsequently the cylinders which replace the CNTs are randomly distributed in the polymer such that homogenization techniques leading to the mechanical properties of the composite.

I. INTRODUCTION

Bedridden patients have a high risk of coming down with decubitus ulcer. This disease is caused by a pressure on the skin and leads to a necrosis of the affected area if left untreated. Detecting the vulnerable areas with the aid of a pressure sensor-matrix and preventing bedridden people from developing decubitus ulcer can lead to an improvement of the healthiness of immobile patients. Considering that a multi-scale model is necessary to investigate the influence of the reinforcement attributes on the mechanical composite properties. In this case the functional layer of the pressure sensors consists of a small fraction of carbon nanotubes (CNTs) and a soft polymer resulting in piezoresistive properties of the composite. Thus, the measured resistance of the composite depends on the pressure between patient and mattress. The goal of this work is the comparison of the developed mechanical multi-scale model of the CNT-polymer-composite with results of FE simulations.

II. MULTI-SCALE MODELING

A hierarchical multi-scale approach is used in this work to model the CNT-polymer-composite and derive their properties. Starting with the replacement of atomic bonds by beams to model single carbon nanotubes and the determination of their equivalent cylinder properties. Subsequently the carbon nanotubes are replaced by the equivalent cylinders and distributed in the polymer matrix such that the properties of the composite can be calculated by different homogenization techniques.

A. Carbon nanotubes

Molecular mechanics approaches are a part of the molecular modeling and are dealing only with the movement of the nuclei [1]. These methods use force fields to describe the potential energy between neighboring atoms during a motion. The atomic bonds of carbon nanotubes can be replaced by elements in the FE software ANSYS representing approximations of these force fields. The linkage between structural and molecular mechanics was found by Li and Chou [2] such that the harmonic approximations of the potential energies are equivalent to the strain energies of a beam for stretching, bending and torsion. This leading to the following relationships between the parameters of a beam and the force field constants

$$\frac{EA}{L} = k_r, \quad \frac{EI}{L} = k_\theta, \quad \frac{GJ}{L} = k_\tau \qquad (1)$$

with the Young's modulus E, the cross-sectional area A, the length L, the area moment of inertia I, the shear modulus G, the torsion constant J and the force field constants for bending k_r, stretching k_θ and torsion k_τ [2]. The equations found by Tserpes and Papanikos [3]

$$d = 4\sqrt{\frac{k_\theta}{k_r}}, \quad E = \frac{k_r^2 L}{4\pi k_\theta}, \quad G = \frac{k_r^2 k_\tau L}{8\pi k_\theta^2} \qquad (2)$$

enable the calculation of the properties of a beam with a circular cross-section and its diameter d. Using this linkage carbon nanotubes can be modeled with beams as atomic bonds and simulated in ANSYS.

B. Equivalent cylinder

In order to reduce degrees of freedom and calculation time, the carbon nanotubes can be modeled as a cylinder. This step is also necessary to enable the meshing of the CNT-polymer-composite for the purpose of doing a finite element simulation in ANSYS and to apply homogenization techniques using the material properties of the equivalent cylinder. Consequently the stiffness of the corresponding cylinder has to be determined. In linear elasticity the fourth order stiffness tensor $\boldsymbol{C}_{ijkl}$ relates the engineering strains ε_{kl} with the stresses $\boldsymbol{\sigma}_{ij}$ of materials

$$\boldsymbol{\sigma}_{ij} = \boldsymbol{C}_{ijkl}\,\varepsilon_{kl}. \qquad (3)$$

Assuming transversely isotropic properties with a symmetry axis in z-direction the stiffness tensor can be simplified

978-1-5386-4483-6/18 $31.00 © 2018 IEEE

to a stiffness matrix consisting of five independent components C_{ij}

$$
\mathbf{C} = \begin{bmatrix}
C_{11} & C_{12} & C_{13} & 0 & 0 & 0 \\
C_{12} & C_{11} & C_{13} & 0 & 0 & 0 \\
C_{13} & C_{13} & C_{33} & 0 & 0 & 0 \\
0 & 0 & 0 & C_{44} & 0 & 0 \\
0 & 0 & 0 & 0 & C_{44} & 0 \\
0 & 0 & 0 & 0 & 0 & \dfrac{C_{11}-C_{12}}{2}
\end{bmatrix}. \quad (4)
$$

The inverse of the stiffness matrix is used to calculate the Young's moduli, the shear moduli and the Poisson's ratios of the material. In order to compute the material properties at least five different load cases of the carbon nanotube have to be determined to calculate the five independent components of the stiffness matrix. In this work the five load cases are based on similar approaches from [4]–[6] and have to be simulated in ANSYS to determine the mechanical properties of the equivalent cylinder. An overview of these five load cases is given in Table I along with the applied boundary conditions of the finite element simulation. Additionally the load cases two to five are depicted in Fig. 1, where the blue carbon nanotubes show the initial condition and the red ones the exaggerated deformation of the respective load case.

The first and second load case stretch the carbon nanotube in axial direction. Thus leading for the radial-coupled nodes in the first simulation to a uniform tapering of the carbon nanotube while the nodes in the second case only have a displacement in axial direction since their in-plane (x,y) positions are fixed. In the third case the carbon nanotube is twisted around its axial axis by a tangential movement of the upper nodes. A compression in radial direction is described by the fourth load case such that all nodes are shifted against the radial direction. Additionally the axial position is fixed prohibiting a movement in z-direction. Load case five induces an in-plane shear in the carbon nanotube by displacing the nodes radially and tangentially dependent on the in-plane position of the respective node.

Finite element simulations of the CNTs for every load case are performed in ANSYS in order to obtain the

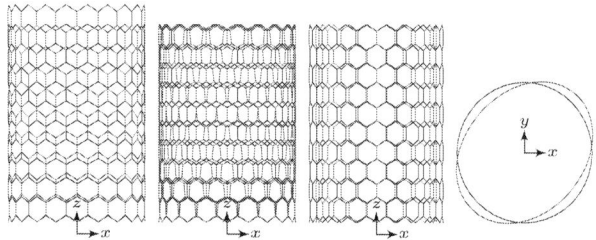

Fig. 1. Displacement of the load cases: axial tension with fixed radial position, axial torsion, radial compression with fixed axial position and in-plane shear

TABLE I
Load cases for FE simulations

Case	Description	Boundary conditions
1	axial tension	$\varepsilon_x = \varepsilon_y \neq 0, \varepsilon_z \neq 0$
2	axial tension with fixed radial position	$\varepsilon_x = \varepsilon_y = 0, \varepsilon_z \neq 0$
3	axial torsion	$\varepsilon_{yz} \neq 0, \varepsilon_{xz} \neq 0$
4	radial compression with fixed axial position	$\varepsilon_x = \varepsilon_y \neq 0, \varepsilon_z = 0$
5	in-plane shear	$\varepsilon_{xy} \neq 0$

reaction forces of the displaced nodes. The reaction force F_i and the respective displacement Δs_i of the i-th node are used to calculate the energy of the load case U_{FE} as

$$
U_{FE} = \frac{1}{2} \sum_{i=1}^{n} F_i \, \Delta s_i. \quad (5)
$$

This energy has to be equal to the strain energy of the equivalent cylinder to obtain an identical behavior for similar strains. In linear elasticity the strain energy of a solid can be calculated with its volume V as

$$
U = \frac{V}{2} \boldsymbol{C}_{ijkl} \varepsilon_{ij} \varepsilon_{kl}. \quad (6)
$$

The strains of the solid $\boldsymbol{\varepsilon}$ are derived from the boundary conditions of the respective load cases and from the node displacements obtained via the ANSYS simulations. Considering (5) and (6) the five independent components of the transversely isotropic stiffness matrix C_{ij} can be calculated using the energies and strains obtained from the ANSYS simulations. Thus leading to

$$
\vec{U}_{FE} = \frac{V}{2} \cdot \begin{bmatrix}
2\varepsilon_x{}^2 & 2\varepsilon_x{}^2 & 4\varepsilon_x\varepsilon_z & \varepsilon_z{}^2 & 0 \\
0 & 0 & 0 & \varepsilon_z{}^2 & 0 \\
0 & 0 & 0 & 0 & \varepsilon_{yz}{}^2 + \varepsilon_{xz}{}^2 \\
2\varepsilon_x{}^2 & 2\varepsilon_x{}^2 & 0 & 0 & 0 \\
\dfrac{\varepsilon_{xy}{}^2}{2} & -\dfrac{\varepsilon_{xy}{}^2}{2} & 0 & 0 & 0
\end{bmatrix} \cdot \begin{bmatrix}
C_{11} \\ C_{12} \\ C_{13} \\ C_{33} \\ C_{44}
\end{bmatrix} \quad (7)
$$

where $\vec{U}_{FE}$ is a column vector containing the five strain energies from the ANSYS simulations. Furthermore the matrix rows consists of the strains for every load case. Due to the transversely isotropic properties the strains in x- and y-direction are partially equal such that the matrix can be simplified according to (7). The solution of (7) can be computed via a left multiplication with the inverse of the strain matrix and a division by the half volume of the cylinder. Thus the stiffness parameters of the cylinder are calculated and can be used to determine their Young's moduli, shear moduli and Poisson's ratios.

In this work hollow and solid cylinders are considered as equivalent cylinders for the carbon nanotubes. For sake of simplicity a uniform strain for radial compression and in-plane shear is assumed for the hollow cylinder. Additionally the wall-thickness of 0.34 nm is used for both cylinders in the calculation.

978-1-5386-4483-6/18 $31.00 © 2018 IEEE

C. Homogenization

The CNT-polymer-composites are modeled via a random distribution of the equivalent cylinders in the polymer which also prohibits an overlapping of the single cylinders. Subsequently the average orientation tensor of the nanotubes is calculated. This tensor provides information about the distribution of the CNTs and is necessary for several homogenization techniques. For the sake of visibility the tensor indices in this section are omitted such that C represents the fourth order stiffness tensor C_{ijkl}. Different homogenization techniques can be applied to the CNT-polymer-composite to calculate its properties.

The Reuss homogenization is a simple approach considering only the fraction of matrix and filler material. Thus the Reuss approach computes the volume average of the compliance tensors as

$$\overline{S} = v_m S_m + v_f S_f, \tag{8}$$

$$v_m + v_f = 1, \tag{9}$$

with the compliance tensor of the matrix S_m and the filler S_f [7]. The volume fraction of filler and matrix represents v_f respective v_m. Neglecting the nanotube orientation is the main disadvantage of this approach since the composite properties depend only on the filler fraction.

Micromechanical homogenization techniques like Mori-Tanaka or Eshelby take the orientation of the carbon nanotubes into account. Therefore the strain concentration tensor A_f is necessary and can be computed as

$$A_f = \left[I + E_f C_m^{-1} (C_f - C_m) \right]^{-1} \tag{10}$$

with the Eshelby tensor E_f and the identity tensor I [5]. The Eshelby tensor depends on the filler geometry and Poisson's ratio of the polymer. The Mori-Tanaka homogenization of the composite is obtained by

$$\overline{C} = C_m + v_f \left\langle (C_f - C_m) A_f \right\rangle \left[v_m I + v_f \left\langle A_f \right\rangle \right]^{-1} \tag{11}$$

where the volume average is denoted by the angle brackets [8]. These homogenization techniques can be used to calculate the stiffness matrix of the CNT-polymer-composites. The material properties can also be computed via ANSYS simulations to verify the results.

III. SIMULATION RESULTS

The equivalent cylinder properties are calculated for CNTs with radii from 0.4 nm up to 2.3 nm and four different aspect ratios (1, 2, 5, 10) using the force field parameters from [3] and strains of one percent for every load case. For the homogenization eight and sixteen (30,30)-CNTs with an aspect ratio of ten are embedded in a soft isotropic polymer ($E = 500$ kPa, $\nu = 0.49$) to reach a CNT volume fraction of 0.5 % and 1 %. In this work the ten CNT-polymer-composites per fraction are obtained via Monte-Carlo simulations where the carbon nanotubes are distributed randomly in the polymer. The Eshelby tensor was calculated using a spheroid inclusion according to [5].

In addition to the micromechanical homogenization using (8)-(11) the composite properties are obtained via average strains and stresses from the ANSYS simulations to verify the results. Therefore the hollow and solid equivalent cylinders are used partially in the simulations. To reach a reasonable computation time of the ANSYS simulations the number of carbon nanotubes is limited to a small amount of eight and sixteen.

A. Equivalent cylinder

Fig. 2 shows the Young's and shear moduli of the equivalent hollow and solid cylinders for the simulated carbon nanotubes. The Young's modulus of the hollow cylinder in z-direction is near 1050 GPa while the one of the solid cylinder decreases with increasing radius. A similar behavior is observable for the other material parameters in Fig. 2. Thus the moduli of the hollow cylinder remain almost constant while the properties of the solid one decreases for larger radii. The difference of the moduli between solid and hollow cylinder is caused by the volume in (7) such that the increasing radius leads to a bigger increase of the solid volume compared to the hollow one. The Poisson's ratios in this case are equal for both cylinders and remain nearly constant: $\nu_{xy} = 0.33$, $\nu_{xz} = \nu_{yz} = 0.06$ and $\nu_{zx} = \nu_{zy} = 0.02$. Similar results of the parameters E_z and G_{yz}/G_{xz} of the hollow cylinder are obtained by [2], [3].

B. CNT-polymer-composite

The homogenization result of the CNT-polymer-composites with 0.5 % nanotubes is shown in Fig. 3 where nearly the same shifts of the average Young's modulus are obtained by the Mori-Tanaka (MT) approach using solid and hollow cylinders and the finite element (FE) simulations with the solid cylinders. The Reuss approach underestimates the shift while the FE results of the hollow cylinder overestimates it. Additionally the results of the

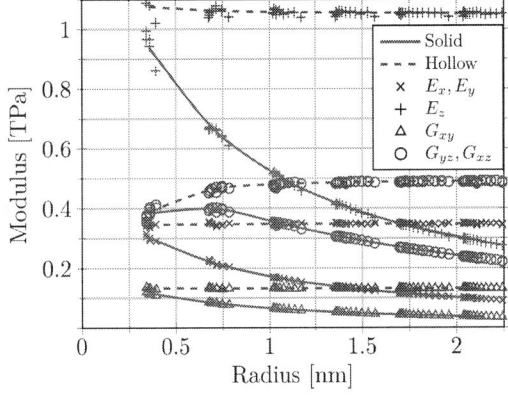

Fig. 2. Young's and shear moduli of equivalent solid and hollow cylinders for different carbon nanotubes with an aspect ratio of ten

978-1-5386-4483-6/18 $31.00 © 2018 IEEE

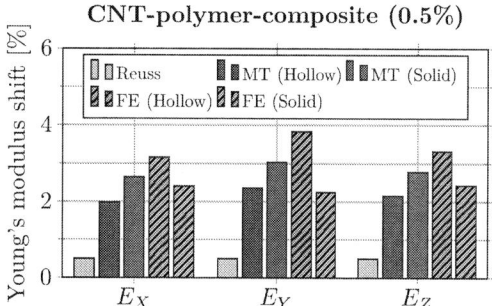

Fig. 3. Average Young's modulus shift of ten CNT-polymer-composites with eight (30,30)-CNTs (0.5 %)

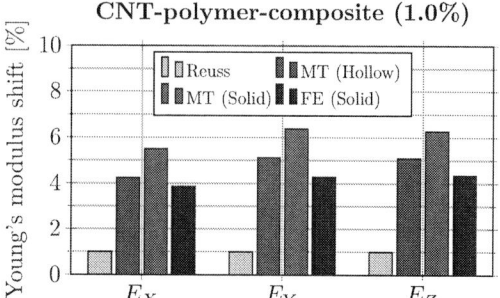

Fig. 5. Average Young's modulus shift of ten CNT-polymer-composites with sixten (30,30)-CNTs (1.0 %)

Eshelby method are neglected in this case since the small filler fractions lead to the same results like Mori-Tanaka. The different results of the FE simulations with hollow and solid cylinders are caused by the small wall thickness of the (30,30)-CNT compared to its radius. Thus the element size of the mesh in ANSYS induces this problem for thin walled cylinders like in this case. Considering this, the result of five additional FE simulations with hollow and solid cylinder using (10,10)-CNTs with a smaller radius is shown in Fig. 4. It's visible that the Young's modulus shift of the FE simulations and the Mori-Tanaka approach are nearly the same such that the hollow cylinder is not applicable in ANSYS if the radius is much bigger than the wall thickness.

The homogenization result of the CNT-polymer-composites with 1.0 % nanotubes is shown in Fig. 5. In this case the Young's modulus shifts agree with the previous findings except that Mori-Tanaka using solid cylinders overestimates the change of modulus. The difference between the homogenization techniques can be caused by the element size of the mesh in ANSYS and the unequal Youngs' modulus of equivalent cylinder and polymer.

IV. CONCLUSION

A finite element approach for the multi-scale modeling of CNT-polymer-composites is presented in this paper. It starts with the computation of the equivalent cylinder of

a carbon nanotube with beam elements using an energy approach. Regarding solid cylinders this method performs well while the assumptions of the hollow one can cause errors in the ANSYS simulations. After that, these cylinders are embedded in the polymer matrix such that homogenization approaches and FE simulations lead to nearly the same properties of the composite. The time-consuming FE simulations are necessary to validate the results of the homogenization since there are no measurements of the CNT-polymer-composites which are essential for future work to evaluate the simulation results.

ACKNOWLEDGMENT

This research has been funded by the European Social Fund (ESF) in the framework of SenseCare (project number 100270070).

REFERENCES

[1] B. R. Gelin, *Molecular Modeling of Polymer Structures and Properties*. HANSER GARDNER PUBL, Jan. 11, 1994.
[2] C. Li and T.-W. Chou, "A structural mechanics approach for the analysis of carbon nanotubes," *Int. J. Solids Struct.*, vol. 40, no. 10, pp. 2487–2499, May 2003.
[3] K. Tserpes and P. Papanikos, "Finite element modeling of single-walled carbon nanotubes," *Composites Part B: Engineering*, vol. 36, no. 5, pp. 468–477, Jul. 2005.
[4] L. Shen and J. Li, "Transversely isotropic elastic properties of single-walled carbon nanotubes," *Phys. Rev. B*, vol. 69, 045414 (1–10), 4 Jan. 2004.
[5] G. Odegard, T. Gates, K. Wise, C. Park, and E. Siochi, "Constitutive modeling of nanotube–reinforced polymer composites," *Compos. Sci. Technol.*, vol. 63, no. 11, pp. 1671–1687, Aug. 2003.
[6] Y. Liu and X. Chen, "Evaluations of the effective material properties of carbon nanotube-based composites using a nanoscale representative volume element," *Mech. Mater.*, vol. 35, no. 1-2, pp. 69–81, Jan. 2003.
[7] R. Hill, "Elastic properties of reinforced solids: Some theoretical principles," *J. Mech. Phys. Solids*, vol. 11, no. 5, pp. 357–372, Sep. 1963.
[8] Y. Benveniste, "A new approach to the application of mori-tanaka's theory in composite materials," *Mech. Mater.*, vol. 6, no. 2, pp. 147–157, Jun. 1987.

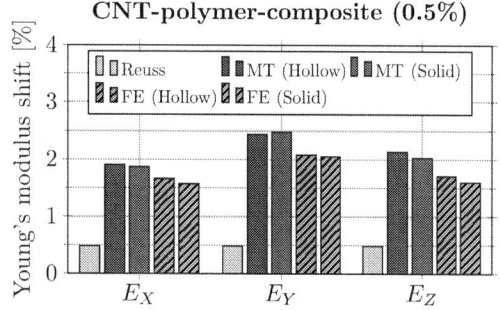

Fig. 4. Average Young's modulus shift of five CNT-polymer-composites with ten (10,10)-CNTs (0.5 %)

978-1-5386-4483-6/18 $31.00 © 2018 IEEE

Modeling of High Total Ionizing Dose (TID) Effects for Enclosed Layout Transistors in 65 nm Bulk CMOS

Aristeidis Nikolaou[1], Matthias Bucher[1], Nikolaos Makris[1], Alexia Papadopoulou[1],
Loukas Chevas[1], Giulio Borghello[2,4], Henri D. Koch[3,4], Federico Faccio[4]

[1]School of Electrical and Computer Engineering, Technical University of Crete, 73100 Chania, Greece
[2]DPIA, Università degli Studi di Udine, 33100 Udine, Italy
[3]SEMi, Université de Mons, 7000 Mons, Belgium
[4]EP Dept., CERN, 1211 Geneva, Switzerland
E-mail:anikolaou@isc.tuc.gr

Abstract— High doses of ionizing radiation drastically impair the electrical performance of CMOS technology. Enclosed gate layout remains an effective means to reduce this impact. Nevertheless, high total ionizing dose (TID) effects remain strong. The paper presents an effective approach to analytically model high TID effects in both NMOS and PMOS transistors with enclosed-gate layout in 65 nm commercial CMOS.

Keywords— Compact modeling, EKV model, enclosed gate MOSFETs, high energy physics, high total ionizing dose, radiation, space applications.

1. Introduction

Electronics operating in extreme conditions like space environments and high-energy physics (HEP) systems, are expected to present significant performance degradation due to their exposure to ionizing radiation. The forthcoming update of the Large Hadron Collider (LHC) at CERN, aims to increase the rate of collisions (luminosity) by a factor of 10; as a result, electronics in the innermost locations of the detector, closer to the collision sites, are expected to be exposed to a cumulative ionizing dose up to 1Grad [1]. Commercial non radiation-hardened CMOS processes, in 130 and 65nm nodes, because of the thin gate oxides, are proven to be advantageous in suppressing TID related degradation effects [2]. Additionally, MOS transistors with enclosed-gate (EG) layout, due to the absence of shallow trench isolation (STI) field oxide, are expected to exhibit enhanced resilience to high TID [3], as well as improved mismatch and low frequency noise characteristics [4].

2. Physical Process of TID Effects

High-energy incident photons, electrons or protons are capable of ionizing atoms creating electron-hole (e-h) pairs [5]. When a MOSFET is exposed to high-energy ionizing radiation, e-h pairs are generated at random sites in the SiO_2 lattice. Quantitatively, $1rad(SiO_2)$ of total absorbed dose generates $\sim 8.1 \cdot 10^{12}$ pairs/cm^3 [6].

Holes that escape from initial recombination will move through the SiO_2 lattice by polaron hopping as a result of shallow trap sites [7]. Close to the Si/SiO_2 interface and inside the oxide, oxygen vacancies (lattice defects) acting as trapping centers will capture a significant number of holes. Oxide trapped charges Q_{ot} are positive for both n- and pMOSFETs and will cause negative voltage shift in either case [8].

Additionally, radiation introduces the formation of traps at the Si/SiO_2 interface. Holes traversing the oxide lattice and oxide charge buildup mechanism described above, will release H+ (Hydrogen) ions that will drift toward the Si/SiO_2 interface forming electrically active interface traps Q_{it}. In a pMOSFET interface traps are predominately positive causing negative threshold voltage shift. Conversely, an n-channel device is mainly affected by negatively charged traps and therefore the threshold voltage will be positively shifted [9]. Table 1 lists individual features of oxide and interface traps.

978-1-5386-4483-6/18 $31.00 © 2018 IEEE

In standard CMOS processes the use of STI oxides can significantly boost the positive charge build up due to TID at the edges of the STI oxide. This charge accumulation can invert the p-type surface forming an n-type region underneath the oxide. As the surface inverts, parasitic conduction paths can significantly increase the leakage current. Since radiation charge build-up is primarily positive, the effect is more pronounced in nMOS transistors [5].

Table 1. Q_{ot} and Q_{it} Impact on V_{TH}

	Sign		V_{TH} shift	
	Q_{ot}	Q_{it}	Q_{ot}	Q_{it}
nMOS	+	-	-	+
pMOS	+	+	-	-

3. Experimental Setup

Standard threshold voltage enclosed gate (EG) nMOS transistors of 65 nm bulk CMOS process, with tetragonal layout, were exposed to radiation levels up to 500 Mrad(SiO$_2$) at room temperature of 25°C. Irradiation experiments were carried out at CERN using a 10 keV X-ray source at a high dose rate ~9 Mrad/h. During the irradiation, devices are biased under worst-case conditions ($V_G=V_D=V_{DD}=1.2$ V). At each targeted TID, irradiation is stopped and an identical DC characterization protocol is followed. Measured IV characteristics comprise transfer characteristics in linear ($V_{DS}=0.02$ V) and saturation regimes ($V_{DS}=1.2$ V), and output characteristics. Note that in this work, possible annealing effects are not investigated.

The examined EG transistors are operated with the drain in the center and have their widths varying according to their channel length. The effective W/L ratio can be approximated by [3],

$$(W/L)_{eff}=8/\ln(D_2/D_1). \qquad (1)$$

In Fig. 1 the available devices within the W-L space are depicted.

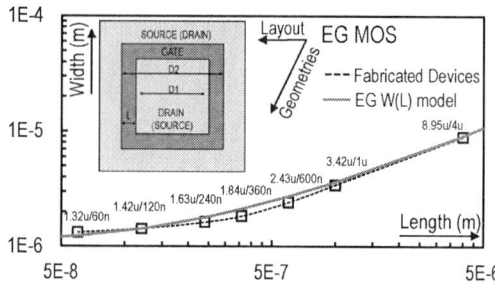

Fig. 1. Available fabricated EG-&ST MOS devices vs. geometry (width, length). For EG devices widths varying according to channel lengths. Typical layout of a tetragonal EG MOS and W/L ratio approximation described by (1) (line) are also depicted.

4. Modeling Approach

In the current section, a simplified analytical model proposed in [10] will be used in order to capture the TID effects in both n- and p-type EG MOSFETs. Inversion coefficient IC is a numerical measure of the inversion level of the channel and is defined as IC=max(i_f, i_r), where i_f and i_r are the normalized forward and reverse current components forming the drain current of the transistor. All currents are normalized to I_{spec} according to,

$$I_{spec}=I_0\frac{W}{L}=\frac{I_F}{i_f}=\frac{I_R}{i_r}=\frac{I_D}{i_d}, \qquad (2)$$

where $I_0=2n\mu C_{ox}U_T^2$ is the technology current. In the latter, $U_T=kT/q$ is the thermal voltage, n is the slope factor, μ is the carrier mobility and C_{ox} is the SiO$_2$ capacitance per unit area. The basic voltage-inversion charge equation is defined as,

$$v_p-v_{s,d} \approx \frac{V_G-V_{TO}-nV_{S,D}}{nU_T}=2q_{s,d}+\ln(q_{s,d}), \qquad (3)$$

where v_p is the normalized pinch-off voltage and $q_{s,d}$ the normalized mobile charge densities at the source and drain end of the channel, respectively.

Parameter λ_c introduced at [11] accounts for velocity saturation (VS) according to,

$$\lambda_c=\frac{L_{sat}}{L}, \qquad (4)$$

where L_{sat} is a characteristic length for velocity saturation. When velocity saturation is present drain current saturates when q_d reaches a saturated value $q_d=q_{dsat}$.

978-1-5386-4483-6/18 $31.00 © 2018 IEEE

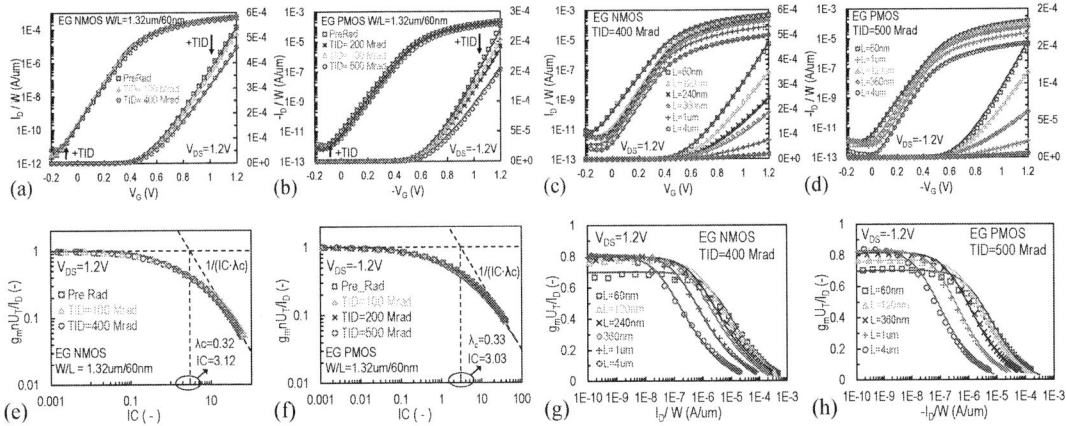

Fig. 2. Transfer characteristics (a, b) and normalized transconductance efficiency vs. IC (e, f), for short-channel (L=60nm) EG n- and pMOSFETs, at increasing TID levels. Normalized drain current I_D/W vs. gate voltage V_G (c, d) and transconductance-to-current-ratio (g, h) for several channel lengths after maximum TID exposure. Markers: measurements, lines: model.

Hence,

$$i_{dsat}=IC=q_s^2+q_s-q_{dsat}^2-q_{dsat}. \qquad (5)$$

Following [10], in terms of λ_c drain current is given by

$$i_{dsat}=\frac{2q_{dsat}}{\lambda_c}. \qquad (6)$$

By using (5) and (6), the following can be obtained,

$$2q_s=\sqrt{(\lambda_c IC+1)^2+4IC}-1. \qquad (7)$$

Equations (3), (4) and (7) constitute an EKV-type model that can predict accurately degradation effects resulting from high TID exposure in saturated transistors.

A set of parameters (I_0, V_{TO}, n, λ_c) is extracted for every geometry and TID level as follows: Following [12], in the widest- and –longest device available, $I_D=I_0$ when $g_mU_T/I_D=0.618(g_mU_T/I_D|_{max})$. Then, I_{SPEC} can be calculated for the rest of the samples using (2). Parameters V_{TO} and slope factor n are extracted in weak inversion, the former by fitting the model for every individual transfer characteristic and the latter from the maximum g_mU_T/I_D plateau. Parameter λ_c is extracted in strong inversion. In each case leakage current is adapted to the observed leakage level.

In Fig. 2(a, b) the described model is adapted to the transfer characteristics of short channel EG n- and pMOSTs (L=60 nm) after exposure at increasing TID levels. In Fig. 2 (e, f), the behaviour of the model over the measured normalized transconductance efficiency for the same devices and TID levels is shown. In Fig. 2(c, d, g, h) transfer characteristics and transconductance-to-current-ratio resulting from maximum TID exposure, for all available channel lengths and both types of transistors, are presented. In all cases both measurements (markers) and model (solid lines) are shown.

In Fig. 3, the extracted model parameters versus channel length, for both EG n- and pMOSTs, for all the different TID conditions are demonstrated. The λ_c parameter, also shown in the same figure, is dependent, essentially, on channel length, and only marginally on TID. Equation (4) shows the typical dependence of λ_c on channel length. In this work a different model is proposed,

$$\lambda_c(L) = \lambda_0+\frac{L_b}{L}. \qquad (8)$$

Parameters λ_0, L_b are listed in Table 2. Interestingly, the velocity saturation effect is, basically unaffected by TID for both n- and p-type EG MOSFETs. Typical L_{sat}/L dependence of λ_c can be observed towards shorter channel lengths. The rest of the parameters follow expected variations [13] after exposure to high TID.

978-1-5386-4483-6/18 $31.00 © 2018 IEEE

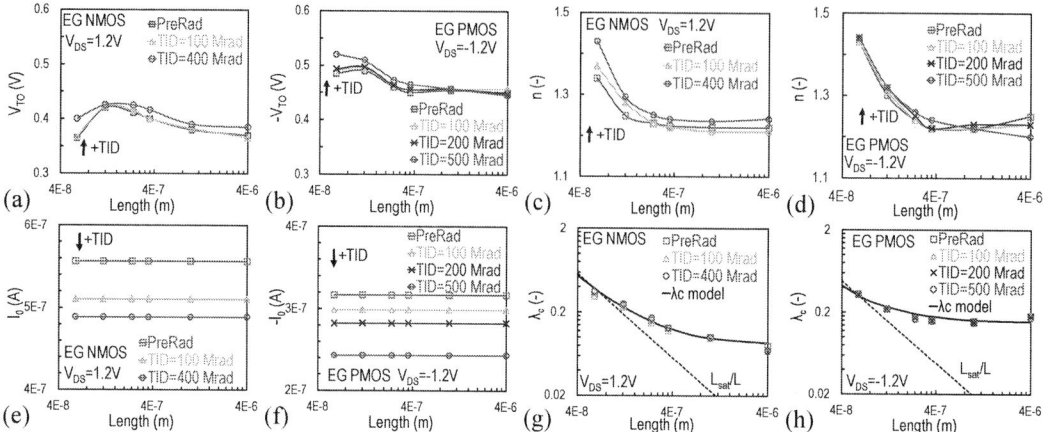

Fig. 3. Extracted parameters V_{TO}, n and I_0 vs. channel length, at increasing TID levels, for EG n- (a, c, e) and p-(b, d, f) type MOSFETs. Velocity saturation related parameter λ_c vs. L at various TID levels (g, h) is also depicted. Markers: extracted λ_c value from measurements, lines: proposed model (8), dashed lines: L_{sat}/L.

Table 2. List of λ_c model parameters

Parameters	Units	EG NMOS	EG PMOS
λ_0	-	0.08	0.15
L_b	m	18n	10n
L_{sat}	m	19.9n	19.2n

5. Conclusions

In the present work, an analytical approach has been demonstrated to modeling saturated drain current, transconductance, and transconductance-to-current ratio of enclosed-gate NMOS and PMOS transistors up to 500 Mrad TID from short to long-channel. The proposed charge-based model covers velocity saturation effects, while parameters such as threshold voltage, slope factor, specific and leakage current, are adapted over channel length.

References

[1] L. Ratti e.a., "Front-end channel in 65 nm CMOS for pixel detectors at the HL-LHC experiment upgrades," IEEE Trans. Nuclear Science, 64(2), 789, Feb. 2017.

[2] F. Faccio e.a., "Radiation-induced short channel (RISCE) and narrow channel (RINCE) effects in 65 and 130 nm MOSFETs," IEEE Trans. Nuclear Science, 62(6), 2933, Dec. 2015.

[3] W. Snoeys e.a.,"Layout techniques to enhance the radiation tolerance of standard CMOS technologies demonstrated on a pixel detector readout chip," Nucl. Instr. Meth. Phys. Res. A, 439, 349, 2000.

[4] M. Bucher e.a., "Variability of low frequency noise and mismatch in enclosed-gate and standard nMOSFETs," IEEE ICMTS, Grenoble, France, Mar. 2017.

[5] J. R. Schwank e.a., "Radiation effects in MOS oxides," IEEE Trans. Nuclear Science, 55(4), 1833, Aug. 2008.

[6] F. B. McLean e.a., Basic mechanisms of radiation effects in electronic materials and devices, Harry Diamond Laboratory, 1987, Tech. Rep. HDL-TR-2129.

[7] R. C. Hughes, "Time-resolved hole transport in a-SiO₂," Phys. Rev. B, Condens. Matter, 15(4), 2012, 1977.

[8] C.-M. Zhang e.a., "Characterization of gigarad total ionizing dose and annealing effects on 28-nm bulk MOSFETs," IEEE Trans. Nuclear Science, 64(10), 2639, Oct. 2017.

[9] P. J. McWhorter e.a., "Simple technique for separating the effects of interface traps and trapped-oxide charge in metal-oxide semiconductor transistors," Appl. Phys. Lett., 48(2), 133, 1986.

[10] C. C. Enz, E. A. Vittoz, Charge-based MOS transistor modeling. John Wiley 2006.

[11] C. Enz e.a., "Low-power analog/RF circuit design based on the inversion coefficient." European Solid-State Circuits Conference (ESSCIRC), pp. 202-208, 2015.

[12] A. Bazigos e.a., "An adjusted constant-current method to determine saturated and linear mode threshold voltage of MOSFETs," IEEE Trans. Electron Devices, 58(11), 3751, Nov. 2011.

[13] M. Bucher e.a., "Total ionizing dose effects on analog performance of 65 nm bulk CMOS with enclosed-gate and standard layout," IEEE ICMTS, 166, Austin, Texas, Mar. 2018.

Analytical analysis of the plasmonic enhancement of resonance energy transfer in the vicinity of a spherical nanoparticle

Titus Sandu*, Catalin Tibeica, Oana T. Nedelcu, Mihai Gologanu

National Institute for Research and Development in Microtechnologies-IMT,
126A, Erou Iancu Nicolae Street, Bucharest, Romania
E-mail: titus.sandu@imt.ro

Abstract—The enhancement factor of intermolecular energy transfer in the vicinity of a plasmonic spherical nanoparticle is calculated analytically. In contrast to other treatments, the present calculations exploit the knowledge of spectral properties of the electrostatic operator for spherical geometry. Some numerical calculations and further discussions are also provided.

Keywords—Plasmonics; boundary integral equation method; electrostatic operator; finite element method; Förster resonance energy transfer.

1. Introduction

The energy transfer between an excited donor molecule and other close by acceptor molecule is very important for the control and understanding of numerous photophysical/photochemical processes from photosynthesis to fluorescence probing in biotechnology [1]. The models for this process depend on the separation distance between donor and acceptor. For small distances (less than 2 nm) the treatment is quantum mechanical [2, 3], while for distances between 2 and 10 nm there is a nonradiative energy transfer via electrostatic dipole–dipole interaction also known as Förster resonance energy transfer (FRET) [2, 4]. FRET is a quite weak process which behaves as the sixth power of the separation distance, hence its enhancement is more than needed in various applications. The coupling between light and the collective excitations of the free electrons in metals may lead to strong fields and confinements below diffraction limit in the vicinity of metallic nanostructures with many sensing applications [5]. Strong fields may imply strong coupling with atomic systems [6-8]. The enhancement of FRET may be obtained with planar metallic structures [9] as well as with various nanoparticles: spheres and spheroids [10] nanodiscs [11], nanorods [12], etc. Analytical treatments of plasmonic enhancement of FRET were invoked in [10]

for spheres and spheroids or for shelled spheres [13] and spheroids [14]. In [13] the calculations are based on Bergman's approach [15], which is an operator approach with linear operators defined in the whole space. In this work we use a close related method, a boundary integral equation approach, which is also an operator method but the operators are defined on surfaces rather than the whole space [16-19]. Our approach is based on well-behaved operators, hence the calculations are faster [18]. In addition to that, for spherical shape there are known all spectral properties of this surface operators [20], consequently, as it will be seen in the next section, the calculations are straightforward. Moreover, the plasmonic enhancement terms can be singled-out in the final expression of the total enhancement factor of FRET. The paper has the following structure: in section 2 we present the analytical treatment, in section 3 we perform some numerical estimations and we conclude our work in section 4.

2. Analytical Treatment of the Plasmonic Enhancement of FRET

A. The model

In a classical picture, the molecules (the donor and the acceptor) are associated with point-like dipoles, d_D and d_A, respectively. The transfer of energy between the donor (D) and the acceptor (A) is governed by the dipole-dipole interaction [2,3,4]. The model for the enhancement energy transfer considers D and A as harmonic point-like dipoles interacting with each other directly or via the metallic nanoparticle (Fig. 1) [10]. The total electric potential for an arbitrary position has four terms

978-1-5386-4483-6/18 $31.00 © 2018 IEEE

$$\Phi(\boldsymbol{r}) = \Phi_A(\boldsymbol{r}) + \Phi_D(\boldsymbol{r}) + \Phi_{Aind}(\boldsymbol{r}) + \Phi_{Dind}(\boldsymbol{r}) \quad (1)$$

The first two terms are the electric potentials generated by A and D and the third and fourth terms are the electric potentials of the charge induced on the nanoparticle. The electric field at the acceptor site is

$$\boldsymbol{E}_A = -\nabla\left[\Phi_D(\boldsymbol{r}_A) + \Phi_{Dind}(\boldsymbol{r}_A) + \Phi_{Aind}(\boldsymbol{r}_A)\right] \quad (2)$$

Thus the interaction energy for the acceptor in the presence of both the donor and the nanoparticle is the following

$$U_A = -\boldsymbol{d}_A \cdot \boldsymbol{E}_A = U_{AD} + U_{ADind} + U_{AAind}. \quad (3)$$

The enhancement factor for the FRET process in the presence of the metallic nanoparticle is simply

$$|A|^2 = \left|1 + \frac{U_{ADind}}{U_{AD}}\right|^2. \quad (4)$$

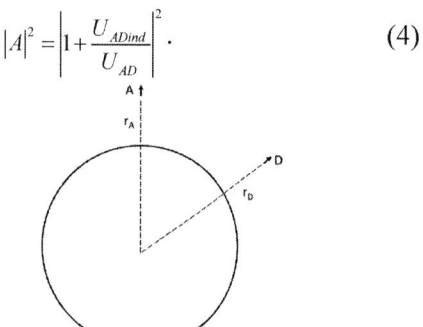

Fig. 1 The model and the geometry for the donor (D) and acceptor (A) nearby a spherical metallic nanoparticle. The A and D dipoles are oriented normal to the spherical surface.

B. Electrostatic Calculations

In the electrostatic limit we will need to know the change in the total electric field $\boldsymbol{E}$ occurred by the presence of a metallic nanoparticle. We assume that the initial charge distribution has temporal evolution of the form $exp(j\omega t)$ and generates an electric field $\boldsymbol{E}_0$ coming from a potential, i. e., $\boldsymbol{E}_0 = -\nabla\Phi_0$. The charge density induced on the surface of the nanoparticle is [16-19]

$$\sigma = \sum_k \frac{1}{\frac{1}{2\lambda} - \chi_k} u_k \langle v_k | \boldsymbol{n}\boldsymbol{E}_0(\boldsymbol{r}) \rangle \quad (5)$$

where: $\lambda = (\epsilon_1 - \epsilon_0)/(\epsilon_1 + \epsilon_0)$, ϵ_1 is the dielectric permittivity of the nanoparticle, ϵ_0 is the dielectric permittivity of the embedding medium, u_k and χ_k are the eigenfunctions and the eigenvalues of the electrostatic

operator defined on Σ of the nanoparticle as [10,12]

$$\hat{M}[\sigma] = -\frac{1}{4\pi}\int_{r'\in\Sigma}\sigma(\boldsymbol{r}')\frac{\partial}{\partial\boldsymbol{n}_r}G(\boldsymbol{r},\boldsymbol{r}')d\Sigma_{r'}, \quad (6)$$

and $\langle v_k | \boldsymbol{n}\boldsymbol{E}_0(\boldsymbol{r}) \rangle$ are the scalar products between the corresponding eigenfunctions v_k of the adjoint operator $M^\dagger$ and the dot product $\boldsymbol{n}\cdot\boldsymbol{E}_0$, with $\boldsymbol{n}$ the normal to the surface Σ of the nanoparticle. In Eq. (6)

$$G(\boldsymbol{r},\boldsymbol{r}') = \frac{1}{4\pi}\frac{1}{|\boldsymbol{r}-\boldsymbol{r}'|} \quad (7)$$

is the free-space Green function and $\partial/\partial\boldsymbol{n}_r$ is the normal derivative. The total electric field is then given finally by

$$\Phi(\boldsymbol{r}) = \Phi_0(\boldsymbol{r}) + \int_{r'\in\Sigma}G(\boldsymbol{r},\boldsymbol{r}')u(\boldsymbol{r}')d\Sigma_{r'}. \quad (8)$$

As we recently analyzed [20], due to its symmetry, in spherical coordinates the free-space Green function (7) has an explicit form provided by the separation of variables method

$$G(\boldsymbol{r},\boldsymbol{r}') = \sum_{l,m}\frac{1}{2l+1}\frac{r_<^l}{r_>^{l+1}}Y_{lm}^*(\theta',\varphi')Y_{lm}(\theta,\varphi), \quad (9)$$

where Y_{lm} are the spherical harmonics. Eq. (9) is a well-known result that may be found in any textbook treating classical electrodynamics. What was less known is the fact that a separated form like Eq. (9) generates also all the eigenvectors and the eigenvalues of (6) not only for spherical shape but also for other shapes like spheroidal, ellipsoidal, etc. [20]. For sphere the electrostatic operator is symmetric. Thus, for a sphere of radius a one has

$$u_{lm}(\theta,\varphi) = v_{lm}(\theta,\varphi) = \frac{Y_{lm}(\theta,\varphi)}{a}, \quad (10)$$

$$\chi_l = \frac{1}{2(2l+1)}. \quad (11)$$

The electric potential of a dipole $\boldsymbol{d}$ located at $\boldsymbol{r}'$ in a point $\boldsymbol{r} < \boldsymbol{r}'$ (near field) may be evaluated with Eq. (9) as

$$\Phi_d(\boldsymbol{r}) = \boldsymbol{d}\cdot\nabla G_{r'}(\boldsymbol{r},\boldsymbol{r}')$$
$$= d\sum_{l,m}\frac{l+1}{2l+1}\frac{r^l}{r'^{l+1}}Y_{lm}^*(\theta',\varphi')Y_{lm}(\theta,\varphi). \quad (12)$$

Moreover, any induced charge density on a sphere of radius a with its center at the origin has the following expansion

$$\sigma = \sum_{lm} \sigma_{lm} \frac{Y_{lm}(\theta, \varphi)}{a} \qquad (13)$$

Using Eqs. (5), (6), and (9-13), the induced charge density of a donor of a dipole strength d_D positioned at r_D has the following spherical components

$$\sigma_{D,lm}(\theta, \varphi) = d_D l(l+1) \times$$
$$\frac{\epsilon_1 - \epsilon_0}{(1/2 - \chi_l)\epsilon_1 + (1/2 + \chi_l)\epsilon_0} \frac{a^l}{r_D^{l+2}} Y_{lm}(\theta_D, \varphi_D) \qquad (14)$$

Now, using Eqs. (8), (9), and (13) it can be calculated the interaction energy on an acceptor with a dipole d_A positioned at r_A as

$$U_{ADind} = d_A d_d \sum_{lm} \frac{l+1}{2l+1} \frac{a^{l+1}}{r_A^{l+2}} \sigma_{D,lm}(\theta_D, \varphi_D) Y_{lm}(\theta_A, \varphi_A), \qquad (15)$$

while U_{AD} is the interaction energy between two dipoles. When the acceptor and the donor are on the opposite sites with respect to the sphere ($\theta_A = 0$, $\theta_D = \pi$) the enhancement factor takes the form

$$|A|^2 = \left| 1 + \frac{(r_A + r_D)^2}{2a^3} \sum_l (-1)^{l+1} \frac{(l+1)^2 (\epsilon_1 - \epsilon_0)}{\epsilon_1 + \epsilon_0 (l+1)/l} \left(\frac{R^2}{r_A r_D} \right)^{l+2} \right|^2. \qquad (16)$$

Eq. (16) are the same as those found in the literature [10,13], but our calculations are more straightforward and systematic.

3. Numerical Results

The FRET between a donor molecule and an acceptor molecule is also studied in the vicinity of a spherical metallic nano-particle with a finite element method (FEM). The nanoparticle is modeled as a sphere with radius of 25 nm, made by Ag with dielectric function described by a Drude model $\epsilon_1(\omega) = \epsilon_0 \left(\epsilon_\infty - \omega_p^2 / (\omega(\omega + i\delta)) \right)$ with the following parameters: $\varepsilon_\infty = 5$, $\omega_p = 9.5$ eV, and $\delta = 0.15$ eV. The donor molecule is modeled as a point dipole with radial orientation with respect to the sphere and placed at a distance of 30 nm from the center of coordinates, or 5 nm from the nano-particle's surface. The acceptor molecule is symmetrically placed in

respect to the center of nano-particle. The numerical simulations were performed by using the AC/DC electrostatic solver, in frequency domain, of COMSOL Multiphysics software.

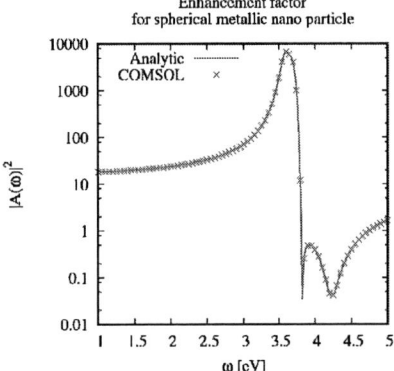

Fig. 2 Analytic (red solid line) versus FEM (symbols) calculations of FRET enhancement factor as function of frequency expressed in energy units.

The computed enhancement factor as function of frequency (in eV) is represented in Fig. 2. There is a pretty good match between analytical [Eq. (16)] and FEM calculations. We found out a maximum value of the enhancement factor of 6673 at 3.6 eV. In order understand the role of the nanoparticle in the FRET enhancement we analyze the electric field configuration for two representative energies: 0.5 and 3.6 eV.

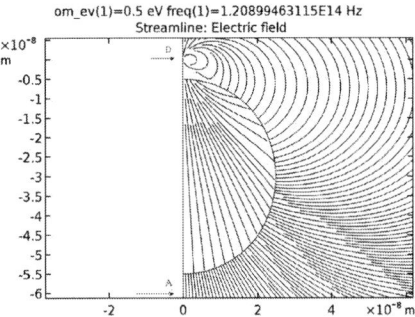

Fig. 3 Streamlines of the electric field in the off-resonance regime at 0.5 eV. The arrows represent the donor and the acceptor.

In the off-resonance regime at 0.5 eV (Fig. 3) the role of the nanoparticle is to enhance the electric field generated by the donor dipole. However, in the on-resonance regime at 3.6 eV (Fig. 4) the role of the nanoparticle is dramatically changed. The nanoparticle becomes a very large and extended dipole that enhances the electric field at the acceptor site.

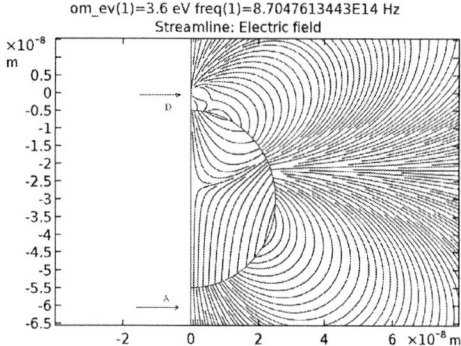

Fig. 4 Streamlines of the electric field in the on-resonance regime at 3.6 eV.

4. Conclusions

In this work we calculated analytically the enhancement of intermolecular energy transfer in the presence of a nearby spherical metallic nanoparticle. In contrast to other treatments, our treatment seems to be more straightforward since it exploits the knowledge of spectral properties of the electrostatic operator for spherical geometry. Our numerical calculations indicate that at large enhancements the nanoparticle becomes a large and extended dipole that positively affects the FRET process.

Acknowledgments. This research was supported by the institutional 2018 CORE-Programme financed by the Ministry of Research and Innovation..

References

[1] L. Stryer, "Fluorescence Energy Transfer as a Spectroscopic Ruler", Annu. Rev. Biochem., **47**, pp. 819-846, 1978.

[2] I. L. Medintz and N. Hildebrandt, "FRET - Förster Resonance Energy Transfer: From Theory to Applications", John Wiley, Weinheim, Germany, 2013.

[3] D. L. Dexter, "A Theory of Sensitized Luminescence in Solids", J. Chem. Phys., **21**, pp. 836-850, 1953.

[4] T. Förster, "Transfer mechanisms of electronic excitation", Discuss. Faraday Soc., **27**, pp. 7-17, 1959.

[5] N. J. Halas, S. Lal, W. -S. Chang, S. Link, and P. Nordlander, "Plasmons in Strongly Coupled Metallic Nanostructures", Chem. Rev., **111**, pp. 3913–3961, 2011.

[6] T. Sandu, V. Chihaia, and W. P. Kirk, "Dynamic squeezing in a singlemode boson field interacting with two-level system", J. Lumin., **101**, pp. 101, 2003.

[7] T. Sandu, "Dynamics of a two-level system coupled with a quantum oscillator: The very strong coupling limit", Phys. Rev. B, **74**, pp. 113405, 2006.

[8] T. Sandu, "Dynamics of a quantum oscillator strongly and off-resonantly coupled with a two-level system", Phys. Lett. A, **373**, pp. 2753, 2009.

[9] J. I. Gersten, "Fluorescence resonance energy transfer near thin films on surfaces," Plasmonics, **2**, pp. 65–77, 2007.

[10] J. I. Gersten and A. Nitzan, "Accelerated energy transfer between molecules near a solid particle," Chem. Phys. Lett., **104**, pp. 31–37 1984.

[11] F. Reil, U. Hohenester, J. R. Krenn, and A. Leitner, "Förster-type resonant energy transfer influenced by metal nanoparticles," Nano Lett. **8**, pp. 4128–4133, 2008.

[12] Y. C. Yu, J. M. Liu, C. J. Jin, and X. H. Wang, "Plasmon-mediated resonance energy transfer by metallic nanorods," Nanoscale Res. Lett. **8**, pp. 209, 2013.

[13] M. S. Shishodia, B. D. Fainberg, and A. Nitzan, "Theory of energy transfer interactions near sphere and nanoshell based plasmonic nanostructures", Proc. of SPIE Vol. 8096, pp. 80961G, 2011.

[14] H. Y. Chung, P. T. Leung, and D. P. Tsai, "Enhanced intermolecular energy transfer in the vicinity of a plasmonic nanorice," Plasmonics, **5**, pp. 363–368, 2010.

[15] D. J. Bergman, "The dielectric constant of a composite material-A problem in classical physics", Phys. Rep., **43**, pp. 377-407, 1978.

[16] T. Sandu, D. Vrinceanu and E. Gheorghiu, "Surface plasmon resonances of clustered nanoparticles", Plasmonics, **6**, pp. 407-412, 2011.

[17] T. Sandu, "Eigenmode decomposition of the near-field enhancement in localized surface plasmon resonances of metallic nanoparticles", Plasmonics **8**, pp. 391-402, 2013

[18] T. Sandu, G. Boldeiu and V. Moagar-Poladian, "Applications of electrostatic capacitance and charging", J. Appl. Phys., **114**, pp. 224904, 2013.

[19] T. Sandu, Near-Field and Extinction Spectra of Rod-Shaped Nanoantenna Dimers, Proc. of the Romanian Academy A, **15**, pp. 338-345, 2014.

[20] R.C. Voicu, T. Sandu, "Analytical results regarding electrostatic resonances of surface phonon/plasmon polaritons: separation of variables with a twist", Proc. R. Soc. A, **473**, pp. 2016079, 2017.

Session **SD**

SEMICONDUCTOR DEVICES

978-1-5386-4483-6/18 $31.00 © 2018 IEEE

Ψ-MOSFET Configuration for DNA Detection

Licinius Benea*, Melania Banu**, Maryline Bawedin*, Cécile Delacour***, Monica Simion**, Mihaela Kusko**, Sorin Cristoloveanu*, Irina Ionica*

*Univ. Grenoble Alpes, CNRS, Grenoble INP, IMEP-LAHC 38016 Grenoble, France
beneal@minatec.inpg.fr
**National Institute for Research and Development in Microtechnologies – IMT Bucharest, 126A Erou Iancu Nicolae Street, 077190, Bucharest, Romania
melania.banu@imt.ro
***Néel Inst., CNRS, 38042 Grenoble, France

Abstract— This work proposes a novel method for DNA detection by using the Ψ-MOSFET configuration. Systematic measurements of the drain current vs. gate voltage revealed an important shift of the characteristics corresponding to the charge of the biochemical species attached to the top surface of the device. The results were validated by fluorescent scanning. The advantages of this method are its simplicity and sensitivity.

Keywords—DNA detection, SOI, Ψ-MOSFET, field-effect, surface functionalization

1. Introduction

DNA biosensors, as well as microarrays, employ a chemically-modified surface on which single-stranded DNA (ssDNA) sequences called probes are attached [1]. The hybridization reaction consists in the recognition of DNA probe sequences with their complementary target DNA and the base pairing through hydrogen bonds. Depending on the transducing system, the measurement of the hybridization can be optical, electrical, electrochemical, ion-sensitive or mass-sensitive [2]. In DNA microarray technology, the hybridization is based on detecting the fluorescent signals coming from the labelled DNA target molecules [3]. Field effect transistor (FET) biosensors are to be acknowledged, since they enable a fast and label-free recognition [4]. The working principle of the FET sensor is based on the measurement of a conductance variation that is induced by the presence of charges to be detected close to the channel of the transistor. The pseudo metal oxide semiconductor FET (Ψ-MOSFET) can be adapted for this aim.

The Ψ-MOSFET is essentially a characterization technique used to determine the different parameters of a Silicon On Insulator (SOI) material such as the threshold voltage, the mobility or the interface trap density [5-7]. This technique uses the bulk silicon substrate as a back-gate and the buried oxide (BOX) as a gate dielectric. The gate voltage (V_G) induces a channel at the interface between the silicon film and the BOX. The conduction between source and drain is ensured by either holes or electrons depending on the V_G polarity. The coupling effect between the top free surface charge and the channel [8] leads to threshold voltage V_T and flat-band voltage V_{FB} variations. For intentionally modified SOI surfaces, these V_T/V_{FB} shifts proved to be convenient for chemical detection, making the Ψ-MOSFET a detection device [9,10]. Provided that the channel is very close to the surface (i.e. for thin silicon films), this device has a high sensitivity. This work aims to demonstrate that the Ψ-MOSFET can be successfully used as a DNA detector. The functionalization methods and the extraction of the electrical parameters V_T and V_{FB} are shown in section 2. Section 3 describes the experimental results.

2. Materials and Methods

A. Functionalization protocol

The utilized SOI substrates had a 145 nm thick BOX and a 70 nm silicon film. Islands of 5x5mm² were patterned by lithography and Reactive Ion Etching and cleaned for 10 min in Piranha solution (H_2SO_4:H_2O_2 3:1 v/v) in order to generate hydroxyl groups on the surface.

978-1-5386-4483-6/18 $31.00 © 2018 IEEE

The solution-phase silanization was achieved by immersing the SOI samples for 2 h in 2.5% hydrolysed (3-Aminopropyl)triethoxysilane (APTES), obtained by mixing water with ethanol in 1:19 v/v ratio. At the end of the silanization process, the biochips were rinsed with ethanol and deionized water, dried under N_2 stream and thermally treated for 30 min at 110 °C. The APTES-modified SOI substrates were further subjected to 5% glutaraldehyde (GAD) functionalization for 4 h, in order to have aldehyde active moieties.

B. Immobilization and hybridization of HPV-specific oligonucleotides

The studies regarding the probes' tethering and hybridization on functionalized SOI supports involved the use of sequences corresponding to L2 gene fragment from HPV 16, acquired from Biomers [11].

Table 1: Probe and target sequences.

HPV 16 specific probe	5'-NH₂-C6- TGGGAGGCCTTGTT CCCAATGGA-3'
Control probe (noncomplementary sequence)	5'-NH₂-C6- CTAGGAATTGCGGG AGGAAAATGGG-Cy3- 3'
Complementary sequence	5'- TCCATTGGGAACA AGGCCTCCCA-3'

The Cy3-labelled noncomplementary sequence was employed to certify by microarray scanning (GeneTAC UC4, Genomic Solutions, USA) the presence of DNA probes attached onto the surface, as an indicator of efficient functionalization and oligonucleotide attachment protocols.

C. Electrical characterization of the Ψ-MOSFET

The setup consists in two tungsten carbide probes (used as source and drain) positioned at a distance of 1 millimetre and having a pressure controlled system (**Fig. 1**). The electrical measurements of the detectors were conducted using an HP4155 Analyser. The drain bias was set at 100mV,

corresponding to the linear regime of the pseudo-transistor.

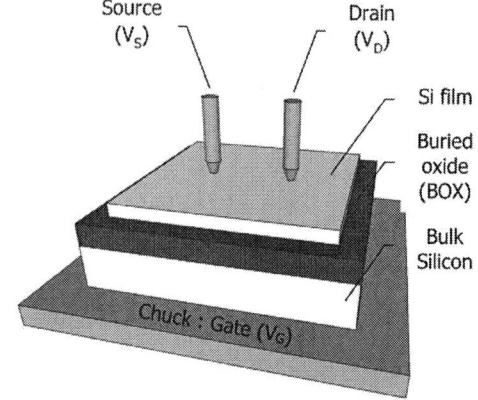

Fig. 1: Experimental setup for the Ψ-MOSFET configuration.

Fig. 2 shows a typical I_D-V_G curve, obtained before any surface treatment. The inversion regime is set for $V_G > V_T$, while an accumulation channel appears for $V_G < V_{FB}$.

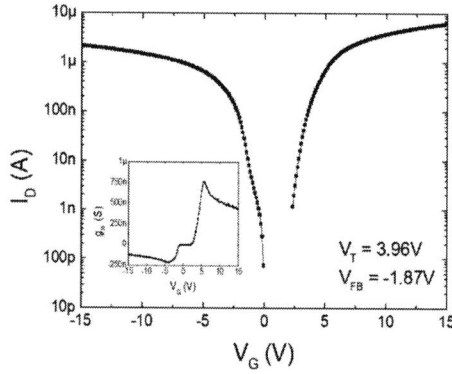

Fig. 2: Drain current and transconductance (inset) vs. gate voltage characteristics of the Ψ-MOSFET.

In order to avoid transient effects, V_G is scanned "slower" by using a hold time (time between measurements steps) which varies between 0.1s and 1s.

The I_D - V_G characteristics in the linear regime can be described by the classical MOS transistor equation [12]:

$$I_D = f_g C_{OX} V_D \frac{\mu_0}{1+\theta\left(V_G - V_{T,FB}\right)} \left(V_G - V_{T,FB}\right) \quad (1)$$

f_g is the geometric factor, C_{OX} is the buried oxide capacitance per unit area, V_T is the threshold voltage, V_{FB} is the flat band voltage (acting as a "threshold" voltage for

978-1-5386-4483-6/18 $31.00 © 2018 IEEE

the accumulation channel), μ_0 is the low-field mobility, and θ factor describes the vertical field mobility reduction.

The transconductance (**Fig. 2** - inset) can be evaluated using the following equation:

$$g_m = \frac{\partial I_D}{\partial V_G} = f_g C_{OX} V_D \frac{\mu_0}{\left[1 + \theta\left(V_G - V_{T,FB}\right)\right]^2} \quad (2)$$

In order to linearize the V_G dependence in equations 1 and 2, the Y-function is used to extract V_T and V_{FB}:

$$Y = \frac{I_D}{\sqrt{g_m}} = \sqrt{\mu_0 f_g C_{OX} V_D} \left(V_G - V_{T,FB}\right) \quad (3)$$

The carrier mobility is determined from the slope and the threshold/flat-band voltage is given by the intercept with the horizontal V_G axis.

3. Results

The fluorescence scanning result after the hybridization step is displayed in **Fig. 3**. The fluorescent spots in the active area can be distinguished from the background, demonstrating suitable substrate functionalization and DNA attachment protocols.

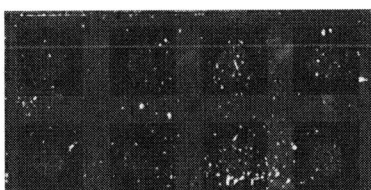

Fig. 3: Immobilization results on an SOI substrate functionalized with APTES and GAD.

Before the electrical measurements of the functionalized samples we evaluated the average and standard variations of V_T and V_{FB} for 19 bare SOI samples (shown in **Table 2**).

Table 2. Average and standard deviation of the threshold voltage and flat-band voltage.

	V_T (V)	V_{FB} (V)
Average	4.08	-1.74
Standard Deviation	0.19	0.34

V_T and V_{FB} were then extracted after each functionalization step; 6 devices were initially used for each step in order to determine the variation induced by each chemical/biological species. Note that, for a

part of the samples, the BOX was leaking after the chemical treatment, therefore the corresponding data is not represented.

The ΔV_T and ΔV_{FB} shifts in characteristics were calculated as a difference between the value after the respective step and the initial value of the same bare SOI device. The variation of the threshold voltage (**Fig. 4**) shows that the immobilization and hybridization steps induce uniform positive shifts of +0.31V and +0.3V, respectively. This corresponds to the negative charge of the DNA sugar phosphate backbone.

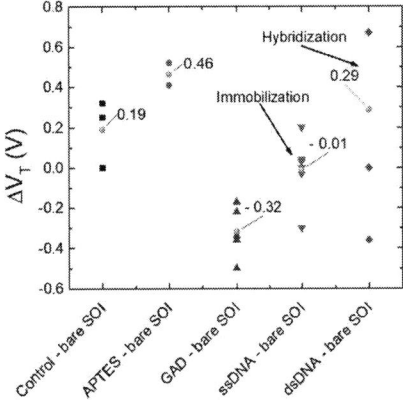

Fig. 4: Variation of the threshold voltage for different functionalization steps. Mean values represented in orange dots.

The same observation can be made in the case of the flat-band voltage (**Fig. 5**). The shift induced by the DNA detection is though lower: +0.04V for the immobilization step and +0.11V for the hybridization.

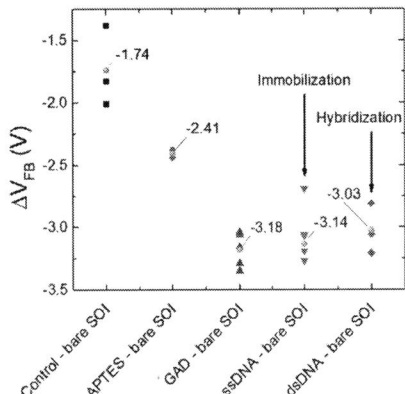

Fig. 5: Variation of the flat-band voltage for different functionalization steps. Mean values represented in orange dots.

It is important to notice that while the

tendency of the shift is clear, the variation is greater than the standard deviation in the case of the threshold voltage, but lower for the flat-band voltage.

Unlike for DNA, the effect of the APTES is less clear. This functionalization step is expected to leave a positive charged surface [13]. The V_{FB} shift is in agreement with the positive charge, but not the V_T. A more detailed analysis is needed to understand this effect.

4. Conclusion

This research attempted to assess the viability of the Ψ-MOSFET device as a DNA detector. We demonstrated the correlation between the charges added at the top surface of the device and the shift in the threshold voltage and the flat-band voltage. It is a promising detection method due to its simplicity and sensitivity. The article presented a proof of concept of this type of sensor. Future work will involve a more detailed experimental part and a larger number of samples, in order to define the sensor figures of merit (limit of detection, behaviour for different concentrations, selectivity, etc.) and benchmark with other electronic devices.

Acknowledgments. This work was supported by the AGIR-POLE CLEPS project funded by the University of Grenoble Alpes, Campus France PHC Brancusi (project 38 395QA) and the European projects WayToGoFast and REMINDER. The authors thank SOITEC for samples and inspiring comments.

The authors also acknowledge the support of the Romanian Ministry of Research and Innovation through PHC Brancusi Bilateral project – BIS-SOI – contract no. PNIII P3-3.1-PM-EN-FR-2016 and for project 67PCCDI 30/03/2018 within PN-III-P1-1.2-PCCDI-2017-0820.

References

[1] R. Monošík, M. Streďanský and E. Šturdík, *"Biosensors - classification, characterization and new trends"*, Acta Chim. Slovaca. **5**, pp. 109–120, 2012.

[2] A. Sassolas, B.D. Leca-Bouvier and L.J. Blum, *'DNA biosensors and microarrays'*, Chem. Rev., **108**, pp. 109–139, 2008.

[3] S.L. Lai, C. H. Chen and K. L. Yang, *"Enhancing the fluorescence intensity of DNA microarrays by using cationic surfactants"*, Langmuir, **27**, pp. 5659-5664, 2011.

[4] S. Cheng, S. Hideshima, S. Kuroiwa, T. Nakanishi and T. Osaka, *"Label-free detection of tumor markers using field effect transistor (FET)-based biosensors for lung cancer diagnosis"*, Sens. Actuators B Chem., **212**, pp. 329–334, 2015.

[5] S. Williams, S. Cristoloveanu and G. Campisi, *"Point contact pseudo-metal/ oxide/ semiconductor transistor in as-grown silicon on insulator wafers"*, Mater. Sci. Eng. B, **12**, pp. 191–194, 1992.

[6] S. Cristoloveanu, M. Bawedin and I. Ionica, *"A review of electrical characterization techniques for ultrathin FDSOI materials and devices"*, Solid-State Electron., **117**, pp. 10–36, Mar. 2016.

[7] S. Cristoloveanu, D. Munteanu and M.S. Liu, *"A review of the pseudo-MOS transistor in SOI wafers: operation, parameter extraction, and applications"*, IEEE Trans. Electron Devices, **47**, pp. 1018–1027, 2000.

[8] G. Hamaide, F. Allibert, H. Hovel and S. Cristoloveanu, *"Impact of free-surface passivation on silicon on insulator buried interface properties by pseudotransistor characterization"*, J. Appl. Phys., **101**, pp. 114513-1 - 114513-7, Jun. 2007.

[9] C. Fernandez, N. Rodriguez, C. Marquez and F. Gamiz, *"Determination of ad hoc deposited charge on bare SOI wafers"*, Ultimate Integration on Silicon (EUROSOI-ULIS), 2015 Joint International EUROSOI Workshop and International Conference, pp. 289–292, 2015.

[10] I. Ionica, A.E.H. Diab and S. Cristoloveanu, *"Gold nanoparticles detection using intrinsic SOI-based sensor"*, Nanotechnology (IEEE-NANO), 11th IEEE Conference, pp. 38–43, 2011.

[11] http://www.biomers.net/.

[12] S. Cristoloveanu and S. Li, *"Electrical characterization of silicon-on-insulator materials and devices"*, Kluwer Academic Publishers, 1995.

[13] E.T. Vandenberg, L. Bertilsson, B. Liedberg, K. Uvdal, R. Erlandsson, H. Elwing and I. Lundström, *"Structure of 3-aminopropyl triethoxy silane on silicon oxide"*, J. Colloid Interface Sci., **147**, pp. 103–118, 1991.

Interface trap effects in the design of a 4H-SiC MOSFET for low voltage applications

G. De Martino, F. Pezzimenti, F. G. Della Corte

Department of Information Engineering, Infrastructures and Sustainable Energy, DIIES
Mediterranea University of Reggio Calabria - Reggio Calabria 89122, Italy
E-mails: giuseppe.demartino@unirc.it, fortunato.pezzimenti@unirc.it,
francesco.dellacorte@unirc.it

Abstract—The current-voltage characteristics of a 4H-SiC MOSFET dimensioned for a breakdown voltage of 650 V are investigated by means of a numerical simulation study that takes into account the defect state distribution at the oxide-semiconductor interface in the channel region. The modelling analysis reveals that, for these low-voltage devices, the channel resistance component plays a key role in determining the MOSFET specific ON-state resistance (R_{ON}) under different voltage biases and temperatures. The R_{ON} value is in the order of a few $m\Omega \times cm^2$.

Keywords—4H-SiC; power devices; ON-state resistance; numerical simulations; defects states.

1. Introduction

Silicon carbide (SiC) -based devices are worldwide recognized as well suited for high-power, high-frequency, and high-temperature applications. In particular, thanks to the excellent material electronic properties, namely high critical electric field and thermal conductivity [1], important industrial sectors such as photovoltaic (PV) and automotive might gain undoubted advantages by using, for example, MOSFETs in SiC as power optimizers also for low voltage ratings. SiC, in fact, shows high mechanical strength and reliability that allow a fully functional of these devices for several years in all operating conditions. However, since SiC is a relatively new technology, the deployments of intensive modeling efforts are needed to gain a better understanding of the device performance.

In this paper, we investigate the main electrical characteristics of a 4H-SiC MOSFET, namely drain current (I_D), specific ON-state resistance (R_{ON}), and breakdown voltage (BV_{DS}), by using a TCAD 2D physical simulator. Detailed results on the existing tradeoff between these fundamental device parameters were presented in [2]. There, however, for the sake of simplicity,

SiO$_2$/SiC interface trap effects were neglected as entry data for modelling. These effects are an unavoidable technological issue and, by developing further the simulations setup, it was clear that an explicit defect concentration at the oxide-semiconductor interface in the channel region has an apparent detrimental effect on the device forward current behavior, and therefore on the R_{ON} value.

2. Device Structure and Parameters

The schematic cross-section of the considered 4H-SiC MOSFET half-cell is shown in Fig. 1. The proposed structure is, in principle, compatible with a process-run based on doping by ion implantation.

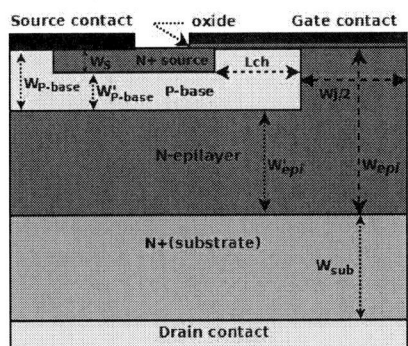

Fig. 1. Schematic cross-sectional view of the MOSFET half-cell. The drawing is not in scale.

The geometrical parameters and doping concentrations of the different MOSFET regions are summarized in Table 1. The device length along the z-direction was set to 1 μm by default. The simulated MOSFET footprint area was 6.5 μm^2.

The device R_{ON} can be expressed as [1]:

$$R_{ON,sp} = R_{N+} + R_{ch} + R_{acc} + R_{JFET} + R_{epi} + R_{sub} \quad (1)$$

978-1-5386-4483-6/18 $31.00 © 2018 IEEE   147

where R_{N+} is the contribution of the N^+-source, R_{ch} is the channel resistance, R_{acc} is the accumulation layer resistance, R_{epi} is the epi-region resistance, R_{sub} is the substrate resistance, and R_{JFET} is the resistance of the depletion layer between the P-base region and the N-epilayer. R_{n+} and R_{sub} are generally negligible because they are localized in heavily doped regions; R_{ch} and R_{acc} mainly depend on the gate bias level; R_{JFET} and R_{epi} are determined by the geometry and doping concentration of the epilayer. This region, in particular, strongly influences the R_{ON} and BV_{DS} values. A good trade-off between the parameters N_{epi} and W_{epi} is therefore required as highlighted from the breakdown expression for abrupt junction p-i-n devices in punch-through conditions [1]:

$$BV_{DS} = W'_{epi}\left(E_C - \frac{qN_{epi}W'_{epi}}{2\varepsilon_s} \right). \qquad (2)$$

Here, E_C = 1.5 MV/cm is the critical electric field assumed in this work, ε_s is the semiconductor dielectric constant, and q is the electron charge. For a fixed W_{epi} value, it follows that the lower is the desired BV_{DS}, the higher should be N_{epi}, with consequent advantages in terms of a low R_{ON}. For example, for the device in Table 1 a BV_{DS} close to 650 V is expected and this value was verified during the simulations.

TABLE 1. MOSFET parameters.

Source thickness, W_S (μm)	0.5
Channel length, L_{ch} (μm)	1
Base junction depth, $W_{P\text{-}base}$ (μm)	1.3
Interspace, $W'_{P\text{-}base}$ (μm)	1
Base-to-base distance, W_j (μm)	5
Epilayer junction depth, W_{epi} (μm)	8.7
Base-to-substrate distance, W'_{epi} (μm)	0.5
Substrate thickness, W_{sub} (μm)	100
Oxide thickness (nm)	80
N^+-source doping (cm^{-3})	10^{18}
P-base doping (cm^{-3})	10^{17}
N-epilayer doping (cm^{-3})	10^{16}
N^+-substrate doping (cm^{-3})	10^{19}

Finally, for low-voltage MOSFETs (VB_{DS} < 1 kV), the R_{ch} component of R_{ON}

plays a key role and its value is determined by the effective carrier mobility (electrons) into the inversion layer.

3. Physical Models

The numerical simulation analysis was performed by using the Silvaco-ATLAS physical simulator [3] to solve the Poisson's equation and the carrier continuity equations for a finely meshed device structure. In particular, around the P-N junctions and within the channel region below the SiO$_2$/4H-SiC interface a mesh spacing down to 25 nm was imposed.

The defect density of states (DoS) can be modelled as a sum of four terms, namely two band tail states localized near the conduction and valence band-edges, and two deep states in the mid-gap, each acting either as donor-like or acceptor-like level [3,4]. More in detail, we used the following expressions for the tail state densities:

$$D_{T,C}(E) = D_{T,C}^0 \exp\left(\frac{E - E_C}{U_C} \right) \qquad (3a)$$

$$D_{T,V}(E) = D_{T,V}^0 \exp\left(-\frac{E - E_V}{U_V} \right) \qquad (3b)$$

where $D^0_{T,C}$ and $D^0_{T,V}$ are the conduction and valence band-edge intercept densities, and U_C and U_V are the characteristic energy decays. At the same time, the mid-gap state densities $D_{G,C}(E)$ and $D_{G,V}(E)$ are modelled as two Gaussian distributions

$$D_{G,C}(E) = D_{G,C}^0 \exp\left[-\left(\frac{E - E_{GC}}{W_C} \right)^2 \right] \qquad (4a)$$

$$D_{G,V}(E) = D_{G,V}^0 \exp\left[-\left(\frac{E - E_{GV}}{W_V} \right)^2 \right] \qquad (4b)$$

where E_{GC} and E_{GV} are the energy values of the defect concentration peaks localized in the bandgap, i.e. $D^0_{G,C}$ and $D^0_{G,V}$, and the terms W_C and W_V account for the spectral width of the Gaussian distributions.

978-1-5386-4483-6/18 $31.00 © 2018 IEEE

The carrier recombination rate is calculated as four terms in the form of

$$\int_{E_V}^{E_C} \left(\frac{D(E) \cdot \bar{n} \cdot \bar{p}}{\bar{n} + e_n + \bar{p} + e_p} \right) \cdot dE \qquad (5)$$

where the density $D(E)$ is one of terms $D^0_{T,C}$, $D^0_{T,V}$, $D^0_{G,C}$, and $D^0_{G,V}$ introduced above, and $\bar{n} = \sigma_n n \cdot v_{th}$ and $\bar{p} = \sigma_p p \cdot v_{th}$. Here, v_{th} (1.34×10^7 cm/s) is the thermal velocity, and σ_n and σ_p are the capture cross sections for electrons and holes, respectively. Finally, e_n and e_p are the trap emission rates:

$$e_n = \sigma_n v_{th} N_c \cdot \exp\left(\frac{E - E_C}{kT} \right) \qquad (6a)$$

$$e_p = \sigma_p v_{th} N_v \cdot \exp\left(\frac{E_v - E}{kT} \right). \qquad (6b)$$

The capture cross sections have to be properly defined both for acceptor-like and donor-like traps. Their values were fixed as in [5] and the fundamental DoS parameters used during the simulations are summarized in Table 2 [3,5].

TABLE 2. DoS parameters.

$D^0_{T,C}$, $D^0_{T,V}$ (cm^{-3})	1.0×10^{14}, 4.0×10^{20}
$D^0_{G,C}$, $D^0_{G,V}$ (cm^{-3})	5.0×10^{10}, 1.5×10^{18}
$E_{G,C}$, $E_{G,V}$ (eV)	1.0, 0.4
U_C, U_V (eV)	0.033, 0.05
W_C, W_V (eV)	0.1, 0.1

The other key physical models include the band-gap temperature dependence and apparent band-gap narrowing, the incomplete ionization of dopants, the Shockley–Read–Hall and Auger recombination processes, the impact ionization, and the concentration and temperature dependent carrier lifetime and carrier mobility. This simulation setup is presented in detail in recent authors papers addressed to the study of 4H-SiC-based devices [6-9]. In addition, it is supported by experimental results on implanted p-i-n diodes [7,9] in a wide range of currents and temperatures.

4. Results and Discussion

In order to assess the impact of traps at the SiO$_2$/SiC interface on the MOSFET current capabilities, several simulations were performed. The device current-voltage characteristics (I_D-V_{DS}) at room temperature are shown in Fig. 2 for three different values of V_{GS}, starting from $V_{GS} = 8$ V that we can assume as the threshold voltage V_{TH}.

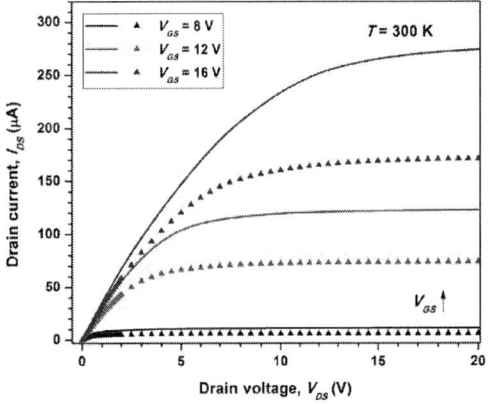

Fig. 2. MOSFET I_{DS}-V_{DS} characteristics for a device with trap (symbols) and no-trap (solid lines) effects at $T = 300$ K.

The same I_D-V_{DS} characteristics were simulated at $T = 473$ K as shown in Fig. 3. This is a typical operation temperature for SiC-based power devices.

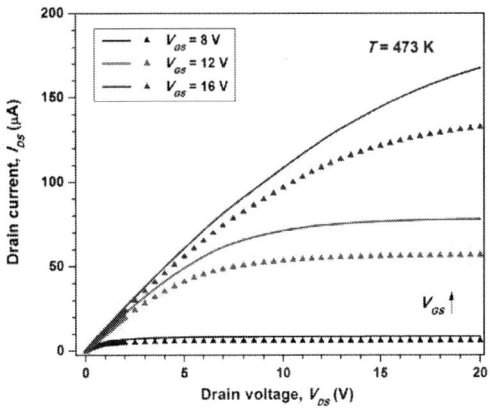

Fig. 3. MOSFET I_{DS}-V_{DS} characteristics for a device with trap (symbols) and no-trap (solid lines) effects at $T = 473$ K.

From these results, it is clear that the interface defects have a strong impact in determining the MOSFET performance at any temperature, in particular, for increasing V_{GS}. In particular, our attention was focused on the

978-1-5386-4483-6/18 $31.00 © 2018 IEEE

R_{ON} value calculated assuming an operating point in the triode region for $V_{DS} = 1$ V similarly to [2]. The R_{ON} behaviors as a function of V_{GS} at $T = 300$ K and 473 K are shown in Fig. 4. One datasheet value of R_{ON} at $T = 300$ K is also reported for comparison [10].

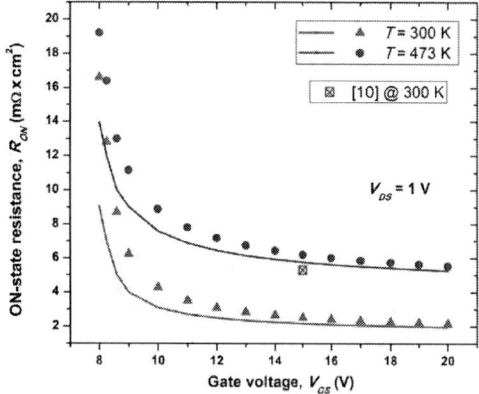

Fig. 4. R_{ON} behaviors as a function of V_{GS} for $V_{DS} = 1$ V with trap (symbols) and no-trap (solid lines) effects.

As expected, introducing the interface defect effects, the MOSFET R_{ON} increases for a fixed temperature especially for low V_{GS} voltages. This increase, however, tends to reduce for higher values of V_{GS}. More in detail, we calculated the percentage variation of R_{ON} vs. V_{GS} as shown in Fig. 5.

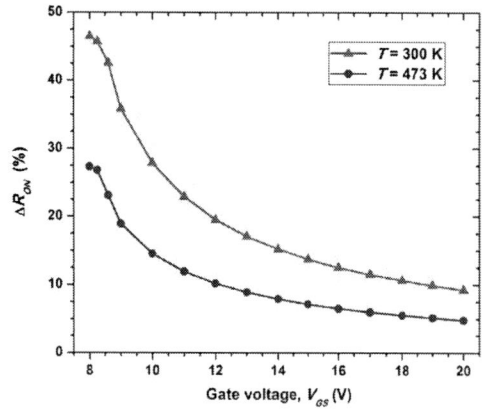

Fig. 5. Percentage variation of R_{ON} vs. V_{GS} introducing trap effects in the simulations.

It is worthwhile noting that, in presence of a defect state distribution, the percentage variation of R_{ON} appears less severe for $T = 473$ K in the whole explored V_{GS} range. Moreover, a V_{GS} higher than about $2V_{TH}$ aids

to limit ΔR_{ON} in the order of 10%.

5. Conclusion

In this work, the role of an explicit defect state concentration at the SiO$_2$/SiC interface in determining the R_{ON} value of a 4H-SiC power MOSFET has been investigated. Numerical simulation results with trap and no-trap effects have been compared at different temperatures. The percentage variation of R_{ON} can be as high as 40% for low gate voltages at $T = 300$ K.

References

[1] B. J. Baliga, *Fundamentals of Power Semiconductor Devices*, Springer, New York, 2008.
[2] G. De Martino, F. Pezzimenti, F. Della Corte, G. Adinolfi, and G. Graditi, *"Design and numerical characterization of a low voltage power MOSFET in 4H-SiC for photovoltaic applications"*, in Proc. IEEE Ph.D. Research in Microelectronics and Electronics - PRIME, 2017, pp. 221–224.
[3] *ATLAS Users Manual*, Silvaco Inc. (2010).
[4] V. Afanasev, M. Bassler, G. Pensl, and M. Schulz, *"Intrinsic SiC/SiO₂ interface states"*. Physica Status Solidi (a), 162(1), pp. 321–337, 1997.
[5] Dimitriadis, N. Archontas, D. Girginoudin and N. Georgoulas, *"Two dimensional simulation and modeling of the electrical characteristics of the a-SiC/c-Si (p) based, thyristor-like, switches"*, Microelectron. Eng., 133, pp. 120-128, 2015.
[6] F. Pezzimenti, *"Modeling of the steady state and switching characteristics of a normally-off 4H-SiC trench bipolar-mode FET"*, IEEE Trans. Electron Devices, 60, pp. 1404-1411, 2013.
[7] M. L. Megherbi, F. Pezzimenti, L. Dehimi, A. Saadoune, and F. G. Della Corte, *"Analysis of the forward I-V characteristics of Al-implanted 4H-SiC p-i-n diodes with modeling of recombination and trapping effects due to intrinsic and doping-induced defect states"*, J. Electron. Mater., 47, pp.1414-1420, 2018.
[8] F. Pezzimenti, and F. G. Della Corte, *"Design and modeling of a novel 4H-SiC normally-off BMFET transistor for power applications"*, in Proc. Mediterranean Electrotechnical Conference - MELECON, 2010, pp. 1129-1134.
[9] F. Pezzimenti, L. F. Albanese, S. Bellone, and F. G. Della Corte, *"Analytical model for the forward current of Al implanted 4H-SiC p-i-n diodes in a wide range of temperatures"*, in Proc. IEEE Int. Conf. Bipolar/BiCMOS Circuits and Technology Meeting,- BCTM, 2009, pp. 214-217.
[10] CREE model C3M0280090D (900V). [Online]. 2018. Available: http://www.cree.com

978-1-5386-4483-6/18 $31.00 © 2018 IEEE

High Pillar Doping Concentration for SiC Superjunction IGBTs

H. KANG* and F. UDREA*

*Electrical Engineering Department of University of Cambridge, 9 JJ Thomson Avenue, CB30FA, Cambridge, U.K.
hk428@cam.ac.uk, fu@eng.cam.ac.uk

*Abstract—This paper is a theoretical study of the optimum doping concentration for the n and p pillars of a superjunction IGBT. As the concentration of the pillar for a silicon-carbide superjunction device increases up to 10 times higher than that of silicon, unipolar drift current in each pillar can be predominant over the bipolar action. The increased doping concentration effectively reduces the potential drop in the pillar for the on-state conduction.
Keywords— PiN Diode, Superjunction, IGBT.*

1. INTRODUCTION

The Superjunction (SJ) Insulated Gate Bipolar Transistor (IGBT) has the potential to achieve a higher current density and a faster reverse recovery than the conventional IGBT [1]–[3]. However, one basic question has not been addressed: what level of the doping concentration for the pillar should be used in a SJ IGBT. Silicon carbide (SiC), the concentration of the pillars for a SiC SJ can be significantly enhanced (> 1.0×10^{16} cm^{-3})[4]–[6] and this increased doping concentration would contribute quasi-unipolar drift current.

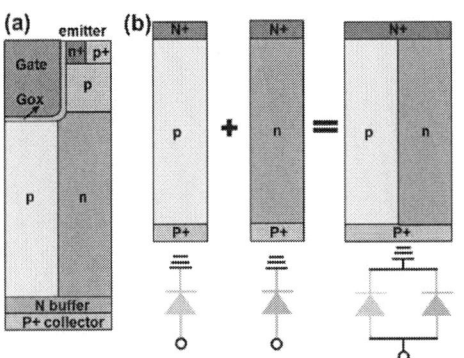

Fig. 1 (a) A schematic superjunction IGBT structure, (b) the suggested inner circuit model for a superjunction IGBT in this study.

To tackle the question, an inner circuit model for a SJ IGBT is provided in Fig. 1. A SJ IGBT system can be regarded as double P-i-N (or P^+-N/P-N^+) diodes. The inversion

layer of the MOSFET will be the N^+ emitter for the diodes. The SJ IGBT's on-state current will be the sum of each P-i-N diode and, therefore, the behavior of each diode should be studied first. This study will investigate the behavior of a SJ IGBT with respect to the pillar length, lifetime and current density. (The influence of the buffer layer will be ignored.)

2. Tendency of Optimum Concentration

The total current density (J_T) in each diode is the sum of the drift current (J_{drift}) and the diffusion current ($J_{diffusion}$). According to arithmetic-geometric mean, the total current will have the following relationship:

$$J_T = J_{drift} + J_{diffusion} \geq 2(J_{drift} \times J_{diffusion})^{1/2}. \quad (1)$$

By developing equation (1), the optimum concentration ($N_{D.min}$) for P^+-N-N^+ and P^+-P-N^+ diodes where total current density is minimum can be found:

$$N_{D.min} \propto n_i \exp\left(\frac{qV_{p+}}{2kT}\right) \times \sqrt{\frac{L_H}{L_D}}, \quad (P^+\text{-N-}N^+) \quad (2\text{-}1)$$

$$N_{D.min} \propto n_i \exp\left(\frac{qV_{N+}}{2kT}\right) \times \sqrt{\frac{L_H}{L_D}}, \quad (P^+\text{-P-}N^+) \quad (2\text{-}2)$$

Where, n_i, k, T, L_H, and L_D are intrinsic concentration of a semiconductor, Boltzmann constant, absolute temperature (300 K) diffusion length at a high-level injection, and length of the pillar. V_{P+} is the applied voltage between P^+ and N in P^+-N-N^+ and V_{N+} is the voltage between N^+ and P in P^+-P-N^+. L_H, was calculated form diffusivity driven by Einstein relationship and doping-dependent lifetime [7]:

$$L_H = \sqrt{D_H \tau_H} \quad (3\text{-}1) \qquad D_H = \frac{kT}{q}\mu \quad (3\text{-}2)$$

978-1-5386-4483-6/18 $31.00 © 2018 IEEE   151

$$\tau_{Hn/Hp}(N) = \frac{\tau_{n0/p0}}{1 + \left(\dfrac{N}{N_{ref}}\right)^{\gamma}} \qquad (3\text{-}3)$$

Where τ_{Hn} and τ_{Hp} are doping dependent lifetimes for electron and hole, respectively (according to Ruff *et al.*, $\tau_{Hn} \approx 5\ \tau_{Hp}$) [8]. N_{ref}, and γ are 3.0×10^{17} cm^{-3}, and 0.3, respectively in this simulation [8].

3. Simulation results

Sentaurus workbench (SWB) was used for this simulation. As can be seen in Fig. 2 ~ 6, with increasing the L_D and decreasing the lifetimes, the $N_{D.min}$ (the concentration having the highest V_{CE}) shifted towards a low N_D. As the current increased from 100 to 500 A/cm^2 (V_{P^+} or V_{N^+} increase), $N_{D.min}$ moved to a higher value. Fig. 6 and 7 show the simulation results of a SJ IGBT. With increasing the concentration of the pillars, the on-state voltages were significantly lowered.

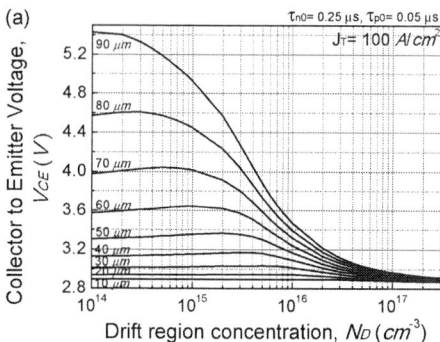

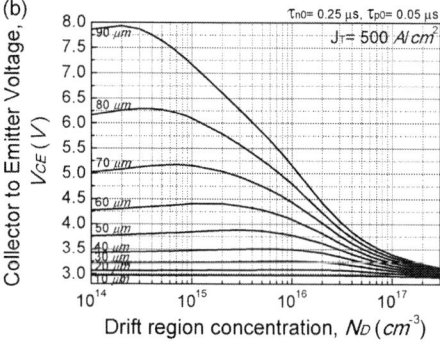

Fig. 3 V_{CE}, with respect to N_D in a P$^+$-N-N$^+$ diode at, $\tau_{n0}=$ 0.25, and $\tau_{p0}=$ 0.05 μs, and (a) $J_T=$ 100 and (b) $J_T=$ 500 A/cm^2.

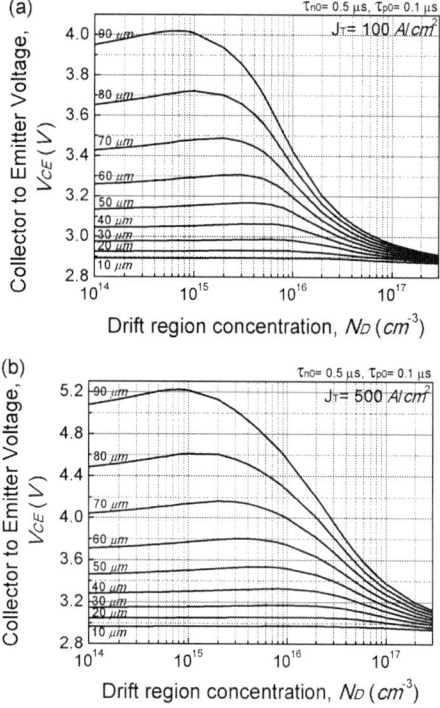

Fig. 2 V_{CE}, with respect to N_D in a P$^+$-N-N$^+$ diode at, $\tau_{n0}=$ 0.5, and $\tau_{p0}=$ 0.1 μs, and (a) $J_T=$ 100 and (b) $J_T=$ 500 A/cm^2.

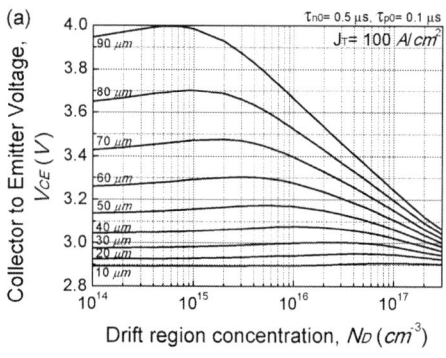

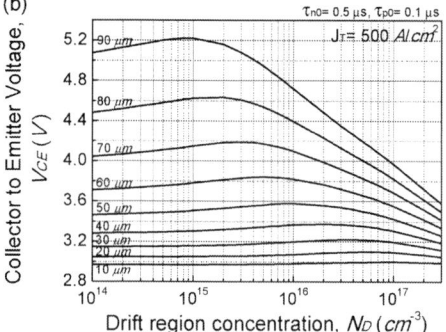

Fig. 4 V_{CE}, with respect to N_D in a P$^+$-P-N$^+$ diode at, $\tau_{n0}=$ 0.5, and $\tau_{p0}=$ 0.1 μs, and (a) $J_T=$ 100 and (b) $J_T=$ 500 A/cm^2.

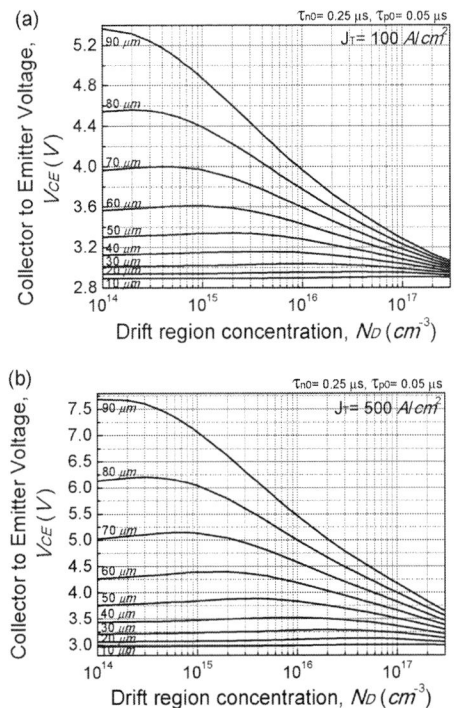

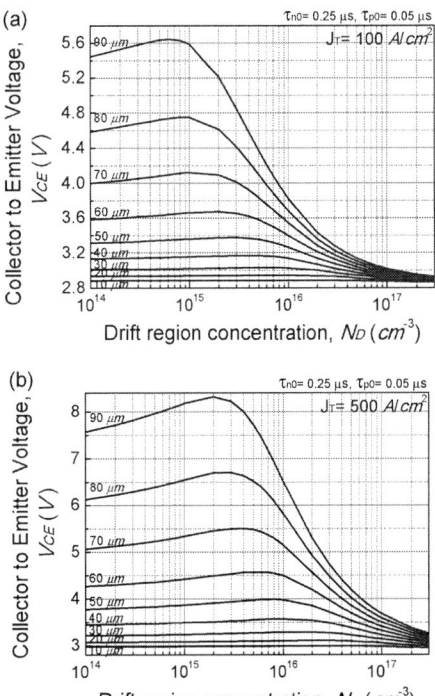

Fig. 5 V_{CE}, with respect to N_D in a P$^+$-P-N$^+$ diode at, $\tau_{n0}= 0.25$, and $\tau_{p0}= 0.05$ μs, and (a) $J_T= 100$ and (b) $J_T= 500$ A/cm^2.

Fig. 7 V_{CE}, with respect to N_D in a SJ diode at, $\tau_{n0}= 0.25$, and $\tau_{p0}= 0.05$ μs, and (a) $J_T= 100$ and (b) $J_T= 500$ A/cm^2.

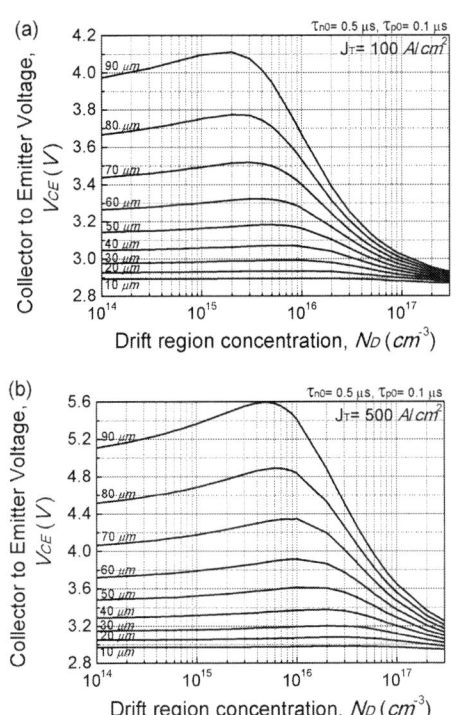

Fig. 6 V_{CE}, with respect to N_D in a SJ diode at, $\tau_{n0}= 0.5$, and $\tau_{p0}= 0.1$ μs, and (a) $J_T= 100$ and (b) $J_T= 500$ A/cm^2.

The $N_{D.min}$ and V_{CE} for the SJ structure were slightly higher than those of P$^+$-P-N$^+$ and P$^+$-N-N$^+$. For example, when $L_D= 90$ μm, the $N_{D.min}$ and V_{CE} at 500 A/cm^2 were 2.0×10^{14} cm^{-3} ($V_{CE}= 7.9$ V) for P$^+$-N-N$^+$, 1.0×10^{14} cm^{-3} ($V_{CE}= 7.7$ V) for P$^+$-P-N$^+$, and 2.0×10^{15} cm^{-3} ($V_{CE}= 8.3$ V) for SJ, respectively (See Fig. 3(b), 5(b), and 7(b)). As shown in Fig. 8, unlike each diode structure, the SJ presented a symmetrical carrier profile with a low electron density in the middle of the pillar ($\sim$ 45 μm) and, due to the low density, a large amount of potential was consumed in the middle of the pillar. When the two diodes are connected as a SJ structure, the injected minority carriers can diffuse into the other side of the pillar. According to Fig. 9, some amounts of the holes injected from the P$^+$ region into the n-pillar diffuse into the p-pillar. This weakens the conductivity modulation in the n-pillar near P$^+$ but results in a higher conductivity modulation in the p-pillar. Electrons showed the same behavior

978-1-5386-4483-6/18 $31.00 © 2018 IEEE

near N^+. Therefore, to make a SiC SJ IGBT as low on-state voltage as possible, the concentration of the pillar should be as high as possible, especially for the device with a long enough pillar length ($L_D > 50 \ \mu m$).

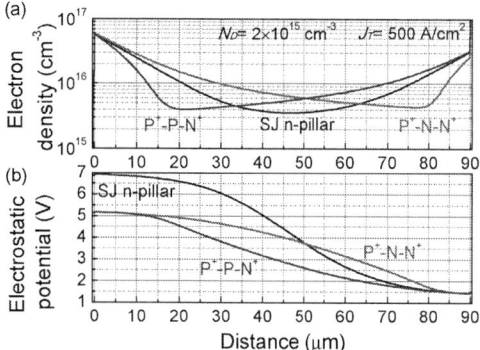

Fig. 8 Profile of (a) electron density and (b) electro static potential in each diode and in the n-pillar of the superjunction. L_D= 90 μm. τ_{n0}= 0.25, τ_{p0}= 0.05 μs, and J_T= 500 A/cm^2.

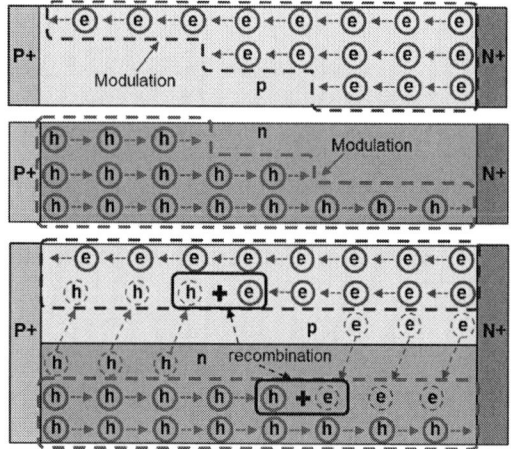

Fig. 9 Schematic illustration of minority carrier movement in each diode and in a superjunction.

4. Conclusion

The on-state behavior of a 4H-SiC SJ IGBT with respect to the concentration of the pillar was investigated. Owing to the significantly increased concentration in the pillar, the on-state voltage could be effectively lowered. Therefore, 4H-SiC SJ IGBT should adopt the pillar concentration as high as possible.

Acknowledgments.

This research is supported by On Semiconductor corporation as a part of a future power electronics technology.

References

[1] M. Antoniou, F. Udrea, and F. Bauer, "Optimisation of SuperJunction Bipolar Transistor for ultra-fast switching applications," in *Proceedings of the 19th International Symposium on Power Semiconductor Devices and IC's*, 2007, pp. 101–104.

[2] M. Antoniou, F. Udrea, F. Bauer, and I. Nistor, "The Soft Punchthrough Superjunction Insulated Gate Bipolar Transistor: A High Speed Structure With Enhanced Electron Injection," *IEEE Trans. Electron Devices*, vol. 58, no. 3, pp. 769–775, Mar. 2011.

[3] F. Bauer, I. Nistor, A. Mihaila, M. Antoniou, and F. Udrea, "Superjunction IGBT filling the gap between SJ MOSFET and ultrafast IGBT," *IEEE Electron Device Lett.*, vol. 33, no. 9, pp. 1288–1290, Jul. 2012.

[4] H. Kang and F. Udrea, "True Material Limit of Power Devices -Applied to 2-D Superjunction MOSFET," *IEEE Trans. Electron Devices*, vol. 65, no. 4, pp. 1432–1439, Mar. 2018.

[5] H. Kang and F. Udrea, "Material Limit of Power Devices--Applied to Asymmetric 2-D Superjunction MOSFET," *IEEE Trans. Electron Devices*, pp. 1–7, May 2018.

[6] F. Udrea, G. Deboy, S. Member, and T. Fujihira, "Superjunction Power Devices, History, Development, and Future Prospects," *IEEE Trans. Electron Devices*, vol. 64, no. 3, pp. 713–727, Jan. 2017.

[7] D. J. Roulston, N. D. Arora, and S. G. Chamberlain, "Modeling and measurement of minority-carrier lifetime versus doping in diffused layers of n+-p silicon diodes," *IEEE Trans. Electron Devices*, vol. 29, no. 2, pp. 284–291, Feb. 1982.

[8] M. Ruff, H. Mitlehner, and R. Helbig, "SiC devices: physics and numerical simulation," *IEEE Trans. Electron Devices*, vol. 41, no. 6, pp. 1040–1054, Jun. 1994.

978-1-5386-4483-6/18 $31.00 © 2018 IEEE

Surface Recombination Evaluation in Bipolar Junction Transistors by Combined Electro-Optical Method

Viorel Banu *, Josep Montserrat**, Xavier Jordá**, Philippe Godignon**

* D+T Microélectronica A.I.E., Campus UAB, 080193 Bellaterra-Barcelona, Catalunya, Spain
E-mail: viorel.banu@imb-cnm.csic.es
** IMB-CNM, CSIC, Campus UAB, 080193 Bellaterra-Barcelona, Catalunya, Spain
** **E-mail: Josep.Montserrat@imb-cnm.csic.es, xavier.jorda@imb-cnm.csic.es,**
philippe.godignon@imb-cnm.csic.es

Abstract— This paper describes an original method for very short minority carrier lifetime evaluation of semiconductor junctions that is aimed to be mainly used for the surface recombination studies of bipolar junction transistors made on silicon carbide, where an important surface recombination between base and emitter occurs. The presented method is based on combined electro-optical measurements and it was prior tested and calibrated using silicon bipolar transistors.

Keywords—minority carrier lifetime; surface recombination; bipolar junction transistor; silicon carbide; optical excitation.

1. Introduction

The progresses in manufacturing high quality silicon carbide (SiC) substrates from the last decade enabled a wide variety of SiC power devices that became a presence on the market substituting the silicon power devices in the new projects.

However, SiC bipolar junction transistors (BJT) are still not an important presence on the market. Even if the old issues linked to the staking faults were already solved, major problems created by the surface recombination are not yet entirely overcome.

The BJTs are important devices that are useful not only in signal or power applications, but are also present in integrated circuits, insulated gate bipolar transistors (IGBT), semiconductor controlled rectifiers (SCR), or TRIACs, that should be also transferred on the silicon carbide semiconductor technology.

In this paper we present an original method for evaluating the effective minority lifetime for the base to emitter junction of silicon carbide BJT.

The main part of this work is dedicated to the method's study and calibration performed on silicon BJTs. The Si BJTs behave a relatively low surface recombination and electrical evaluation of the both base-collector and base emitter junctions are possible.

2. Minority carrier lifetime measurement

The electrical approach of minority carrier lifetime was detailed in [1], [2] and [3].

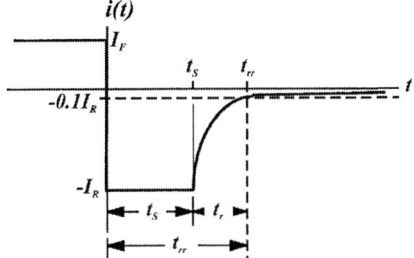

Fig. 1 Switching current response of a p-n junction.

The relation between the electrons lifetime τ_n into the p type base of p-n junction and the storage time t_s defined at the point when the voltage drop across the diode become zero, is described by (1):

$$ erf\sqrt{\frac{t_s}{\tau_n}} = \frac{1}{1+\frac{I_R}{I_F}} \qquad (1) $$

Where I_F is the conduction current through the diode prior to switching, I_R is the reverse recovery flow current during the storage phase between t=0 and t=t_s, (see Fig. 1). After t=t_s phase, the current flow drops exponentially to zero, in fact to the leakage current of the junction. The time t=t_{RR} defined at the reverse current point i(t_{RR})=0.1I_R is called reverse recovery time and is a parameter specified for the fast rectifiers.

Fig. 2a illustrates the schematic of the test circuit used for electrical measurement of the minority carrier lifetime in semiconductor junctions. Before t=0 the diode is forward biased through V$^+$ and R1 and the current flow is I_F (Fig. 1). After t=0+ the diode is reverse biased through V$^-$ and R2. The current flow is negative as shown in Fig. 1. Experimental voltage waveforms using the schematic from Fig. 2a are shown in Fig 2b. In order to precisely define the t_s time, the voltage of the points A and K from Fig. 2a were plotted together, the time t_s being measured at the intersection of V(A) and V(K) curves, i.e. when the voltage drop across the diode is zero. The current flow through the diode is obtained by dividing the voltage characteristic V(K) by the value of R_{Shunt}.

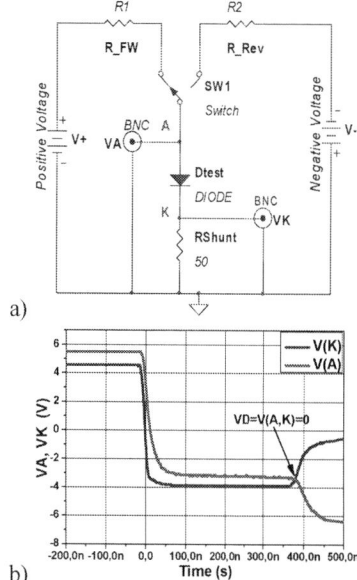

a)

b)

Fig. 2 a) Circuit schematic for switching test; b) switching waveforms of the anode A and cathode K of the p-n junction.

For the BJT characterization, we have separately measured the lifetime of the base-collector junction and the base-emitter junction as for a p-n junction.

For the method's calibration, the minority carrier lifetime of several commercial 1N1711 npn BJTs having different designs was measured. The micrograph captures of the tested designs are shown in Fig. 3.

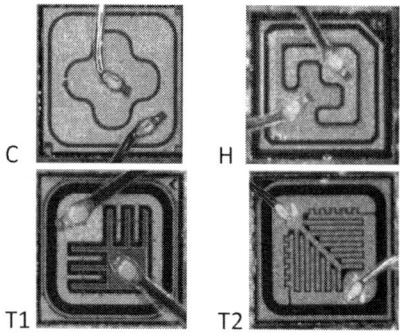

Fig. 3 Layout format of tested 1N1711 npn BJTs.

In the absence of surface recombination, the lifetime values τ_{BC} and τ_{BE} of both junctions should be equals. Different values between τ_{BC} and τ_{BE} is the signature of a certain degree of surface recombination; because the bulk lifetime of the base is the same, the measurement was done on the same base epitaxial layer having two junctions, a surface BE junction and a buried BC junction. Thus, the BE junction is subject to surface recombination. Further we'll consider the

base-collector measured lifetime as reference for the calculation of base-emitter effective lifetime.

In Fig. 4 a plot of measured τ_{BE} versus τ_{BC} is presented. The transistors having lower τ_{BE} values than τ_{BC} indicate the presence of surface recombination. The highest mismatch of τ_{BE} versus τ_{BC} occurs in the case of design T2 which have the highest periphery to surface ratio (perimeter governed surface recombination) [5]. More details about surface recombination can be also found in [7] and [8].

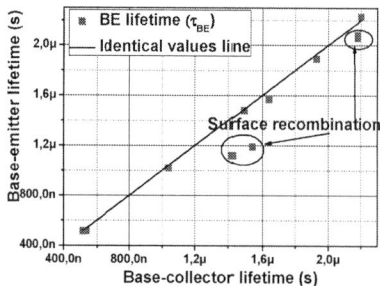

Fig. 4 Base-emitter versus base-collector lifetime

3. Gummel plot

The Gummel plot is another electrical method to evidence the surface recombination [4]. The method

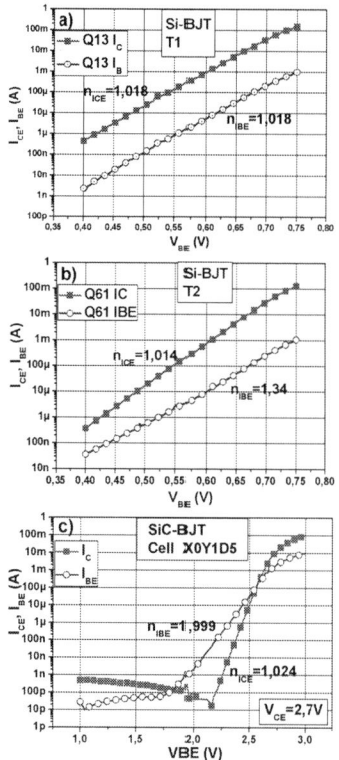

Fig. 5 Gummel plots: a) Si BJT design T1; b) Si BJT design T2; c) SiC BJT

978-1-5386-4483-6/18 $31.00 © 2018 IEEE   156

consists in plotting together the collector current I_C and base-emitter current I_{BE} versus the base-emitter voltage V_{BE} at a constant V_{CE} slightly above the specific junction threshold value (>0.7V for Si and > 2.5V for SiC).

Fig. 5 presents three Gummel plots, a) and b) for silicon BJT and c) for SiC BJT. In the absence of surface recombination, the ideality factors n_{BC} and n_{BE} of the collector current I_C and base current I_{BE} should be close to the value 1. In Fig. 5a the ideality factors of both I_C and I_{BE} are 1, so there is no surface recombination. In Fig. 5b the ideality factor n_{BE}=1.34 indicating the presence of surface recombination. In Fig. 5c is presented the Gummel plot of silicon carbide BJT. In this case n_{BE}=2 indicating a dominant surface recombination.

4. Lifetime response to light excitation

Under low voltage reverse bias of the junction and strong illumination (Fig. 6), a photocurrent due to the generation of electron-hole pairs occur as described by the equation (2), [6].

$$I_{Photo}=qG_L\,(L_p+L_n+W)A_j \quad (2)$$

Where G_L is the generation rate of electron-hole pairs due to absorbed light (per unit time and volume), q is the electric charge of electron, L_p is the diffusion length of holes, L_n is the diffusion length of electrons, Aj is the cross-sectional area of the p-n junction and W the depletion region width of the junction.

$$L_{p=}\sqrt{D_p\tau_p} \quad and \quad L_{n=}\sqrt{D_n\tau_n} \quad (3)$$

Where D_p and D_n are the diffusivity of holes and electrons, τ_p and τ_n the lifetime of holes and electrons.

$$D_p = \frac{kT}{q}\mu_p \quad and \quad D_n = \frac{kT}{q}\mu_n \quad (4)$$

Where μ_p and μ_n are the hole and electrons mobility, k the Boltzmann's constant and T is the absolute temperature. We can write now:

$$I_{Photo}=qG_L\left(\sqrt{\frac{kT}{q}\mu_p\tau_p} + \sqrt{\frac{kT}{q}\mu_n\tau_n}+W\right)A_j \quad (5)$$

Because usually $L_n \gg W$ and $L_p \gg W$. W could be neglected i.e. $W \cong 0$. Taking into account that:

$$\mu_p = B_m \times \mu_n \quad and \quad \tau_p = B_t \times \tau_n \quad (6)$$

Where B_m and B_t are proportionality factors between the holes to electrons mobility, respectively holes to electrons lifetime in the same semiconductor material layer. Thus we can write having B as global proportionality factor:

$$I_{Photo}=qG_L B \sqrt{\tau_n}A_j \quad (7)$$

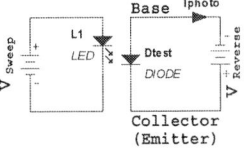

Fig. 6 Light sweep schematic of the p-n junction

Considering G_L as variable input, the sweep of G_L gives a linear function having the slope m that contain precious information about the minority electrons in the P type base of the transistor:

$$I_{Photo}=mG_L \quad (8)$$

Where $\quad m=qB A_j\sqrt{\tau_n} \quad (9)$

We can also write:

$$\frac{I_{Photo2}}{I_{Photo1}} = \frac{m_2}{m_1} = \frac{A_{j2}}{A_{j1}}\sqrt{\frac{\tau_{n2}}{\tau_{n1}}} \quad (10)$$

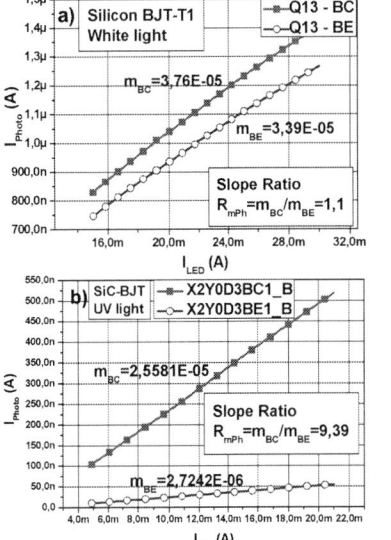

Fig. 7 Photocurrent experimental slopes versus light intensity for a) Silicon BJT without surface recombination (white-LED) and b) SiC BJT with strong surface recombination (UV-LED)

Note in Fig. 7 the BC to BE slopes difference between Si-BJT with no surface recombination and SiC-BJT that behave a strong surface recombination.

For the same light intensity excitation G_{L1}=G_{L2} we obtain:

$$\frac{m_{BC}}{m_{BE}} = B_A\sqrt{\frac{\tau_{BC}}{\tau_{BE}}} \quad (11)$$

B_A is a proportionality factor related on the geometrical dimensions of the base-collector and base-emitter junctions.

5. Base-emitter lifetime evaluation by electro-optical measurements

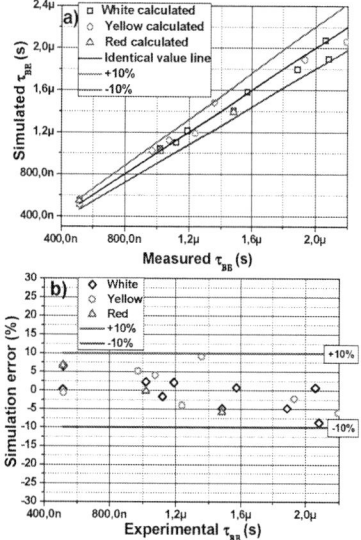

Fig. 8 Simulated τ_{BE} values starting fom τ_{CE} and photocurrent slopes.

The mean photo-current slope ratio of all tested Si BJTs is $R_{mPh}=m_{BC}/m_{BE} =1.0975$ with a standard deviation $s_{d1}= 0.018332$ and the mean ratio $(\tau_{BC}/\tau_{BE})^{1/2}=1,015$ with a standard deviation $s_{d2}= 0.01069$.

The correction factor is $B_A=1.0975/1.015=1.08$ as results by applying (11).

Taking as reference the measured base–collector junction lifetime τ_{BC} in the same base epitaxial layer, we have calculated the base–emitter junction effective lifetime using the photo-current correction factor $B_A=1.08$. The results are presented in Fig. 8. Fig. 8a shows the calculated values of τ_{BE} starting from the measured τ_{BC} for white, yellow, and red light between +10% and -10% dispersion lines. Fig 8b illustrate the errors of calculated base to emitter junction lifetime τ_{BE} relative to measured values. It is obvious that the method of calculating τ_{BE} junction lifetime starting from electrical measured τ_{BC} junction lifetime and the photo-current slope ratio between BC and BE gives 10% accuracy.

We can now apply this method to evaluate the effective base-emitter junction lifetime τ_{BE} of SiC BJT that is strongly affected by the surface recombination and cannot be measured by the standard electrical approach. We can evaluate it using the equation (11).

For a measured $\tau_{BC}=250ns$ of SiC BJT we obtain with the photo-current slope values from Fig. 7b a BE lifetime $\tau_{BE}=3.3ns$. Obviously, such a small effective lifetime value is difficult to be evaluated by the standard electrical approach.

6. Conclusions

A novel method for evaluating the effective lifetime of the base-emitter junction in SiC bipolar transistors was presented. The method combines the electrical measurement of the base-collector junction in the SiC transistor with the photo-current response of the both BC and BE junctions under reverse bias. The slope ratio of the two photo-current characteristics versus light intensity are proportional with the square root of the lifetime ratio of BC lifetime and BE lifetime. The method is able to easily evaluate effective lifetime of the SiC BE junction in order of ns.

Acknowledgments. This work has been partially supported by the Spanish Ministry of Economy, Industry and Competitiveness through "High Voltage devices for green power electronics: technology/processing trends", HiVolt-Tech project (TEC2014-54357-C2-1-R) cofounded by the EU-ERDF (FEDER), and by AGAUR funds (2014-SGR-1596).

References

[1] R. H. Kingston, *"Switching Time in Junction Diodes and Junction Transistors,"* Proc. *IRE*, **42**, 829 (1954).

[2] M. Byczkowski, J. R. Madigan, *"Minority Carrier Lifetime in p-n Junction Devices"* Journal of Applied Physics **28**, 878 (1957); doi: 10.1063/1.1722879.

[3] W. H. Ko, *"The reverse transient behavior of semiconductor junction diodes,"* IRE Transactions on Electron Devices, **vol. 8**, no. 2, pp. 123-131, March 1961; doi: 10.1109/T-ED.1961.14719.

[4] W. Liu and J. S. Harris, *"Diode ideality factor for surface recombination current in AlGaAs/GaAs heterojunction bipolar transistors,"* in *IEEE Transactions on Electron Devices*, **Vol. 39**, no. 12, pp. 2726-2732, Dec 1992; doi: 10.1109/16.168749

[5] P.G. Neudeck, *"Perimeter governed minority carrier lifetimes in 4H-SiC p$^+$n diodes measured by reverse recovery switching transient analysis"*, Journal of Elec Materi (1998) Vol. 27, (4) pp 317–323; doi:10.1007/s11664-998-0408-5.

[6] A. S. Grove, *"Physics and technology of semiconductor devices"*, John Wiley & Sons, 1967, pp.172-180.

[7] D. J. Fitzgerald, A. S. Grove, *"Surface recombination in semiconductors"*, Surface Science, **Vol. 9**, Issue 2,1968, pp. 347-369; doi: 10.1016/0039-6028(68)90182-9.

[8] L.E. Black, *"Surface Recombination Theory"*, In: New Perspectives on Surface Passivation: Understanding the Si-Al2O3 Interface; pp.15-28 (2016) Springer Theses (Recognizing Outstanding Ph.D. Research). Springer, Cham

Session IC 2

INTEGRATED CIRCUITS 2

978-1-5386-4483-6/18 $31.00 © 2018 IEEE

Low Power and Low Area CMOS Capacitance Multiplier

Gabriel Bonteanu, Arcadie Cracan
"Gheorghe Asachi" Technical University of Iasi
E-mail: gbonteanu@etti.tuiasi.ro, acracan@etti.tuiasi.ro

Abstract—*A low power and low area voltage mode CMOS capacitance multiplication technique is presented. The multiplication factor is conveniently given by the transconductance ratio of two transistors, thereby improving the immunity to process and temperature variations. A low power wide range adjustable relaxation oscillator is presented as application.*

Keywords—*MOS capacitor; capacitance multiplier; adjustable capacitor; low power; relaxation oscillator.*

1. Introduction

Capacitance multipliers are essential elements in the design of analog integrated circuits due to the need for the capacitive elements to use a small area on the chip. As a result, circuit designers have focused their attention on developing techniques that amplify the capacitive effect without compromising other circuit performance.

Two variants of capacitance multipliers are observed in practice [1]: current mode and voltage mode. The first category is based on current scaling to amplify the capacitive effect, as Fig. 1a shows. Voltage applied to the base capacitor will cause a $i_c = sCv_i$ current. This current will be detected and amplified by the parallel connected current controlled current source, so that the current

absorbed by the entire structure will be $K + 1$ times greater than that through the capacitor. In this way a $K + 1$ times capacitor is simulated. As illustrated in Fig. 1b, the second category is based on voltage scaling [2] to amplify the capacitive effect: a current injected into the base capacitor will cause a $v_c = i_c/sC$ voltage to be produced. This voltage will be detected and amplified by the series connected voltage controlled voltage source, so that the voltage across the entire structure will be $K + 1$ times lower that the voltage across the base capacitor. In this way a $K + 1$ times capacitor is simulated.

$$Z_{eq} = \frac{1}{s(K+1)C} \quad (1)$$

A typical transistor level implementation [1] of the current mode capacitance multiplier is shown in Fig. 2a. In this structure the current-controlled current source is implemented by the $M_1 - M_2$ current mirror: the current through the base capacitor is detected by the M_2 current mirror input transistor and amplified by the M_1 output transistor. The form factor of transistor M_1 is designed to be K times larger than that of M_2.

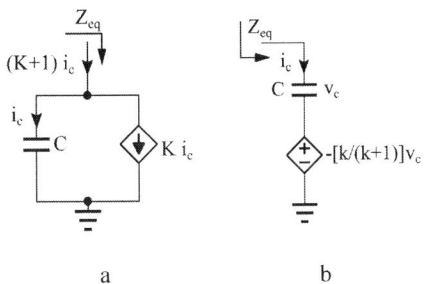

Fig. 1. Capacitance multiplication principles: (a) current mode and (b) voltage mode

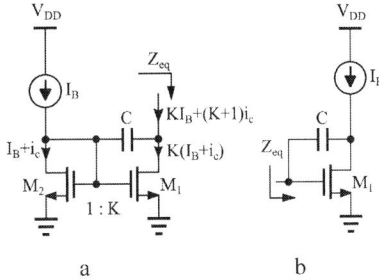

Fig. 2. Typical transistor level implementations: (a) current mode and (b) voltage mode

978-1-5386-4483-6/18 $31.00 © 2018 IEEE

As relationship (2) shows, the equivalent capacitance at the input of the circuit in Figure 2a is $K + 1$ times multiplied but the series connected $1/g_{m2}$ resistor limits the high frequency performance. Also, the M_1 equivalent output resistance degrades the low frequency behavior of the structure: the resistive element becomes dominant.

$$Z_{eq} = \frac{1}{g_{m1}+g_{m2}} + \frac{1}{s(K+1)C} \qquad (2)$$

The typical implementation of the voltage mode multiplier shown in Fig. 2b is based on the Miller effect [1]. The equivalent capacitance at the input of the structure is given by:

$$C_{eq} = (1 + A_v)C = (1 + g_m r_o)C \qquad (3)$$

where $A_v = g_m r_o$ is the low frequency voltage gain and r_o is the transistor's output resistance (considering an ideal current source). Although this structure makes the implementation of high value multiplication factors possible due to high gain at low frequencies, the fact that this gain varies with the process and temperature restricts the application range of this circuit solution.

Innovative methods that enhance or combine voltage/current modes are illustrated in [1, 3], but the multiplication factor is influenced by process and temperature variations.

2. Circuit description

A. The proposed capacitance multiplier

The proposed circuit for capacitance multiplication is illustrated in Fig. 3. The principle used is the Miller multiplication, but in order to better control the voltage amplification and thus the multiplication factor, the transistor responsible with the voltage gain is loaded with the impedance of a diode connected identical type transistor. A common mode control circuit, not shown in Fig. 3, is required to control the drain voltage for M_1 and M_{1b} transistors. Neglecting the transistor output resistance, the voltage gain will be given by the relationship (4) which

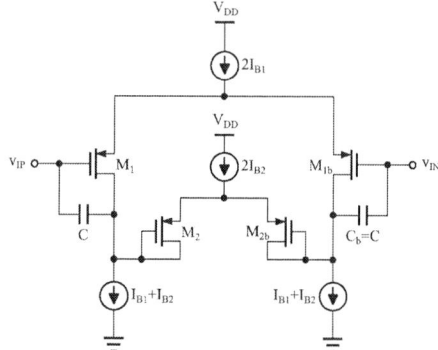

Fig. 3. Proposed circuit for capacitance multiplication

indicates a good immunity to process and temperature variations.

$$A_v = \frac{g_{m1}}{g_{m2}} = f\left(\frac{I_{B1}}{I_{B2}}\right) \qquad (4)$$

B. Frequency analysis

Considering ideal transistors, the equivalent half circuit depicted in Fig. 4 can be used for frequency analysis. The input impedance of the structure can be computed as:

$$Z_i = \frac{V_i}{I_i} = -\frac{V_{sg1}}{I_c} = \frac{g_{m2}+sC}{sC(g_{m1}+g_{m2})} \qquad (5)$$

or

$$Z_i = \frac{1}{g_{m1}+g_{m2}} + \frac{1}{sC(1+\frac{g_{m1}}{g_{m2}})} \qquad (6)$$

The equivalent model for Fig. 4 circuit is a series connection of a $R_{eq} = 1/(g_{m1} + g_{m2})$ resistor and a $C_{eq} = C(1 + g_{m1}/g_{m2})$ capacitor. The capacitance multiplication factor will be higher as M_1's transconductance will be higher and that of M_2 smaller. One can note that the expression of the equivalent input impedance for the proposed structure is equivalent to the one corresponding to the Fig. 2a current mode implementation.

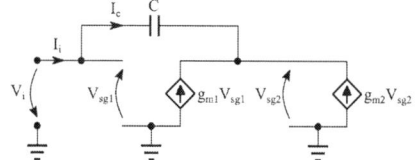

Fig. 4. Equivalent half circuit for frequency analysis considering ideal transistors

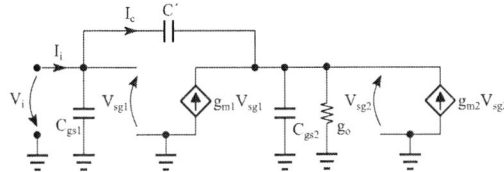

Fig. 5. Equivalent circuit for frequency analysis considering non-ideal effects for the transistors

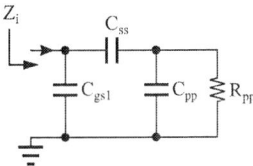

Fig. 6. Input impedance equivalent model for the proposed circuit

Taking into consideration the non-ideal effects of the gate-source and gate-drain capacitances as well as the drain-source resistance for the MOS transistors, the equivalent circuit for the frequency analysis of the structure in Fig. 3 is illustrated in Fig. 5, where $C' = C + C_{gd1}$, $g_o = g_{o1} + g_{o2} + g_{ocs}$ and g_{ocs} is the output conductance of the transistor implementing the current source. One can write:

$$Z_i = \frac{V_i}{I_i} = \frac{1}{sC_{gs1}} \; // \; Z_i' \qquad (7)$$

where

$$Z_i' = \frac{V_i}{I_c} = -\frac{V_{gs1}}{sC'(V_{gs2}-V_{gs1})} \qquad (8)$$

Further processing the Z_i' impedance expression, we get:

$$Z_i' = \frac{C'+C_{gs2}}{C' \cdot C_{gs2}} \frac{s+\frac{g_{m2}+g_o}{C'+C_{gs2}}}{s\left(s+\frac{g_{m1}+g_{m2}+g_o}{C_{gs2}}\right)} \qquad (9)$$

Using the following result:

$$\frac{s+\alpha}{s(s+\beta)} = \frac{1}{\beta}\left(\frac{\alpha}{s} + \frac{\beta-\alpha}{s+\beta}\right) \qquad (10)$$

the expression of Z_i' becomes:

$$Z_i' = \frac{1}{sC_{ss}} + \frac{\frac{1}{C_{pp}}}{s+\frac{1}{C_{pp}R_{pp}}} \qquad (11)$$

where:

$$C_{ss} = \left(1 + \frac{g_{m1}}{g_{m2}+g_o}\right)C' = (1+K)C' \qquad (12)$$

$$C_{pp} = \frac{1}{1+\frac{g_{m1}}{g_{m1}+g_{m2}+g_o}\frac{C_{gs2}}{C'}}C_{gs2} = \frac{1}{1+\frac{K}{K+1}\frac{C_{gs2}}{C'}}C_{gs2} \qquad (13)$$

$$R_{pp} = \frac{1}{g_{m1}+g_{m2}+g_o}\left(1 + \frac{K}{K+1}\frac{C_{gs2}}{C'}\right) \qquad (14)$$

The last relations show that Z_i can be replaced with the simplified circuit in Fig. 6. The multiplication factor is given by $K_v = g_{m1}/(g_{m2}+g_o)$. By neglecting the non-ideal effects, $C_{gs1,2} = C_{gd1} = 0$ and $g_{o1,2} = 0$, we can observe the correspondence of the last relations with those obtained for the structure built with ideal transistors.

Relationship (12) shows that g_o is important for the multiplication factor if $g_{m2} < g_o$, introducing a limitation effect of K adjustability. A similar analysis made on the structure proposed in [1] and illustrated in Fig. 2a, leads to an equivalent model similar to that in Fig. 5 but which contains an additional element: an equivalent parallel input resistance corresponding to the drain-source resistance of M_1. Therefore, comparing the current mode implementation in Fig. 2a with the structure proposed in this paper we can say that the circuit in Fig. 3 will have a capacitive behaviour at low frequencies, while the one in Fig. 2a will have a resistive one.

3. The relaxation oscillator

As an application for the capacitance multiplier proposed in the previous section, a single comparator relaxation oscillator as in [4] has been designed using the equivalent input capacitance of the structure in Fig. 3. The block diagram of this oscillator is illustrated in Fig. 7 and the oscillation frequency expression is given by (15). The comparator in Fig. 7 structure is implemented with a simple NMOS differential stage that has a PMOS current mirror as load. Given the poor input dynamic range of voltage mode multipliers the threshold voltages at the comparator input have to be close to the supply middle.

978-1-5386-4483-6/18 $31.00 © 2018 IEEE

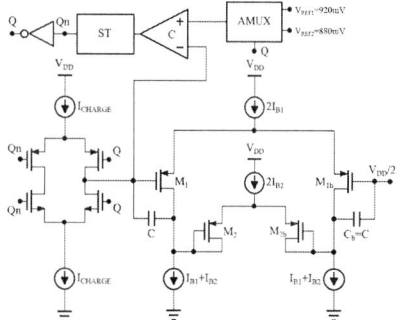

Fig. 7. Relaxation oscillator

$$f \approx \frac{I_{CHARGE}}{2C_{eq}\Delta V_{REF}} \quad (15)$$

4. Simulation results

The proposed circuit was simulated using the models of an AMS 180nm CMOS process. In order to meet the requirements of a low-power application, the g_{m1} transconductance has been tuned by varying the I_{B1} current in the $[10n \dots 990nA]$ range. The g_{m2} transconductance has been tuned by a complementary varying $I_{B2} = 1uA - I_{B1}$ current. Fig. 8 shows the input impedance magnitude and phase for different values of I_{B1}. The proposed structure equivalent input capacitance and the oscillation frequency versus the I_{B1} bias current are illustrated in Fig. 9 and Fig. 10.

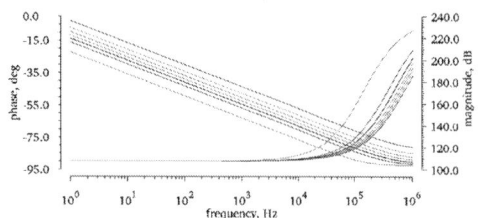

Fig. 8. Magnitude an phase of the input impedance for different values of tune current

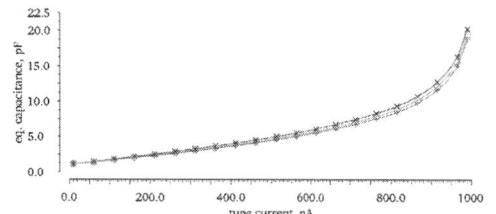

Fig. 9. Equivalent input capacitance versus tune current: (○) nominal, (x) worst speed and (+) worst power

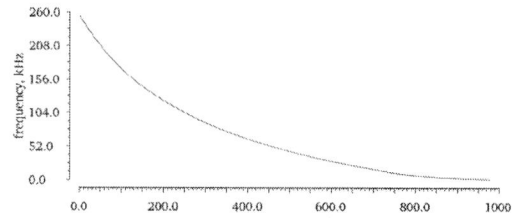

Fig. 10. Oscillation frequency versus tune current

5. Conclusions

In this paper a low power and low area voltage mode CMOS capacitance multiplication technique was presented. The circuit exhibits a good immunity to process variations and the low frequency behavior is capacitive. As an application a relaxation oscillator was presented. There is a good agreement between the theoretical expectations and the simulation results. The proposed capacitance multiplier tune range is $1p \dots 20pF$, the total current consumption is $5uA$ and the oscillation frequency can be tuned in the $35Hz \dots 250kHz$ range by varying the bias current. The proposed circuit has been simulated using the models of an AMS 180nm CMOS process.

References

[1] G.A. Rincon-Mora, "*Active capacitor multiplier in Miller-compensated circuits*", IEEE Journal of Solid-State Circuits, **35**, (1), pp. 26-32, 2000

[2] G. Bonteanu, "*A Review of Capacitance Multiplication Techniques*", 10th International Conference on Electronics, Computers and Artificial Intelligence (ECAI), 2018, *in press*

[3] I. Padilla-Cantoya, „*Capacitor Multiplier With Wide Dynamic Range and Large Multiplication Factor for Filter Applications*", IEEE Trans. on Circuits and Systems - II: Express Briefs, **60**, (3), pp. 152-156, 2013

[4] G. Pristavu, A.G. Vasilica, M. Apostolescu, G. Brezeanu, "*Fully-integrated oscillator in CMOS technology*", CAS 2010 Proceedings, pp. 517-520, 2010

Regulation Mechanism for Dickson Charge Pumps Using Charge Recycling and Adiabatic Charging

A. Florin Bîzîitu, B. Liviu Goraş

*A. Infineon Technologies Romania
E-mail: florin.biziitu@infineon.com
**B. „Gh. Asachi" Technical University of Iaşi and Institute of Computer Science,
Romanian Academy, Iaşi Branch, Romania
E-mail: lgoras@etti.tuiasi.ro

Abstract — Integrated charge pumps optimized for driving resistive loads are usually using a feedback control loop for improving the power efficiency over the entire specified load range. We propose a new circuit topology that combines charge recycling with a feedback control loop based on a regulated clock buffer supply that is also capable of adiabatic charging. The clock buffer supply regulation, as well as the adiabatic charging mechanism are both employed via a single on-chip fast transient response voltage regulator.

Keywords — regulated charge pumps, charge recycling, power efficiency, multi-step charging, adiabatic charging.

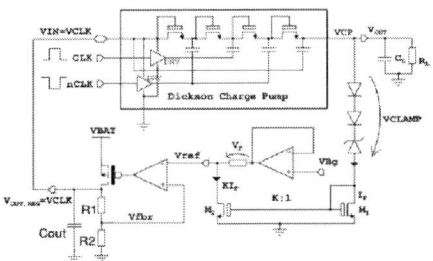

Fig. 1 Regulated Dickson CP

1. Introduction

Multiple regulation mechanisms exist for improving the output voltage accuracy and the power efficiency of fully integrated charge pumps (CPs). Regulated CP topologies based on variable clock frequency are more often implemented [1], but have the disadvantage of a load or input voltage dependent EMI spectrum.

Regulation methods based on the adaptive control of the CP clock buffer supply voltage as a function of the load current are used less frequently, but generally offer better EMC performance due to a more predictable emission profile regardless of the load current value.

First proposed in [2], charge recycling is a technique used to substantially reduce the performance impact that the parasitic capacitance has on the power efficiency of integrated charge pumps.

Adiabatic charging and in particular multi-step charging is a method that can further improve the energy efficiency of CP circuits. The term adiabatic refers to energy transfer that occurs between system components with minimal losses to the surrounding environment.

In this paper charge recycling techniques are applied to a basic Dickson CP core using passive charge transfer switches (diode connected NMOS devices) while multi-step (adiabatic) charging is employed by an additional modulation applied to the supply voltage of the clock buffers. The adiabatic charging modulation is applied in addition to the adaptive control of the clock supply as a function of the load current (the main CP regulation loop).

The major advantage of our proposed topology is that both the main CP regulation loop and the adiabatic charging are implemented with the same sub-circuit: a fully integrated, fast transient response voltage regulator.

2. Basic Concepts

A. Regulated Dickson Charge Pump

In order to improve the power efficiency, to limit the

amount of radiated emissions, as well as the CP output voltage when it is delivering less than the nominal load it was designed for, a regulation mechanism is usually implemented in practical CP designs.

The adaptive clock voltage regulation concept for integrated Dickson CPs is depicted in Fig. 1. The main blocks are a Dickson-type CP core generating the high voltage output VCP, an integrated LDO regulator providing the V_{CLK} supply voltage for the clock buffers as well as the input of the CP core and a circuit which is sensing the output voltage and generating an error signal V_F proportional to the difference between VCP and the reference voltage $VCLAMP$ at which the CP core is clamped. The error signal V_F is fed back and effectively subtracted from the (band gap reference V_{BG} derived) reference input V_{ref} of the CP voltage regulator, modulating the V_{CLK} value as a function of the load current (modeled by the resistive load $R_{L,CP}$).

The V_F voltage drop is generated by I_F multiplied by the M1, M2 current mirror ratio K. If the CP core has not reached the desired regulation voltage, the current through the clamp is zero and $Vref=Vbg$ leading to the maximum intended value for $VCLK$ (the core is running at maximum supply).

B. Charge recycling mechanism

The basic concept behind recycling the charge stored in the parasitic capacitors associated with the main pumping capacitors is illustrated in Fig. 2. First the clock buffers inside the CP core need to be replaced by 3-state clock buffers (with a dedicated control signal that allows for high impedance output).

Instead of switching the bottom plates of the pumping capacitors from one state to the next (low to high and vice versa) in the rhythm of the clock, for a very short time interval at the end of every semi-period, the clock buffers are driven to high impedance and their outputs are short circuited by means of an additional switch (transistor). Charge sharing between the parasitic capacitors present at the buffer outputs occurs during this phase and has two consequences:

978-1-5386-4483-6/18 $31.00 © 2018 IEEE   165

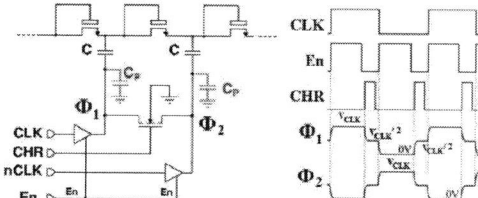

Fig.2 The charge recycling mechanism

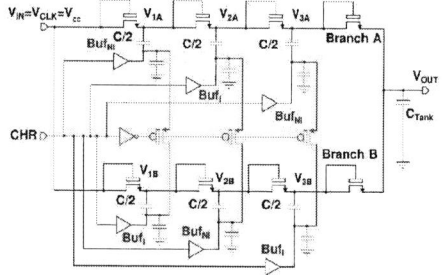

Fig.3 Dual branch charge recycling CP with *CHR* signal driven buffers

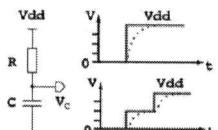

Fig.4 The adiabatic charging mechanism

- the parasitic capacitor on the buffer output about to transition from low to high will only need to be charged from $V_{CLK}/2$ to V_{CLK} during the next low to high transition.

-only half of the charge initially stored on the parasitic capacitor present on the buffer about to transition from high to low will be wasted as it is discharged from $V_{CLK}/2$ to ground.

In order to achieve optimal output impedance (maximum output current capability) for a given spent silicon footprint, our actual design uses a charge pump core that is composed of two pumping branches that work in anti-phase, delivering charge to the output every semi-period. The charge sharing is done in this case between the corresponding stages having the same index number in each pumping branch instead of using neighboring stages in the same branch as described in [4].

In order to have only one clock tree and a better phase correlation between the *CHR* signal (that controls the charge sharing switches) and the CP clock driving the pumping capacitors, only the *CHR* signal (having a frequency of $2 \times f_{CLK}$) is distributed to both the charge sharing switches and the clock buffers. The clock signal is regenerated locally in each buffer. Inverting (*Buf_I*) and non-inverting (*Buf_NI*) buffers are used in the CP core (actual implementation in Fig. 3) all of them driven by *CHR* signal.

C. The adiabatic charging mechanism

In order to evaluate how much energy is consumed from the CP core supply rail to charge the pumping capacitors each clock cycle, we can model the charge transfer switch and output stage of the clock buffer as an equivalent resistor through which the charge flows. Looking at Fig. 4 and

expression (1) we easily come to the conclusion that for each clock cycle, half of the energy drawn from the supply rail is dissipated on this equivalent charging resistance of the pumping capacitor. Minimizing the value of the equivalent resistor will only increase the amplitude of the charging current and decrease the charging time but will not have any effect on the amount of dissipated energy.

Equations (1) and (2) show that there is a difference in the energy delivered by the supply rail of the of the CP core if the pumping capacitors were to be charged in multiple steps instead of one step.

$$E_{source} = Q \cdot Vdd \tag{1}$$

$$E_{source} = \frac{1}{2}Q \cdot \frac{1}{2}Vdd + \frac{1}{2}Q \cdot Vdd = \frac{3}{4}Q \cdot Vdd \tag{2}$$

$$E_{source} = \frac{M+1}{2 \cdot M}Q \cdot Vdd \tag{3}$$

In order to minimize losses it would be therefore extremely useful if we could implement a multi-step charging mechanism for the CP core. Equation (3) shows that the amount of energy drawn from the source decreases as the number of intermediate charging steps M increases. Practical implementations should satisfy a compromise between the optimal number of intermediate charging steps and circuit complexity.

Together with a simple form of charge recycling, adiabatic charging of the pumping capacitors was already reported in [2] and [3]. Three-step charging was implemented in [3]. The voltage levels for each charging step were generated by dedicated regulators for each supply rail (charging step). The connection between the supply rail of the clock buffers and the output of the regulators generating the intermediate voltage levels was done via MOS switches driven by higher frequency phases from the main clock generator. A careful phase relationship must be maintained between the main CP clock the signal controlling the charge recycling phase and the signals driving the switches that connect the intermediate voltages during the multistep charging process.

3. Proposed concept

The main disadvantage of the solutions reported in [2] and [3] is the large silicon footprint required for implementing the voltage regulators and switches corresponding to each of the intermediate charging steps.

The silicon footprint can be drastically reduced if the charging steps are generated with a single, very fast transient response voltage regulator that is also part of the global CP regulation loop in which the clock buffer supply level is varied as a function of the CP load current.

Adiabatic charging can be easily combined with charge recycling for a significant increase in power efficiency. Given that the charge sharing mechanism used for charge recycling is already charging the pumping capacitors up to *VCLK/2*, the adiabatic charging steps should be chosen at *3VCLK/4* and *VCLK* for a simple two-step charging approach.

The number of charging steps dictates how fast the regulator supplying the clock buffers must be. The LDO must have a transient response time that is less than half of the CP clock frequency putting a limit on how high the main CP clock frequency can be. Achieving a small LDO transient response time is only possible by maximizing the LDO bandwidth that is achievable with the existing quiescent current budget. Taking into consideration the reasons mentioned above, we decided to use a simple two-step charging approach.

The *3VCLK/4* voltage step necessary for adiabatic charging is generated by dynamically modifying the resistor divider ratio of the on-chip LDO supplying the clock buffers during each CP clock cycle. This is the same LDO that is used as a variable clock buffer supply depending on the CP load current. The *Clk_STEP* clock signal of twice the CP clock frequency (2 x f_CLK) is used for the purpose of rapidly switching the output of the LDO from *3VCLK/4* to the *VCLK* value. This ensures that every CP clock semi-period is divided into two equal duration quarter periods. In one quarter of a CP clock period the output of the clock supply LDO will be *3VCLK/4* while on the remaining quarter it will jump to VCLK.

Switching from *3VCLK/4* to *VCLK* happens regardless of the effective value of *VCLK* which is as previously discussed, a function of the CP load current.

The transistor level schematic of the LDO is presented in Fig. 6. The output of the regulator is the supply for the CP core which is modeled here as resistor R_L. The input differential pair is implemented with minimum channel length low voltage transistors with a high W/L aspect ratio. R_{S1} and R_{S2} are source degeneration resistors. R_{gain} is chosen to be smaller than the impedance seen at node X. This means that the effective mid-band voltage gain of the g_m stage (equation 4) can simply be approximated as $\sim R_{gain}/R_{S1,2}$.

$$A_{V1} \approx \frac{g_{m1,2} \cdot \left[\left(r_{o5,6}(1 + g_{m1,2}\, r_{o1,2}) + r_{o1,2} \right) || r_{o3,4} || R_{gain} \right]}{1 + g_{m1,2} R_{S1,2}} \quad (4)$$

Our main goal was to achieve the highest LDO bandwidth possible with the minimum quiescent current. This is approximately equivalent to pushing the 0dB cross-over frequency of the LDO open loop transfer function at high frequencies while also maintaining the system stable. The LDO consists of a simple g_m stage driving directly the PLDMOS (p-type lateral double-diffused metal oxide semiconductor) pass device via resistor R_{gain}.

Transistor sizing and the high level of bias current (50µA) flowing through each branch of the differential input stage ensure that the mirror pole, p_4 at node Y is beyond the 0dB cross-over frequency and does not play any important role in bandwidth decrease or stability considerations.

The effect of the p_2 pole in the gate of the PLDMOS device is mitigated with the z_1 zero introduced by R_Z and C_Z, even if p_2 is slightly mobile as a function of the load current because of the Miller effect on the pass device. To assess in more detail the position of other poles we can write the transfer function from the gate of M_{pass} to the output of the LDO. The bypass capacitor C_{fb} is significantly (one hundred times) smaller than the rest of the capacitors, so that, in order to simplify the calculations, we did not take it into consideration in the transfer function $H_{pass}(s)$.

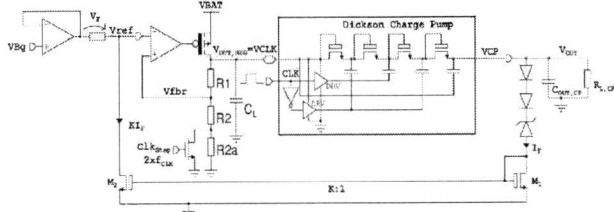

Fig.5 Proposed Regulated CP concept using charge recycling and adiabatic charging

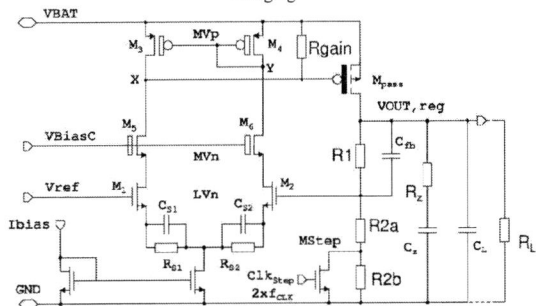

Fig.6 Fast transient response LDO

$$\omega_{p2} \approx \frac{1}{R_{gain}C_{e,pass}}, \qquad C_{e,pass} = C_{SG,pass} + C_{Miller,pass} \quad (5)$$

$$H_{pass}(s) \approx \frac{R_L(1 + sR_ZC_Z)}{s^2 R_L R_Z C_L C_Z + s(R_L C_Z + R_L C_L + R_Z C_Z) + 1} \quad (6)$$

$$\omega_{z1} \approx \frac{1}{R_Z C_Z} \quad (7)$$

We observe that adding R_Z and C_Z alongside C_L at the LDO output has actually introduced two load current (R_L) dependent poles in the transfer function: the dominant pole p_1 and pole p_3. For low load current values (high R_L), the two poles are widely split apart with p_3 residing after the 0dB cross-over frequency. With increasing load current values (decreasing of R_L) p_1 moves higher in frequency while p_3 will move towards the origin decreasing the phase margin of the LDO. The effect of p_3 on the LDO open loop phase is thus corrected with the z_2 zero generated by the bypass capacitor C_{fb}. The effect is clearly visible by comparing the gain and phase diagrams of Fig 7. The flat region of the phase diagram present at low load currents (R_L=1MOhm) in Fig. 7 (two poles p_1 and p_2 and two zeroes z_1 and z2) disappears at high load currents (R_L=300Ohms) when p_3 moves below the 0dB cross over, closer to p_2. Without the effect of z_2 we would have a poor phase margin. The pole associated with z_2, (p_{Z2}) is always positioned after the 0dB cross-over.

Although not mandatory, the phase margin can be further improved by introduction of z_3 by bypassing the source degeneration resistors $R_{S1,2}$ with capacitors $C_{S1,2}$.

Transistor M_{step} has the purpose of shorting out part of the resistor divider rapidly changing the divider ratio depending on the value of the *Clk_Step* signal: if *Clk_Step* is low then the divider is adjusted for *3VCLK/4* at the LDO output, if *Clk_Step* is high then the LDO should provide the maximum charging step equal to *VCLK*.

$$\omega_{p1} = \frac{-(3R_L + 2R_Z) + \sqrt{9R_L^2 + 4R_L R_Z + 4R_Z^2}}{4R_L R_Z C_L} \quad (8)$$

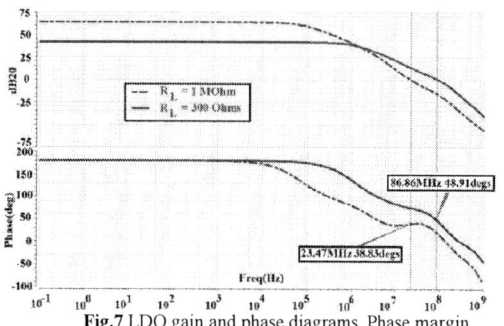

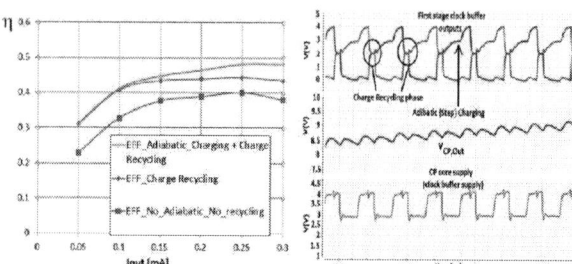

Fig.7 LDO gain and phase diagrams. Phase margin

Fig.8 a. Efficiency comparison **b.** Internal waveforms for CP1(top − clock buffer outputs showing the charge recycling and adiabatic two-step charging phases, middle - CP1 output ripple, bottom − LDO output

$$\omega_{p3} = \frac{-(3R_L + 2R_Z) - \sqrt{9R_L^2 + 4R_L R_Z + 4R_Z^2}}{4R_L R_Z C_L} \quad (9)$$

$$\omega_{p4} \approx \frac{g_{m4}}{C_{gs4} + C_{ds4} + C_{gs3}} \quad (10)$$

$$\omega_{z2} \approx \frac{1}{R_1 C_{fb}}, \qquad \omega_{p,z2} \approx \frac{1}{(R_1 \| R_2) C_{fb}} \quad (11)$$

$$\omega_{z3} \approx \frac{1}{R_S C_S}, \qquad \omega_{p,z3} \approx \frac{1 + g_{m1,2} R_S}{R_S C_S} \quad (12)$$

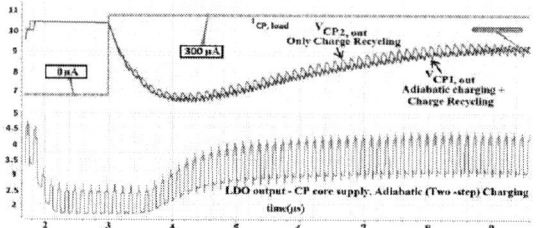

Fig.9 CP1 and CP2 load step (0μA to 300μA) response and ripple comparison (lower graph shows the CP1 core supply with two-step charging)

4. Simulation results

A comparison between three CPs was made in order to assess the effectiveness of the proposed regulation mechanism. All three regulated CPs use the same regulation loop topology (Fig.1) where the clock buffer supply level is a function of the CP load current. The CP1 and CP2 systems are using the same CP core that has charge recycling implemented according to the concept schematic presented in Fig. 3, while CP3 has no charge recycling implemented. All three CPs have $V_{CLK,max}$=4.5V (during ramp-up and full load), n=3 pumping stages, m=2 pumping branches, are driven by a f_{CLK}=4MHz clock signal, with each of the pumping capacitors being C=24pF and α=0.24 as the ratio between the pumping capacitor and the bottom plate parasitic capacitor. All three CPs were targeted to deliver I_{OUT}=250μA at V_{OUT}=10V. The only difference between CP1 and CP2 is that the first one has the adiabatic charging turned on and implicitly receives the Clk_{Step} clock signal, while for the second one adiabatic charging is turned off by always connecting Clk_{Step} to logical 1 (one step charging after the charge recycling phase). Even if CP3 is using the same LDO for supplying the clock buffers, it has both charge recycling and adiabatic charging turned off with only the base regulation loop operational.

The reduction in ripple amplitude is clearly visible for CP1 that has adiabatic charging is enabled, especially at high load currents. In case of CP1 charge is transferred to the output in two sub-cycles corresponding to each of the charging steps. (Fig.8b, Fig.9). However if we examine the efficiency comparison in Fig.8a, it is clearly visible that the adiabatic charging's role decreases as the load current decreases. The efficiency graphs of CP1 and CP2 become indistinguishable for lower load currents. This happens because at low load current levels the pumping capacitors are never depleted completely being completely recharged after the first charging step. The observed behavior is also due to the low loop gain of the external regulation loop that doesn't decrease $VCLK$ aggressively enough for low load current levels.

Increasing the loop gain would require a change in the frequency compensation concept of the CP regulation loop (currently the dominant pole is defined by the CP core output capacitor $C_{OUT, CP}$ together with the CP output impedance).

The improvement efficiency at high loads when using both charge recycling and adiabatic charging (~12.2%, 8% from the charge recycling and 4.2 % from two-step charging) translates into a reduction of CP input current pulse amplitude and lower EMI. Losses always convert to EMI, in the case of a CP system lower efficiency means higher radiated EMI.

5. Conclusion

A new circuit topology that combines charge recycling with a feedback control loop based on a regulated clock buffer supply that is also capable of adiabatic charging was presented.

The main advantage of the proposed topology compared to existing solutions [2] [3] is that the clock buffer supply regulation, as well as the adiabatic charging mechanism are employed via a single on-chip fast transient response voltage regulator which was also described in detail.

References

[1]. L. Aaltonen and K. Halonen, *"On-chip charge-pump with continuous frequency regulation for precision high-voltage generation"*, Proc. of the 2009 Ph.D. Research in Microelectronics and Electronics.

[2]. C.Lauterbach et al" *Charge sharing concept and new clocking scheme for power efficiency and electromagnetic emission improvement of boosted charge pumps"*, IEEE JSSC, Vol. 35, No. 5, May 2000

[3]. Steve Ngueya, W & Mellier, Julien & Ricard, Stephane & Portal, J.M. & Aziza, Hassen. (2017*). " High voltage recycling scheme to improve power consumption of regulated charge pumps.* „2016 IEEE 8th International Memory Workshop (IMW), Paris, 2016, pp. 1-4

[4]. F. Bîzîitu and L. Goraş, *"Improving IC power efficiency by implementing charge recycling in Dickson charge pumps with multiple pumping branches,"* 2017 (CAS), Sinaia, 2017, pp. 187-190.

[5]. J. F. Dickson, *"On-chip high-voltage generation in MNOS integrated circuits using an improved voltage multiplier technique,"* IEEE JSSC, vol. SC-11, no. 3, pp. 374–378, June 1976.

Application Specific Integrated Circuit (ASIC) for an energy efficient impulse radio ultra-wideband transceiver. Testing and statistic assessment

Nicolae Varachiu*, Bilal Benamrouche**, J.-L. Noullet**, A. Rumeau**, Daniela Dragomirescu**

* National Institute for R&D in Microtechnology IMT-Bucharest
E-mail: nicolae.varachiu@imt.ro
** LAAS-CNRS, Universite de Toulouse, CNRS, INSA, Toulouse, France
E-mail: daniela@laas.fr

Abstract—This paper presents the set-up, test results and statistic evaluation for a lot of five ASIC prototypes of an impulse radio ultra-wideband transceiver (IR-UWB), realized in ST Microelectronics CMOS 65 nm technology. The main purpose of our undertaken is to provide an energy efficient device: having five functional prototypes we assessed the manufacturing process stability in respect with average power consumption, at different data rates.

Keywords— ASIC design and manufacturing; energy efficient IR-UWB Transceiver; power measurment technique; statistics for manufacturing evaluation

1. Introduction

Energy efficient devices are, for many years, an important required characteristic for all new implemented ones; especially when they are autonomous (i.e. having battery as energy source), low power consumption being crucial for an efficient long-term functioning.

Staring from the research deployed in [1] and [2] in [3] was proposed an IR-UWB transceiver for wireless sensors network, using *clock-gating* solution (Fig.1) in order to improve the energy efficiency.

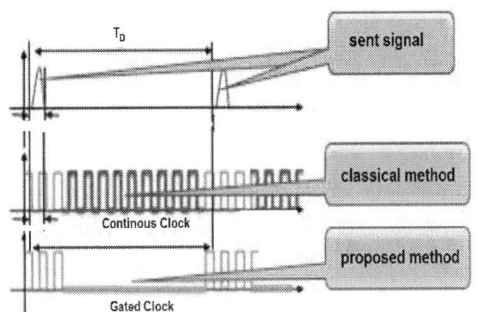

Fig. 1 Clock-gating solution versus continuous clock

The designed ASIC was implemented in LAAS Toulouse using the technology CMOS 65 nm from ST Microelectronics (Fig. 2)

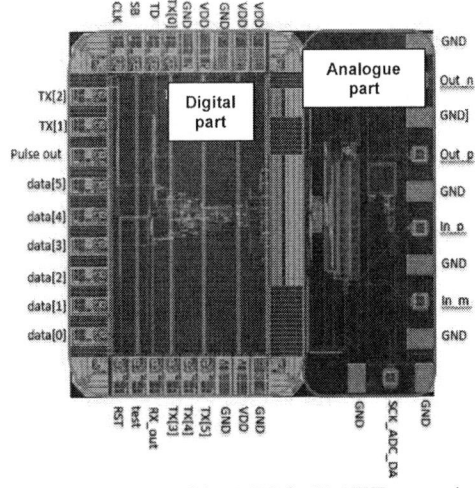

Fig. 2 Final layout of the ASIC for IR-UWB transceiver realized in ST Microelectronics CMOS 65 nm technology

2. Simulation

Using the mask design CADENCE platform, we performed the simulation for average power consumption of ASIC, both for proposed solution based of clock-gating versus the classical solution of continuous clock as in (Fig. 3). We get 2500 Kbps (Kb/s) as the point limit of data rate: above this value, the clock-gating technique-based solution is no more energy efficient than the classical one.

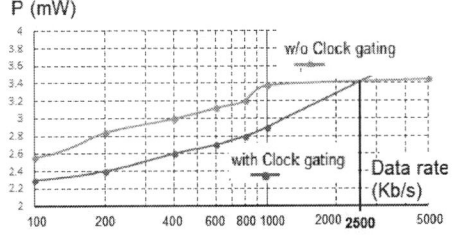

Fig. 3 Power consumption in respect with data rate

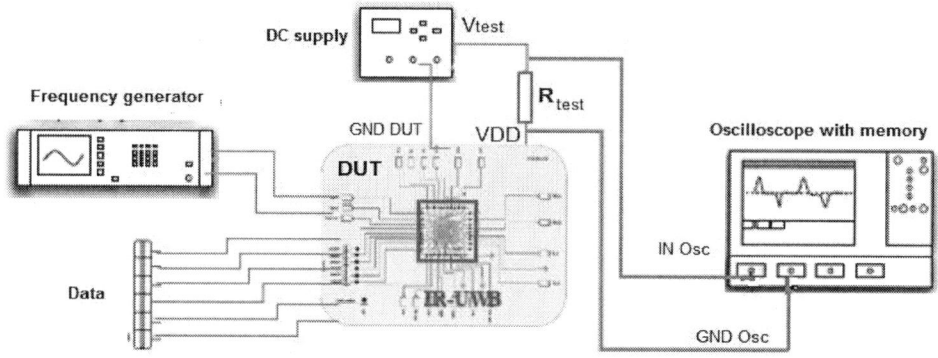

Fig. 4 Set-up for average power consumption measurement

3. Experimental set-up and background

For electrical devices the energy consumed in the time interval from zero to T is the integral of the product of voltage and current, both in respect of current time t, having as limits 0 and T:

$$E = \int_0^T u(t)i(t)dt \tag{1}$$

In order to estimate the *energy efficiency* for different approaches and to effectively characterize particular solutions and implementation, we could not use directly the measurement of energy, as it depends also of the period of time that we observe the device in function, but a "normalized" in respect of time measure. So, in order to compare the energy efficiency of a different solutions and to further measure the realized devices, we further consider the average power consumption P_{ave}:

$$P_{ave} = \frac{1}{T_D} \int_0^{T_D} u(t)i(t)dt \tag{2}$$

where T_D is the period time corresponding to data rate ($=1/T_D$), i.e. the period of time between two packets of signal (Fig. 1).

If we have a periodical signal, the average power over one period of time is a constant: so, *this is a consistent measure that we will use further to estimate the energy efficiency of our manufactured circuits.*

Specifically, we introduced in (2) VDD for voltage, i.e. the practical constant supply voltage applied to device under test (DUT),

as shown in the set-up and measurement approach description (Fig. 4); we use by now simply P as notation for the average power:

$$P = \frac{1}{T_D} \int_0^{T_D} VDD\, i(t)dt \tag{3}$$

We apply V_{test} − 1.2 V and we use a R_{test} = 1 Ohm, same physical resistor for all tests, in order to not increase measurement errors. Because the maximum current consumption of ASIC (DUT) is less than 10 mA, the voltage over R_{test} is less than 10 mV, so we could consider with an error less than 0.1% (precision) that VDD = V_{test} = 1.2 V.

By using the oscilloscope with memory (LeCroy SDA 813Zi) we recorded the effective curve representing the voltage u(t) over R_{test}; from Ohm law u(t) = R_{test} * i(t). So, computing the area under the memorized curve for one period T_D and by considering formula (3) we determine the average power P in all further tests.

4. Tests, results and discussions

We performed average power consumption tests for five ASIC functional prototypes (DUT in Fig 4), all manufactured in LAAS Toulouse, using the technology CMOS 65 nm from ST Microelectronics.

Starting from the customer specifications (Voice of Customer) for Structural Health Monitoring in commercial aviation, we provided more test points in the main range of 100 to 1000 Kb/s. The results are presented in the Tabel 1.

978-1-5386-4483-6/18 $31.00 © 2018 IEEE

Table 1 Average power consumption (mW) of five ASIC prototypes for different data rates

Data rate / Prototype #	1 Kb/s	10 Kb/s	100 Kb/s	200 Kb/s	400 Kb/s	600 Kb/s	800 Kb/s	1 Mb/s	5 Mb/s
1	3.1	3.22	3.36	3.4	4	3.95	3.87	3.9	4.1
2	2.97	3.12	3.31	3.44	3.65	4	3.98	3.85	4.05
3	3.1	3.22	3.35	3.38	3.9	3.75	3.85	3.83	4.14
4	3.2	3.12	3.3	3.42	3.8	3.7	3.9	3.89	4
5	2.9	3.17	3.33	3.43	3.7	3.68	3.91	3.92	4.08

The main purpose of our undertaken is to provide energy efficient devices, so, after the electrical functionality was checked for all five prototypes, we further measure and determine the average power for all five prototypes at mentioned data rates, from recorded data with the memory oscilloscope, as presented in section 3. To asses the stability of the manufacturing process. we use the Individual Values and Moving Range (I-MR) Charts, having a number of five samples as a minimum accepted for this kind.

I-MR Charts works best with data that is normally distributed (4). So, we provided the Anderson-Darling tests for normality for the average power consumption measured values, as presented in section 3, for all five prototypes tested at the mentioned data rates.

It is using the null hypothesis testing; a *p-value* is the probability of obtaining the result at least as extreme as the one that was actually observed, assuming that the null hypothesis (= normal distribution of data) is true. Results are presented in bellow Table 2:

Table 2 The p-value for the five ASIC prototypes

Data rate (Kb/s)	1	10	100	200	400
p-value	0.537	0.164	0.157	0.289	0.148
Data rate (Kb/s)	600	800	1000	5000	
p-value	0.336	0.231	0.886	0.912	

As largely accepted by statisticians, a threshold of 0.1 for p-value is used as minimum value: i.e. over that value the null hypothesis could not be rejected.

As in Table 2, all p-values are over 0.1, so *we could consider the normality for measured data and further build I-MR charts,* one for each data rate value, in each chart being represented all five prototypes as input

observations on x-axis. The control limits, i.e. UCL (=Upper Control Limit) and LCL (=Low Control Limit) are computed from data (4), using the general accepted formulas starting from Xbar (average of individual values), MRbar (average of Moving Range) and applicable constants.

All nine I-MR Charts are in control, no special causes are present: all represented points are inside the control limits. In Fig. 5 are represented the I-MR Charts for the average power consumption corresponding to data rates of 100, 200, 400, 600, 800 and 1000 Kb/s, i.e. the customer specifications for Structural Health Monitoring in commercial aviation. **So, our designed manufacturing process that provided the five prototypes is a stable one from the perspective of average power consumption.**

We provide in Table 3 a performance comparison among different solutions for impulse radio ultra-wideband transceivers, two of them, IEEE 802.15.4 (ZigBee) and IEEE 802.11 (WiFi), being benchmarks in the field.

We could observe the advantages of our proposed and implemented solution in respect with maximum average power consumption and/or circuit dimensions.

Acknowledgments. The main work was performed in the framework of the European Union project *CHIST-ERA SMARTER (**S**mart **M**ultifunctional **A**rchitecture and **T**echnology for **E**nergy aware wireless senso**R**s)*, and in UEFISCDI Romanian grant *Sensors and networks of sensors, from laboratory experiments to industrial applications*, PN-III-P1-1.1-MC-2017-1851

Table 3 Performance comparison

Type	Frequency band	Data rate (Kb/s)	Maximum average power consumption (mW)	Circuit dimension	Fabrication technology
[5]	960 MHz	630-1300	NA	4.52 mm²	0.13 µm CMOS
IEEE 802.15.4 [6]	NA	250	27	NA	NA
IEEE 802.11 [7]	NA	11000	2900	NA	NA
This solution [3]	1 GHz	10-5000	3.9 @ 1Mb/s 4.14 @ 5 Mb/s	1 mm²	**ST Microelectronics CMOS 65 nm**

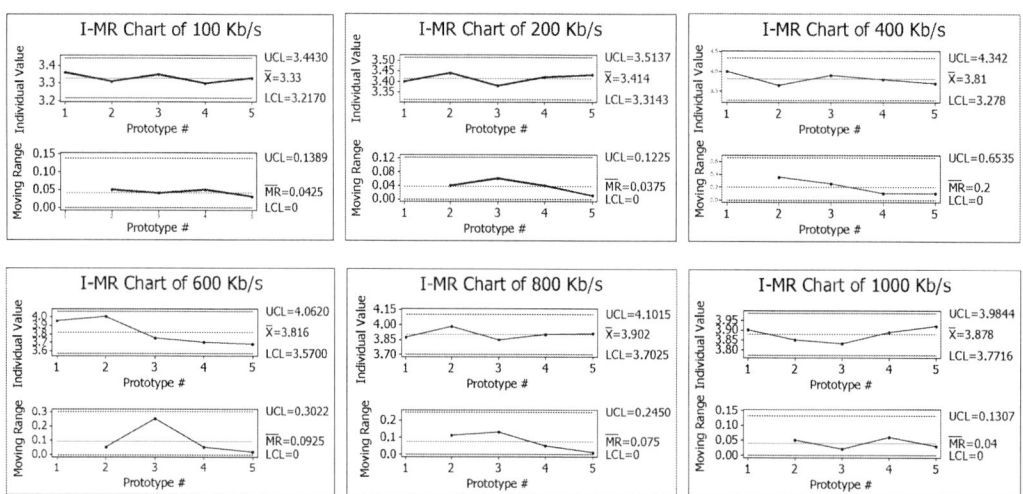

Fig. 5 I-MR Charts for average power consumption, from 100 to 1000 Kb/s data rates

References

[1] Lecointre, A.; Dragomirescu, D.; Plana, R., *"Largely reconfigurable impulse radio UWB transceive"* ELECTRONICS LETTERS Volume 46 Issue 6 pp. 453-U102 Mar. 2010

[2] Jatlaoui, Mohamed Mehdi; Dragomirescu, Daniela; Ercoli, Mariano; et al., *"Wireless communicating nodes at 60 GHz integrated on flexible substrate for short-distance instrumentation in aeronautics and space"* INTERNATIONAL JOURNAL OF MICROWAVE AND WIRELESS TECHNOLOGIES Volume 4 Issue 1 Special Issue SI pp. 109-117, Feb. 2012

[3] B. Benamrouche, A. Rumeau and D. Dragomirescu, *"Ultra-low power IR-UWB transceiver for wireless sensors network"*, Proceedings of CAS 2017 -International Semiconductor Conference, Sinaia, Romania, pp. 285-288, Oct. 2017

[4] Quintin Brook, *Lean Six Sigma & Minitab* (4th Edition) OPEX Resources Ltd, 2014

[5] Verhelst, M; Van Helleputte, N; Gielen, G; Deha, "A reconfigurable 0.13 mm CMOS fully integrated IR-UWB receiver for communicationand sub-cm ranging" in Solid-State Circuits ConferenceCC 2009, pp 250-251, 2009

[6] Somayazulu, V.S. et al, *"Design challenges for very high data UWB systems"*, Asilomar Conference on Signals, Systems and Computers, pp. 717-721, 2002

[7] Aedudodla et al, *"Timimg acquisition in ultra-wide band communications system"* IEEE Transaction on Vehicular Technology, vol. 54, no.5, Sept. 2005

Wide Dynamic Range Current Mirror

Arcadie Cracan*, Gabriel Bonteanu*
*"Gheorghe Asachi" Technical University of Iasi
E-mail: acracan@etti.tuiasi.ro

Abstract—A novel circuit solution for implementing a known low-voltage current mirror topology is presented. It demonstrates a close to the rail input and output voltages (one saturation voltage at the input and two saturation voltages at the output) while being able to support an input current variation of orders of magnitude. The structure uses dynamic biasing in order to keep the transistors in the active region.

Keywords—wide range; high swing; rail-to-rail; current mirror; current amplifier.

1. Introduction

The current mirror is an essential analog building block. With the continuing trend in lowering the power supply, an important feature of a current mirror is being able to operate close to the power and ground rails. In the basic current mirror structure, as shown in Fig. 1a, the input voltage is given by the gate-to-source voltage of the input diode. The basic current mirror demonstrates a high input voltage (a gate-to-source voltage) and a poor matching of the drain-to-source voltages of the master and slave transistors.

In order to lower the input voltage a level shifter can be used, as presented in [1]. The drain potential is shifted relative to the gate potential by using a floating voltage source such that the drain-to-source voltage is brought closer to V_{DSsat}. To improve the matching of the drain-to-source voltages of the master and slave transistors a generalized

regulated cascode technique is used [2]. An amplifier regulates the potential of the gate of the cascode transistor such that the drain-to-source voltage of the mirror slave is kept equal to the drain-to-source voltage of the mirror master. The series feedback increases the output resistance of the current mirror.

Applying the two methods for the basic current mirror a generic improved current mirror is obtained, as shown in Fig. 1b. A wide variety of implementations for amplifiers A_1 and A_2 exist in the literature [3]–[6]. In this paper we present a current mirror structure that can operate from a power supply as low as $V_{GS} + 2V_{DSsat}$ providing a close to the ground rail input voltage of a V_{DSsat} and a close to the ground rail minimum output voltage of $2V_{DSsat}$.

2. The proposed wide dynamic range current mirror

The proposed current mirror employs flipped voltage follower (FVF) type structures in order to implement the amplifiers A_1 and A_2 of the generic current mirror in Fig. 1b, as shown in Fig. 2, where the ideal I_B current sources are implemented with simple CMOS current mirrors from a reference current source.

The transistor M_{12} is in an FVF configuration. It has a constant current, set by I_B, so the drain-to-source voltage of the input transistor M_1 is given by

$$v_{DS1} = v_{GS13} - v_{GS12} \qquad (1)$$

The transistor M_{11} is also in an FVF configuration, so the drain-to-source voltage of the output transistor M_2 is given by

$$v_{DS2} = v_{GS13} - v_{GS11} \qquad (2)$$

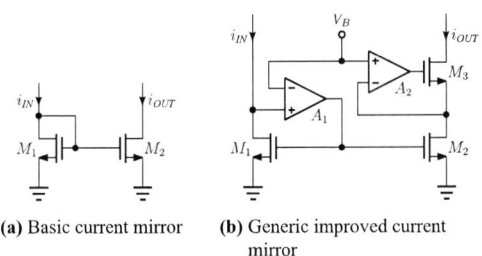

(a) Basic current mirror   **(b)** Generic improved current mirror

Fig. 1 Basic and improved current mirrors

978-1-5386-4483-6/18 $31.00 © 2018 IEEE   173

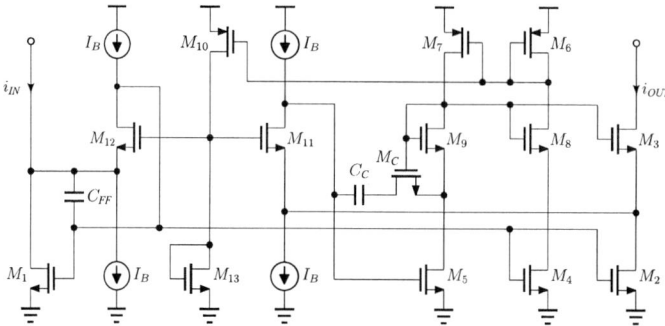

Fig. 2 Proposed wide dynamic range current mirror

Because the transistors M_{11} and M_{12} have the same form factor and are biased by equal currents, it follows that $v_{GS11} = v_{GS12}$, so the drain-to-source voltages of the input and output transistors are equal $v_{DS1} = v_{DS2}$ The M_{13} transistor has been sized for a minimum drain-to-source voltage of the output M_2 transistor, as described in [7].

Compared to the solution in [4], the proposed circuit has a dynamic biasing circuitry. The input current is used to generate the gate voltage of the transistors M_3, M_{11} and M_{12}. The transistors M_{11} and M_5 are part of the voltage amplifier that regulates the voltage at the gate of the cascode transistor M_3 (the amplifier A_2 in Fig. 1b). Unlike the circuit in [4], the transistor M_5 is biased dynamically so that its gate-to-source voltage is equal to the gate-to-source voltage of the input transistor M_1. That makes the drain-to-source voltage of the transistor M_{11} be equal to the drain-to-source voltage of the transistor M_{12}, which achieves a better matching between these two transistors.

A. Small-signal analysis

The small-signal analysis can be made on the two relatively independent parts of the current mirror: the input stage – from the vol-

tage of the M_1 transistor – and the output stage – from the gate-to-source voltage of the M_1 transistor to the output current i_{out}. The input stage is represented simplified in Fig. 3. In this figure G_D and G_G represent the total conductance at the drain and at the gate of the transistor M_1, respectively. The capacitors C_D and C_G represent the total capacitance at the drain and at the gate of the transistor M_1, respectively. The conductance g_{o12} is the output conductance of the transistor M_{12}.

B. Input stage small-signal analysis

By applying KCL at the input and output nodes one can write:

$$i_{in} = (G_D + g_{o12} + g_{m12} + sC_{FF} + sC_D)v_{in} + \\ + (g_{m1} - g_{o12} - sC_{FF})v_{gs1} \quad (4)$$

$$(g_{o12} + g_{m12} + sC_{FF})v_{in} = (G_G + g_{o12} + sC_G + \\ + sC_{FF})v_{gs1} \quad (5)$$

Considering that $g_{m12} \gg g_{o12}$, $g_{m12} \gg G_D$ and $C_{FF} \gg C_D$ (the drain-to-source capacitance of a transistor is typically negligible and the drain junction capacitance is typically small enough), the relationships can be simplified to

$$i_{in} = (g_{m12} + sC_{FF})v_{in} + (g_{m1} - sC_{FF})v_{gs1} \quad (6)$$

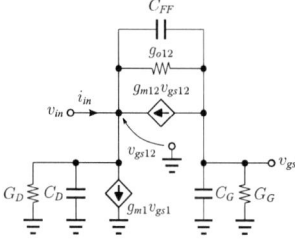

Fig. 3 Input stage equivalent small-signal circuit

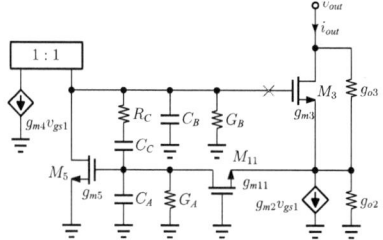

Fig. 4 Output stage equivalent small-signal circuit

$$(g_{m12} + sC_{FF})v_{in} = (G_G + g_{o12} + sC_G + \\ + sC_{FF})v_{gs1} \quad (7)$$

It follows that

$$i_{in} = (g_{m1} + G_G + g_{o12} + sC_G)v_{gs1} \quad (8)$$

and considering that $g_{m1} \gg G_G$ and $g_{m1} \gg g_{o12}$ we finally obtain

$$v_{gs1} = 1/(g_{m1} + sC_G)i_{in} \quad (9)$$

It can be observed from the expression of the overall input transconductance (v_{gs1}/i_{in}) that, given the previous assumptions, the C_{FF} capacitance has little influence on the expression (9).

Using relationships (4) and (5) and the previous assumptions the input impedance, $Z_{in} = v_{in}/i_{in}$ can be written as

$$Z_{in} = \frac{G_G + g_{o12} + sC_{FF} + sC_G}{(g_{m12} + sC_{FF})(g_{m1} + sC_G)} \quad (10)$$

which at low frequency has the expression

$$Z_{in}|_{s \to 0} = \frac{1}{g_{m1}} / \frac{g_{m12}}{G_G + g_{o12}} \quad (11)$$

that shows that the input impedance $1/g_{m1}$ of a simple diode-conected transistor is lowered in this circuit by the gain of the M_{12} cascode stage.

C. Output stage small-signal analysis

The simplified equivalent small-signal circuit of the output stage is represented in Fig. 4. The transistors M_2 and M_4 have been represented as voltage controlled current sources. In order to simplify the analysis we can assume that the $g_{m4}v_{gs1}$ component is zero (its effect in the output current is approximately $g_{m4}v_{gs1}G_A/g_{m5}$, which, assuming $G_A/g_{m5} \propto 1/100$, is small enough to be neglected).

By breaking the loop at the "x" marked in Fig. 4 the open loop circuit transfer function can be written as

$$T(s) \approx -\frac{g_{m3}}{g_{m11} + g_{m3}} \cdot \frac{g_{m11}}{G_A} \cdot \frac{g_{m5}}{G_B} \cdot \frac{1 + sC_C(R_C - 1/g_{m5})}{1 + s\alpha + s^2\beta + s^3\gamma} \quad (12)$$

where $\alpha \approx \frac{g_{m5}}{G_B}\frac{C_C}{G_A}$, $\beta = \frac{C_AC_B + C_AC_C + C_BC_C}{G_AG_B} + C_CR_C\left(\frac{C_A}{G_A} + \frac{C_B}{G_B}\right)$ and $\gamma = \frac{C_AC_BC_CR_C}{G_AG_B}$. This open loop transfer function expression is typical for Miller-type compensation. The dominant pole has the approximate expression

$$p_d \approx -\frac{1}{\alpha} = -\frac{G_A}{C_C\frac{g_{m5}}{G_B}} \quad (13)$$

while the non-dominant poles, considering them widely separated, can be expressed as

$$p_{n1} \approx -\frac{\alpha}{\beta} \quad p_{n2} \approx -\frac{\beta}{\gamma} \quad (14)$$

where it has been assumed that $p_{n1} \ll p_{n2}$. The transfer function also exhibits a zero that can be either in the left half-plane or in the right half-plane depending on the sign of $R_C - 1/g_{m5}$. The resistance R_C is implemented using a triode-region NMOS transistor (M_C) such that the value of R_C is correlated to $1/g_{m5}$ and is designed such that this zero cancels the first non-dominant pole and enhances the phase margin.

3. Simulation results

To assess the performance of the proposed circuit several simulations have been made using the AMS $0.18\mu m$ CMOS PDK models.

A. DC results

Fig. 5 presents the dependence of the current mirror output current on the output voltage. It can be observed that for a value of $I_{OUT} = 500\mu A$ the output voltage can be as low as $V_{OUTmin} \approx 350mV$. Fig. 6 shows the output conductance for different levels of output voltage. For an output resistance greater than $100M\Omega$ the output voltage can be as low as $V_{OUTmin} \approx 470mV$.

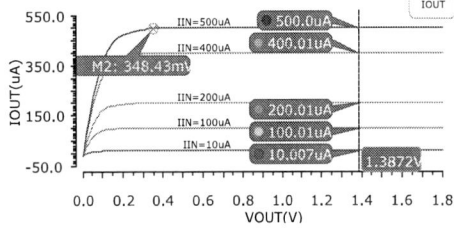

Fig. 5 Output characteristics $I_{OUT}(V_{OUT})$

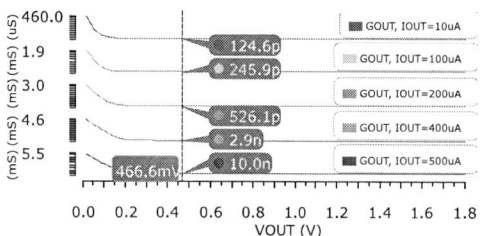

Fig. 6 Output conductance $\partial I_{OUT}/\partial V_{OUT}$ for different output currents

B. Small-signal AC results

Fig. 7 depicts the variation of the output stage loop transfer function phase margin as a function of the output current. It can be observed that in the considered output current range $(10\mu A - 500\mu A)$ the chosen compensation capacitor $(C_c = 1pF)$ is sufficient to obtain a phase margin (PM) of at least 60°.

C. Transient results

Fig. 8 shows the normalized step response of the current mirror (i_{OUT}/step value). The simulation has been performed for the following step values: $10\mu A$, $100\mu A$, $200\mu A$, $400\mu A$ and $500\mu A$. As the step value is increased, the delay at the output also increases.

For a sinusoidal input current of 1kHz the simulated level of total harmonic distortions (THD) is less than 0.004%, as shown in Table 1.

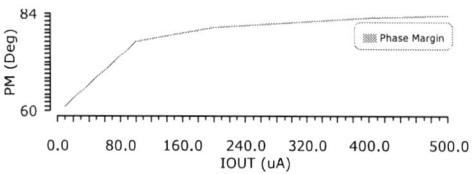

Fig. 7 Output stage loop transfer function phase margin for different output currents

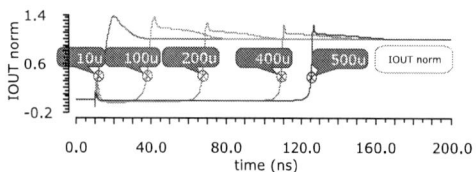

Fig. 8 Normalized step response (i_{OUT}/step size) for different step sizes

Table 1. i_{OUT} THD for several DC levels and AC amplitudes of an input sinusoidal current

I_{IN} (DC)	I_{in} (AC)	i_{OUT} THD
20μA	10μA	0.0038%
100μA	90μA	0.0037%
260μA	250μA	0.0037%

4. Conclusions

A novel wide range current mirror has been presented that can achieve close to the rails operation, while maintaining a high output resistance. The innovation consists of a dynamic biasing circuit and also a compensation scheme that uses a triode biased MOS the resistance of which depends on the DC level of the input signal. The structure is able to achieve a very low level of total harmonic distortions of less than 0.004%.

References

[1] Y. Cong and R. L. Geiger, "*Cascode current mirrors with low input, output and supply voltage requirements,*" in Proceedings of the 43rd IEEE Midwest Symposium on Circuits and Systems (Cat.No.CH37144), **1**, pp. 490–493, 2000.

[2] T. Serrano and B. Linares-Barranco, "*The active-input regulated-cascode current mirror,*" IEEE Transactions on Circuits and Systems I: Fundamental Theory and Applications, **41**(6), pp. 464–467, Jun 1994.

[3] V. Prodanov and M. Green, "*CMOS current mirrors with reduced input and output voltage requirements,*" Electronics Letters, **32**(2), pp. 104–105, Jan. 1996.

[4] A. Torralba, R. G. Carvajal, J. Ramirez-Angulo, and E. Munoz, "*Output stage for low supply voltage, high-performance CMOS current mirrors,*" Electronics Letters, **38**(24), pp. 1528–1529, Nov 2002.

[5] N. Hassen, H. B. Gabbouj, and K. Besbes, "*Low-voltage high-performance current mirrors: Application to linear voltage-to-current converter,*" International Journal of Circuit Theory and Applications, **39**(1), pp. 47–60, 2011.

[6] G. K. Chakravarthy, N. Laskar, S. Nath, S. Chanda, and Baishnab, "*Flipped voltage follower based high dynamic range current mirror,*" in 2017 Devices for Integrated Circuit (DevIC), pp. 311–315, 2017.

[7] P. J. Crawley and G. W. Roberts, "*High-swing MOS current mirror with arbitrarily high output resistance,*" Electronics Letters, **28**(4), pp. 361–363, Feb 1992

Session IC-S 1

INTEGRATED CIRCUITS 1
Student papers

978-1-5386-4483-6/18 $31.00 © 2018 IEEE

A High Performance Mixed-Voltage Digital Output Buffer

Anca Mihaela Dragan[1,2], Andrei Enache[1,2], Alina Negut[1], Adrian Macarie Tache[1], Gheorghe Brezeanu[2]

[1]ON Semiconductor Romania

AncaMihaela.Dragan@onsemi.com, Andrei.Enache@onsemi.com, Alina.Negut@onsemi.com,
AdrianMacarie.Tache@onsemi.com

[2]University "Politehnica" of Bucharest, Romania

Gheorghe.Brezeanu@dce.pub.ro

Abstract— A digital push-pull output buffer is designed and implemented in a 0.18μm CMOS EEPROM process. The buffer acts as an interface between an internal low voltage and an external, higher level voltage. The circuit can operate in a wide range of power supply voltages, from 1.6V to 5.6V, at data rates of up to 20 Mbps. These performances were achieved through topology changes to a classic digital buffer.

Keywords— *output buffer; push pull stage; wide voltage range; high data rate.*

1. Introduction

Signal and power integrity are crucial in Very Large Scale Integration (VLSI) systems. Modern trends in deep sub-micron circuit designs, such as high operating frequencies, short rise/fall delays and a wide range of supply voltages, are some of the most desired targets to achieve [1].

Certain portable devices, including wireless hand-sets, mobile phones, tablets, often employ circuitry which runs on two or more different voltage levels. For instance, ICs utilized within such portable devices may run internal circuitry at a lower level voltage, while being required to interface with other circuits at higher voltages [2]. Thus, a buffer is required between the internal circuitry and external signals.

Output buffers for digital communication lines are very popular because of their high stability and large drive capability. Furthermore, output buffer circuits are required to switch as fast as possible, in order to be able to operate at high data transmission rates. One important target of an output buffer is to shift from one voltage level to another quickly [3].

Output buffers can be designed as an open drain or as a push pull stage. The advantage of using the open drain configuration is the wired AND/OR capability, but the disadvantages are a slow rise (AND) /fall (OR) time and the usage of an external resistor. On the other hand, the push

pull configuration offers high speed response and the capability to source or sink current [4].

This paper presents the implementation of a high-speed output buffer based on a push pull topology in a 0.18μm CMOS EEPROM process with low and high voltage (20V) transistors. This buffer can operate at data rates of up to 20 Mbps (switching frequency of 10 MHz), for a wide range of power supply voltages, 1.6V to 5.6V.

2. The standard output buffer circuit

The block schematic of a typical push-pull output buffer is shown in **Fig 1**. The buffer has a low voltage side (LV), supplied from an internal voltage (V_{INT}), and a high voltage part (HV), which uses the power supply voltage (V_{DD}). The circuit has a LV input data signal, a driver for selectively switching the transistors of the output stage (P_1, N_1) and a data output terminal which is connected to a HV digital communication line. The buffer also has an enable control signal (*EN*), which, when low, indicates that neither transistor should be driving the output (high output impedance – hiZ).

The driver contains level shifter circuits, in order to drive the gates of the output transistors at the same voltage as the external line (V_{DD}).

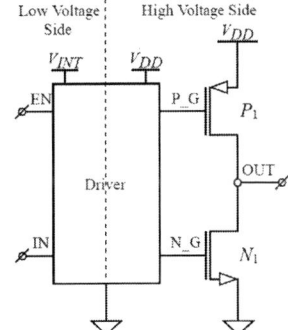

Fig. 1. Output buffer block schematic.

978-1-5386-4483-6/18 $31.00 © 2018 IEEE

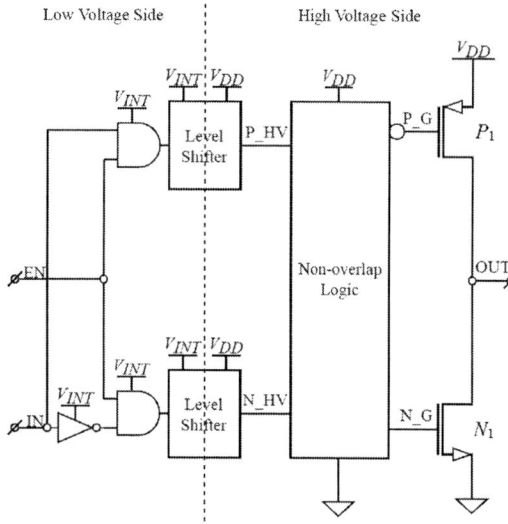

Fig. 2. Detailed schematic of the standard output buffer.

A more detailed schematic of the standard output buffer is given in **Fig. 2.** The LV side receives the input data signal (*IN*), which is gated with the enable signal *EN*. The level shifters allow the change of the gated signals logic "1" level, providing HV outputs (*P_HV, N_HV*).

The Non-overlap Logic block is driven by the level shifters and generates the *P_G* and *N_G* signals which have a "break-before-make" non-overlap. This has an important role for cross current and ground bounce issues [5].

A buffer of the type shown in **Fig. 2** was designed and implemented in a 0.18μm CMOS EEPROM process. Its operation was tested through HSPICE simulations, with the resulting waveforms shown in **Figs. 3,4**, for external (and power supply) voltages V_{DD} = 1.6V and V_{DD} = 5.6V, respectively. In the simulations, a capacitive load of 30pF was considered. In both cases, the data (*IN*) is transmitted at a speed of 10Mbps, which corresponds to a signal with a maximum frequency of 5 MHz.

The operation of the level shifters can be observed in **Fig.4**, where V_{DD} = 5.6V is higher than the internal voltage $V_{INT} \approx 1.9$ V. The LV input is changed into HV signals (*P_HV* and *N_HV*). For V_{DD} < 1.9V (**Fig. 3**), the LV and HV supply voltages are equal ($V_{DD} = V_{INT}$). Thus, the level shifters no longer change the voltage level, but only contribute to the buffer delay (especially due to their high output rise time). This is one of the limitations of this topology.

The logic "1" pulse is wider for *P_G* than *N_G* (as seen in **Figs. 3,4**), which means the output

transistors never conduct at the same time ("break-before-make"). Due to the reduced speed of the high voltage (20V) transistors available in the process, the Non-overlap Logic block introduces an unnecessarily large delay when operating at low voltages (V_{DD} = 1.6V – **Fig. 3)**, increasing overall delay. This is another disadvantage of the basic topology from **Fig. 2**.

On the waveforms from **Fig. 3** and **Fig. 4**, the following timing parameters were measured: t_r – *OUT* signal rising time (10% to 90% of V_{DD}), t_f – *OUT* signal falling time, t_{plh} – propagation delay of low to high transition from *IN* to *OUT* (50% of V_{INT} to 50% of V_{DD}), and t_{phl} - propagation delay of high to low. The average value of the current consumed by the HV side (powered by V_{DD}) was also measured. The variation with temperature (from -40°C to 125°C) and process corner was investigated for both V_{DD} voltages and the results are given in **Table 1**.

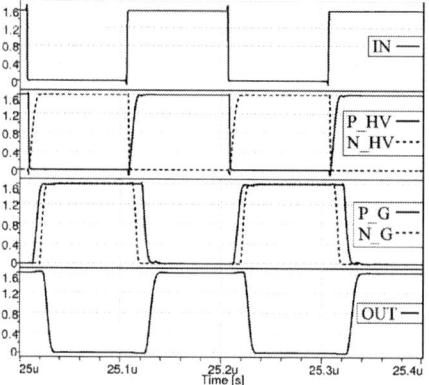

Fig. 3. Standard digital output buffer waveforms for V_{DD} = 1.6V.

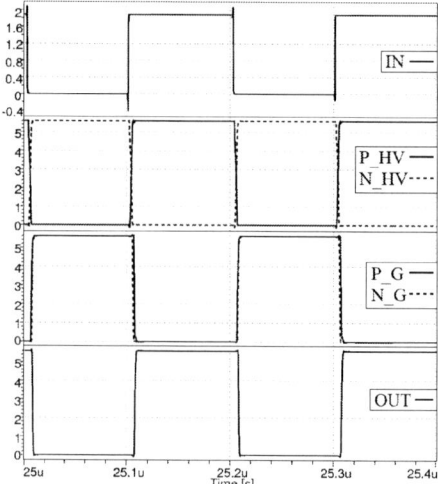

Fig. 4. Standard digital output buffer waveforms for V_{DD} = 5.6V.

978-1-5386-4483-6/18 $31.00 © 2018 IEEE

Table 1. Basic output buffer dynamic parameters

Parameter	Power Supply	Value		
		Min	*Typ*	*Max*
t_r [ns]	1.6V	6.2	8.0	9.3
	5.6V	1.2	1.4	1.8
t_f [ns]	1.6V	5.2	6.7	7.5
	5.6V	0.9	1.2	1.5
t_{plh} [ns]	1.6V	19.8	25.5	28.1
	5.6V	5.5	6.7	7.5
t_{phl} [ns]	1.6V	16.8	20.8	24.8
	5.6V	4.4	5.9	6.2
I_{DD} @5MHz [mA]	1.6V	0.342	0.347	0.356
	5.6V	1.26	1.28	1.31

Acceptable rise and fall times and propagation delays are obtained for $V_{DD} = 5.6$ V. In contrast, at 1.6 V, the rising and falling times are up to ~6 times as large, while the propagation delays are up to ~4 times greater. This propagation delay is as high as ~30% of the bit time at 10Mbps. This can be an issue especially for two-wire communication (clock and data), where the data being sent has to switch within a certain time from an input clock edge.

3. The proposed buffer topology

In order to deal with the limitations described in the previous section, improvements were made to the standard topology from **Fig. 2**. The proposed output buffer is shown in **Fig. 5**.

A significantly lower propagation delay from *IN* to *OUT* is obtained, as demonstrated by simulation from **Figs. 6, 7**, for $V_{DD} = 1.6$V and 5.6V, respectively, and an output load of $C_L = 30$ pF. For these simulations, a bit-rate of 20 Mbps was considered, equivalent to a switching frequency of 10 MHz, in order to illustrate the increased speed of the proposed buffer.

The buffer's delay was reduced through several changes. Firstly, a faster n-MOS transistor, with a lower threshold voltage, was available in the used process. This transistor was used instead for the N_1 device from **Fig. 2**. The threshold of N_1 is low enough that it can be driven with a LV signal. In this manner, the level shifter for the n-MOS, a significant delay contributor, was eliminated. In addition, the delay is further reduced, due to the removal of all HV devices from the n-MOS drive path.

The level shifter is still necessary for P_1, since its gate needs to be driven with V_{DD} in order for it to be OFF. This transistor is driven by the P_2, N_2 stage. The effect of the level shifter delay was diminished by using as N_2 the low threshold n-MOS (same as N_1) and driving its gate with a LV signal (N_D). Thus, the level shifter's slow rising slope no longer has any contribution to P_1 turn-on delay. This improvement can be seen in **Fig. 6**, where P_G falls very quickly.

A small rising delay was added for the N_D signal. In this manner, N_D rises while P_D is halfway through switching, so P_2 is basically OFF (due to its high threshold voltage) when N_2 turns ON. This non-overlap is also maintained at 5.6V, as can be seen in **Fig. 7**.

The effect of the non-overlap delay was reduced through the fact that P_1 turning ON is not conditioned by N_1 turning OFF. This is permissible since N_1 switches quickly compared to P_1. N_1 turn ON conditioning is done with the N_EN signal, which is obtained by shifting P_D back to LV. In this way, the propagation delay was reduced, while maintaining the previous "break-before-make" non-overlap (P_G, N_G signals – **Figs. 6, 7**).

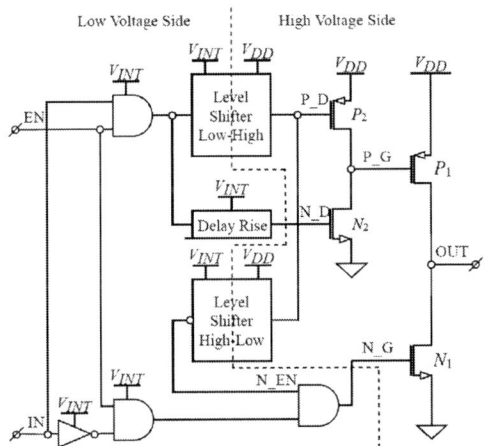

Fig. 5. Detailed schematic of the improved output buffer.

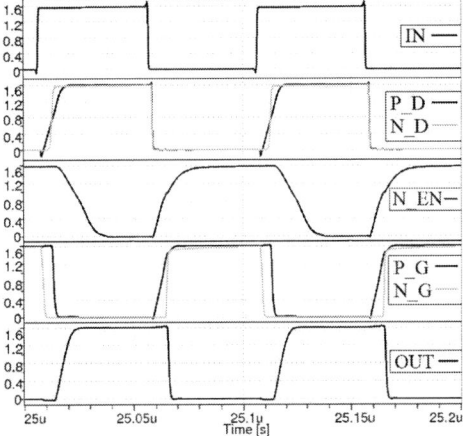

Fig. 6. Improved buffer waveforms for $V_{DD} = 1.6$V.

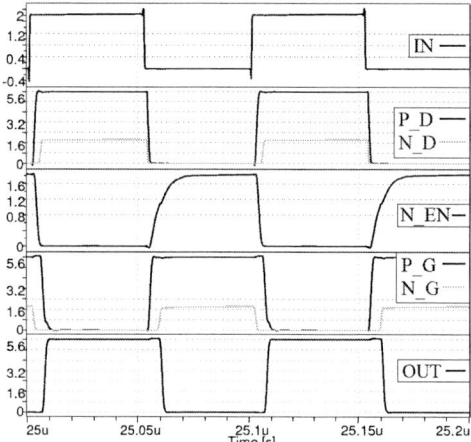

Fig. 7. Improved buffer waveforms for V_{DD} = 5.6V.

Table 2 shows the dynamic parameters of the improved buffer and their spread with temperature and process corner. Due to the faster n-MOS output transistor, the fall time at 1.6V is around 5 times lower than the values from **Table 1**. The rise time is also improved ~1.3 times, due to P_G switching faster (because N_2 is faster).

The most significant improvement, however, is the decrease of the propagation delay at 1.6V, resulting in a buffer that is twice as fast as the previous implementation. This, in turn allows it to operate at twice as high bit-rates (20 Mbps vs. the original 10 Mbps).

Compared to the basic circuit, the current drawn from V_{DD} by the improved buffer (at 5MHz) is reduced by ~15..20%. This is due to the simplified drive of the output n-MOS, powered by the LV supply, V_{INT}.

A DC analysis of the improved output buffer was carried out and the results are given in **Table 3**. The output transistors P_1, N_1 were designed to allow for current loads I_{OL} = 3 mA and I_{OH} = 3 mA for power supply voltages V_{DD} = 1.6…5.6V.

Table 2. Improved output buffer dynamic parameters

Parameter	Power Supply	Value		
		Min	Typ	Max
t_r [ns]	1.6V	5.7	6.7	8.6
	5.6V	1.4	1.6	1.9
t_f [ns]	1.6V	1	1.2	1.4
	5.6V	1.2	1.4	1.8
t_{plh} [ns]	1.6V	8.9	11.0	14.2
	5.6V	5.6	6.7	8.5
t_{phl} [ns]	1.6V	6.9	9.0	12.5
	5.6V	6.2	7.7	10.5
I_{DD} @5MHz [mA]	1.6V	0.295	0.297	0.303
	5.6V	1.027	1.031	1.036

Table 3. DC parameters for output buffer

Parameter	Power Supply	Value		
		Min	Typ	Max
V_{OL} [mV] @ I_{OL} =3mA	1.6V	85.5	100	140
	5.6V	14.4	18.5	25.7
V_{OH} [V] @ I_{OH} =3mA	1.6V	1.43	1.47	1.50
	5.6V	5.56	5.57	5.58

In **Table 3**, fairly low values of V_{OL} and V_{OH} can be seen, which show that the buffer has a high current drive capability. However, the increased size of the output transistors means that they also have higher capacitances, which raises delay (hence the need for a faster topology).

4. Conclusions

An improved digital push-pull output buffer was designed and implemented in a 0.18µm CMOS process with low and high voltage transistors. The circuit is an interface between internal LV logic and a HV communication line.

The buffer was demonstrated, through simulations, to operate at supply voltages from 1.6V to 5.6V, allowing for loads of up to 3 mA.

As a result of the topology changes made, the maximum delay of the buffer was reduced by ~50%, from 28.1 ns to 14.2 ns at 1.6V. The output rise time is reduced ~1.3 times, while the falling edge is around 5 times faster, compared to the basic topology. These improvements allow for a higher data rate (20 Mbps instead of the initial 10 Mbps). Moreover, a reduced current consumption of the HV side (by up to 20%) was observed.

References

[1] V. L. Le, T. T. H. Kim, "*An Area and Energy Efficient Ultra-Low Voltage Level Shifter With Pass Transistor and Reduced-Swing Output Buffer in 65-nm CMOS*", IEEE Transactions on Circuits and Systems-II:Express Briefs, vol.65, no. 5, May 2018.

[2] C. C. Wang, R. C. Kua, J. W. Liu, "*0.9V to 5V Bidirectional Mixed-Voltage I/O Buffer With an ESD Protection Output Stage*", IEEE Transactions on Circuits and Systems-II, vol. 57, no. 8, Aug. 2010.

[3] M. Lanuzza, P. Corsonello, S. Perri, "*Fast and Wide Range Voltage Conversion in Multisupply Voltage Designs*", IEEE Transactions on Very Large Scale Integration Systems, vol. 23, no. 2, Feb. 2015.

[4] J. Zhao, S. Gao, "*Power Efficient Push-Pull Buffer Circuit, System and Method for High Frequency Signals*", Patent US8319530B2, Nov. 27, 2012.

[5] A. Todri et al., "*A Study of Path Delay Variations in the Presence of Uncorrelated Power and Ground Supply Noise*", IEEE 14th Internal Symposium on Design and Diagnostics of Electronic Circuits and Systems, May. 2011.

Duty Cycle Adjustment for the Low Cost High Frequency Charge/Discharge CMOS Oscillator

Alexandru Mihai Antonescu, Lidia Dobrescu, Dragoş Dobrescu

"Politehnica" University of Bucharest, Faculty of Electronics, Telecommunications and Information Technology
andu.antonescu@gmail.com, lidia.dobrescu@electronica.pub.ro, dragos1.dobrescu@yahoo.com

Abstract— A new technique for adjusting the duty cycle in low cost 70MHz charge/discharge based oscillator topology is proposed. Added circuitry is optimized in order to maintain the frequency variation of the initial oscillator topology (without duty cycle adjustment) for a supply voltage range between 1.6V and 2V. The circuit uses different bias currents for each stage and it is implemented using Cadence design suite. It features reduces sensitivity to supply voltage range of the output frequency and low duty cycle variation.

Keywords— duty cycle; bias; charge/discharge oscillator; low voltage.

1. Introduction

Low cost oscillators are versatile topologies and have the following prerequisites: low area of implementation, reduced current consumption and ease of design. They include the current starved ring oscillator [1] and RC oscillators such as relaxation oscillators [2] and they can be tuned to reach hundreds of MHz, usually by changing the bias current or certain reference voltage. Applications and large scale designs use them: driving standalone circuits such as sensors [3] phase locked loops [4] and clock systems, often based on logic gates and they have a high supply voltage sensitivity on both frequency and duty cycle due to switching point variation on the logic gates.

An improved oscillator schematic that lowers the supply voltage sensitivity on frequency is analyzed in [5]. Logic gates (inverters) with compensated switching point are also used. This solution, in spite of its good results has become too elaborate, consuming current and area for only one oscillator compensation. Compensated switching point inverters are used.

Another frequency compensation method is presented in [6]. This combines two

currents, using a specific ratio, providing a current proportional to VDD.

One of the common design requirements for oscillators is to have a duty cycle close to 50%. This is highly dependent on the symmetry of the schematic, biasing precision and topology used. For example, it is easier to achieve 50% duty cycle for the ring oscillator topology than other low cost topology analyzed in [5].

2. Proposed Oscillator Schematic

In Fig. 1 the schematic of the low cost high frequency charge/discharge oscillator is shown. The main difference from [5] is the circuit will be biased with two currents, I_{bias1} and I_{bias2} using P1-P2 and P3-P4 PMOS current mirrors (P2 and P3 provide trimming for the output frequency). The circuit encloses two delay stages. By adding inverter I3, the circuit will have an odd number of inverting stages, providing the oscillation condition.

Inverters I1, I2 and I3 are identical. I4 and I5 isolate the circuit output from capacitive loads. Delays are generated using a "current starved" capacitor charge, fast discharge technique. P5 and P6 FETs control the charging current and N1, N2, are providing the fast discharge, by pulling the capacitors to ground. Equation 1 provides the circuit propagation delay as a sum of two half period delays, t_{pLH} and t_{pHL}. The two half

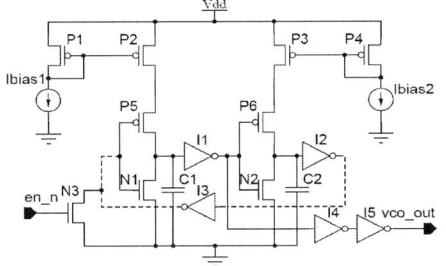

Fig. 1 Low cost Charge/Discharge Oscillator

period delays are computed using (2) and (3) where the charging and discharging delays of each individual stage are used.

$$t_p = t_{pLH} + t_{pHL} \qquad (1)$$

$$t_{pLH} = t_{chgstage2} + t_{dchgstage1} + t_{invLH} \qquad (2)$$

$$t_{pHL} = t_{dchgstage2} + t_{chgstage1} + t_{invLH} \qquad (3)$$

Each of the delays computed in (4) and (5) is based on capacitors controlled charging (t_{chgcap}) or fast discharge ($t_{dchgcap}$) and (t_{invLH}) (t_{invHL}) inverters delays.

$$t_{dchgstage} = t_{dchgcap} + t_{invHL} + t_{invLH} \qquad (4)$$

$$t_{chgstage} = t_{chgcap} + t_{invLH} + t_{invHL} \qquad (5)$$

In equation 6 the current controlled capacitor charging time is estimated, where C is the capacitor value, V_{sp} the inverter switching point and I_{bias} the external current (I_{bias1} for stage 1 and I_{bias2} for stage 2).

$$Q = CV_{sp}; \; I - Q/T \rightarrow t_{chgcap} = C \cdot V_{sp}/I_{bias} \qquad (6)$$

The discharge time in (7) equals 5% of the initial voltage across the capacitor at the end of the discharge phase (R_{DSN} corresponds to the on resistance of N1 or N2 transistors).

$$t_{dchgcap} = 4 \cdot R_{DSN} \cdot C \qquad (7)$$

The half period delays can be estimated using (8) and (9).

$$t_{pLH} \approx t_{chgstage2} = C \cdot V_{sp}/I_{bias2} \qquad (8)$$

$$t_{pHL} \approx t_{chgstage1} = C \cdot V_{sp}/I_{bias1} \qquad (9)$$

The circuit propagation time and duty cycle can be computed using (10) and (11).

$$t_p = C \cdot V_{sp}(1/I_{bias1} + 1/I_{bias2}) \qquad (10)$$

$$duty\_cycle = t_{pLH}/t_p = I_{bias1}/(I_{bias1} + I_{bias2}) \quad (11)$$

The bias circuit that is used for trimming I_{bias1} and I_{bias2} currents is depicted in Fig. 2.

The bias trimming circuit will keep both output currents approximately equal at the center trimming value. In this case duty cycle will be close to 50%. The trimming is done by the two PMOS current mirrors. By increasing the trimming one step, a client

will be added to the P1-P2 current mirror and an identical client will be subtracted from P3-P4; this way, the duty cycle of the oscillator signal will increase. Summing the two output currents, a constant value is obtained regardless of the chosen trimming.

The bias for the two PMOS current mirrors is provided by doubling the reference current, I_{bias}. N4:N6 current mirror uses N1:N3 cascade. I_{bias1} and I_{bias2} currents are provided to the oscillator using two auto-bias NMOS current mirrors N7:N10 and N11:N14. The device threshold doesn't use cascode for the trimming current mirror because of the low voltage requirements. Each generated bias current has range between $0-2I_{bias}$. The two generated currents dependence on reference are described in (12) and (13) where X is the trimming factor, between 0 and 2. The half period propagation delays become the ones in (14) and (15).

$$I_{bias1} = I_{bias} \cdot X \qquad (12)$$

$$I_{bias2} = I_{bias} \cdot (2-X) \qquad (13)$$

$$t_{pLH} \approx C \cdot V_{sp}/(I_{bias} \cdot X) \qquad (14)$$

$$t_{pHL} \approx C \cdot V_{sp}/[I_{bias} \cdot (2-X)] \qquad (15)$$

The circuit propagation time, frequency and duty cycle are stated in (16), (17) and (18), at half supply voltage switching point ($Vsp = VDD/2$).

$$t_p = 2 \cdot C \cdot V_{sp}/[I_{bias} \cdot X \cdot (2-X)] \qquad (16)$$

$$F = I_{bias} \cdot X \cdot (2-X)/(4 \cdot C \cdot VDD) \qquad (17)$$

$$duty\_cycle = X \cdot I_{bias}/(2 \cdot I_{bias}) = X/2 \quad (18)$$

According to biasing scheme, Ibias1 and Ibias2 show a low supply voltage influence. Fig. 3 depicts a comparison between computed and simulated bias currents. (Ibias1+Ibias2)/2 indicates that the current trimming step remains constant. The frequency variation is reverse proportional

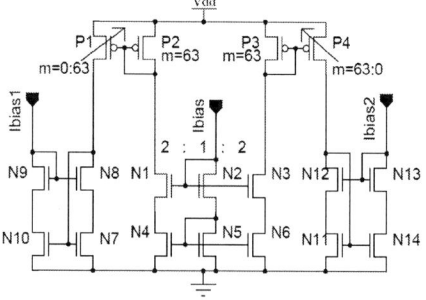

Fig. 2 Current trimming circuit for adjusting the duty cycle of the Low Cost Charge/Discharge Oscillator

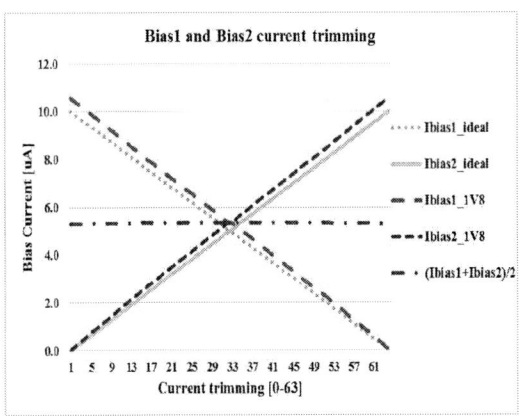

Fig. 3 Bias1 and Bias2 current trimming

and has the same magnitude as the supply voltage (VDD) variation due to the inverter switching point variation.

Low duty cycle variation (less than 10-15%) will weakly affect the frequency; for higher variation a frequency compensation technique is needed.

3. Simulation Results and Discussion

The circuit is implemented in a low voltage technology with a feature size lower than 0.5um. Simulations include supply voltage variation for each duty cycle trimming. Table 1 shows simulation results. Table 2 points out the frequency and duty cycle variation for the designed supply voltage range. As to be expected, the central value has a slight offset and is obtained for trimming 35. This trimming value will be used as central value and the frequency and

Table 1. Frequency Variation Due to Duty Cycle Variation

Supply Voltage [V]	Duty cycle trimming [0-63]	25	26	27	28	29	30	31	32	33	34
1.6	Freq. [MHz]	70.6	71.8	72.8	73.7	74.7	75.5	76.2	76.2	76.6	76.8
	Freq. Var. [%]	-8.5	-6.8	-5.6	-4.4	-3.2	-2.1	-1.2	-1.2	-0.6	-0.4
	Duty. [%]	64.7	63.0	62.6	60.7	59.0	57.6	56.0	55.1	54.2	52.2
1.8	Freq. [MHz]	69.1	70.4	71.5	72.4	73.4	74.2	75.1	75.4	75.5	75.8
	Freq. Var. [%]	-9.4	-7.6	-6.3	-5.0	-3.8	-2.6	-1.5	-1.1	-0.9	-0.6
	Duty [%]	66.0	65.5	63.7	62.1	60.3	58.5	57.3	57.0	55.3	53.6
2	Freq. [MHz]	66.5	68.4	69.2	70.2	71.3	71.8	72.4	72.4	72.6	73.5
	Freq. Var. [%]	-9.8	-7.2	-6.1	-4.8	-3.2	-2.5	-1.8	-1.8	-1.5	-0.2
	Duty [%]	66.9	65.8	63.8	63.1	61.8	59.7	58.5	57.3	55.0	53.4
Supply Voltage [V]	Duty cycle trimming [0-63]	35	36	37	38	39	40	41	42	43	44
1.6	Freq. [MHz]	77.1	76.9	76.9	76.2	75.5	74.9	73.8	72.2	70.5	68.9
	Freq. Var. [%]	0.0	0.3	0.3	1.1	2.1	2.8	4.3	6.4	8.6	10.6
	Duty. [%]	50.5	48.8	47.0	44.8	42.8	41.4	39.7	37.2	35.5	33.9
1.8	Freq. [MHz]	76.2	75.8	75.9	75.0	74.9	73.6	72.3	71.1	69.2	67.7
	Freq. Var. [%]	0.0	0.6	0.4	1.6	1.8	3.5	5.1	6.8	9.3	11.2
	Duty [%]	51.5	49.1	47.4	44.8	43.1	40.8	37.8	35.9	33.7	32.2
2	Freq. [MHz]	73.7	73.5	73.3	73.2	72.3	70.9	70.2	68.8	67.6	65.6
	Freq. Var. [%]	0.0	0.3	0.5	0.7	1.9	3.8	4.8	6.7	8.2	11.0
	Duty [%]	51.7	49.5	47.5	45.1	42.3	40.4	38.0	36.3	33.7	30.4

duty cycle variations will be computed

Table 2. Supply Voltage Frequency and Duty Cycle Variation

Duty cycle Set	Measured Frequency [MHz]					Measured Duty cycle [%]				
	min	max	center [1.8V]	delta	delta [%]	min	max	center [1.8V]	delta	delta [%]
70	62.4	66.2	65.3	3.8	5.9	68.1	71.3	70.3	3.2	4.6
65	68.4	71.8	70.4	3.4	4.9	63.0	66.1	65.5	3.1	4.7
60	71.1	74.8	73.4	3.8	5.2	59.0	61.8	60.3	2.7	4.5
55	72.6	76.8	75.5	4.2	5.6	53.9	55.3	55.3	1.4	2.5
50	73.7	77.1	76.2	3.4	4.5	50.0	51.9	51.5	1.9	3.6
45	73.1	76.4	75.0	3.3	4.4	44.1	45.2	44.8	1.1	2.5
40	70.9	75.0	73.6	4.1	5.6	39.7	41.4	40.8	1.8	4.3
35	68.8	72.4	71.1	3.7	5.2	35.4	37.3	35.9	1.9	5.4
30	63.5	66.9	65.7	3.4	5.2	28.6	31.9	30.0	3.3	11.0

reporting to it.

A maximum frequency variation of 5% is depicted in green and 11% in yellow. A duty cycle range of 40%-60% is achievable with a 5% frequency influence. For a 35%-65% duty cycle, the frequency influence is about 12%. In both cases, the frequency is lower compared with the 50% duty cycle simulation as in equation 17 but it seems to have about tree time higher magnitude. Fig. 4 depicts the frequency variation over the supply voltage range for different duty cycles and Fig. 5 the supply voltage sensitivity of the duty cycle.

In both cases higher spread is obtained for the 30% and 70% setting, especially for the duty cycle. The frequency variation remains under 6% between 1.6V and 2V supply voltage. The frequency spread proves to be almost haft compared to the circuit that has no duty cycle adjustment [5].

4. Conclusions and Future Work

The proposed circuit provides a trimming

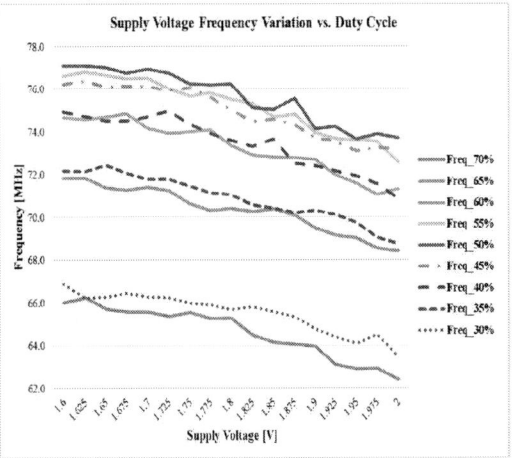

Fig. 4 Supply Voltage Frequency Variation vs. Duty Cycle trimming

technique for the duty cycle, using two bias currents and a biasing circuitry. For the 30%-70% duty cycle range the frequency spread is reduced to half compared to the oscillator with no duty cycle control circuitry [5]. This is the effect of the supply voltage influence on the bias block [7] primarily due to the difference in output resistances of the current mirrors [9].

From the layout point of view, this difference can increase if symmetry and matching are not well take care of [9]. Adjusting the circuit for a duty cycle different than 50% lowers the central frequency for that particular trimming. This behavior is partly explained by the added term, $X(2-X)$, in equation 17. This can bring nonlinearity to schematic [10].

For duty cycles non-equal with 50%, the contribution to the overall delay of the oscillator stage of the inverters and capacitor discharge time is higher for the stage with lower delay and lower for the one with higher delay. This is another key factor for lowering the frequency when adjusting the duty cycle.

To counteract this influence, the reference current I_{bias} needs to be trimmed in order to get higher current for each non 50% duty cycle setting. Trimming only I_{bias} will provide same ratio between the two oscillator currents I_{bias1} and I_{bias2}, thus not affecting the duty cycle setting.

The used topology, based only on controlled charge of the capacitor is more versatile than the ring oscillator because of the two stages differently biased. A topology

that uses both charge and discharge capacitor delays will need to use a biasing topology that changes the current synchronized to each half period of the signal, which is hard to achieve especially for high speed circuitry.

The biasing circuit brings and added current consumption of about 20uA, which represents about 25% of the total oscillator current consumption.

Acknowledgments. The paper represents the results of the first author PhD thesis.

References

[1] R. Jacob Baker, "Circuit design, Layout and Simulation", IEEE Press Series on Microelectronic Systems

[2] Y. Tokunaga, S. Sakiyama, A. Matsumoto, "An on chip CMOS relaxation oscillator with voltage averaging feedback", IEEE Journal of Solid-State Circuits, 2010

[3] M. Rinaldi, C. Zuniga, B. Duick, "Use of a single multiplexed CMOS oscillator as direct frequency read-out for an array of eight AlN Contour-Mode NEMS Resonant Sensors", SENSORS, 2010 IEEE

[4] Y. Wang, P. K. Chan, K. Ho Li, "A Compact CMOS Ring Oscillator with Temperature and Supply Compensation for Sensor Applications", VLSI (ISVLSI), 2014 IEEE Computer Society Annual Symposium

[5] A. Antonescu, L. Dobrescu, "A low cost high frequency charge/discharge CMOS oscillator", 9th International Conference on Electronics, Computers and Artificial Intelligence (ECAI), 2017

[6] A. Antonescu, D. Dobrescu, D. Dobrescu, "Self-bias frequency compensation technique for low cost charge/discharge oscillators", International Semiconductor Conference (CAS), 2017

[7] Yannis Tsividis, "Operation and Modeling of the MOS Transistor, Second edition", Oxford University Press

[8] P. Gray, R. Meyer, P. Hurst, S. Lewis, "Analysis and Design of Analog Integrated Circuits", John Wiley & Sons, Inc.

[9] A. Hastings, "The Art of Analog Layout", Prentice Hall, 1997

[10] A. Rusu, "Non-Linear Electrical Conduction in Semiconductor Structures" Romanian Academy Publishing House, 2000

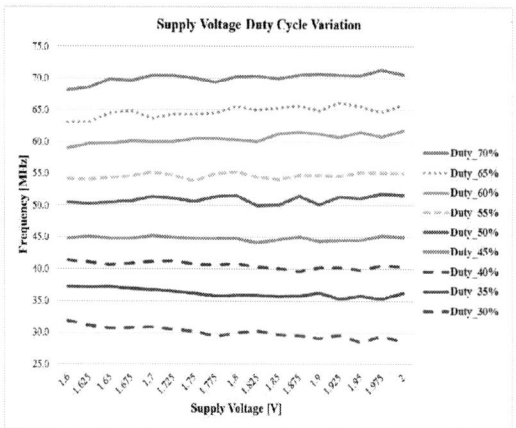

Fig. 5 Supply Voltage Frequency Variation vs. Duty Cycle trimming

Resistor Based Temperature Sensor Using Active Inductor Oscillator

Savinescu Viorel-Stefan[*], Nica Ioan-Alexandru[*], Liviu Goras[**]
*"Gheorghe Asachi" Technical University
Iasi, Romania
E-mail: stefan.savinescu@yahoo.com, nc_ionutz@yahoo.com
**Institute for Information Technology Iasi, Romania
E-mail: lgoras@etti.tuiasi.ro

Abstract— In this communication we present a temperature sensor based on the conversion of an on chip resistor variation with temperature into the period of a harmonic oscillator built with an active inductor resonator. The nonlinearity behavior of active inductor resonator is suppressed using an automatic amplitude control (AAC) loop which in turn makes the period of oscillation linear. An inaccuracy of ±0.34°C/±0.1°C after a first/second order polynomial fitting at a conversion speed of 30μs. The resolution for this time of conversion is 0.25°C.

This innovative principle can be applied to any harmonic oscillator based on gyrator active inductor resonator.

Keywords—Temperature sensor, active inductor, peak detector, oscillator amplitude control, low area.

1. INTRODUCTION

The high density of devices in modern integrated chips require integrated sensors to sense temperature in order to keep it at an acceptable level. To integrate several temperature sensors on the same chip, it is obviously necessary that they occupy an area as small as possible, to absorb low power, to have a high enough conversion rate and a reasonable (moderate) accuracy. From thermo-resistances or thermocouples to integrated CMOS sensors, technological development made possible new methods for temperature measurements and conversion into digital form [1].

Sensors based on resistance dependence on temperature have been also very common. Fortunately (for such applications) CMOS technology resistors have a rather high temperature coefficient (from 0.15% to 0.3%/°C, a poly and nwell resistor, respectively in standard CMOS). However, technological dispersion can be up to 15-20% which, together with the nonlinear dependence of the temperature makes calibration necessary. According to [2], an accuracy of ±1°C in 3σ for a temperature variation in the military domain (-55°-125°C) has been reported using a single calibration point and a LUT to obtain a linear relation between temperature and digital output.

The aim of this paper is to present the design of a temperature sensor based on an on chip resistance variation, which is converted into the time domain by means of an oscillator realized with one of the simplest active inductor to prove the principle. Using this innovative principle of conversion, a temperature inaccuracy of ±0.34°C/±0.1°C after a first/second order polynomial fitting.

Using more complex transconductor topologies to build active inductor resonators, better performances can be obtained at the cost of higher power consumption and a larger area.

2. ARCHITECTURE

The architecture of the temperature sensor is presented in Fig.1.

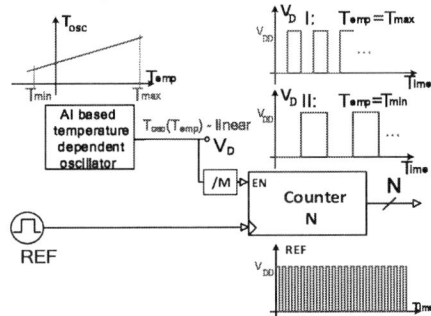

Fig.1 Block schematic of the temperature sensor

An oscillator, whose period is linear dependent on temperature, drives a divider whose output signal represents the EN (active 'High') of a counter. The divide ratio (M) is imposed by the desired resolution. In our design, we aimed to a resolution of 0.3 degrees resulting in a divide factor of 4096.

If an accurate and stable oscillator is used to drive the clock input of the counter, the digital output will be in a linear dependence as shown in (1):

$$Dout = \frac{M * Tosc(T_{emp})}{2Tref} \qquad (1)$$

978-1-5386-4483-6/18 $31.00 © 2018 IEEE

3. CIRCUIT IMPLEMENTATION OF ACTIVE INDUCTOR BASED TEMPERATURE DEPENDENT OSCILLATOR

The temperature sense element is represented by a resistor as show in Fig.2. Using the information from the AMS – 0.18µm design manual, the diffusion resistance was chosen due to the linear dependence on temperature and the highest temperature sensitivity. The temperature dependence of the resistor is modeled with relation (2), where the influence of end resistance and other second order effects (e.g. voltage dependence) are not taken into account because they are not a major contributor in our design:

$$R(T) = R(T_0)(1 + TC(T - T_0)) \qquad (2),$$

where $R(T_0) = 9K\Omega$ and $TC = 1340$ ppm/°C.

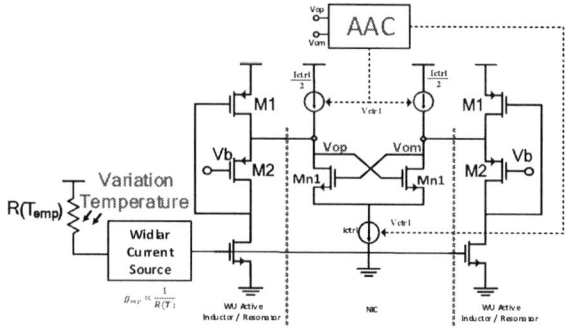

Fig.2 Block schematic of the temperature sensor

The key of conversion from temperature variation to period, is represented by the link between the biasing block of the active inductor based oscillator that includes the sense element, (1), and the fact that the oscillator presented in Fig.1 is based on the gyrator principle.

The bias circuit from Fig.2 is presented in [3] and is known as CMOS Widlar current source. It can be shown that the transconductance (gm's) of a pMOS devices that is biased with it, will be defined like in (3):

$$gm = \frac{\alpha}{R(T)} \qquad (3),$$

where in this case the resistor that was used, R(T), is described by (2) and α is a constant.

Using a gyrator-based oscillator that is implemented with transconductors, [4], it can be shown that (4) describes the frequency of oscillation:

$$f_0 = \frac{1}{2\pi}\sqrt{\frac{gm_1 gm_2}{C_1 C_2}} \qquad (4)$$

The gyrator is implemented using the configuration formed by M_1 and M_2 transistors, [5], due to the simplicity of circuits that implies low area and low power.

If gm_1 and gm_2 are chosen equally to $gm_{1,2}$ and respectively, C_1 and C_2 to $C_{gs1,2}$, the relation (4) is define like in (5), using the parameters of M_1, M_2 transistors

$$f_0 = \frac{1}{2\pi}\frac{gm_{1,2}}{C_{gs1,2}} \qquad (5)$$

Due to the parasitic elements, the gyrator implemented using [5], needs a negative resistance to maintain oscillations, that is implemented using M_{n1} transistors from Fig.2.

Relation (5) is based on small signal analysis, so an amplitude loop control is needed to maintain this condition. In Fig.2, the loop control is represented by AAC block together with the negative resistance [6].

Combining relations (2), (3) and (5) and considering that all transistors are in strong inversion, the relation (6) is obtained

$$T_0 = 2\pi\beta CR(T_{amb})(1 + TC(T - T_{amb})) \qquad (6),$$

where β is a constant that depends on transistors aspect ratio.

Relation (6) shows that the period of oscillation is linear depended on temperature. Introducing (6) in (1), the digital output will be:

$$Dout = \frac{2\pi\beta MCR(Tamb)(1 + TC(T - Tamb))}{2Tref} \qquad (7)$$

In reality, the linearity of conversion suffers due to the non-ideal and nonlinear effects of the circuits and to the dispersion of the devices. Some of the most important effects will be discussed in next section and circuit techniques and boundaries in design will be presented that can be used to obtain the desired performance of the sensor.

4. IMPACT OF THE NONIDEALITIES, NONLINEARITIES EFFECTS AND OF THE DISPERSION TECHNOLOGY

From layout and design perspective, one of the most important aspects is the matching of p-type

978-1-5386-4483-6/18 $31.00 © 2018 IEEE

transistors from Widlar current source and Wu active inductor resonator.

If, from the point of view of layout, the symmetry is the most important aspect to be achieved, the biasing of the transistors from Widlar current source and Wu active inductor in strong inversion region represent the rule of thumb in the design phase. The biasing condition can be accurate represented using inversion coefficient. How it is defined in [7], the inversion coefficient needs to be larger or equal to 10 for transistors to be in strong inversion. In our design, the transistors from Widlar current source are designed at an inversion coefficient around 20 and in active inductor, $M_{1,2}$ around 10 and 30 respectively. If the conditions mentioned above are not met, the relation between oscillation period and temperature will not be linear like in (6).

Another important aspect that need to be satisfied in order that relation (6) to be true, is to maintain the Wu active inductor in linear range. To show the impact of amplitude value on the linearity of $T_{osc}(T)$, a parametrical simulation has been done as shown in Fig.3:

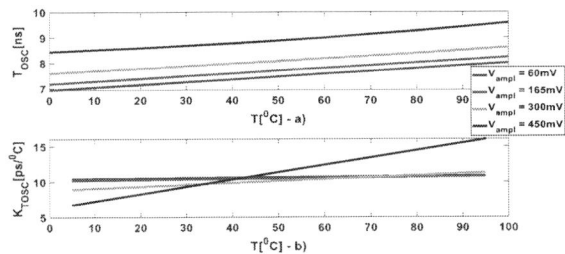

Fig.3 Parametric analysis: a) Period of oscillations; b) Slope factor of period curve

As can be seen from Fig.3 –b), the smaller the amplitude, the more linear $T_{osc}(T)$ is. An amplitude of 60mV has been chosen. This value represent a trade-off between sensor linearity and sensitivity of AAC circuit.

The relation (5) represent the resonance frequency using a simple model for active inductor [5]. Considering the second effects of transistor and knowing that the amplitude control circuit and negative resistance load active inductor resonator, the period of oscillation will deviate from (6), Fig.3, but nonlinearity introduced in the overall sensor will be small, lower than 0.14°C as shown in Fig.4.

In an ideal case, the temperature sensor should be insensible to supply voltage variation. In this implementation, a low sensitivity is achieved by using the Widlar bias circuit. From Fig.5, it is observed that a good accuracy can be achieved even for a supply voltage of 1.6.

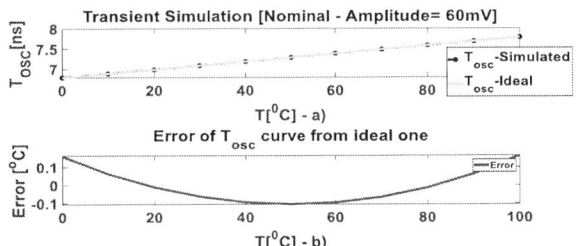

Fig.4 a) Comparison of period oscillation simulation with ideal one (6); b) Resulting error

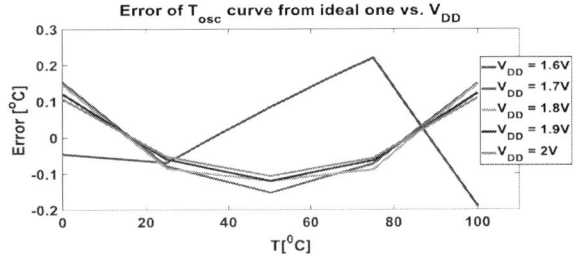

Fig.5. Resulting error vs. V_{DD}

5. SIMULATION RESULTS

To analyze the impact of technological spread on linearity of sensor, transient simulations using MC analysis were performed. The setup of this analysis is based on local and global mismatch, 100 samples per temperature. The period of oscillation versus temperature is shown in Fig.6:

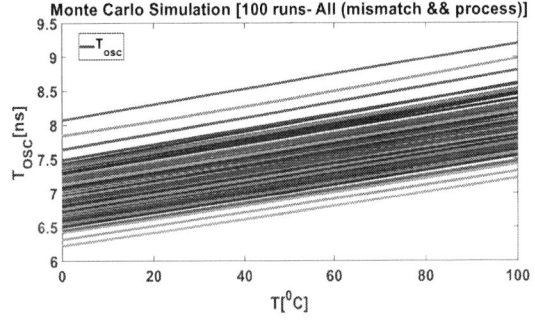

Fig.6 Monte Carlo Simulation: Period of oscillation vs. Temperature

Using the statistical information from Fig.6 and relation (2), the maximum counting number is determined to be 3768. After exhaustive simulations, it was determined that the circuit never leaves the

978-1-5386-4483-6/18 $31.00 © 2018 IEEE

linear region and no overflow will occurred if a 12 bit counter is used. Considering this, trimming operation is not necessary and the calibration will be performed in digital domain using a polynomial fitting. This technique offers a trade-off between complexity and accuracy.

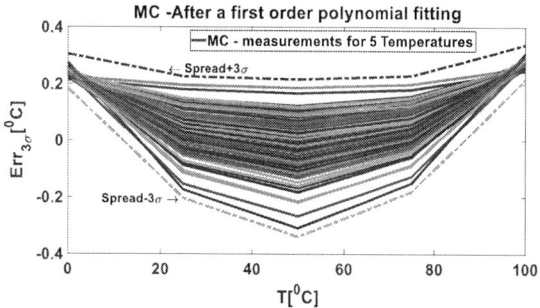

Fig.7 Inaccuracy after a first order polynomial fit

In this work, a first and a second order polynomial fit were used. The results are presented in Fig.7, 8 for a first and second order respectively. As can be seen, due to the weakly nonlinearity, a first order polynomial fit, using data measurement at 0 and 100°C, achieved an error smaller than ±0.34°C in ±3σ. An important remark represents the fact that the errors after a first order fit, has a parabolic shape, this means that a better error can be obtained if a second order fit is used, but a three-point measurement is necessary. After this fit an error smaller than ±0.1°C in ±3σ.

At a 30μs time conversion, a resolution of 0.25°C is obtained, at a current consumption of 400μA. Using a reset technique, a lower time of conversion and a low power is achieved.

Table 1 summarizes the performance of the active inductor oscillator based sensor and compares it with that of other resistor-based sensors.

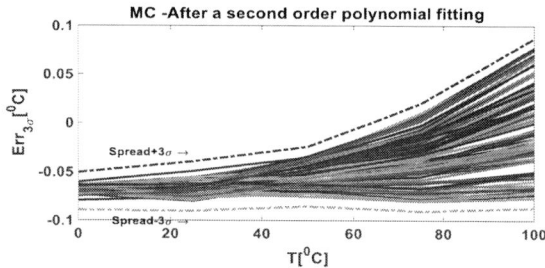

Fig.8 Inaccuracy after a second order polynomial fit

6. CONCLUSIONS

In this paper, a new principle of temperature conversion into oscillation period was presented. The temperature sensor is made with a Widlar current source and a gyrator based oscillator into 0.18μm-AMS CMOS technology. A temperature inaccuracy of ±0.34°C/±0.1°C after a first/second order polynomial fitting. A resolution of 0.25°C into 30μs time conversion.

Table 1. Performance summary

Parameter	This Work	[2]	[8]
Technology	0.18μm	0.18μm	0.18μm
Temperature range	0-100°C	-45-125°C	-40-85°C
Inaccuracy (trim points)	±0.1°C (3)	±1°C(1)/ ±0.4°C(2)	±0.15°C (3)
Resolution (T_{conv})	0.25°C (30μs)	0.01°C (100μs)	0.006°C (100msec)
Power consumption	400μA	43μA	20μA

REFERENCES

[1] S. Mahdi Kashmiri, Sha Xia, and Kofi A. A. Makinwa, "A temperature-to-digital converter based on an optimized electrothermal filter", IEEE Journal Of Solid-State Circuits , VOL. 44, NO. 7, 2009

[2] C.-H. Weng et al., "A CMOS thermistor-embedded continuous-time delta-sigma temperature sensor with a resolution FoM of 0.65pJ/0C". IEEE J. Solid State Circuits 50(11), 2491–2500 (2015)

[3] J. M. Steininger, "Understanding wide-band MOS transistors", IEEE Circuits and Devices, Vol. 6, No. 3, pp. 26–31, May 1990

[4] http://pallen.ece.gatech.edu/Academic/ECE_6440/Summer_2003/L130-VCO-I(2UP).pdf

[5] Yue Wu, Xiaohui Ding, Mohammed Ismail, and Håkan Olsson, "RF bandpass filter design based on CMOS active inductors", IEEE Transactions On Circuits And Systems-II. VOL. 50, NO. 12, December 2003

[6] Faramarz Bahmani, and Edgar Sánchez-Sinencio, "A stable loss control feedback loop for VCO amplitude tuning", IEEE Transactions On Circuits And Systems I: Regular Papers, VOL. 53, NO. 12, December 2006

[7] David M. Binkley, "Tradeoffs and optimization in analog CMOS design", 2008 John Wiley & Sons Ltd

[8] Mina S., Kianoush S., Kofi A.A. Makinwa, "A Resistor-Based Temperature Sensor for MEMS Frequency References", IEEE ESSCIRC 2013 - 39th European Solid State Circuits Conference - Bucharest,Romania

LDO with a Dual Complementary Buffer Architecture

Mihai Dicianu
POLITEHNICA University of Bucharest,
Bucharest, Romania
E-mail: mihaidicianu@gmail.com

Vlad Ionescu
Infineon Technologies Romania SCS,
Bucharest, Romania
E-mail: vlad.ionescu@infineon.com

Claudius Dan
POLITEHNICA University of Bucharest,
Bucharest, Romania
E-mail: claudius.dan@upb.ro

Abstract—This paper presents a 5V LDO architecture with a buffered error amplifier. This is achieved by using a functional block called a dual complementary buffer which consists of two buffers, one using a NMOS output transistor, the other a PMOS output transistor. The main advantage of this architecture is the rail-to-rail output voltage swing of the buffer, improving performance in both the tracking and the regulating operating regions of the voltage regulator. Simulation results show load regulation of 4.47uV/mA and line regulation of 3.92uV/V. The maximum input voltage is 40V and the maximum load current is 200mA. The LDO was simulated using a 0.8um BiCMOS process.

Keywords—LDO; dual complementary buffer; regulation loop.

1.Introduction

An LDO (low dropout voltage regulator) is a linear voltage regulator that can still operate when the voltage difference between the input and the output is low (as low as hundreds of millivolts). To achieve low dropout voltage the series PMOS pass transistor is designed to have large dimensions in order to reduce its on resistance $R_{sd,on}$. However, this will raise the gate capacitance up to tens, sometimes hundreds of picofarads.

Due to the fact that, in a classic LDO architecture, the gate of the series pass transistor is driven directly by an OTA (operational transconductance amplifier) [1] having high output impedance, the pole associated with the gate node will be placed at very low frequencies (Hz, tens of Hz). This low-frequency pole has a negative effect on the bandwidth of the LDO and the frequency compensation circuitry.

To move the gate associated pole to higher frequencies, even beyond the UGF (unity gain frequency), one must either reduce the gate capacitance of pass transistor (not feasible for LDOs) or reduce the impedance driving the gate node. This implies the necessity to use a voltage buffer to drive the series pass PMOS. By design, such buffer should have a low output resistance and an extended output voltage range.

Simple buffer stages, such as emitter follower for bipolar transistors and source follower for MOS transistors have the disadvantage of being unable to drive the output voltage close to one of the power supply rails. In addition, in the case of bipolar transistor buffers, nonlinear current gain β variation with temperature and biasing current will degrade the circuit's performance.

The N-type buffers (NPN or NMOS) have the output voltage limited near the positive supply by V_{BE} or V_{GSN}. Using this type of buffer will degrade the performance of the LDO in the regulation region where the pass transistor, for very low to no output current, operates in the sub-threshold region and needs to be completely turned off.

The P-type buffers (PNP or PMOS) are limited to an output voltage higher than negative supply by $|V_{BE}|$ or $|V_{GSP}|$. This will degrade the LDO's performance in the tracking region, as the gate of the series PMOS can't be driven to the negative supply of the LDO (in order to reduce $R_{sd,on}$).

978-1-5386-4483-6/18 $31.00 © 2018 IEEE

In order to fulfill the rail-to-rail voltage swing requirement of the buffer that drives the series pass transistor this work proposes the dual complementary buffer, an alternative to already existing topologies [2,3,4].

Section 2 presents the LDO architecture and the buffer concept. The schematics and operation of the buffer are displayed in Section 3. The proposed LDO is simulated and the results are shown in Section 4. Section 5 presents the conclusions and possible future developments of this LDO architecture.

2. The LDO and dual buffer concept

A. The LDO

Fig. 1 displays the proposed LDO architecture. The series regulating element is a PMOS transistor and the error amplifier is a symmetrical OTA. A Brokaw bandgap voltage reference is used for the V_{REF} block.

The LDO also includes a pre-regulator, which supplies internal circuit blocks with a lower and filtered supply voltage $V_{INTERNAL\_SUPPLY}$, and a PTAT (proportional to absolute temperature) current source from which all biasing currents are derived.

The pole associated with the gate node has the following frequency dependence:

$$f_{p,gate} = 1/(2\pi R_{o,buffer} C_{gate}) \qquad (1)$$

where $R_{o,buffer}$ is the output resistance of the buffer and C_{gate} is the gate capacitance of transistor M_{PASS}. The buffer is designed so that it exhibits a low output resistance when the LDO operates in the regulation region.

The frequency of the pole associated with the output node is

$$f_{p,out} = 1/(2\pi (R_{o,pass} \parallel R_L) C_O) \qquad (2)$$

where $R_{o,pass}$ is the output resistance of the pass transistor, R_L the load resistance of the LDO and C_O the output capacitance.

The frequency compensation circuitry consists of a Miller network (R_C and C_C) and the feed-forward capacitor C_{FF}. The gate node was avoided when considering compensation in order to minimize the capacitance of the node.

B. The dual complementary buffer

Due to the need of a full swing buffer, the circuit presented in Fig. 2 has been designed. The circuit proposes a complementary use of the two buffers, biasing only one block at a certain time using the control signal that indicates whether the LDO is in the tracking or regulation region.

The P-type buffer needs to drive the gate close to the positive power supply voltage in low load current conditions. Also, because the regulation region poses the risk of oscillations of the output voltage, the output resistance of the buffer needs to be as low as possible in order to push the gate pole to higher frequencies. To handle fast transients, the P-type buffer will be biased with a current that is proportional to the load in order to faster charge or discharge the gate capacitance.

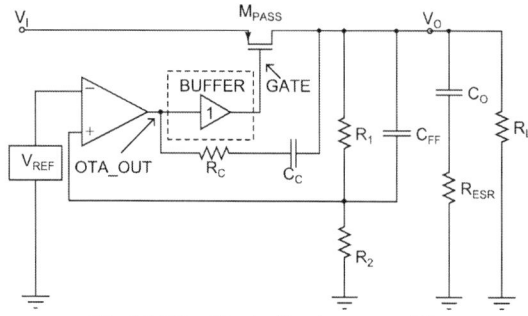

Fig. 1 LDO with a buffered error amplifier

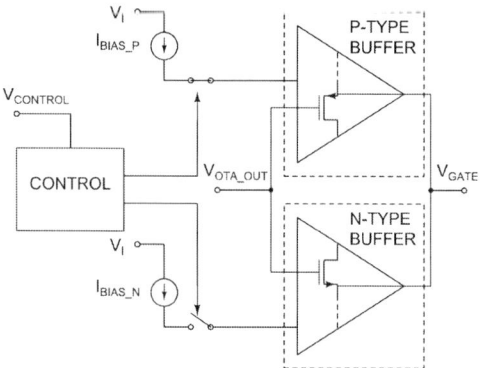

Fig. 2 Dual complementary buffer

3. Dual complementary buffer schematics

In Fig. 4 the electrical schematic of the buffer is shown. The P-type buffer, used in the regulation region, is able to drive the gate node to V_I. The N-type buffer, used in the tracking region, is able to drive the gate node to ground. Fig. 3 displays which of the buffers is enabled by the control circuitry based on the $V_{CONTROL}$ signal. The regions of operation of the LDO are also shown.

The P-type buffer uses a super-source follower topology [4] which implements a local negative feedback loop through transistor M5 in order to reduce its output resistance

$$R_{o,buffer} = r_{o5}/(1 + g_{m5}(r_{o5} + g_{m4}\,r_{o4}\,r_{o5})) \quad (3)$$

where r_{ox} and g_{mx} are the output resistance and transconductance of Mx, respectively.

The block is biased by a current source delivering $I_{BIAS\_P} = I_0 + kI_{LOAD}$, where I_0 is a PTAT current, independent of the LDO's load, k a subunitary constant and I_{LOAD} the current through the LDO's load. The RC groups of R1, C1, R2, C2 are used to maintain proper biasing during negative load current jumps.

The N-type buffer is a simple NMOS source follower (transistor M9), biased by $I_{BIAS\_N} = I_0$. A more complex topology for this circuit is not required because it only needs to drive the gate node to ground in the tracking region.

The control block consists of two inverting stages in series (M11 and M12 in common-source configuration) which drive

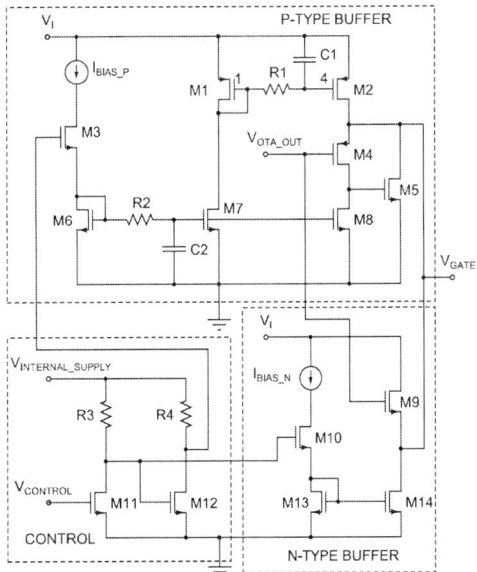

Fig. 4 Schematic of the dual complementary buffer

switches M3 and M10. The second stage was used in order to reduce the loading on the OTA pin which supplies $V_{CONTROL}$.

4. Simulation results

Table 1 shows the performance of the presented circuit in comparison to LDOs with a similar buffered architecture. The circuit was simulated (results shown in Fig. 5) using an output capacitor of 1µF with low ESR (10mΩ).

Table 1. Performance comparison

	2007 [3]	2009 [4]	2018 [5]	This work
Nominal output voltage [V]	1.8	3.3	2.8	5
Full load current [mA]	200	300	50	200÷500
Maximum input voltage [V]	5.5	5.5	3.3	40
Line regulation [mV/V]	2	1.7	23.4	0.0039
Load regulation [µV/mA]	170	6	310	4.47
Quiescent current at no load [µA]	20	130	36.1	40
Temperature range [°C]	NA	NA	NA	-40÷175
Output capacitor [µF]	1	1	0.01	1

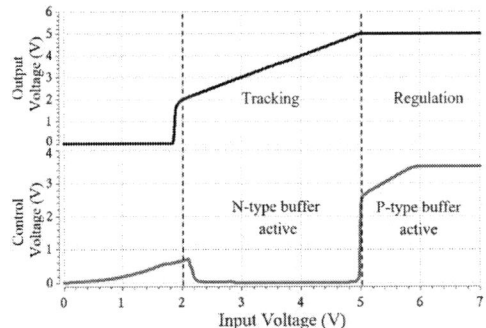

Fig. 3 Output and control voltage vs input voltage, I_{LOAD} = 1 mA

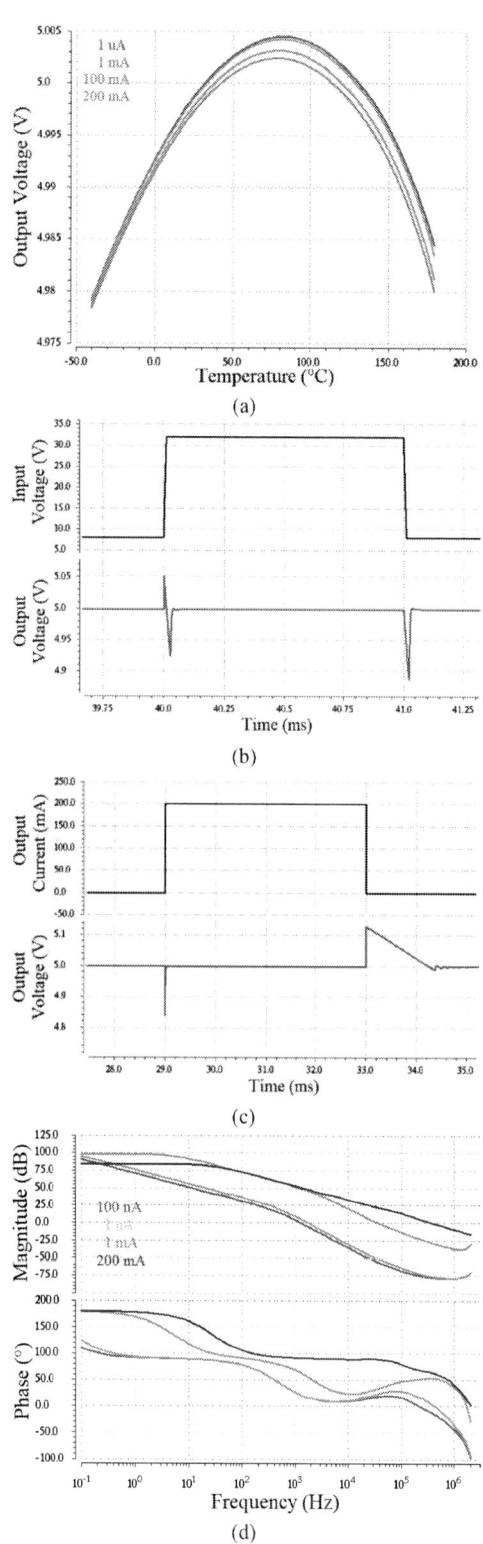

Fig. 5 Simulation results: (a) Output voltage vs temperature, (b) Line jump, (c) Load jump and (d) Bode plots

Fig. 5 (a) indicates the relation between the output voltage and temperature for four load currents. Fig. 5 (b) shows a line jump simulation where the input voltage increases from 8V to 32V with a 2V/µs ramp. The load jump analysis in Fig 5. (c) uses a jump of the load current from 100µA to 200mA with rise and fall times of 1µs. Bode plots for different load currents are presented in Fig. 5 (d) from which it can be determined that the LDO is stable from no load to full load current.

With degradation to the load step response, the full load current can reach 500mA.

5. Conclusions

The simulated results show a very good stability of the output voltage with respect to temperature, input voltage and load current. The proposed LDO can be suitable for the automotive industry due to the extended input voltage and operating temperature ranges. Future developments of this architecture can include adaptive compensation techniques and load dependent gain stages in order to improve the transient response of the presented LDO.

References

[1] C. Raducan, M. Neag, "*Capacitorless LDO with Fast Transient Response Based on a High Slew-Rate Error Amplifier*", 2015 International Semiconductor Conference, 2015

[2] C. Stanescu, C. Dinca, R. Iacob, "*Soft-Start Low Voltage CMOS LDO*", 2013 International Semiconductor Conference, 2013

[3] M. Al-Shyoukh, H. Lee, R. Perez, "*A Transient-Enhanced Low-Quiescent Current Low-Dropout Regulator With Buffer Impedance Attenuation*", IEEE Journal of Solid-State Circuits, 2007

[4] J. Choi, et al., "*Design of LDO linear regulator with ultra low-output impedance buffer*", International SoC Design Conference, 2009

[5] M. Khan, M. Chowdhury, "*Capacitor-less Low-Dropout Regulator (LDO) with Improved PSRR and Enhanced Slew-Rate*", IEEE International Symposium on Circuits and Systems, 2018

Session IC-S 3

INTEGRATED CIRCUITS 3
Student papers

978-1-5386-4483-6/18 $31.00 © 2018 IEEE

I/O Blocks Reliability for an SRAM-Based FPGA when Exposed to Ionizing Radiation

V. M. PLACINTA[1,2], L. N. COJOCARIU[1], C. RAVARIU[2]

[1]Horia Hulubei National Institute for R&D in Physics and Nuclear Engineering,
Department of Elementary Particle Physics,
Reactorului 30, RO-077125, Bucharest-Magurele, Romania
[2]Polytechnic University of Bucharest,
Faculty of Electronics, Telecommunications and Information Technology,
Splaiul Independentei 313, RO-060042, Bucharest, Romania
E-mail: vlad-mihai.placinta@cern.ch

Abstract—In this paper we present a survey of radiation induced failures in the Input/Output blocks of an SRAM-based Field Programmable Gate Array (FPGA), using a ring oscillator-based measurement technique. This study has been done on Xilinx's KINTEX-7 FPGA, while exposed to ion and X-rays beams. Two types of failures have been identified, amplitude and duty cycle failures, and the cross-section values were estimated to be approximately $0.6 \cdot 10^{-5}$ cm²/device for the amplitude failures and $1.6 \cdot 10^{-5}$ cm²/device for the other ones.

Keywords—radiation effects; I/O blocks; single event upset; ring oscillator; FPGA.

1. Introduction

The reliability of the reconfigurable logic devices, mostly focused on the Field Programmable Gate Arrays (FPGAs), has become a major topic, and efforts are done to qualify them as potential replacement of the Application Specific Integrated Circuits (ASICs) for use in safety-critical applications such as space and accelerator environments. Such environments are harsh with a broad range of background radiation as those in Large Hadron Collider (LHC) experiments at CERN, e.g. Large Hadron Collider beauty (LHCb) experiment. Because of their high logic density and low price compared with an ASIC, or with a radiation hardened by design (RHBD) FPGA, the commercial of the self (COTS) FPGAs are the most investigated devices for use in such applications. Exposing the device under test (DUT) to radiation source and monitoring its response is a straightforward way to validate the DUT's radiation tolerance. The gathered data is used then to establish and extrapolate the DUT behavior and its reliability for a given radiation environment.

During the second-long shutdown (LS2) of the LHC, scheduled for 2019, the upgrade program of the LHCb detector [1] and its sub-detectors is foreseen to start. The photo-detection system of the Ring Imaging Cherenkov (RICH) [2] sub-detectors will be upgraded to work at a 40 times increased trigger rate than its current operation [3]. The digital readout architecture of the front-end trigger board includes a communication board with an SRAM-based FPGA from KINTEX-7 family. The Xilinx's 7-series are manufactured using TMSC's high performance and low power (HPL) process with a 28 nm high-k metal gate (HKMG) technology node [4].

When using SRAM-based FPGAs in such experiments the main concerns are focused to its reliability to cumulative effects and single event effects.

The cumulative effects cause a degradation of the device with time as trapped charges are accumulated, leading to a decrease in performance along with a modification of its electrical parameters when the total dose is higher than its tolerated dose. In the CMOS technology the transistors can tolerate a certain amount of ionizing radiation without any issues. With more dose accumulated the transistors will start to be affected and parameters like leakage current and threshold voltage will be modified for the same switching conditions.

Single Event Effects (SEEs) are failures triggered by the crossing of a single energetic and high-Z particle passing through the device in active layers while losing its energy through ionization. The energy loss causes a localized charge to be deposited into semiconductor or dielectric layers and if its value is larger than a critical charge value given for a specific technology node, then the SEE occurrence is probable. The SEEs are classified as soft errors if the device can recover without a power cycle, hence these errors are single-event upsets (SEUs), where one or more bits are flipped. Hardware errors are the other class of SEE which can be non-destructive if are being mitigated properly, being called single event latchups (SELs) or destructive resulting in a permanent damage of the device. Single event gate rupture (SEGR) and single event burnout (SEB) are the most common destructive effects occurring in semiconductor devices when exposed to radiation.

SRAM-based FPGAs contains a board range of resources like: user Flip-Flops, embedded RAM, distributed memory, configuration memory (CRAM), PLLs, and I/O blocks (IOBs). All of them are susceptible to radiation induced-failures, hence they need to be proper tested and qualified.

The reliability of the IOBs has become a major concern as they may perturb the operation of other components embedded in a larger system installed in a harsh environment. They can be affected by SEUs in the FPGA's configuration memory where an IOB can be reconfigured from input to output (or vice versa) or by changing its attributes (slew rate, drive strength etc.).

This paper presents the methodology and the investigation of radiation-induced failures in the IOBs of a KINTEX-7 FPGA using ion and X-ray beams. Several research groups have investigated the reliability of the IOBs in other SRAM-based FPGAs either by using radiation or by emulating the behavior with a proper test bench [5-7].

2. DUT and its Setup

The smallest device from KINTEX-7 family has been chosen for testing, XC7K70T. This device has the following features: 240 DSP slices, 82000 user Flip-Flops, 65.6 k logic cells, 4.86 Mb Block RAM (BRAM), 8 GTX transceivers, 300 IOBs and 18884576 bits of CRAM [8]. Being manufactured in a flip-chip technology, the thermal interface material and substrate from the top side of the package had to be thinned from about 250 to about 60 µm in order to allow the ions to reach to the active layer.

DUT's response to ionizing radiation has been monitored using a custom test bench, specially designed for these tests in which all the DUT activity along with its electrical parameters are monitored and saved in ASCII files for later analyses [9].

The IOBs reliability was tested by using the vendor's I/O buffer primitives [10], shown in **Fig. 1**, without any additional logic or any other connections on the PCB.

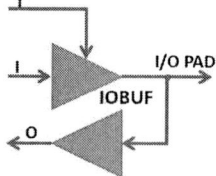

Fig.1. The I/O buffer architecture of the 7-series [10]

Based on these structures, 4 ring oscillators were implemented using 5 from all 6 I/O banks of the FPGA. The I/O blocks were configured as delay elements in a buffer configuration in which the data is shifted from first to the last element. The output from the last element is inverted using a NOT gate, and the result is connected at the input of the first delay element. This architecture, presented in **Fig. 2**, meets the minimum condition to generate a self-oscillation while not using any additional logic resources of the DUT.

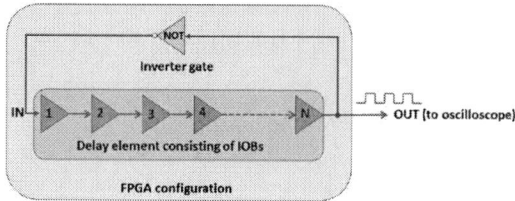

Fig.2. The ring oscillator architecture used for testing

Equation (1) defines the RO oscillation frequency:

$$F_{RO} = \frac{1}{2*T*n} \tag{1}$$

where T is the total propagation delay and n is the number of delay elements which in our case the IOBs and NOT gates used.

Each of the four ring oscillators (ROs) has a fixed frequency which varies with the number of I/O used, IO bank type [10] in our case high performance [HP] or high range [HR], and with the voltage applied to each I/O bank, as shown in table 1. The entire firmware with only the ROs use about 0.19 % essential bits [11] of total CRAM size.

Table.1. Resource utilization of the ring oscillator architectures

Nr.	I/O used	IO BANK	Type	VCC [V]	Freq [MHz]
RO1	20	13	HR	1.5	~4
RO2	35	15&16	HR	1.5&1.8	~10
RO3	20	33	HP	1.8	~10
RO4	18	34	HP	1.5	~11

The ROs oscillation signals, seen in **Fig. 3**, were monitored over 5 m of coaxial cable using an oscilloscope. Parameters like frequency, duty cycle, and amplitude were recorded and saved in ASCII files using a graphical user interface (GUI) designed using LabVIEW™ and connected to the oscilloscope.

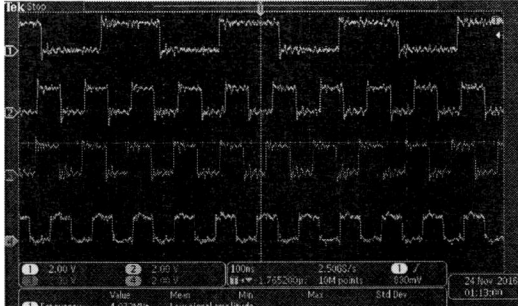

Fig.3. The ROs waveforms (oscilloscope snapshot)

3. Test Beam Results

A. Ion Beam Results

The SIRAD facility of the INFN Legnaro National Laboratories (LNL) from Italy provided ^{28}Si ion beams with a 157 MeV kinetic energy and a linear energy transfer (LET) of 8.59 MeV · cm^2 / mg [12].

The DUT was placed in a vacuum chamber on a sample holder, as shown in **Fig. 4**, and all of its connections including power, communication and control signals were linked via vacuum pass through connectors with the monitoring system placed outside of the chamber. This operation had to be done due to facility constrains, as beam pipe vacuum is protected and proton/ion energy loss in air is reduced.

Fig.4. The DUT prepared to be irradiated and placed in front of the beam

The DUT was irradiated continuously for 3186 seconds while accumulating a total fluence of 5 · 10^5 particles/cm^2. Tests and measurements were performed before, during and after irradiation to establish the DUT's response for the given fluence accumulated.

During irradiation, besides the corruption of the CRAM monitored with the pre-verified solution provided by vendor [13], failures in the ROs were seen. Most of them were failures due to positive and negative shifts in frequency and duty cycle while a small fraction of them was due to complete loss of the oscillation. All failures were recovered either by CRAM monitor [13] or by a fully reconfiguration of the device. The failures are classified as amplitude (Amp.), frequency (Freq.) and duty cycle (Duty) failures and presented in table 2 for each RO tested.

Table.2. RO failures recorded and classified for each RO tested

Nr.	Amp failures	Freq failures	Duty failures	Total failures
RO1	0	0	0	0
RO2	3	0	2	5
RO3	0	0	2	2
RO4	0	0	4	4

The frequency failures are large positive or negative shifts of the nominal frequency. Most of these failures are due to changed attributes occurred in some IOBs of the RO's delay element or by local charge deposition near the IOB resulting in modifications of the transistors proprieties. Slew rate and drive strength attributes can be the candidates for this type of failures because every change in each of them can contribute to timing modifications. In our case these can be seen as an increasing or decreasing of the total propagation delay in the RO, hence changes in the RO's oscillating frequency. During irradiation we saw minor shifts of the frequencies of each RO, but the values are too small to perform a detailed analysis in time domain (e.g. FFT).

However, the duty cycle failures which can be seen in **Fig. 5**, are failures occurring only during half of RO's total oscillation period. These are due to transient changes of the IOBs attributes or transient effects in the IOB itself which cause only half of the period to be affected.

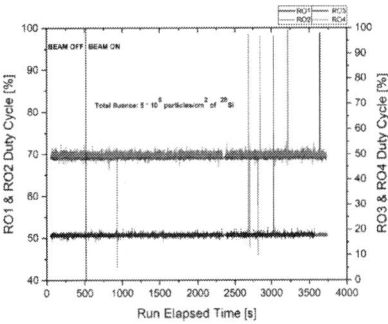

Fig.5. Duty cycle stability failures seen in the ROs during irradiation

The amplitude failures are total failures of the ROs due to complete loss of the oscillation, hence at least one IOB from the entire chain was affected.

IOBs resilience in presence of ionizing radiation proved to be good comparative with the results of other tested resources from the same device published by our group [14-15].

The DUT's cross-sections (σ) for the IOBs failures have been calculated for the amplitude and duty cycle failures for the total fluence which was accumulated by the DUT, of $5 \cdot 10^5$ particles/cm^2:

$$\sigma_1 = 0.6^{+0.58}_{-0.32} \cdot 10^{-5} \text{ cm}^2/\text{device (68\% CL)}$$
$$\sigma_2 = 1.6^{+0.79}_{-0.55} \cdot 10^{-5} \text{ cm}^2/\text{device (68\% CL)}$$
$$\sigma_1 = 0.6^{+1.15}_{-0.47} \cdot 10^{-5} \text{ cm}^2/\text{device (95\% CL)}$$
$$\sigma_2 = 1.6^{+1.55}_{-0.91} \cdot 10^{-5} \text{ cm}^2/\text{device (95\% CL)}$$

where σ_1 is the value for the RO amplitude failures, σ_2 is the value for the RO duty cycle failures, and CL is the Confidence Level and its associated interval.

B. X-ray Beam Results

With 50 KeV photons provided by SIRAD's X-ray machine [16] we irradiated the DUT up 200 krad. Besides a small current increasement in the power rails we did not see any failures either in the CRAM, or the IOB logic.

4. Conclusions and Future Plans

The reliability of the KINTEX-7 FPGA was studied by our group in order to establish its tolerance limits for a given environment. The IOBs have been studied due to their important role of connecting the FPGA's internal logic with other electronic components, all of them being embedded in a larger system.

The only observed IOB SEUs have a small cross-section compared with CRAM SEUs and might be caused by the latter. The severity of the IOB SEUs is not dangerous so far and can be mitigated if properly monitored.

By now, the IOBs were tested with protons, ions and X-ray beams at facilities from Switzerland, Germany and Italy and the detailed results will be published in scientific journals after the data analysis is completed.

Acknowledgments. This study and all the materials used to achieve it, were supported by the Ministry of National Education (MEN) and the Institute of Atomic Physics Bucharest (IFA) under grants 7/16.03.2016, and national project "NUCLEU" through grant number PN 16 42 01 03.

References

[1] The LHCb Collaboration, *"The LHCb Detector at the LHC"*, Journal of Instrumentation, 3, S08005, 2008.

[2] The LHCb Collaboration, *"LHCb Particle Identification Upgrade Technical Design Report"*, CERN/LHCC 2013-022, 2013.

[3] The LHCb Collaboration, *"The upgrade of the LHCb trigger system"*, Journal of Instrumentation, vol. 9, 2014.

[4] J. Hussein, M. Klein and M. Hart, *"Lowering Power at 28 nm with Xilinx 7 Series Devices"*, Xilinx White Paper, WP389, 2015.

[5] F. Z. Tazi, C. Thibeault, Y. Savaria, S. Pichette and Y. Audet, *"On Extra Delays Affecting I/O Blocks of an SRAM-based FPGA due to Ionizing Radiation"*, IEEE Transactions on Nuclear Science, vol.6, pp. 3138-3145, 2014.

[6] F. Z. Tazi, C. Thibeault and Y. Savaria, *"Detailed Analysis of Radiation-Induced Delays on I/O Blocks of an SRAM-Based FPGA"*, 2016 IEEE Canadian Conference on Electrical and Computer Engineering (CCECE), 2016.

[7] N. Rollins, M. J. Wirthlin, M. Caffrey and P. Graham, *"Reliability of Programmable Input/Output Pins in the Presence of Configuration Upsets"*, Los Alamos National Laboratory internal document, LA-UR-02-3163, 2002.

[8] Xilinx, *"Kintex-7 FPGAs Data Sheet: DC and AC Switching Characteristics"*, Xilinx Data sheet DS182, v2.16, 2017.

[9] L. N. Cojocariu, V. M. Placinta and L. Dumitru, *"Monitoring system for testing the radiation hardness of a KINTEX-7 FPGA"*, AIP Conference Proceedings, 1722, pp 140009, 2016.

[10] Xilinx, *"7 Series FPGAs SelectIO Resources"*, Xilinx User Guide, UG471, 2018.

[11] Xilinx, *"Soft Error Mitigation Using Prioritized Essential Bits"*, Xilinx Application Note, XAPP538, 2012.

[12] J. Wyss, D. Bisello and d. Pantano, *"SIRAD: an irradiation facility at the LNL Tandem accelerator for radiation damage studies on semiconductor detectors and electronic devices and systems"*, Nuclear Instruments and Methods in Physics Research A, col. 462, pp. 426-434, 2001.

[13] Xilinx, *"Soft Error Mitigation Controller v4.1 LogiCore Ip Product guide"*, Xilinx Product Guide, PG036, 2017.

[14] V. M. Placinta and L. N. Cojocariu, *"Radiation Hardness Studies and Evaluation of SRAM-Based FPGAs for High Energy Physics Experiments"*, in proceedings of Topical Workshop on Electronics for Particle Physics 2017, 2018.

[15] L. N. Cojocariu and V. M. Placinta, *"Ion Beam Irradiation Effects in KINTEX-7 FPGA Resources"*, accepted for publishing in Romanian Journal of Physics, 2018.

[16] D. Bisello, A. Candelori, A. Litovchenko, E. Noah and L. Stefanutti, *"X-ray radiation source for total dose radiation studies"*, Radiation Physics and Chemistry, vol. 71, pp.713-715, 2004.

Fault Impact Assessment for Automotive Smart Power Products in an Electric Power Steering Application

Jonas Stricker*, Clemens Kain†, Andi Buzo†, Jerome Kirscher†, Linus Maurer*, Georg Pelz†

Email: stricker.external@infineon.com

†) Infineon Technologies AG, *) Bundeswehr Universität München

Abstract—The paper presents a methodology to propagate the consequences of random hardware faults in automotive smart power products to the application level. To accomplish this, the random hardware faults on chip level are assessed through fault injection into circuit simulations and are collapsed to come up with the relevant fault modes of a certain chip block. Then, these fault modes are propagated to the application level by injecting them into application simulations. The above is accomplished in an automated, seamless flow, which supports the engineering judgment in safety analysis by simulation results. The viability of the proposed approach is shown along a real-life example application (electric power steering) and a related smart power function (current measurement in the three phases). [1]

Keywords: Requirement Verification, Fault Clustering, Fault Collapsing, Fault Propagation, Electric Power Steering

I. INTRODUCTION

Ensuring automotive safety is a task for the complete automotive value chain. How to accomplish this is defined in the ISO 26262 as the absence of unreasonable risk [1]. The acceptable risk is defined in automotive safety integrity levels (ASIL A ... ASIL D). For the levels ASIL C and ASIL D, the ISO 26262 recommends to support the safety assessment with simulations anyway [2]. On the other hand, to assess the consequence of a random hardware fault, injecting faults into the related simulations is very helpful.

In this context, it makes sense to link the chip failure modes with the application simulation. The above requires the fault collapsing of the high number of basic circuit faults (wire opens, wire shorts, etc.) into a much lower number of chip failure modes [3]. Injecting the chip failure modes into application simulations, will propagate these failure modes into the application and helps to evaluate the related consequences. This cascading of simulations is indispensable, as system simulations employing low level models (or even circuitry) for the chip(s) end up in excessive runtime. At application level, faulty and fault free

application responses are compared to assess the fault's impact.

This methodology has been applied to the current measurement as a function of a chip in the context of an electrical power steering (EPS) application. The impact of the current sensor's faults are clustered in current sensor failure modes. This leads to a reduction of 20x for the application simulations.

The paper is organized as follows. In section 2 we discuss the state of the art. Section 3 describes the proposed method. The demonstrator, an EPS system, is introduced in section 4, while section 5 provides the results. Finally, the paper is concluded in section 6.

II. STATE OF THE ART

This section discusses the three main sections: Fault impact judgment, fault modeling and fault simulation & fault clustering.

A. Fault impact judgment

Fault impact evaluation is typically done by engineering judgment, which is quickly applied and lens itself for less complex fault behavior [4]. Lack of this method is the abstraction of a complex fault behavior and the automation for a larger number of faults.

B. Fault modeling

Fault modeling implies the simulation of a fault usually through a wire open/short in the system [5]. The fault affected behavior can be separated from the normal behavior of the model. A transistor fault of a current sensor for example has an impact on the output voltage. The relation of the input (current to measure) to the output (voltage representing the current) can be described through a model. While most faults of a current sensor are linear, the fault behavior can be described through a linear regression model (LRM) [6] [7]. Advantage of this method is the consideration of the mutual impact of the input factors and the easy description through a linear formula.

---

[1] This research project is supported by the German Government, Federal Ministry of Education and Research under the grant number 16ES0356-61.

978-1-5386-4483-6/18 $31.00 © 2018 IEEE

C. Fault simulation and fault clustering

The chip is simulated faulty and fault free to get the transient behavior of the chip. To cluster the transient fault behavior of the chip, the silhouette clustering algorithm can be used [3] [8]. It clusters the faults based on the similarity in the transient signals. Needed is a clustering based on the application performance, where this algorithm clusters based on the similarity of the transient behaviors.

III. METHOD DESCRIPTION

The proposed method combines the phases from chip level simulation to application level simulation. It is illustrated in figure 1 and it consists of four phases: chip simulation, fault clustering, fault modeling and application simulation.

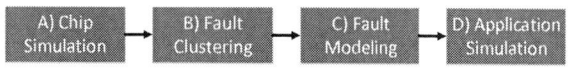

Figure 1. Phases of the Method

A. Chip simulation

At the chip simulation phase, the chip is simulated fault free in its test bench. Afterwards the faults under consideration are successively injected at chip level and simulated. Typical faults are opens and shorts. The fault behavior is dependent on the environmental conditions of the chip. While these can vary at application level over the time, the fault at chip level has to be simulated with varying environmental conditions which might occur at chip level. The transient responses of each fault are saved for further steps.

B. Fault clustering

The reduction to the most representative fault modes is done with the silhouette clustering algorithm [3] [8]. To accomplish this the transient time signal is clustered based on the Formula 1, where y_i is the time signal of each fault, $a(y_i)$ the average dissimilarity of y_i to all other y_i and $b(y_i)$ is the minimum dissimilarity over all clusters. The output s represents how well a certain fault fits to a cluster. From each cluster the most representing fault is chosen which has the shortest distance to the center point of the cluster. These faults are selected for fault modeling.

$$s(y_i) = \frac{b(y_i) - a(y_i)}{max(a(y_i), b(y_i))} \quad (1)$$

C. Fault modeling

To propagate the fault behavior, the fault has to fit to the fault description. For this reason the description has to be accurate enough to describe the fault impact and the complexity of the fault description has to be limited, otherwise the simulation time might become prohibited. To do so, the fault behavior is separated from the normal chip behavior to limit the complexity of the fault mode.

Figure 2 demonstrates how the fault is implemented at application level. To accomplish this, $f_{fault} = f_{meas} - f_{ideal}$ need to be calculated for each fault under consideration, where f_{meas} is the transient behavior of the chip simulation. $f_{fault} + f_{ideal}$ is the modeled fault affected behavior at application level which is called f_{sim}. The fault impact f_{fault} is linearized through a linear regression model with a feedback loop of its response. This assumption holds for many practical examples. A first order linear model is represented at Formula 2,

$$y(t) = b_0 + \sum_{i=1}^{n} b_i x_i + b_r y(t-1) + \epsilon \quad (2)$$

where y is the response variable, x_i the input variables, b_0, b_i, b_r are the coefficients and ϵ the normally distributed error. These coefficients have to be calculated with the least squares method to reduce the ϵ [9]. This function represents the fault in dependence of the input. For instance, if the current sensor fault has an offset of $1V$ and a faulty slope of $0.1V/A$ then $b_0 = 1$ and $b_1 = 0.1$. Limitations are the abstraction of a non linear behavior and the abstraction of noise, while this oscillates at the output with a static input.

D. Application simulation

The application is first simulated without any fault impact to measure the fault free response. Afterwards the modeled fault is added, which is shown at figure 2, and simulated successively for each fault mode. The analysis shows the impact of a chip fault on the responses of an application, which can violate the requirements of the application.

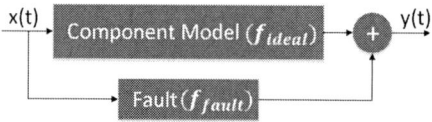

Figure 2. Fault Integration at Application Level

IV. APPLICATION DEFINITION

To apply the previous described methodology the electric power steering (EPS) application and the current sensor chip are chosen. This two models are described at the following sections.

A. Application description

The EPS application consists of several electrical and mechanical components. Figure 3 represents a general overview of the EPS [10] [11]. Its task is the assistance of torque for steering, which is enabled through an electric motor. The added torque is strongly dependent on the car speed. For lower car speeds, for instance parking, a higher assist of the motor is needed, while at higher car speed, the assist is close to zero. To support the driver by steering, several sensors measure the position and rotational velocity of the steering wheel. Together with the car speed, the motor position and currents the motor voltages are calculated at the controller. The motor converts the electrical power to the mechanical one, which is attached at the steering shaft.

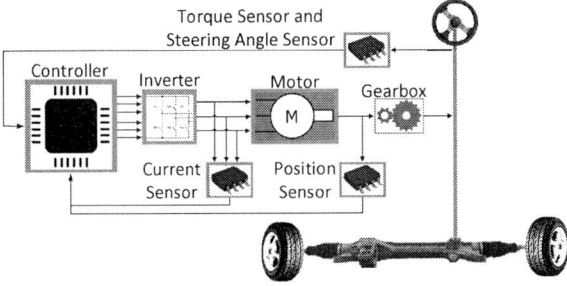

Figure 3. EPS Application Overview

B. Component description

To show the fault impact of a semiconductor component, the current sensor chip is selected. An example is shown at figure 4. This measures the current in the corresponding phase, which is done through a shunt measurement in the phase and an amplifier [12]. The gate voltage (GV) switches the MOSFET on/off with a frequency of 20 kHz. While the MOSFET is switched on the shunt voltage corresponds to the current at the phase, while at off phase the shunt voltage is zero. Component failures under consideration are opens and shorts of the transistors, resistances, diodes and capacitors at one section of the amplifier.

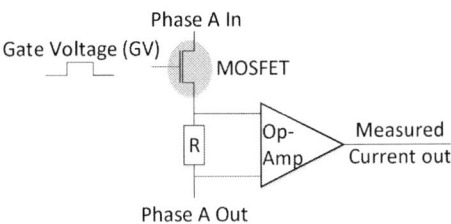

Figure 4. Current Sensor for one Phase

To check if and how the chip failure influences the performance, the following responses have been measured.

Torque Ripple(T_R): The definition of T_R is a vibration at the steering wheel which can be described like in table I. The frequency spectrum of the simulation $T_{ideal}(f)$ and $T_{real}(f)$ are filtered with a low pass filter at 20 Hz. $T_{Max}(f)$ is the frequency spectrum of the maximal assisted torque.

Precision of wheel angle ϕ_W: ϕ_W is the precision how close the wheel angle $\phi_{real}(t)$ is to the wheel angle of the ideal simulation $\phi_{ideal}(t)$. This behavior is described in table I:

These responses have to be in the requirement limits described in table I.

Table I
RESPONSE DEFINITION AND REQUIREMENTS

Response definition	Requirements
$T_R[\%] = \dfrac{\sum_{f=0}^{20} \lvert T_{real}(f) - T_{ideal}(f)\rvert * 100\%}{T_{Max(f)}}$	$0\% \leq T_R \leq 0.7\%$
$\phi_W[^{\circ}] = rms(\phi_{real}(t) - \phi_{ideal}(t))$	$0^{\circ} \leq \phi_W \leq 0.3^{\circ}$

V. RESULTS

We show an automated flow from chip simulation to application one to reduce unreasonable risk and to stress the safety mechanisms. To accomplish this the EPS application, is stressed with faults at the current sensor. The results are spitted to the sections: Component simulation & fault clustering, fault modeling and application simulation.

A. Component Simulation & fault clustering

Each transistor, resistance, diode and capacitor at one section of the current sensor amplifier is afflicted with two faults (open and short). This results to 86 faults which are automatically injected one after the other at the chip level.

To ensure that the fault model of figure 2 represents the fault under different environmental conditions, the fault is modeled for four use cases (UC) which are shown in table II. At this UCs, the phase current to measure and the GV duration shown at figure 4 are varied. The measured faulty and fault free responses of each UC are used to abstract the fault to a fault model. All fault models are compared with the fault model of Test 1 which is used for further steps. The comparison shows a maximal difference of 5% which is at this UC considered as good to proceed.

Table II
USE CASES UNDER CONSIDERATION

Test No:	Phase Current:	Pulse Duration Length of GV:
Test1	20A	25μs
Test2	20A	25μs
Test3	40A	5μs
Test4	40A	5μs

Based on the methodology, the faults are clustered into 4 fault modes. Figure 5 shows the fault responses and their grouping. Compared with the fault free chip behavior (magenta), the fault mode 1 (FM1) (black) is the closest, followed by FM2, FM3 and FM4. FM3 and FM4 have a significant offset to the fault free behavior and some noise affected faults. The chosen representative fault mode is shown at figure 6.

B. Fault Modeling

The faults under consideration are propagated through a linear first order model, which is described at section III. All representative faults and their modeled fault affected

978-1-5386-4483-6/18 $31.00 © 2018 IEEE

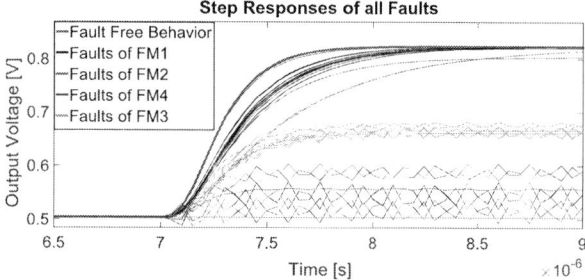

Figure 5. Transient behavior of all faults under consideration

behavior are illustrated at figure 6. Differences between the modeled behavior and the representative of the mode are especially at the beginning of the step. It has to be ensured that this difference is limited, otherwise the error of abstraction is too big to proceed with the method. In addition it has to be ensured that this difference has a low impact on the application responses.

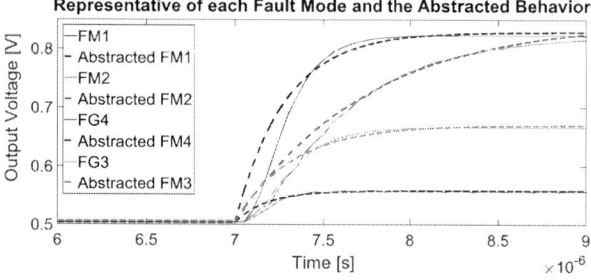

Figure 6. Comparison between the representative of the fault mode and the modeled fault

C. Application Simulation

Table III illustrates the application performances T_{R_r} and ϕ_{W_r} with the impact of the modeled fault affected behavior. The fault modes 1&2 have a torque ripple impact which is close to zero and meet the requirements at table I, while fault modes 3 and 4 don't meet the requirements. It seems that the offset between fault free and faulty transient signal has a big impact. This table shows which faults of the chip violate the application needs and which faults are less critical.

A comparison between the fault propagation and a direct chip implementation with fault at application level is impossible in reasonable time, while one run would need ∼4 months. This is caused by the complexity and timing of the chip.

To ensure that the performance of the represented fault mode (T_{R_r}, ϕ_{W_r}) have a similar application response behavior than all group members (T_{R_i}, ϕ_{W_i}), all faults were successively implemented at the application. Table III shows the maximal absolute response difference of all group members to its representative ($max(|T_{R\_r} - T_{R\_i}|), max(|T_{R\_r} - \phi_{W\_i}|)$). Fault mode 4 has a higher

variation, which is caused by the large transient signal difference shown at figure 5. While this fault mode is not even close the the requirements limits, this variation is uncritical. For future appliance, it is helpful to model the fault modes less representative to verify the application performance variance at one fault mode merges the requirements.

Table III
CONFORMITY OF THE GROUPS

Responses: Groups:	T_{R_r}:	ϕ_{W_r}:	$max(|T_{R\_r} -T_{R\_i}|)$	$max(|\phi_{W\_r} -\phi_{W\_i}|)$
FM1(Black):	0.04%	0°	0.05%	0.0056°
FM2(Red):	0.17%	0.04°	0.32%	0.019°
FM3(Green):	21%	2°	2.07%	0.204°
FM4(Blue):	42%	3.25°	11.84%	0.66°

VI. SUMMARY

The paper proposes a methodology for propagating low level faults as triggered by random hardware faults in microelectronics into automotive applications to assess the related consequences. In this way safety assessment starts at the set of all potential random hardware faults of a chip function (or a complete chip) and seamlessly tracks the related consequences of chip failure modes into application consequences. This substantially supports the current engineering judgment in today's safety analysis. Fault collapsing is employed on this way, to reduce the number of faults to a tractable amount. The above is demonstrated along a real-life electric power steering application.

REFERENCES

[1] Christian Giesselbach,Juergen Mottok,Vera Gebhardt,Gerhard M. Rieger *Funktionale Sicherheit nach ISO 26262: Ein Praxisleitfaden zur Umsetzung*, dpunkt.verlag, 2013.

[2] Peter Loew, Roland Pabst, Erwin Petry *Funktionale Sicherheit in der Praxis*, dpunkt.verlag, 2010.

[3] Oezlem Karaca, Jerome Kirscher, Arnaud Laroche, Andreas Tributsch, Linus Maurer and Georg Pelz *Fault Grouping for Fault Injection Based Simulation of AMS Circuits in the Context of Functional Safety*, Synthesis, Modeling, Analysis and Simulation Methods and Applications, 2016.

[4] James Parkin *Engineering Judgement and Risk*, Thomas Telford, 2000.

[5] R.J.A. Harvey, A.M.D. Richardson, H.G. Kerkhoff *Defect oriented Test Development Based on Inductive Fault Analysis*, IEEE International Mixed Signal Testing Workshop, Grenoble,1995.

[6] Frank E. Harrell , Jr. *Regression Modeling Strategies*, Springer, 2015.

[7] Douglas C. Montgomery, Elizabeth A. Peck, G. Geoffrey Vining *Introduction to Linear Regression Analysis*, John Wiley & Sons, 2012.

[8] Renato Cordeiro de Amorim, Christian Hennig *Recovering the number of clusters in data sets with noise features using feature rescaling factors*, Information Sciences 324, 2015.

[9] D. Montgomery *Design and Analysis of Experiments*, John Wiley & Sons, 2005.

[10] Z. Qun, H. Juhua *Modeling and Simulation of Electric Power Steering System*, Pacific-Asia Conference on Circuits, Communications and System, 2009.

[11] X. Chen, T. Yang, X. Chen, K. Zhou *A Generic Model-Based Advanced Control of Electric Power-Assisted Steering Systems*, IEEE Transactions on Control Systems Technology, 2008.

[12] Muhammad H. Rashid *Microelectronic Circuits: Analysis and Design*, Cengage Learning, 2017.

978-1-5386-4483-6/18 $31.00 © 2018 IEEE

Message Recovered: A Robust Fault Detection and Reporting Method for Galvanically Isolated IGBT Gate Drivers

Ines Hurez *,**, Ted Chen***, Florin Vlădoianu**, Vlad Anghel **, Gheorghe Brezeanu*

* University "Politehnica" of Bucharest, Romania
**ON Semiconductor Romania
*** ON Semiconductor USA
ines.hurez@onsemi.com

Abstract—This paper presents a fault detection and reporting technique for galvanically isolated Insulated Gate Bipolar Transistor (IGBT) gate drivers. This technique provides robust transmission of Under Voltage Lock Out (UVLO) and Desaturation (DESAT) events. The proposed method was verified by means of simulations and implemented in a standard 0.25μm CMOS BCD technology, as part of a galvanically isolated IGBT gate driver. Experimental results highlight proper reporting of UVLO and DESAT faults.

Keywords—galvanic isolation, gate driver, Under Voltage Lock Out, Desaturation, IGBT

1. Introduction

In the last decade, there has been an increasing demand for galvanically isolated gate drivers (GIGD) in a wide field of applications, such as inverters for photovoltaic arrays, variable frequency motor drives etc [1]. Galvanic isolation (GI) has become an essential requirement in order to separate the voltage domains between two or more system modules/sub-systems, hence guaranteeing safety and realiability in harsh environments. The modules need to be galvanically isolated since one of them is subject to considerable fluctuations in the ground level and the occurrence of high surge voltage/ current [2].

Traditionally, GI was obtained by means of optocouplers and/or discrete transformers but these solutions are not suitable for cost and area restricted applications, respectively [3][4]. Currently, on chip galvanically isolated modules exploit architectures that use an inductive or capacitive isolation element (IE) and are based on oscillators and voltage rectifiers [2][3].

One important set of applications for on chip galvanic isolation is represented by IGBT gate drivers. In such a driver, data received on a low voltage module is transmitted to a high voltage module which drives the IGBT, while maintaining galvanic isolation between the two sub-systems [2]. Several methods of transmitting information across the galvanic barrier are described in state-of-the-art literature [2][3].

This paper presents a technique for detecting and transmitting fault signals that can occur in a galvanically isolated IGBT gate driver. The fault signals analysed represent UVLO (Under Voltage Lock Out) events and the desaturation of the IGBT (DESAT), and their transmission across the isolation barrier is based on a robust method.

2. Principle of Operation

The concept schematic employing the proposed technique is presented in **Fig. 1** and the ideal waveforms of the pins of the circuit are given in **Fig. 2**. The architecture consists of two galvanically isolated modules - low voltage module (LVM – connected to the microcontroller) and high voltage module (HVM – connected to the IGBT) that are able to exchange information by means of an IE. IGBT gate signal is transferred on a forward path (FP - from LVM to HVM) and fault signals are reported on the feedback route (FR – from HVM to LVM).

The LVM depicted in **Fig. 1** consists of the conceptual schematic for UVLO (UVLO1) detection and reporting. COMP1 is used to detect the occurrence of this error. The supply voltage of LVM, VCC1, must operate above a preset threshold, V_{UV1}, in order to ensure the correct transmission of the control signal and reliable decoding of the fault signals sent from HVM.

978-1-5386-4483-6/18 $31.00 © 2018 IEEE

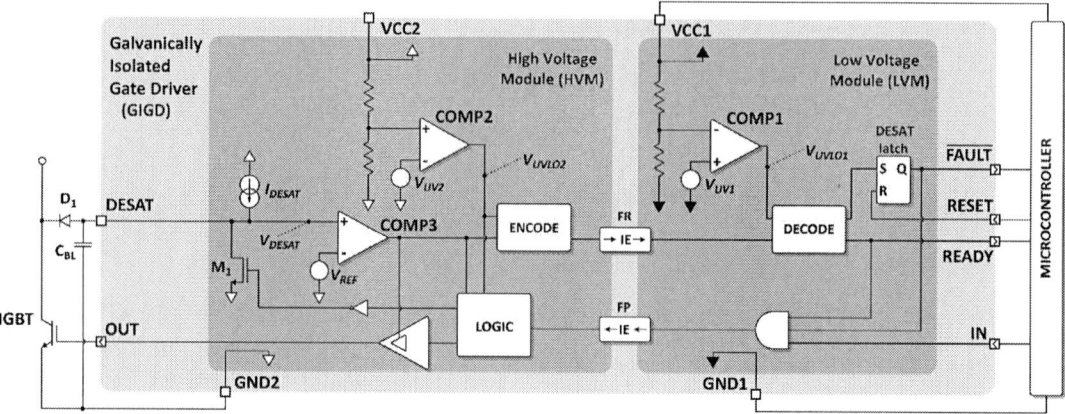

Fig. 1 Concept of GIGD with proposed technique included

In HVM, depicted in **Fig. 1**, are included conceptual implementations of specific UVLO (UVLO2) and DESAT detection. The former fault is detected using COMP2. When the IGBT is turned ON, the command voltage, $V_{OUT,}$ follows the VCC2 variations (as illustrated in **Fig. 2**) and it can lead to the occurrence of insufficient gate voltage if the supply voltage decreases, in general, by more than 20%. In this situation, the collector-emitter voltage is increased in order to provide the same current. Hence, the conduction losses increase too [4].

A DESAT event is detected using an external group formed by D_1 and C_{BL} (as pictured in **Fig. 1**). A comparator (COMP3) monitors the V_{CE} voltage of the IGBT, a current source (I_{DESAT}) charges C_{BL}, and a switch (M_1) offers a path for I_{DESAT}, as well as the possibility to discharge C_{BL}, during IGBT OFF state. When the IGBT receives a turn ON command, the current source starts charging C_{BL} for period of time t_{BL} [4]:

$$t_{BL} = C_{BL} \frac{V_{REF}}{I_{DESAT}} \qquad (1)$$

After t_{BL}, diode D_1 is forward biased and COMP3 monitors V_{DESAT}:

$$V_{DESAT} = V_{CE} + V_F \qquad (2)$$

The value of C_{BL} is chosen in order for COMP3 to start checking V_{DESAT} after V_{CE} reaches the saturation value, $V_{CE,sat}$. When V_{CE} increases excessively (usually at least four times the value of $V_{CE,sat}$) under high current conditions, IGBT desaturation occurs, and the output of COMP3 switches high, as depicted in **Fig. 2** [4].

The outputs of COMP2 and COMP3 are fed into a LOGIC block which turns OFF the IGBT in case one of the monitored faults takes place. In order to communicate the correct state of the GIGD and the IGBT to the microcontroller, the fault signals are sent into an ENCODE block on HVM. The low voltage module decodes the received signals and reports them consequently to the microcontroller. Furthermore, the error signals transferred from HVM are used to stop the transmission of the control signal (by the AND gate of LVM) in case of a desaturation or UVLO2 condition.

The UVLO2 signal is active low so as to ensure correct transmission of the error to the microcontroller. Therefore, when VCC2 is within the operating voltage range, the ENCODE block generates a periodic signal which is sent across the galvanic barrier from HVM to LVM, thus confirming the normal operation of the HVM. When the signal stops being trasmitted, LVM interprets it as the occurance of an UVLO2 event.

Inside the ENCODE block, a DESAT error takes priority over an UVLO2 check and is transmitted to LVM. This is because the IGBT desaturation can be a destructive process if it is not acted upon when it is detected.

The role of the DESAT latch is to prevent further turning ON of the IGBT until the microcontroller resets it. As an additional measure of protection, when a DESAT event is detected, the IGBT is not quickly turned

OFF. Instead, a soft shutdown (SSD) function is activated (highlighted in **Fig. 2**) that helps prevent high switching over-voltages that could destroy the IGBT [4].

Output signals READY and $\overline{\text{FAULT}}$ are used to communicate to the microcontroller the operation mode of the system. If both READY and $\overline{\text{FAULT}}$ are in high state, the system is in normal operation and control signals received on the IN pin are transmitted to the OUT pin. An UVLO event on either HVM or LVM is reported on READY by changing its state to low. At the same time, the command signal is no longer propagated to OUT. $\overline{\text{FAULT}}$ will remain in high state until a desaturation event occurs. The gate driver prohibits further turning ON of the IGBT until the microcontroller resets the latched DESAT flag using the RESET pin.

3. Simulation Analysis

The concept of the technique presented was designed and simulated in a Cadence environment using a 0.25um CMOS technology and BSIM3v3 models. The simulated plots are shown in **Fig. 3**. A good agreement with the ideal waveforms from **Fig. 2** can be observed.

Initially, the supply voltages of the HVM and the LVM exceed their corresponding reference voltage and the READY voltage is pulled high. This reflects GIGD normal operation and the command signal on the IN pin is transmitted to the OUT pin. When VCC1 starts decreasing, the READY and $\overline{\text{FAULT}}$ voltages follow its variation. READY pin is set to 0 at the moment VCC1 is reduced generally by more than 30%. At the same time, the OUT voltage is pulled to ground as a measure of protection for the IGBT. An UVLO2 condition is also reported by changing the state of the READY pin and setting low the OUT voltage.

When V_{DESAT} increases to the limit of desaturation, the $\overline{\text{FAULT}}$ voltage is pulled to ground and the SSD function is activated in order to ensure that the OUT pin changes its state safely. Both pins switch to high state only after a pulse is sent on the RESET pin.

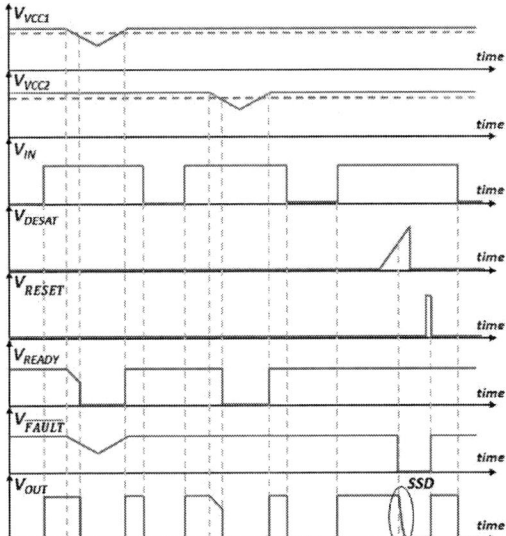

Fig. 2 Ideal plots representing the functionality of the detection and reporting technique

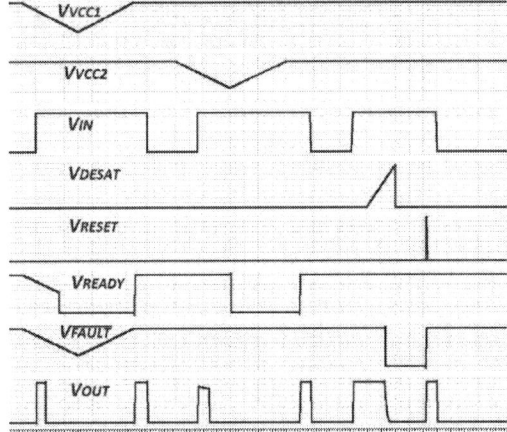

Fig. 3 Simulation results emphasizing the detection and reporting technique functionality.

4. Experimental Results

A galvanically isolated gate driver containing the detection and reporting technique from **Fig. 1** was implemented in a standard 0.25µm CMOS BCD technology. Measurements were performed to validate proper functionality of UVLOx and DESAT reporting. **Fig. 4** shows an oscilloscope screen capture of an UVLO1 event. Initially, VCC1 is above the preset threshold and V_{READY} is high. This corresponds to the normal operation of the gate driver and V_{OUT} is enabled. When an UVLO1 condition is forced, V_{READY} is set low and the control signals from the IN pin are no longer transmitted to the OUT pin.

Moreover, the $\overline{\text{FAULT}}$ pin does not change state. An UVLO2 event was captured in **Fig. 5** using the same sequence of events as described for an UVLO1 error.

An oscilloscope screen capture of a DESAT event is illustrated in **Fig. 6**. After V_{OUT} reaches its nominal voltage, a DESAT condition is forced and the V_{OUT} signal goes low slowly, emphasizing the soft shutdown function. Furthermore, the error is reported by setting $V_{\overline{FAULT}}$ low without influencing V_{READY}. This sequence of events demonstrates proper functionality of the detection and reporting technique implementation.

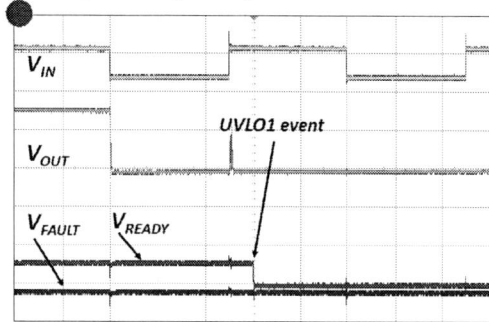

Fig. 4 Oscilloscope screen capture emphasizing UVLO1 condition reporting.

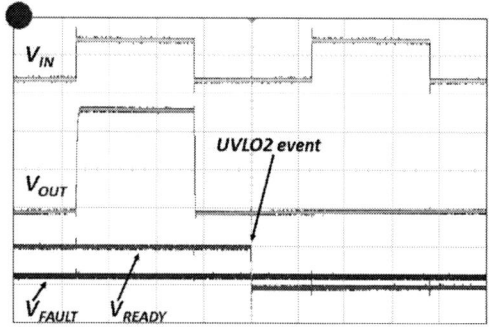

Fig. 5 Oscilloscope screen capture validating UVLO2 reporting.

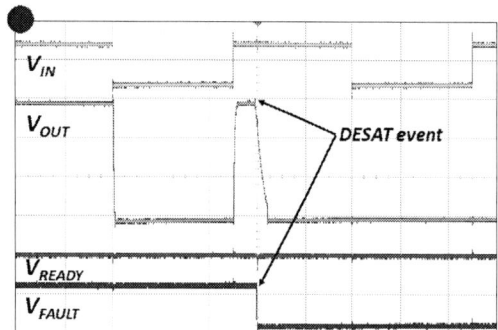

Fig. 6 Oscilloscope screen capture evincing DESAT event reporting.

5. Conclusions

This paper proposed a fault detection and reporting method for galvanically isolated IGBT gate drivers. The fault signals being transmitted from HVM represent an UVLO2 event and the desaturation of the IGBT. In the absence of the former fault, periodic pulses are sent from HVM to LVM confirming that the GIGD circuit operates properly. When VCC2 decreases by more than 20%, the fault is consequently signaled. The LVM has its own UVLO1 detecting block that is reported to the microcontroller. In case of a desaturation event, the fault takes priority on the transmission path because of its destructive potential and a soft shutdown function is activated. While UVLO2 is constantly monitored and dynamically reported, DESAT is latched and prohibits further turning ON of the IGBT until the latch is reset.

The proposed fault detecting and reporting technique was verified by means of simulation and a galvanically isolated gate driver containing the proposed concept was implemented in a 0.25µm CMOS BCD technology. Simulations and measurements were in agreement with the expected behavior.

Considering the silicon validation of the proposed technique based on a robust transmission method, this concept of detecting and reporting fault signals can be recommended for any application with galvanically isolated gate drivers.

References

[1] M. Txapartegi and J. Liao, "*Gate Driver Market and Technology Trends*" Report, Yole Developpement, March 2017.

[2] R. Yun, J. Sun, E. Gaalaas and Baoxing Chen, "*A transformer-based digital isolator with 20kVPK surge capability and > 200kV/µS Common Mode Transient Immunity*" 2016 IEEE Symposium on VLSI Circuits (VLSI-Circuits), Honolulu, HI, pp. 1-2, Sept. 2016.

[3] S. Kaeriyama et al., "*A 2.5 kV Isolation 35 kV/us CMR 250 Mbps Digital Isolator in Standard CMOS With a Small Transformer Driving Technique*" in IEEE Journal of Solid-State Circuits, vol. 47, no. 2, pp. 435-443, Feb. 2012.

[4] A. Volke and M. Hornkamp, "*IGBT Modules*", 3rd ed., Infineon Technologies AG, 2017

Comparison of Level Shifter Architectures: Application to I/O cell

Radu-Valentin Petrica*, **, Mihaela-Daniela Dobre*, **, Philippe Coll*, Florin Draghici**, Gheorghe Brezeanu**

*Microchip Technology Inc., **University Politehnica, Bucharest
Radu.Petrica@Microchip.com, Daniela.Dobre@Microchip.com

Abstract—Novel low-voltage and high-speed level shifter topologies will be presented. The level shifters circuits were designed in 40 nm technology using 1.2V devices and zero-V$_T$ transistors. These techniques will provide functionality near the threshold region. The simulated results were compared with a reference architecture. The resulted level shifters will be integrated in an already tested I/O structure. The results were analyzed in terms of electrical performance and silicon area.

Keywords—native devices; level shifter; I/O strucutre; low voltage.

1. Introduction

Power dissipation is a problem highlighted in recent years [1, 2]. Due to this fact, Systems on Chips (SoCs) tend to consist of significant power domains [2, 3]. Low-power dissipation was achieved by scaling down core supplies, to even below 1V. One of the many challenges of scaling down the core voltage comes in translating the signals from core voltage domains to the outside world; this phenomenon occurs within the I/O pad ring surrounding the core of the chip. In the context of the I/O cells, the level shifters (LS) are the circuits that translate the signal from core voltage, typically of 1.2V, to the outside world voltage, typically of 3.3V.

The purpose of this paper is to analyze various suggested level shifter architectures with better capabilities (lower input/core voltage, higher frequency, improved area compared to already existing level shifters) and integrate the resulted designs in an I/O cell. After the two main sections of this paper, the first one consisting of a comparative analysis between different level shifters and the second with the analysis of the impact after the I/O integration, the resulted I/O

capabilities will be discussed, and conclusions will be drawn.

2. Level shifters comparative analysis

A. Conventional Level Shifter

A conventional level shifter based on a cross-coupled structure is shown in **Fig. 1**.

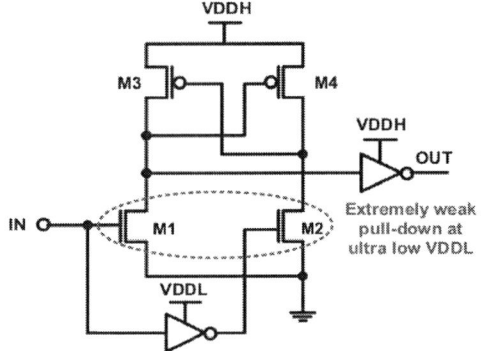

Fig. 1 Conventional Level Shifter [1]

The transistors M1, M2, M3 and M4 are of thick gate oxide to overcome voltage stress.

The transistors M1 and M2 seed voltage on cross-coupled input nodes. The positive feedback connection of M3 and M4 transistors uses this voltage to provide a full-swing VDDH on the output node spot between M3 and M1. VDDH is considered in this case at I/O voltage level. When IN is low, M1 and M4 are turned off, and M2 and M3 are on. At the switching time of IN from low to high, M2 turns off, M1 turns on, M4 switches on and output node switches from low to high [4].

The major drawback of this architecture is the strong competition between pull-down and pull-up networks during switching [1].

978-1-5386-4483-6/18 $31.00 © 2018 IEEE

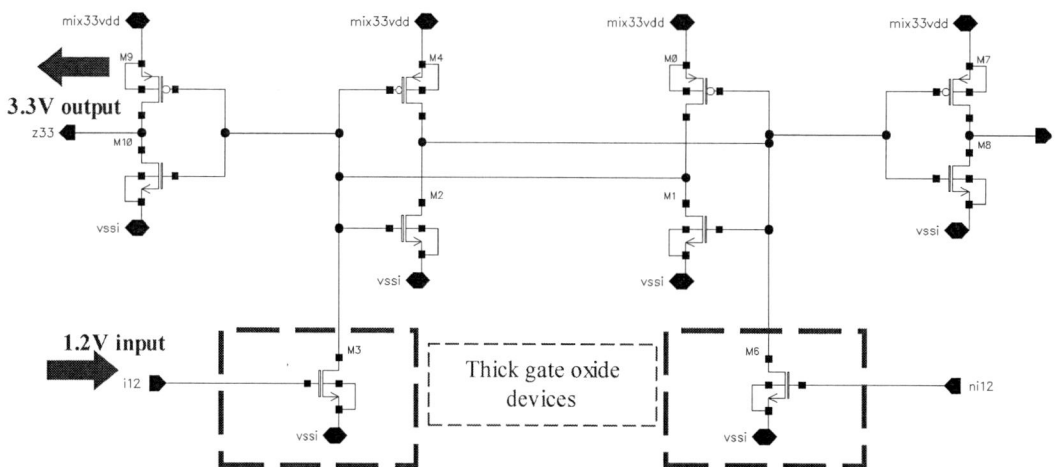

Fig. 2 Reference Level Shifter architecture

B. Reference level shifter

The reference level shifter architecture is displayed in **Fig. 2**. The functionality is similar with the level shifter presented in **Fig. 1** with one exception: the positive feedback is formed by two inverters in this case. The transistors that receive 1.2V signal are thick-gate oxide devices, illustrated with interrupted line boxes. In **Fig. 2**, the I/O voltage is mix33vdd (3.3V), the core signal (1.2V) is applied on the i12 input (ni12 is negated input). The desired signal translated to the I/O voltage is on the z33 output pin along with the inverted version on the zn33 output pin.

According to the spice simulations, the minimum input voltage is 0.81V. The frequency analysis conditions are: a full swing of the output signal, duty cycle within 45:55 and output slopes measured between 10%-90% of the mix33vdd voltage level need to be less than 75% of half of the period. Under these specific conditions, the reference level shifter has a maximum operating frequency of 120MHz. The resulted area is of 180μm² (6μm width x 30μm height). These parameters will be considered a comparison point for the newly implemented architecture in section C.

C. Architecture improvments

The newly suggested level shifter design uses the same architecture as the one provided in **Fig. 2**. The difference consists in placing a stack connection of a native device and a 1.2 volt-device instead of the highlighted thick-gate oxide devices (devices of 3.3V). This technique solves the major drawback of the shifting circuits – the weak pull-down network[1]. **Table 1** summarizes the connection of the considered architectures: both topologies have a connection consisting of a native device (in the upper stack part) and a 1.2V device (in the lower part), which receives the 1.2V input signal. The difference between Topology1 and Topology2 lies mainly in the different geometry of the devices, each with a different purpose.

The threshold voltage (V_T) of the native device along with the 1.2V transistor minimum length are the major advantages of the newly implemented architecture. Even if the V_T of the 3.3V transistors used in the reference level shifter is lower than the 1.2V transistor, the V_{DS} required to place transistor in the saturation region is higher ($V_{DS} > V_{GS} - V_T$).

Table 1 Devices used in the suggested architecture

	Reference	Topology1	Topology2
			Different transistor dimensions
V_T	0.5V (3.3V device)	0.07V (native device)	
		0.65V (1.2V device)	

The goal of the first improved version (Topology1) is a reduced silicon area compared to the reference architecture, by keeping the same input voltage. The second improvement (Topology2) aims ultra-low input voltage, using the same technique, but with different size of the marked devices. **Table 2** comprises the resulted electrical results. In addition, the LS layout width, height and resulted layout dimensions are provided. **Fig. 3** shows the layout representation for all the three circuits.

Table 2 Level shifters performance

LS	Reference	Topology1	Topology2
Vin [V]	0.81 V		0.65 V
Fr [MHz]	120 MHz	170 MHz	110 MHz
Width [μm]	6.03	5.16	8.875
Height [μm]	30.495		
Area [μm²]	183.89	157.36	270.65

Both topologies reach their goals: Topology1 has the area shrunk by 14.24% and frequency capabilities improved by 50MHz; Topology2 permits a lower input voltage (0.65V) with the cost of 47.18% of the added area.

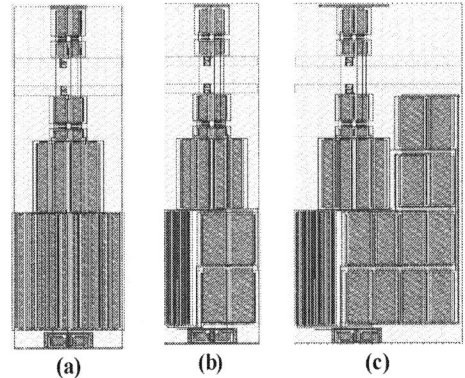

Fig. 3 Layout view for:(a) reference level shifter; (b) topology1; (c) topology2

3. I/O context

Fig. 4 is a schematic representation of a generic output I/O cell. The main composing blocks that work with signals from core voltage domain are: the logic section which is responsible with signal control principles such as enable, pull-up and pull-down etc. and the level shifters involved in the core voltage translation at a standard value of an amplitude

that meets communication requirements with the other peripherals. The pre-driver along with the output driver have a critical role in signal propagation delay: they guide the signal to the bounding PAD which encounters a large external load capacitance, so it needs sufficient current capability. These two blocks operate at I/O voltage level.

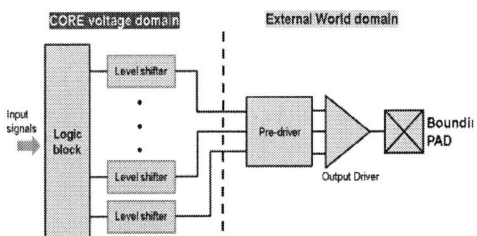

Fig. 4 Basic output I/O structure

The occupied area by the level shifters in an I/O cell depends on the number of input pins from the core side, necessary to meet the required functionality. The challenge is to try to integrate as many level shifters in a designated I/O area, preserving the same complex functionality. The suggested level shifters were integrated in an existing bidirectional reference structure maintaining the same 33.6μm x 170μm area for the I/O cell. **Fig. 5** has the resulted layout for the bidirectional I/O with the three-level shifter variants.

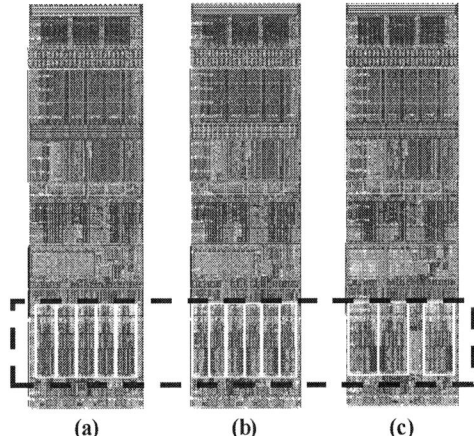

Fig. 5 I/O layout after level shifter integration: (a) five reference level shifters integrated in a bidirectional I/O; (b) integrating six level shifters from topology1; (c) three level shifters with lowest input voltage using topology2 integrated in I/O.

Table 3 contains information about the I/O cell performance and area, resulted after the integration part was performed. The reserved level shifters structures occupy around 18% of the entire I/O layout.

Table 3 I/O pad characteristics after integration

I/O	Reference	Topology1	Topology2
Vin	0.81 V	0.81 V	0.65 V
Fr	60 MHz	90 MHz	10 MHz
Area reserved for LS	18%		
Area occupied by 1 LS	3.218%	2.75%	4.73%
Integrated LS	5	6	3
Total area occupied by LS	16.09%	16.52%	14.21%

The number of integrated level shifter circuits reflects a downward characteristic in terms of area. The decreasing area trend given by Topology1 permits one more level shifter to be integrated in the same space. This has a major impact on the logical block functionality because it allows for a new control mechanism to be added. The results listed in **Table 3** are validated with post-layout simulation waveforms shown in **Fig. 6**. The worst-case scenario - shifting from 0.81V and 0.65V input signals to a maximum 3.6V output signal - is done correctly for all considered LS designs. The output enable signal was represented for functionality purposes and for highlighting one of the core voltage domain control I/O cell pins.

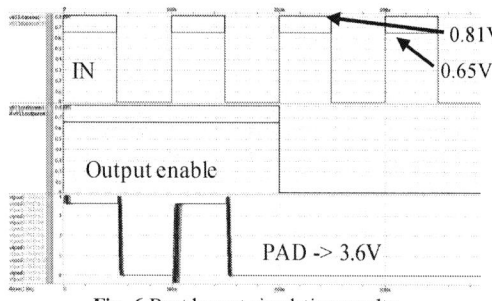

Fig. 6 Post layout simulation results

Parameters such as fall/rise and delay time are essential when a new I/O library is prepared for release. A rigorously characterization was done to understand the eventually timing strengths and limitations of the I/O cell after integration. **Fig. 7** gives an insight on the timing delay results in a graph; Topology1 has similar delays to signals rising to the reference design, which can explain the similar output signal frequencies of 60 MHz and 90 MHz; Topology2 has a delay degradation which explains the lower frequency obtained of 10 MHz.

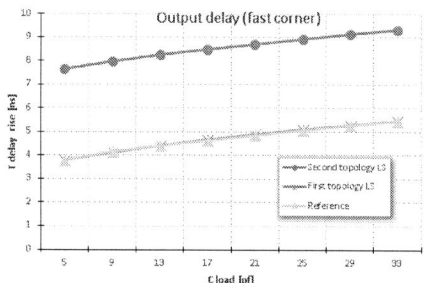

Fig. 7 Delay timing overview

4. Conclusions

This paper brings into attention two novels, ultra-low voltage and high-speed level shifters designed in 40 nm technology, along with their integration into an I/O structure. These suggested topologies use mixed V_T devices that grant robustness and low-power dissipation. The first-level shifter suggested (Topology1) can be used for applications where I/O cells need higher frequency and increased number of digital control signals, whereas the second one (Topology2) allows for ultra-low core voltage applications, down to 0.65V.

References

[1] J. Zhou, C.Wang, X.Liu, X. Zhang, and M. Je, *"An Ultra-Low Voltage Level Shifter Using Revised Wilson Current Mirror for Fast and Energy-Efficient Wide-Range Voltage Conversion from Sub-Threshold to I/O Voltage"*, IEEE, pp.697, 2015

[2] S. Hossain, I. Savidis, *"Bi-directional Input/Output Circuits with Integrated Level Shifters for Near-threshold Computing"*, IEEE, pp.1240, 2017.

[3] J. Zhou, C. Wang, X. Liu, X. Zhang, M. Je, *"A Fast and Energy-Efficient Level Shifter with Wide Shifting Range from Sub-threshold up to I/O Voltage"*, IEEE, pp.137, 2013.

[4] K.H. Koo, J.H. Seo, M.L. Ko and J.W. Kim, *"A new level-up shifter for high speed and wide range interface in ultra deep sub-micron"*, IEEE, 2005.

Session D&IC-S

DEVICES & INTEGRATED CIRCUITS

978-1-5386-4483-6/18 $31.00 © 2018 IEEE

Power Supply Duty Cycling for Highly Constrained IoT Devices

A. Monti*, E. Alata, A. Takacs***, D. Dragomirescu****

* LAAS-CNRS, University of Toulouse, CNRS, Toulouse, France
E-mail: amonti@laas.fr
** LAAS-CNRS, University of Toulouse, CNRS, INSA, Toulouse, France
E-mail: ealata@laas.fr, dragomirescu@laas.fr
*** LAAS-CNRS, University of Toulouse, CNRS, UPS, Toulouse, France
E-mail: atakacs@laas.fr

Abstract—With increasing interest emerging towards the Internet of Things (IoT) area, many new applications require embedded devices to integrate ever more sensors and communication interfaces while keeping hard constraints on battery life. We propose to go beyond classical radio duty cycling by disconnecting unused devices from the power supply, eliminating quiescent power consumption. Generalizing this technique to other sensors, we present a flexible IoT architecture, which we illustrate through an example industrial application and test results from an instrumented prototype.

Keywords—IoT; embedded systems; low power; quiescent current; generic architecture; power gating.

1. Introduction

Internet of Things (IoT) devices are highly constrained embedded systems, as they need to combine several sensors, one or more communication systems (e.g. LoRa radio and GPS module) and a microcontroller (UC) allowing implementation of the communication protocol as well as gathering and processing of sensor data, while keeping the overall energy consumption to a minimum in order to meet strong battery life requirements. Additional constraints such as cost, space or shape requirements are also usually encountered and further increase the associated engineering complexity. Still, the energy consumption remains the most powerful constraint when it comes to designing an IoT device, therefore our work focuses on this particular issue.

2. Proposed Low Power Architecture

A. Motivation

In a good first approximation, a vast majority of IoT devices follow a common activation profile structure. Indeed, typical communication profiles consist of long inactivity periods punctuated by short time slots where wireless transfers are done [1]. As IoT devices are energy autonomous systems that need to operate during a long time span on small finite energy resources, the common strategy is to alternate periodically between low power consumption states (denoted thereafter as sleep mode or state) and active states. Therefore, the two activation profiles are usually synchronized and results in the device being most of the time in sleep mode, while being periodically activated in order to gather and process sensor data, then proceed to a transmission to the IoT infrastructure.

Following this simple assumption, the overall energy consumption of the IoT device can be estimated by the weighed sum of average consumptions in active and sleep states, with respect to the periodic activation profile. We observe that given that the active period is very short, reducing the average energy consumption in the sleep state can significantly improve the battery life of the device. The sleep state energy consumption mostly results from the sum of all currents drawn by active devices such as the microcontroller, sensors and communication systems. During the sleep state, the radio transceiver is not used and is also put to a low power consumption mode; this technique is well described in the literature as radio duty cycling [2-3]. However, even when not used or put to a low power state, every active system consumes energy, usually quantified

978-1-5386-4483-6/18 $31.00 © 2018 IEEE

by its quiescent current. Such currents can range from some µA (e.g. accelerometers) to several tens of mA (e.g. GPS modules), and can even exceed the required minimal sleep state current with respect to a given device lifetime.

B. Electronic Architecture

As a first feature of the presented architecture we propose, in order to eliminate quiescent currents in sleep mode, to isolate groups of devices from the main power supply by introducing a switch on their supply rail. The switch circuit, realized for a standard 3.3V supply voltage, is represented on Fig. 1. The circuit can be controlled by a single microcontroller pin configured as push-pull, and does not change state when the controlling pin switches to high impedance, allowing the microcontroller to be put in sleep mode without altering the switch's state. Indeed, when the switch is on, a feedback through R1 keeps Q2 conducting even when the driving pin is in a high impedance state. This topology differs from commercially available load switches such as TI's TPS22860 or Vishay's SiP32431 by the bistable characteristic.

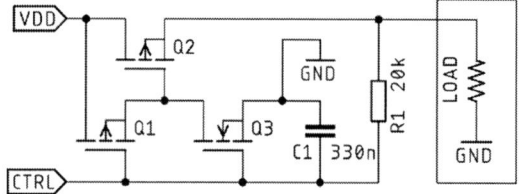

Fig. 1 Switch circuit for a 3.3 Volts supply voltage

Next, we note that increasing the number of integrated sensors modules requires the use of a higher-end microcontroller featuring many interface peripherals, mainly General Purpose Input Outputs (GPIOs), Serial Peripheral Interface (SPI) or Inter Integrated Circuits (I^2C) and high processing power. However, using a high pin count, fully featured microcontroller implies a higher power consumption even in sleep mode. We thus propose to integrate two smaller microcontrollers to further reduce power consumption and decrease integration

complexity. The first microcontroller, denoted thereafter as the master UC, shall control power switches, interfacing to the main communication radio (e.g. LoRa or Sigfox) and implementation of the wireless protocol. On the other hand, the second microcontroller, denoted as the sensor UC, provides interfaces to sensors and additional modules as well as processing power to handle sensor data. Interfacing of the two microcontrollers is realized using a SPI bus, preferred over I^2C for reduced power consumption.

Active devices used infrequently are thus organized in groups we denote as Power Supply Domains (PSDs), each of which is isolated from the main power supply. As a general guideline, we propose two PSDs, the first including the main radio circuitry (used at a rate fixed by the wireless protocol), the second containing the sensor UC and associated sensors (potentially used at a different rate than the main radio).

The overall architecture is summarized on Fig. 2.

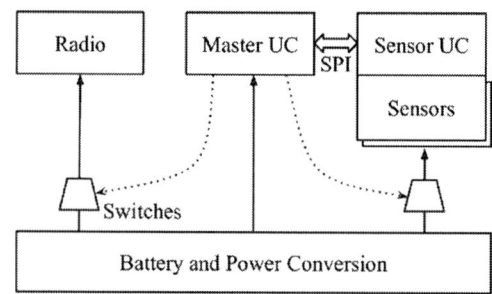

Fig. 2 Summary of the proposed architecture

C. Main Benefits and Limitations

This simple electronic architecture presents several advantages, the first of all being the capability to integrate many infrequently used sensors while minimizing sleep mode power consumption by eliminating quiescent currents. Integration of sensors or modules that are not designed for low power applications (such as GPS modules) is also possible, at the condition that they are only used a fraction of the device's operating cycle.

978-1-5386-4483-6/18 $31.00 © 2018 IEEE

Elimination of quiescent currents is a common practice in the low-power SoC community, which is usually referred to as power gating [4]. However, the proposed architecture being built using discrete electronics, it is simpler to integrate and deploy and therefore targets different engineering problems. In that sense, our approach is similar to the one described in [5].

From an engineering perspective, the architecture is hardware generic (as no assumptions were made on integrated sensors and modules) but also scalable, because it allows integration of as many PSDs as needed, possibly with several sensor UCs. The main limitation of the proposed technique is the quiescent current imputable to the power switches in their off state, which is entirely due to the FET's leakage current. For standard components such as FDV303N and FDN340P, and for a 3.3V drain-source voltage, this value is on the order of 250 nA.

On the software side, the dual microcontroller setup provides room for parallelism and load balancing, which makes a versatile architecture and allows a wide range of applications from simple IoT devices to intensive processing sensor platforms.

3. Hardware Realizations and Validation

A. Example Industrial Application

We illustrate the proposed architecture through a hardware prototype tailored for an example industrial application, which was an IoT object aiming to help find lost domestic animals. It consisted of several elements: the embedded device as a collar fixed to the pet, a base station in the user's home and a cloud infrastructure. The device has two operating modes; when it is close to the base station, activity parameters are periodically monitored.

Component-wise, the device must have integrated a LoRa radio, an accelerometer for activity monitoring, a GPS module, a temperature sensor as well as user interface such as buttons and LEDs. Meeting the lifetime constraint required a sleep mode current several orders of magnitude lower

than the quiescent current of the GPS module, hence the relevance of our architecture.

The prototype is shown on Fig. 3 and features, from left to right and top to bottom, power switches and conversion, a LoRa radio, the GPS module, sensors and user interaction elements, master and sensor UC.

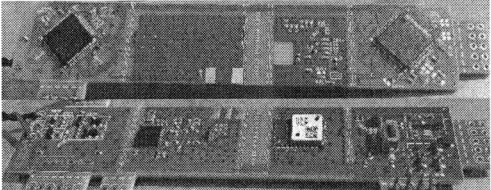

Fig. 3 Prototype for an example industrial application

Software validation of the architecture has shown excellent versatility and dynamic configurability provided by the power switches and dual microcontroller architecture.

B. Instrumentation Prototype

As, for technical reasons, no precise and complete current measurements could be conducted on this prototype, another prototype was designed including several current probes, allowing precise power measurements. We also developed dedicated firmware for both master and sensor UCs highlighting different states and transitions. The prototype is shown on Fig. 4.

Fig. 4 Instrumented prototype for current measurements

As to present a specific example, Fig. 5 depicts a typical IoT device current consumption profile when not using the proposed architecture. In this experiment, only the master UC and LoRa radio were used for the sake of simplicity. In the first part of the curve, both microcontroller and LoRa radio are in low power mode, and the corresponding current consumption is measured to be 2.72 mA. When the master UC wakes up to enter in an active state, the radio is used to send a frame (the next two peaks), a LED is blinked (which corresponds to the last current peak) and the cycle starts again.

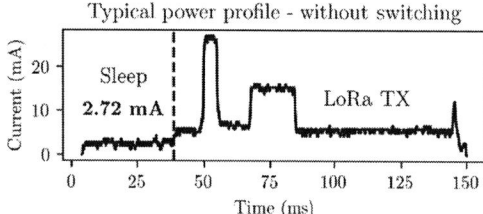

Fig. 5 Reference current measurement for validation

On Fig. 6 we can observe the same operating cycle, but using the proposed architecture to switch off the LoRa radio power supply whenever it is not used. We then measure a sleep mode current consumption of 1.94 mA.

The sleep mode current consumption reduction is measured to be 780 µA for this configuration. We would like to emphasize on this gain rather than on absolute power consumption values, which are artificially increased when running microcontrollers in debugging mode.

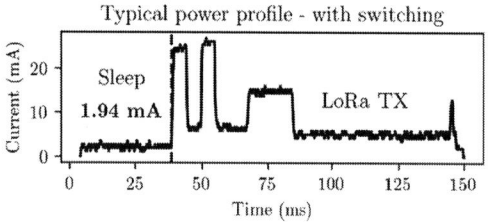

Fig. 6 Example current measurement for validation

We also note an additional current peak when powering the LoRa radio, which corresponds to the chip's booting sequence. However, this additional power consumption is still compensated by the sleep mode current reduction if the sleep state duration is high enough (e.g. greater than 800 ms for this particular example, which is in the correct order of magnitude for targeted IoT applications). This is our main result, showing the current consumption gain with respect to classical radio duty cycling.

4. Conclusion and Future Work

We proposed an energy efficient and simple IoT device architecture allowing the integration of multiple power consuming sensors and modules while providing an extended battery life by putting them in separate power supply domains that can be turned off to eliminate quiescent currents and lower the average energy consumption.

Experiments on an instrumented prototype shown a significant reduction of the sleep mode current consumption when using the discrete power gating topology.

The proposed architecture is generic and scalable, as any number of power supply domains can be added, limiting the number of available sensors and modules only to microcontroller interfacing limits.

References

[1]   I. Demirkol, C. Ersoy and F. Alagoz, *"MAC protocols for wireless sensor networks: a survey"*, IEEE Communications Magazine, **44**(4), pp.115-121, 2006.

[2]   U. Raza, P. Kulkarni and M. Sooriyabandara, *"Low power wide area networks: An overview"*, IEEE Communications Surveys & Tutorials, **19**(2), pp.855-873, 2017.

[3]   G. Anastasi, M. Conti, M. Di Francesco and A. Passarella, *"Energy conservation in wireless sensor networks: A survey"*, Ad hoc networks, **7**(3), pp.537-568, 2009.

[4]   K. Parthiban and S. Sasikumar, *"Performance Analysis of Leakage Current Reduction in Standby Mode of Zigbee SoC Using Active Mode Logic"*, Journal of Computational and Theoretical Nanoscience, **15**(2), pp.525-529, 2018.

[5]   M. Hayashikoshi, H. Noda, H. Kawai, K. Nii and H. Kondo, *"Low-power multi-sensor system with task scheduling and autonomous standby mode transition control for IoT applications"*, Low-Power and High-Speed Chips (COOL CHIPS), 2017 IEEE Symposium, pp.1-3, 2017.

Over-Temperature Protection for a Switched-Capacitor DC-DC Converter with Controlled Charging Current

Cosmin-Sorin Plesa*, Marius Neag*, Cristian Mihai Boianceanu**

*Technical University of Cluj-Napoca, Basis of Electronics Department, 400027, Romania

E-mail: Cosmin.Plesa@bel.utcluj.ro

**INFINEON Technologies, Bucharest, Romania

*E-mail: CristianMihai.Boianceanu@infineon.com

Abstract—This paper presents an over-temperature protection (OTP) circuit for a DC-DC converter based on switching capacitors (SC DC-DC). The circuit was designed by using a two-step approach that encompasses running both electrical and electro-thermal simulations. The die temperature distribution was analyzed for two critical operational scenarios in order to identify the worst case and to provide design data for implementing a robust and precise OTP. Finally, a design example of a SC DC-DC with OTP is presented in some detail, complete with the key electrical and electro-thermal simulation results.

Keywords—over-temperature protection; switched-capacitor converter; thermal shutdown;

1. Introduction

DC-DC converters based on switching capacitors (SC DC-DC) have become real challengers to inductor-based converters for low- to medium- power automotive applications [1]. External inductors are not only expensive but also bulky and susceptible to mechanical vibrations.

Over-temperature protection (OTP) is a basic feature for linear voltage regulators, essential for avoiding physical destruction in fault conditions such as output shorted to ground. The SC DC-DC does not dissipate too much power in normal operation but in some scenarios the overheating could influence the functionality. Therefore, an accurate OTP is necessary to protect the converter against thermal runaway that can appear in certain situations [2].

This paper shows that even for a fairly well-known SC DC-DC topology the power dissipated within some of its circuitry could affect the robustness of the converter. Three design options for implementing the OTP circuit are briefly discussed, then a design

example is presented. Results yielded by both standard, electrical-only, and electro-thermal simulations are shown for two operational scenarios. They demonstrate the need for an OTP circuit and indicate a systematic approach to its design.

2. OTP for a SC DC-DC Converter with Controlled Charging Current

A. Core Schematic and the Case for OTP

The core of a SC DC-DC converter is the switched capacitor array which consists of MOS switches and "flying" capacitors (C_{FLY}) used for storing and transferring energy.

Of the many possible topologies the 2:1 (step-down) or 1:2 (step-up) configuration was chosen for the discussion here, as they are among the most used SC DC-DC converter topologies. In general, they provide good efficiency and require a relatively small die area, due to their employing only one C_{FLY} [3]. Moreover, we focused on the implementation shown in Fig. 1, an one-C_{FLY} SC DC-DC with controlled charging current.

The circuit converts the input voltage, V_{IN}, into a regulated output voltage, V_{OUT} within two phases: C_{FLY} is charged from the input in the first phase (Φ_1) and discharged on the load in the second phase (Φ_2). This is achieved by turning ON switches S_2 and S_4 during phase Φ_1 and switches S_1, S_3 and S_5 during phase Φ_2. The output voltage level is set and maintained by a feedback loop that controls the amount of charge transferred from the input to the output. The transferred charge is sourced by a power current mirror

978-1-5386-4483-6/18 $31.00 © 2018 IEEE

(PCM – transistors M_2 and M_3); the amount of charge transferred is set by the differential error amplifier (EA) that drives transistor M_1 proportionally to the difference between the voltages applied to its inputs: the reference V_{REF} and a fraction of the output voltage brought to the inverting EA input by the resistive divider R_1, R_2 (V_{FB}).

In normal operation conditions the power dissipated on switches S_1 - S_4 is not large enough to cause their overheating. But in certain operational scenarios large voltage drops can appear across the power transistor M_3 within the PCM, then can result in substantial levels of dissipated power. Here are two such critical scenarios:

- Normal start-up but a large input voltage is applied. The PCM will source the maximum possible current in the first phase (Φ_1) because the inputs of the EA are totally unbalanced, as $V_{FB}=0$. Until the steady state is reached (that is, until $V_{FB} = V_{REF}$) a large amount of power is dissipated by the PCM in each phase Φ_1.

- Start-up in shorted-output fault condition (the output shorted to ground). In this scenario the inputs of the EA remain unbalanced all the time and the PCM will try to source the maximum possible current in each phase Φ_1.

The designer should ensure that in all valid operational scenarios the maximum die temperature within the converter – called here $T_{HOTSPOT}$ - remains below a safe value, defined by application and technology limits.

B. Design Options for Circuit Implementation of OTP

Fig. 2 presents a typical circuit implementation of the OTP, usually employed for LDOs [4] but which can be adapted for SC DC-DC converters, as well.

The voltage produced by a temperature sensor, OTP_{SENSOR} – here the base-emitter voltage of a BJT, V_{EB} – is applied to the inverting input of a voltage comparator with hysteresis, implemented by the operation amplifier, the gain stage based on transistor T_1, and the Schmitt Trigger that selects the

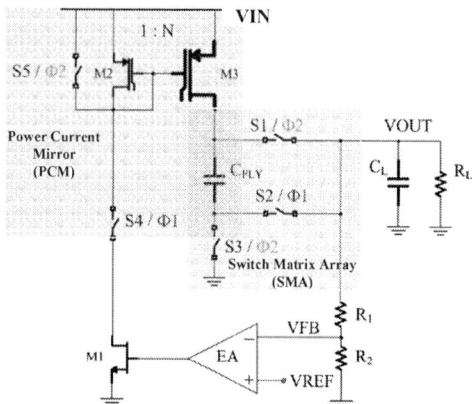

Fig. 1 Schematic of the core of a SC DC-DC converter with controlled charging current.

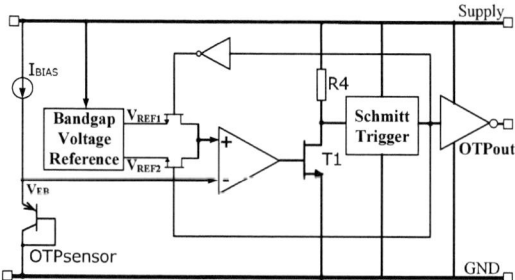

Fig. 2 Simplified schematic of the proposed OTP; only the main functional elements are detailed.

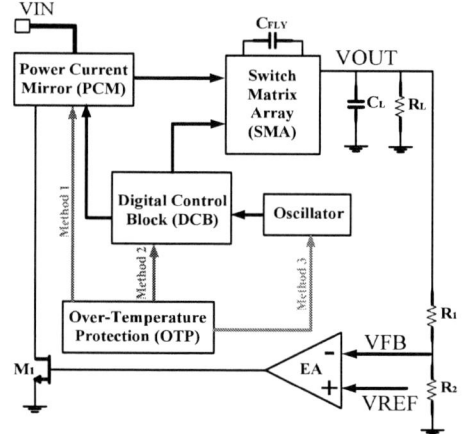

Fig. 3 Block diagram of a SC DC-DC Converter with OTP; three ways of using the OTP signal to shut-down the converter are highlighted.

reference voltage applied to the non-inverting input, V_{REF1} or V_{REF2}. When the temperature of the OTP_{SENSOR} goes above the activation temperature of the OTP circuit - called here T_{HIGH} -, the output OTP_{OUT} will go to logic "1"; when the temperature goes below the de-activation temperature – called here T_{LOW} -, the OTP_{OUT} will trip to logic "0".

For the SC DC-DC topology shown in Fig. 1 the OTP_{OUT} signal can be used to implement the OTP in three different ways, illustrated in Fig. 3:

- Method 1: The OTP$_{OUT}$ signal can be used as a control signal for the PCM. When an OTP event occurs the OTP$_{OUT}$ goes high, and a logic circuit can turn ON the switch S$_5$ and turn OFF the switch S$_4$.

- Method 2: The OTP$_{OUT}$ signal is used as an additional input to the Digital Control Block (DCB) of the converter so that when OTP$_{OUT}$ goes high the DCB will activate phase Φ_2 and keep the converter in that state until the OTP$_{OUT}$ signal goes low, that is, until the temperature of the OTP$_{SENSOR}$ drops below T$_{LOW}$.

- Method 3: The OTP$_{OUT}$ signal can be used as an additional enabling condition for the oscillator that generates the Φ_1 and Φ_2 clocks: the oscillator is turned off as long as OTP$_{OUT}$ stays high.

3. Design Example and Simulation Results

Fig. 4 presents the floorplan of a SC DC-DC converter we designed for a specific application, based on the schematic shown in Fig. 1. A version of the OTP circuit shown in Fig. 2 was also integrated. Method 2 described above was used to implement the OTP: the SC DC-DC converter was kept in phase Φ_2 as long as OTP$_{OUT}$ had a logic "1" value.

Quite often OTP circuits are designed based on electrical-only simulations. But these simulations assume that the die temperature is constant across the entire converter, which is obviously not the case in a real-life integrated circuit (IC); therefore, the designed/simulated OTP activation and de-activation temperatures can be quite far from their actual, measured values. To avoid these errors our design followed a two-step approach:

Step 1 = the circuit was sized by using electrical-only simulations, aiming for T$_{HIGH}$=175°; and T$_{LOW}$=160°;

- Fig. 5 presents results from a temperature sweep simulation; one can derive the activation (T$_{HIGH}$) and the de-activation temperature (T$_{LOW}$) for the OTP.

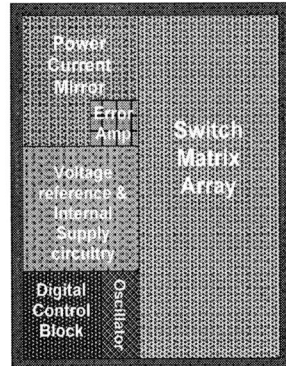

Fig. 4 Floorplan of a SC DC-DC converter with controlled charging current and OTP designed based on Figs.1-3

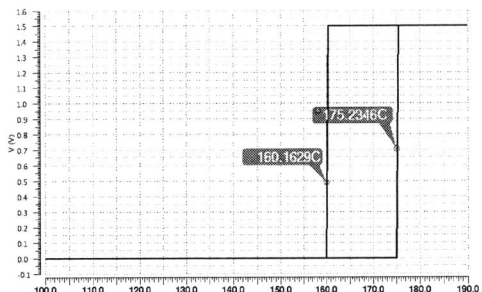

Fig. 5 OTP$_{OUT}$ variation obtained for a temperature sweep

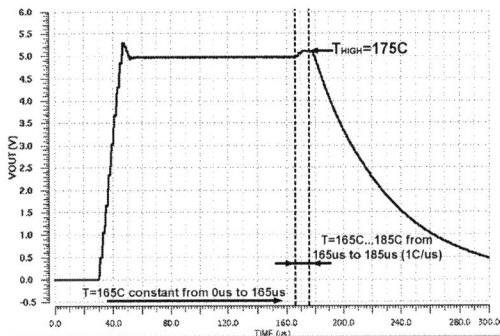

Fig. 6 The SC DC-DC output voltage for a test that involved increasing the die temperature by 1°C/us after t=165us

- Fig. 6 presents the output of the SC DC-DC converter yielded by a transient simulation that also involved changing the die temperature. Until t=165us the die temperature was kept constant, T=165°; afterwards the temperature was increased by 1°C/us. Thus, at around t=175us the OTP activation temperature, T$_{HIGH}$, was reached, OTP$_{OUT}$ went high and the SC DC-DC converter was held in phase Φ_2. Therefore, the output voltage dropped to zero as C$_{FLY}$ discharged through the load.

Step 2 = finely adjust the OTP thresholds based on electro-thermal simulations run as described in [5], but considering the critical

978-1-5386-4483-6/18 $31.00 © 2018 IEEE   221

Table 1. DeltaT[°C] yielded by electro-thermal sims

Test scenario (Section II.A)	DeltaT [°C]	
	$T_{START} = -40°C$	$T_{START} = 150°C$
Normal Start-up	8.5°C	16.5°C
Start-up in fault condition	52°C	65°C

scenarios presented in Section 2.A. The main results are: the maximum temperature that can appear within the IC ($T_{HOTSPOT}$) and the hot-spot location. These data are necessary for deciding the OTP$_{SENSOR}$ location and for calculating the resulting thermal coupling:

$$DeltaT[°C] = T_{HOTSPOT} - T_{OTPSENSOR} \quad (1)$$

The electro-thermal simulations consider the IC layout, the physical description of the package, the ambient temperature (T_{START}) and the power dissipated by the PCM yielded by electrical-only simulations.

Table 1 summarizes results of tests run for the two critical scenarios with T_{START} set to -40°C or 150°C. The worst case DeltaT value was obtained for the "start-up in fault condition" test, with $T_{START}=150°C$: DeltaT= 65°C. Fig. 7 presents electro-thermal simulation results for this case: it shows the variation over time of temperatures $T_{HOTSPOT}$ (red trace) and $T_{OTPSENSOR}$ (black trace). Note that the PCM temperature, from where $T_{HOTSPOT}$ is extracted, varies: in Φ_1 the temperature increases as power is dissipated by the PCM) while in Φ_2 the temperature decreases because the PCM dissipates no power. Overall the PCM temperature increases until the temperature of the OTP$_{SENSOR}$ reaches the T_{HIGH} threshold. The thermal-map at t=150us is presented in Fig. 8 The hotspots appear in the PCM – see Fig.4.

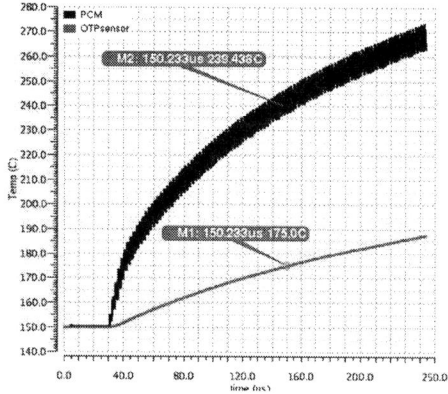

Fig. 7 Temperature variation over time for the OTP$_{SENSOR}$ (red) and for the hotspot within the PCM (black)

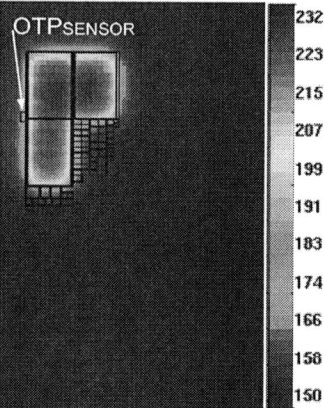

Fig. 8 Temperature map of the SC DC-DC converter generated by the electro-thermal simulation shown in Fig.7 before the OTP event (t=150us).

4. Summary and Conclusions

First, it was demonstrated that a SC DC-DC converter with controlled charging current needs a precise OTP circuit. A design example based on both electrical and electro-thermal simulations was then shown. It involved determining the worst-case difference between the OTP$_{SENSOR}$ and the maximum temperature developed within the IC. This not only allows for fine tuning of the OTP thresholds but also provides data for optimal placement of the OTP$_{SENSOR}$.

Acknowledgments. This work was co-funded by the European Regional Development Fund through the Operational Program "Competitiveness" POC -A1.2.3-G-2015, project "PartEnerIC", contract 19/01.09.2016.

References

[1]. G. Villar Pique, H.J. Berveld, E. Alarcon, "*Survey and benchmark of fully integrated switching power converters: switched-capacitor versus inductive approach*," IEEE Transaction on Power Electronics, vol. 28, no. 9, 2013

[2]. V. W-S. Ng, S. R. Sanders, "*Switched Capacitor DC-DC Converter: Superior where the Buck Converter has Dominated*", PhD Thesis, 2011

[3]. M. Budaes, L. Goras, "*Burst mode switched capacitor voltage converter modeling and design*", Romanian Journal of Information Science and Technology, vol. 11, 2008

[4]. C.-S. Plesa, M. Neag, L. Radoias, "*Design options for thermal shutdown circuitry with hysteresis width independent on the activation temperature*", Advances in Electrical and Computer Engineering, vol.17, no.1, 2017

[5]. M. Pfost, C. Boianceanu, H. Lohmeyer, M. Stecher, „*Electrothermal simulation of self-heating in DMOS transistors up to thermal runaway*", IEEE Transactions on Electron Devices, vol. 60, 2013

Influence of platinum-hydrogen complexes on silicon p+/n-diode characteristics

Jennifer Prohinig[*, **], Fabian Rasinger[*], Hans-Joachim Schulze[***], Gregor Pobegen[*]

*KAI Kompetenzzentrum Automobil- u. Industrieelektronik GmbH, 9524 Villach, Austria

E-mail: jennifer.prohinig@k-ai.at

**Graz University of Technology, Institute of Solid State Physics, 8010 Graz, Austria

***Infineon Technologies AG, 81726 Munich, Germany

Abstract—Deep level impurities in p+/n silicon diodes are investigated using deep-level transient spectroscopy (DLTS). Three different deep levels are observed: two electron traps located 0.23 eV and 0.50 eV below the conduction band as well as a hole trap 0.36 eV above the valence band. The impurities are identified as platinum and platinum-hydrogen related defects. From current voltage (IV) and capacitance voltage (CV) characteristics the generation lifetime and the saturation diffusion current are obtained. A depth profile of the Pt-H complex is calculated by using the reverse IV characteristic. All the measurements are put together in order to show that mid-bandgap traps such as Pt-H complex increase the leakage current in reverse and the non-ideality factor in forward operation.

1. Introduction

The investigation of deep-level traps in semiconductor devices is of high interest due to their potential application as lifetime killers. In 1974 D. Lang [1] proposed a non-destructive, highly sensitive technique for deep-level impurity examination called deep-level transient spectroscopy (DLTS). This method allows the determination of trap parameters such as the trap energy level within the bandgap E_T, the electron (hole) capture cross section σ_n (σ_p) and the trap concentration N_T.

In this paper two similar p+/n Si diodes, in the following referred to as Sample A and Sample B, are investigated using the DLTS technique. Both samples are approximately 40 µm thick, have a phosphorous n-doping concentration N_D of 10^{14} cm^{-3} and a breakdown voltage of about 480 V. The only difference between the two samples is an additional field-stop implanted by using protons [10] in Sample A, which is missing in Sample B. We assign the detected traps to already identified defects in silicon by comparing them to data published in the literature [2, 5, 7, 9]. The influence of the traps on the current voltage (IV) characteristic in forward

and reverse operation is discussed.

2. Experimental results

A. Deep-level transient spectroscopy

DLTS spectra of both p+/n Si diodes are recorded applying a filling pulse V_f of +1 V (forward bias) for 10 ms to include electron as well as hole traps in the measurement and a reverse recovery pulse V_r of -9 V for 100 ms. The filling pulse duration is varied and a duration of 10 ms showed sufficient DLTS signal saturation, which is necessary for complete trap filling. As visible in **Fig. 1,** both diodes show three significant peaks: two electron trap peaks at about 105 K and 240 K and a hole trap peak around 160 K. In the following they are labelled using the type of trap and the corresponding temperature, e.g. E(105) considering the first peak.

It can be concluded that the protons implanted to form the field-stop in Sample A may anneal the defect responsible for the E(240) peak while leaving the other 'main' defects unaffected. Due to the high DLTS signal of peak E(105) its amount is not measureable exactly, as its defect concentration is bigger than the in literature broadly accepted limit in DLTS of $0.1 \cdot N_D$.

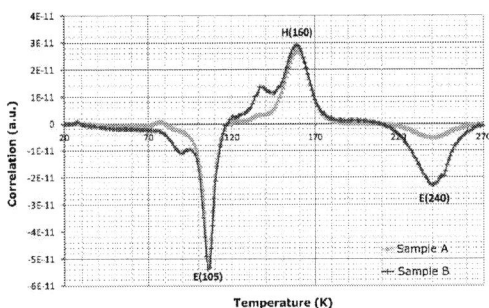

Fig. 1 DLTS spectra of the two investigated p+/n Si diodes calculated using a rectangular correlation function.

Table 1. Trap parameters determined from Arrhenius plots

Property	Level		
	E(105)	*H(160)*	*E(240)*
E_T *(eV)*	$E_C - 0.23$	$E_V + 0.36$	$E_C - 0.50$
σ_n *or* σ_p *(cm²)*	$5 \cdot 10^{-15}$	$2 \cdot 10^{-15}$	$2 \cdot 10^{-16}$
N_T *(cm⁻³)* *Sample A*	$> 0.1 \cdot N_D$	$\sim 9.7 \cdot 10^{12}$	$\mathbf{1.6 \cdot 10^{12}}$
N_T *(cm⁻³)* *Sample B*	$> 0.1 \cdot N_D$	$\sim 1.0 \cdot 10^{13}$	$\mathbf{7.0 \cdot 10^{12}}$
Assignment	Pt	Pt	Pt-H
Reference	$E_C - 0.23$ [2, 7]	$E_V + 0.36$ [9]	$E_C - 0.50$ [5]

Both, the E(105) and H(160) peaks show a shoulder at their low temperature side in the spectrum of Sample B, suggesting further Pt-H$_x$ defects, which appear less pronounced in the spectrum of Sample A.

The calculation of characteristic trap parameters (see **Table 1**) and comparison with [2], [5] and [9] allows an identification of the observed defects. The E(105) and H(160) peak are identified as substitutional Pt defect while the E(240) peak is likely caused by a Pt-H defect complex. As indicated by the different E(240) peak heights the mean concentration N_T of this defect is about four times higher in Sample B compared to Sample A.

B. Current-voltage measurements

To correlate the trap information gained from DLTS with device performance, IV-curves of both diodes are measured. The diode characteristics in forward and in reverse bias are depicted in **Fig. 2** at top and bottom, respectively.

The exponential increase of the forward current below about 0.4 V differs between Sample A and B. By fitting this part of the diode characteristic using the Shockley diode equation for voltages which fulfill $qV >> kT$

$$I = I_S e^{\frac{qV}{\eta kT}} \qquad (1)$$

the ideality factors η of both diodes has been calculated. In (1) I is the diode current, I_S the reverse bias saturation current, q the elementary charge, V the voltage across the diode,

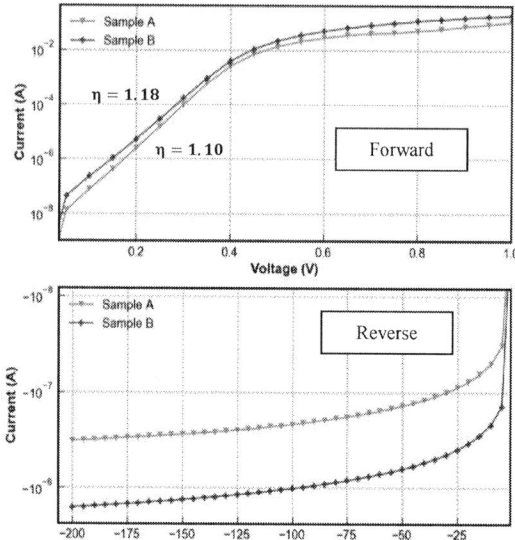

Fig. 2 Forward (top) and reverse (bottom) characteristics.

k the Boltzmann constant and T the temperature. The calculation yields an ideality factor η of 1.10 for Sample A and 1.18 for Sample B, respectively. This means that Sample A shows the more ideal behavior which is due to a lower overall trap concentration. The same trend can be observed in the reverse current characteristic. Sample B shows a reverse current which is significantly higher than the one of Sample A.

C. Capacitance-voltage measurements

The CV curve of both samples is measured. By plotting $1/C^2$ versus the applied voltage (**Fig. 4**) the mean doping concentration can be calculated [3]. Since the CV curves as well as the $1/C^2$ curves are identical, different doping concentrations in our samples as reason for the measured differences in device behavior can be excluded.

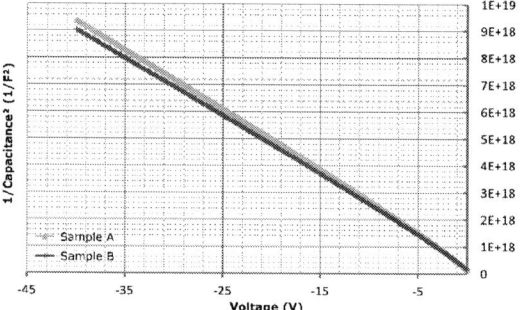

Fig. 3 Capacitance voltage characteristic up to -40 V.

D. Generation lifetime and saturation diffusion current determination

For a pn-junction which obeys the Shockley-Read-Hall theory, the reverse current I_R can be described by the sum of diffusion current I_{diff} and generation current I_{gen}

$$I_R = I_{diff} + I_{gen} \qquad (2)$$

where I_{gen} is defined by

$$I_{gen} = \frac{Aqn_iW_D}{\tau_G} \qquad (3)$$

with the diode area A, the intrinsic carrier concentration of silicon n_i, the depletion width W_D and the carrier generation lifetime τ_G. A method to separate the two current components of I_R is proposed by [4]. It is based on the depletion width dependence of the generation current and requires the IV- and the high frequency CV-curve of the same pn-junction. From the IV curve one calculates the reverse current density J_R and plots it versus the depletion width W_D calculated from the CV curve. This yields a graphical representation of (2) as visible in **Fig. 4**. The carrier generation lifetime τ_G can be derived from the slope of the resulting curve whereas the diffusion current density J_{diff} is given by the y-intercept. The obtained values for the samples observed are listed in **Table 2**.

A detailed description of the separation method can also be found in [6]. It should be noted that this method yields a linear dependence for reverse bias voltages up to around 5 V.

Table 2. Generation lifetime τ_G and saturation diffusion current density J_{diff} are extracted by using the separation method [4].

	τ_G (μs)	J_{diff} (nA/cm²)
Sample A	20.5	3.7
Sample B	5.0	12.6

E. Depth profile of the E(240) defect

From monitoring the reverse current density gradient with respect to depletion width dJ_{gen}/dx, information about the distribution of defect centers responsible for current generation such as the E(240) defect can be obtained [9]. A depth profile of the defect concentration N_T (x) (see **Fig. 5**) is calculated using the relation

$$N_T(x) = \frac{1}{q\xi}\frac{dJ_{gen}}{dx} \qquad (4)$$

with the depletion width x and the emissivity ξ. The emissivities are calculated using the mean trap concentration of the E(240) level from **Table 1** and the derivative of the generation current with respect to the depletion width at bias up to –9 V. For Sample A and Sample B emissivities ξ are calculated as 486 s⁻¹ and 470 s⁻¹, respectively. Both samples show the same trend with respect to depth, a steady increase of the trap concentration, up to a depth of 20 μm. There is a four times lower trap concentration in Sample A (including the field stop implantation on the backside).

While the trap density keeps increasing with depth in Sample B, above 20 μm the

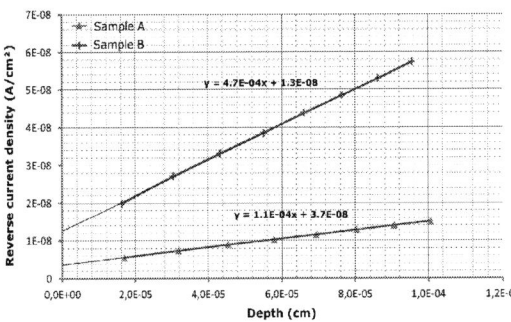

Fig. 4 Graphical representation of (3) using the separation method by [4] up to a reverse bias of -9 V.

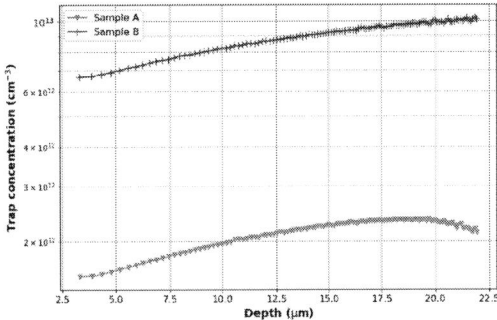

Fig. 5 Calculated depth profile of the E(240) defect center measured from the p+/n-junction [9].

concentration of the Pt-H complex starts to fall in Sample A. This behavior is attributed to the field-stop which may influence the defect concentration of the E(240) defect and create additional defects.

3. Discussion

The evaluation of the DLTS measurements reveals a concentration of 7.0×10^{12} cm^{-3} for the E(240) complex in Sample B which is about four times higher than in Sample A. This defect has been identified as a Pt-H complex with a trap level E_T of E_C–0.50 eV. Since this is about mid-bandgap of Si, this defect can explain the high reverse current as well as the higher non-ideality factor η (see **Fig. 2**).

By using the separation method [4] the generation lifetime as well as the saturation diffusion current has been calculated. Higher trap densities cause a lower τ_G and therefore a higher generation current density J_{gen}, which explains the higher total reverse current in Sample B.

It has been expected that the implantation of protons encourages the formation of Pt-H complexes, thus causes an increase of the associated peaks while reducing the pure Pt peaks in DLTS, but this is not observed in our samples. Considering **Fig. 5** the conclusion is made, that the proton implantation causes several kind of vacancy complexes within the field-stop region which consume hydrogen introduced from the backside and tend to form hydrogen-related donors or other stable higher-order hydrogen-related complexes. Thus the number of hydrogen atoms available for Pt-H complex formation is reduced and this can lower the overall Pt-H defect concentration in Sample A in the drift region.

4. Summary

Investigations of deep level impurities in p+/n Si diodes are performed using DLTS. Three trap levels located at E_C–0.23 eV, E_V+0.36 eV and E_C–0.50 eV are found and are associated with substitutional Pt and Pt-H defects. The influence of single trap levels with high trap concentrations lying at mid-bandgap on device characteristics like forward and reverse current is shown. The depth profile gives additional information about the evolution of the defect concentration of the E(240) trap with respect to depth.

Acknowledgments

This work was funded by the Austrian Research Promotion Agency (FFG, Project No. 863947).

References

[1] D. Lang, „Deep-level transient spectroscopy: A new method to characterize traps in semiconductors,“ *Journal of applied physics*, Bd. 45, Nr. 7, pp. 3023-3032, 1974.

[2] Brotherton, S. D., P. Bradley, and J. Bicknell, „Electrical properties of platinum in silicon,“ *Journal of Applied Physics*, Bd. 50, Nr. 5, pp. 3396-3403, 1979.

[3] Hillibrand, J. and Gold, R. D., „Determination of the impurity distribution in junction diodes from capacitance-voltage measurements,“ *Semiconductor Devices: Pioneering Papers*, pp. 191-198, 1991.

[4] Murakami, Y. and Shingyouji, T., „Separation and analysis of diffusion and generation components of pn junction leakage current in various silicon wafers,“ *ournal of applied physics*, Bd. 75, Nr. 7, pp. 3548-3552, 1994.

[5] Sachse, J. U., Sveinbjörnsson, E. Ö., Yarykin, N. and Weber, J., „Similarities in the electrical properties of transition metal–hydrogen complexes in silicon,“ *Materials Science and Engineering: B*, Bd. 58, Nr. 1-2, pp. 134-140, 1999.

[6] Simoen, E., Claeys, C., and Vanhellemont, J., „Defect analysis in semiconductor materials based on pn junction diode characteristics.“ *Defect and Diffusion Forum*, Bd. 261, pp. 1-24, 2007.

[7] Mantovani, S., Nava, F., Nobili, C., Conti, M., & Pignatel, G., „Thermal diffusion of Pt in silicon from PtSi,“ *Applied Physics Letters*, Bd. 44, Nr. 3, pp. 328-330, 1984.

[8] Hazdra, P., Rubeš, J. and Vobecký, J., „Divacancy profiles in MeV helium irradiated silicon from reverse I–V measurement,“ *Nuclear Instruments and Methods in Physics Research Section B: Beam Interactions with Materials and Atoms*, Bd. 159, Nr. 4, pp. 207-217, 1999.

[9] Kwon, Y. K., Ishikawa, T. and Kuwano, H., „Properties of platinum-associated deep levels in silicon,“ *Journal of applied physics*, Bd. 61, Nr. 3, pp. 1055-1058, 1987.

[10] Niedernostheide, F. J., Schulze, H. J., Felsl, H. P., Hille, F., Laven, J. G., Pfaffenlehner, M., ... and Schustereder, W., „Tailoring of field-stop layers in power devices by hydrogen-related donor formation,“ in *Power Semiconductor Devices and ICs (ISPSD)*, 2016.

Numerical Simulations of Radiation Damage Effects in Active-Edge Silicon Pixel Sensors for High-Energy Physics Experiments

D. Djamai*, E. Leonidas Gkougkousis**, M. Chahdi***, A. Lounis****, S. Oussalah*****

*Laboratoire d'Ingénierie et Sciences des Matériaux Avancés (ISMA), Université Abbes Laghrour Khenchela, Algeria
djamai@lal.in2p3.fr
**Institut de Fisica d'Altes Energies (IFAE), Barcelona, Spain
egkougko@cern.ch
*** Université de Batna1, Algeria
chahdi.mohamed@yahoo.fr
****Laboratoire de l'Accélérateur Linéaire, Université Paris-Sud XI, CNRS/IN2P3, Orsay, France
lounis@lal.in2p3.fr
*****Centre de Développement des Technologies Avancées (CDTA), Algiers, Algeria
soussalah@cdta.dz

Abstract—*High-energy physics experiments at the future CERN High Luminosity LHC (Large Hadron Collider) require highly segmented pixelated sensors of increased geometrical efficiency and the ability of withstanding extremely high radiation damage. The performance of planar n-on-p sensors with active edges is simulated at very high fluences (2×10^{16} n$_{eq}$/cm^2), using a recent three level trap model for p-type silicon material. Precise structural definition is achieved by investigating the doping profile of the devices via the Secondary Ion Mass Spectrometry technique. The breakdown voltage, and hole density distribution are studied as a function of radiation fluences.*

Keywords—Active edge sensor; radiation damage; TCAD simulation.

1. Introduction

Pixel silicon sensors are the most precise instruments for charged particle tracking currently in use at high-energy physics 14 TeV proton-proton collision experiments at CERN[*] large hadron collider (LHC) in Geneva, Switzerland. Closely located to the interaction point, they are required to function in a radiation harsh environment [1]. It is hence necessary for such sensors to demonstrate an increase radiation tolerance [2] as well as to maintain a good performance at high beam luminosity conditions ($l = 10^{35}$ cm^{-2}s^{-1}) for fluences ranging up to 2.2×10^{16} n$_{eq}$/cm$^{2[\dagger]}$ at the inner layers. A major concern is increasing overall sensitive detection area while maintaining radiation tolerance. Several technologies address these issues,

mainly combining a p-bulk process sensor with active edge [3-4] or optimized bias rail geometries.

The main objective of this work is to study a new detector structure implementing n-on-p active edges and analyze the electrical characteristics of the device using Secondary Ion Mass Spectroscopy (SIMS) data as input for the simulation algorithm. Breakdown voltage, leakage current, and charge carrier distributions (holes, electrons) are presented. The ATLAS simulator of Silvaco™ Technology Computer Aided Design (TCAD) software [5] is used for the simulation and study of the electrical characteristics of the pixel device.

This paper is organized as follows. The planar active edge sensor is presented in Section 2. Section 3 presents the TCAD simulator as well as the doping profiles measured in the structure through SIMS techniques. Section 4 presents the radiation trapping model used in this work. The obtained results are presented in Section 5. Finally, some concluding remarks are provided in the last Section.

2. Structure Definition

A schematic cross-section of the simulated device is presented in *Fig 1*. In this work a

---

[*] European Organization for Nuclear Research

[†] All radiation fluences are normalized to the standard 1MeV neutron equivalent fluence.

high resistivity (>10 kΩ-cm) p-type silicon substrate is used in an n-on-p configuration with a total thickness of 150 μm. The distance between the edge and the first pixel is set at 100 μm, while the top layer metal and SiO₂ layer thicknesses were set at 1.5 μm and 600 nm respectively. A layout detail of the structure is plotted in *Fig 2*. The particular geometry considered here was produced by ADVACAM Ltd [6].

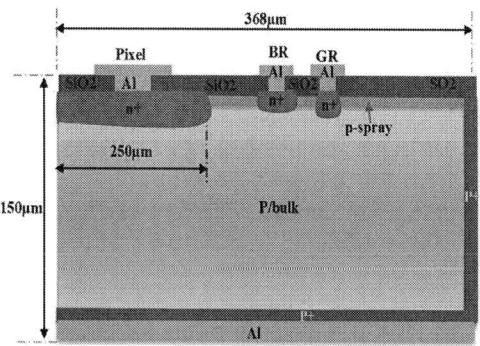

Fig. 1 Schematic cross-section of n-in-p planar active-edge pixel sensor with one guard-ring and bias rail.

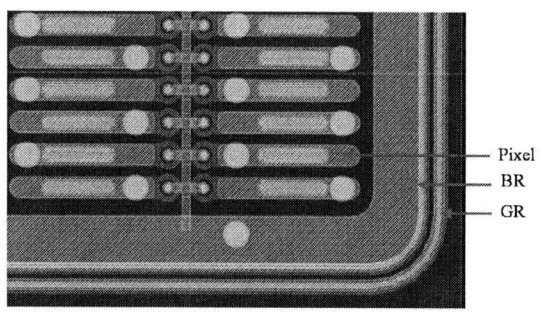

Fig. 2 Layout of a corner of the pixel sensor.

3. TCAD Simulation

Simulations presented in this paper are carried out using the Silvaco™ finite element method implemented within the corresponding TCAD software framework.

A. Theoritical Model

The basic framework for device simulation is provided by coupling the continuity and diffusion equations (*Eq. 1* and *Eq. 2*) to the Poisson's equation (*Eq. 3*).

$$\frac{\partial p}{\partial t} = \nabla . D_h \nabla p + \nabla .(p\mu_h \vec{E}) + G_h - \tau_h \quad (1)$$

$$\frac{\partial n}{\partial t} = \nabla . D_e \nabla n - \nabla .(n\mu_e \vec{E}) + G_e - \tau_e \quad (2)$$

$$-\nabla^2 V = \nabla . \vec{E} = \frac{q(n - p - N_D^+ + N_A^-) - Q_T}{\varepsilon} \quad (3)$$

where p and n denote the carrier concentrations of holes and electrons, respectively, D and μ, their respective diffusion coefficient and mobility. N_D^+ and N_A^- are the ionized donor and acceptor impurity concentrations, respectively. Q_T refers to the charge concentration due to traps and defects. $G_{h,e}$ and $\tau_{h,e}$ correspond to the generation and recombination rates, respectively, for holes h and electrons e respectively.

The system is solved using a set of physical models to obtain the different electrical characteristics of the semiconductor device. The dynamic behavior of traps is described by the Shockley-Read-Hall recombination model, which depends on trap concentration, energy, and carrier cross section [7].

B. SIMS as Technical Tool to Improve the Reliability of Process Simulation

Secondary Ion Mass Spectroscopy (SIMS) [8] is a very powerful tool allowing a detailed characterization of the dopant profile. Measurements presented in this study were conducted at the GEMAC‡ CNRS laboratory using the Cameca IMS-7F system. Further details are available in reference [9].

In *Fig. 3*, the doping profiles implemented in the simulation for the pixel (phosphorus), BR, and GR regions are presented, as obtained through SIMS measurements. At the same figure the profile for the p-spray (boron) areas is also overlapped.

‡Groupe d'Etude de la Matière Condensée. Université de Versailles-Saint Quentin en Yvelines

978-1-5386-4483-6/18 $31.00 © 2018 IEEE

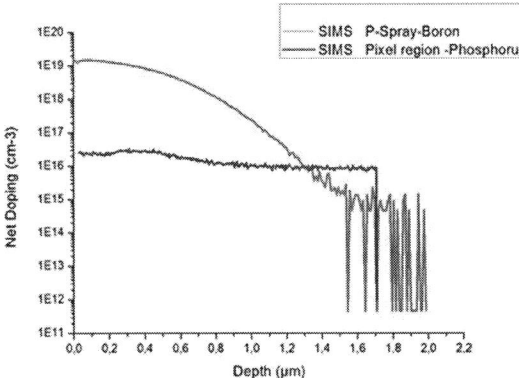

Fig. 3 Doping profiles measured in the active edge sensors provided by "ADVACAM Ltd" through SIMS techniques. The same doping profile have been used in the TCAD simulations.

4. Radiation Damage Modeling

It is well established that exposure of planar pixel sensors to irradiation in the fluence range from 10^{15} n_{eq}/cm^2 to 2×10^{16} n_{eq}/cm^2 significantly affects the electrical properties of the device. The radiation damage model used in this analysis is based on the three level traps model [10], where irradiation generates two acceptor levels, positioned slightly above the mid band gap, and one donor level, located far below the mid band gap. The details of the trap model are presented in *Tables 1* and *2*.

Table 1. Trap model parameters for fluences up to $7 \times 10^{+15}$ n_{eq}/cm^2.

Defect	E(eV)	σ_e (cm^{-2})	σ_n (cm^{-2})	η
Acceptor	Ec-0.42	10^{-15}	10^{-14}	1.6
Acceptor	Ec-0.46	7×10^{-15}	7×10^{-14}	0.9
Donor	Ev+0.36	3.23×10^{-13}	3.23×10^{-14}	0.9

Table 2. Trap model parameters for fluences between $7 \times 10^{+15}$ n_{eq}/cm^2 and $2.2 \times 10^{+16}$ n_{eq}/cm^2.

Defect	E(eV)	σ_e (cm^{-2})	σ_n (cm^{-2})	η
Acceptor	Ec-0.42	10^{-15}	10^{-14}	1.6
Acceptor	Ec-0.46	3×10^{-15}	3×10^{-14}	0.9
Donor	Ev+0.36	3.23×10^{-13}	3.23×10^{-14}	0.9

It is found that silicon dioxide is also damaged by irradiation. In this analysis, we take into account the damage inflicted to the SiO_2 layer by introducing a positive charge

near the interface. For a good quality SiO_2 layer with an initial charge density at the interface layer in the order of 2×10^{11} cm^{-2} (for a non-radiated detector), the result demonstrates that the final amount of interface charge can exceed 3×10^{12} cm^{-2} at saturation level for a heavily irradiated structure [11].

5. Results

To get a better understanding of the phenomenon involved, the Silvaco^{TM} TCAD framework is used to simulate the hole concentration for an active edge sensor biased at -150 V for different fluences as shown in *Fig. 5*. Obtained results show that the non-depleted region increases with the radiation dose used. This region starts from the corner between the active edge and the backside and extends to the pixel region of the structure. As the fluence increases, the entire population of holes diffuses deeper inside the bulk, thus allowing for higher breakdown voltage. The simulated breakdown voltages were -165 V and -235 V for a non- irradiated sensor and irradiated sensor with fluence of 2×10^{16} n_{eq}/cm^2, respectively. This information gives an indication of less probable breakdown risk for low depletion voltages in high radiation level environments.

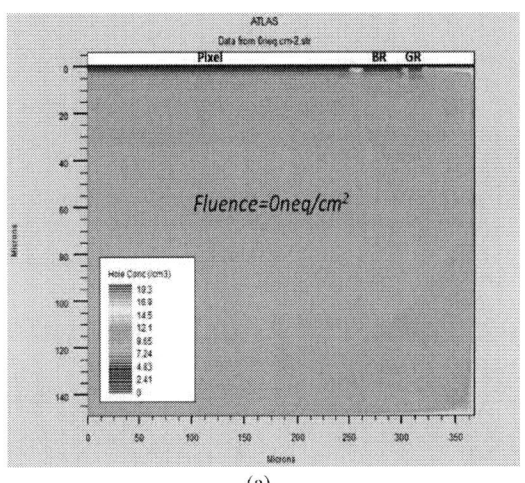

(a)

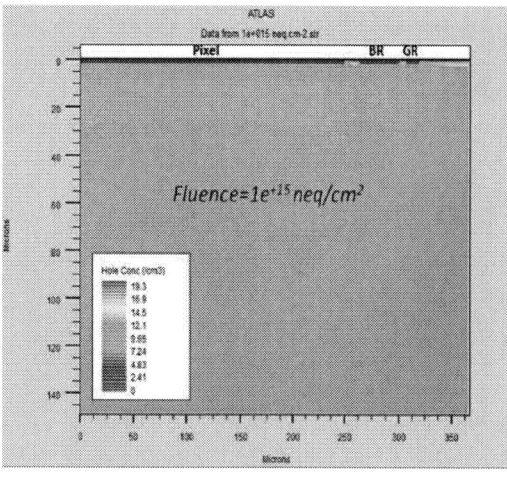

(b)

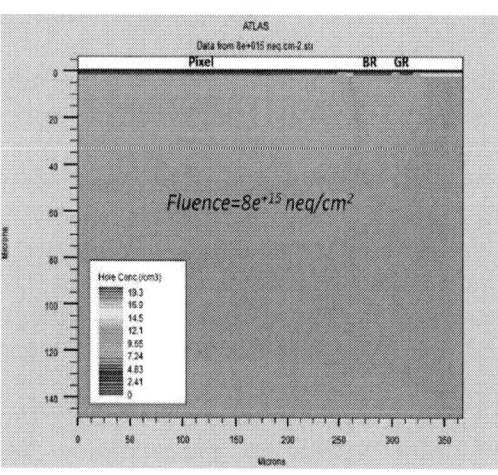

(c)

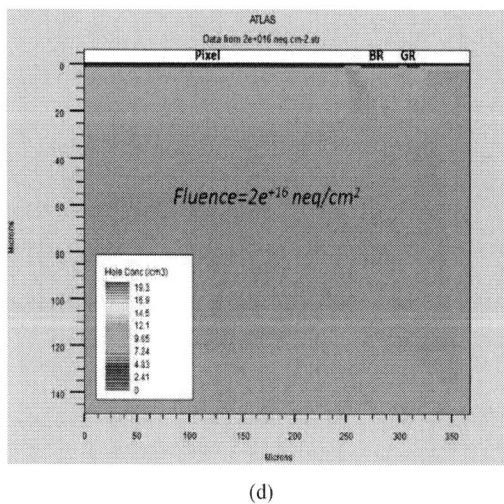

(d)

Fig. 4 Simulated holes distribution for different radiation fluences at -150 V bias voltage, a) unirradiated, b) 10^{15} n_{eq}/cm^2, c) 8×10^{15} n_{eq}/cm^2, d) 2×10^{16} n_{eq}/cm^2.

6. Conclusions

In this work TCAD simulation was used to study the effects of radiation damage in active edge sensors in detail the variation of the active region. Doping profile data for phosphorous and boron spray extracted from SIMS measurements were used for accurate structure definition. It was found that, with increasing radiation fluences, the non-depleted regions start to evolve from the structure's lower corner. Significant effect on edge efficiency is visible from fluences of 10^{15} n_{eq}/cm^2, were edge performance is comparable to a standard non-active edge sensor. This result may help not only to improve the design of edgeless planar pixel sensors but also to avoid risk of breakdown at high radiation doses.

References

[1] L. Evans and P. Bryant, "*LHC Machine*", JINST, vol. 03, S08001, August 2008.

[2] M. Moll, "*Radiation tolerant semiconductor sensors for tracking detectors*", Nucl. Instr. and Meth. A., vol. 565, pp. 202-211, 2006.

[3] G.-F. Dalla Betta et al., "*Design and TCAD simulation of planar p-on-p active edge pixel sensors for the next generation of FELs*", Nucl. Instr. and Meth. A., vol. 824, pp. 384-385, 2016.

[4] M. Mekheldi, S. Oussalah, A. Lounis, and N. Brihi, "*Comparison of electrical performances of two n-in-p detectors with different implant type of guard ring by TCAD simulation*", Results In Physics, Elsevier, Vol. 6, pp. 80-81, Feb. 2016.

[5] Device Simulation Software, SILVACO Inc., Santa Clara, CA, 95054, 2010., http://www.silvaco.fr/products/tcad/index.html.

[6] VTT Technical Research Centre of Finland, P.O. Box 1000, FI-02044 VTT, Finland

[7] Atlas User's Manual, Device Simulation Software, SILVACO Inc., Santa Clara, CA, 95054, 2010.

[8] E. Gkougkousis, PhD thesis, LAL15-426, February 2016

[9] AMETEK, Inc - CAMECA SAS. www.cameca.com

[10] D.Passeri et al., "*Modeling of radiation damage effects in silicon detectors at high fluences HL-LHC with Sentaurus TCAD*", Nucl. Instr. and Meth. A., vol. 824, pp. 443-445, 2016.

[11] M. Benoit, A. Lounis, and N. Dinu, "*Simulation of Radiation Damage Effects on Planar Pixel Guard Ring Structure for ATLAS Inner Detector Upgrade*", IEEE Transactions on Nuclear Science, vol. 56, no. 6, pp. 3236-3243, June 2009.

Investigation of carbon interstitials in the vicinity of the SiO_2/4H-SiC(0001) interface

Hind Alsnani[1], JP Goss[1], Oras Al-Ani[1], SH Olsen[1], PR Briddon[1], MJ Rayson[1] and AB Horsfall[2].

[1]School of Engineering, Newcastle University, United Kingdom

[2]Department of Engineering, Durham University, United Kingdom

E-mail: h.a.t.alsnani2@ncl.ac.uk

Abstract—Native defects, including carbon vacancies and interstitials, are understood to be important defects that degrade device performance. In particular, during oxidation to form SiO_2 dielectric layers, carbon interstitials are thought to be injected into the SiC channels. Using density functional theory, the kinetics of diffusion of interstitial from the SiO_2/4H-SiC(0001) interface has been investigated. The results show that the injection of a carbon interstitial from a site at the 4H-SiC-SiO_2 interface is hindered by an additional 1 eV relative to the migration barrier of the native defect in bulk 4H-SiC.

Index Terms—SiO_2/4H-SiC, Density Functional Theory, Carbon-Interstitial, Diffusion.

I. INTRODUCTION

CARRIER lifetime is a significant parameters to achieve effective conductivity modulation, but this quantitiy is sensitive to temperature, doping and defects that introduce trapping levels [1]. 4H-SiC devices often exhibit relatively low carrier lifetimes, with a typical carrier lifetime in commercially available SiC epilayers of the order of $1\mu s$. This value is dominated by the effect of the $Z_{1/2}$ trap concentration of about $10^{13}\,cm^{-3}$. with a carrier lifetime of over $5\mu s$ desirable for high performance bipolar devices [2]. However, since very long carrier lifetime lead to large switching losses [1], it is highly desirable to have control of the concentration of deep level defects with the active region of a SiC device [3]–[10].

Two of the most significant deep level defects in as-grown n-type 4H-SiC are the $Z_{1/2}$ and $EH_{6/7}$ centres, with trap levels located at 0.56 eV and 1.50 eV, respectively, relative to the conduction band minimum. $Z_{1/2}$ has been observed in a number of experimental studies using techniques including deep level transient spectroscopy (DLTS), and this centre is considered to be the dominant carrier-lifetime killer in MOSFET (metal-oxide–semiconductor field-effect transistors) devices [4]. Microwave photoconductance decay [4], [7], [10] suggests high concentrations of $Z_{1/2}$ reduce the carrier lifetime by more than an order of magnitude [3], [4]. There is compelling evidence [6]–[9] that $Z_{1/2}$ is linked to the concentration of carbon vacancies in the material, which are incorporated as a consequence of the growth conditions.

Introduction of carbon interstitials (C_i) responsible for reduction $Z_{1/2}$ and $EH_{6/7}$ centres has been reported [7], but additional deep levels labelled ON1 and ON2 that lie 0.84 eV and 1.1 eV below the conduction band [3], [4] are generated during the thermal oxidation process of n-type 4H-SiC that acts as the generation process for the carbon interstitials. It is generally agreed from experiment that the controlling processing for devices and the subsequent role of these defects in the active regions of SiC devices is a significant challenge. During oxidation of SiC, the formation of SiO_2 necessitates the removal of carbon, which may occur processes such as the formation and outgassing of volatile species such as CO and CO_2, the trapping of carbon within the oxide, or may result in an increased concentration of carbon within the SiC. Interstitial interstitials C_i are thought [3] to result with high concentration of electrical traps at the SiO_2/4H-SiC interface [11], and may be directly associated with the ON1 and ON2 centres. Thus, migration of C_i in the vicinity of SiO_2/4H-SiC interfaces is likely to have a significant impact on performance of devices such as MOSFETs [4], [9].

In this paper, a quantum-chemical investigation of C_i in the vicinity of a model interface between SiO_2 and the 4H-SiC (0001) face is presented, in which the effect of the interface on C_i diffusion is highlighted The results of density function theory (DFT) simulations show that the barrier to migration of the carbon spilt-interstitial (C-C)$_C$ into the SiC from the interface is increased by 1 eV relative to the comparable bulk process.

II. COMPUTATION METHOD

Simulations employ DFT as implemented in AIMPRO [12], with atom cores modelled using norm-conserving, separable pseudo potentials [13]. Kohn-Sham eigen-functions are expanded using atom-centred Gaussian basis sets [14] consisting of independent sets of s- and p-type Cartesian Gaussian functions centred at atomic co-ordinates, with 4 different widths for Si, C, and O, and 3 for H. 2(4) additional sets of d-type orbitals are included for Si and C (O) to account for polarization. Matrix elements of the Hamiltonian are calculated using a plane wave expansion of the density and Kohn-Sham potential [15] with a cutoff of 150 Ha. Structures are optimized by a conjugate gradients algorithm, and convergence of the total energy taken when forces are less than 10^{-3} atomic units. The final structural optimization step is required also to result in a reduction in the total energy of less than 10^{-5} Ha. All systems are modelled using periodic boundary conditions.

978-1-5386-4483-6/18 $31.00 © 2018 IEEE

For the defects in bulk 4H-SiC 72-atom cells made up from $(3 \times 3 \times 1)$ primitive, 8-atom cells [16]. The calculated bulk lattice-constants of 4H-SiC are $a = 3.06\,\text{Å}$ and $c = 10.04\,\text{Å}$, in excellent agreement with experiment [12].

The structure of SiO_2/4H-SiC interface is modelled with finite thickness slabs separated by vacuum. A SiO_2 layer built on top of the k-site Si-face of a 4H-SiC slab [17], with the bottom, C-face of the slab saturated by hydrogen to remove surface states. The lower surface is held fixed during optimization and diffusion barrier calculations to mimic the long-zone is sampled using Monkhorst-Pack grids [16]. For the bulk reference cells, a grid of $(2 \times 2 \times 2)$ k-points is used, whereas for the interface simulations a -point approximation is employed. Simulations with in-plane sampling show that both the energies and structures are not significantly affected by this approach due to the relatively large in-plane length-scales and the absence of dispersion in the surface normal directions. The migration paths have been calculated using the climbing nudged elastic band method [18].

III. RESULTS AND DISCUSSION

A. C_i in bulk 4H-SiC

Bulk 4H-SiC has two, non-equivalent host sites for each chemical species, which are conventionally labelled k (cubic-like) and h (hexagonal-like) [19]. This means that for each point defect there are at least two distinct forms. Additionally, many point defects will exhibit orientationally distinct forms at each of the h- and k-sites, depending upon the internal structure and orientation relative to the c-axis of the underlying crystal. We have focused upon the lowest energy conformers at each of the h- and k-sites in our determination of the diffusion of C_i in the c-axis direction, this direction being the key for inject of these point defects from the interface layer in our model. The ground state of a single interstitial carbon atom in 4H-SiC bulk is understood to form a split-interstitial configuration [20], although a wide range of structures for carbon in SiC have been explored [21], including non-bonded 2 and sp^3 bonding, respectively. C_i split-interstitials involve both sp^2 and sp^3 hybridizations, depending upon whether the interstitial is located at the h- or k-carbon site. Other interstitial configurations are less energetically favourable or unstable [20], but may represent configurations passed through during diffusion.

We have investigated the motion of C_i diffusing from a k-site to a h-site, and then to a second k-site to represent a complete process in the c-axis direction. The initial k-site and intermediate h-site structures. The minimum energy path between these structures in bulk 4H-SiC yields a barrier of $0.86\,\text{eV}$, and the energies are plotted in Fig. 1. This estimate for the diffusion barrier agrees well with literature values [22]. This process involves the motion of one of the carbon atoms in the initial split-interstitial configuration moving to the adjacent basal-plane. The energies of the split-interstitial configurations at the k- and h-sites do not differ significantly in energy. Based upon the agreement of these energetics with the literature,

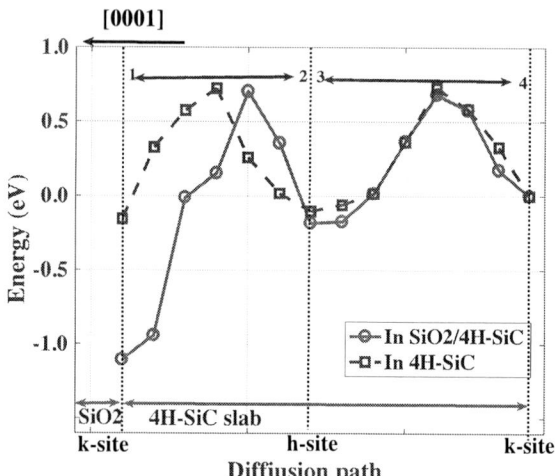

Fig. 1. Minimum energy paths for C_i in bulk 4H-SiC (dashed red) and SiO_2/4H-SiC interfaces (solid blue). To facilitate comparison, the zero on the energy scale is defined as the k-site energy at the end of the process. 1-4 are as indicated in the labeling if Fig. 2.

we have used the same approach to examine diffusion in the vicinity of the SiO_2/4H-SiC interface.

B. $(C\text{-}C)_C$ in SiO_2/4H-SiC

The processes corresponding to those investigated for bulk 4H-SiC have been examined in the SiO_2/4H-SiC system, as shown in Fig. 2. The process is taken to start with the carbon interstitial in the interfacial layer. The final configuration places C_i at a k-site a lattice constant from the interface, but sufficiently far from the lower face of the slab for it to have no significant affect. The final structure (labelled 4 in Fig. 2) is considered to be close to bulk, and we therefore use this configuration to define the zero on the energy scale. In so doing, it can be seen in Fig. 1 that the energy of C_i in the interfacial layer is lower than the corresponding bulk structure by around 1 eV.

Although there is a significant impact upon the energetics, the underlying geometry of the split-interstitials at each site (k and h) is the same as found in bulk 4H-SiC, and in particular the combination of 3- and 4-fold coordination of the carbon atoms in each split-interstitial configuration is as found for the bulk comparators.

Comparison of the bulk and interfacial energy profiles is presented in Fig. 1, in which there are some notable features. The overall energy barrier of the process labelled 1–2, the maximum energy is approximately the same for both interfacial and bulk processes. For the second stage (the process labelled 3–4 in Fig. 2) both the overall barrier shape and the absolute energies match well, supporting the view that the environment beyond the first Si-C bi-layer from the SiO_2/4H-SiC interface is very similar to bulk SiC. It has been reported that there is experimental evidence [23] that interstitial carbon produces trap states, and we have found that the region

978-1-5386-4483-6/18 $31.00 © 2018 IEEE

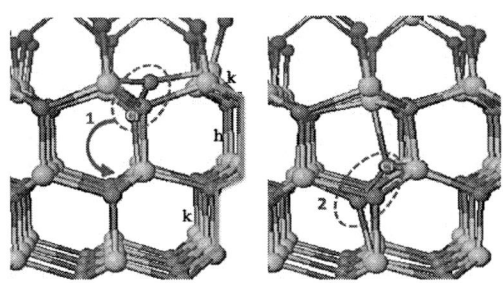

First diffusion process $\xrightarrow{step1} (C-C)^k \xrightarrow{step\ 2} (C-C)^h$

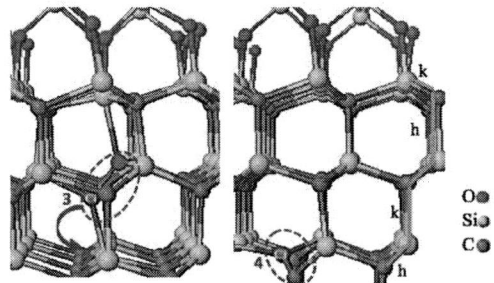

O	●
Si	●
C	●

Second diffusion process $\xrightarrow{step3} (C-C)^k \xrightarrow{step\ 4} (C-C)^h$

Fig. 2. Illustration of C_i diffusion at the $SiO_2/4H$-SiC interface. (1) as initial step, and (2) final step between first and second layers in 4H-SiC slab. while (3) and (4) step; showing the diffusion process of $(C$-$C)_C$ between the initial (3) and final (4) structures.

immediately adjacent to the SiO_2 layer is an energetically favourable environment for C_i. This being the case, the buildup on carbon in this region is expected to give rise to a potentially high concentration of interfacial traps.

IV. CONCLUDING REMARKS

Using a DFT approach, the diffusion processes of C_i in the vicinity of the $SiO_2/4H$- SiC interface has been simulated and the results presented in this paper. We find that the effect upon the carbon-interstitial in terms of geometry and due to the interface is very minor where the interstitial is located at beyond the interface layer, with the diffusion barrier very similar to to that obtained for bulk 4H-SiC. In contrast, the first layer is affected strongly, with the interstitial being energetically stabilised by about 1 eV relative to when in bulk for our model interface. This energetic stabilisation is expected to inhibit the injection of single self interstitials into bulk 4H-SiC, although the processes by which oxidation results I modification of deep traps within the SiC remain to be fully explored and understood. We conclude that the control over the injection of self-interstitials to modify the profile of trap states within the channel, and therefore the carrier lifetime, in devices such as MOSFETs is rate-limited by this initial step.

V. ACKNOWLEDGMENTS

The authors thank the Omm Al-Qura University (KSA) for sponsoring HA's PhD study.

REFERENCES

[1] K. Danno, D. Nakamura, and T. Kimoto, "Investigation of carrier lifetime in 4H-SiC epilayers and lifetime control by electron irradiation," *Appl. Phys. Lett.*, vol. 90, no. 20, 2007.

[2] K. Kawahara, X. T. Trinh, N. T. Son, E. Janzén, J. Suda, and T. Kimoto, "Investigation on origin of $Z_{1/2}$ center in SiC by deep level transient spectroscopy and electron paramagnetic resonance," *Appl. Phys. Lett.*, vol. 102, no. 11, 2013.

[3] K. Kawahara, J. Suda, and T. Kimoto, "Analytical model for reduction of deep levels in SiC by thermal oxidation," *J. Appl. Phys.*, vol. 111, no. 5, p. 053710, 2012.

[4] T. Kimoto, K. Danno, and J. Suda, "Lifetime-killing defects in 4H-SiC epilayers and lifetime control by low-energy electron irradiation," *Phys. Status Solidi*, vol. 245, no. 7, pp. 1327–1336, 2008.

[5] K. Kawahara, "Identification of Deep Levels in SiC and Their Elimination for Carrier Lifetime Enhancement," Ph.D. dissertation, Kyoto University, Japan, 2013.

[6] N. T. Son, X. T. Trinh, L. S. Løvlie, B. G. Svensson, K. Kawahara, J. Suda, T. Kimoto, T. Umeda, J. Isoya, T. Makino, T. Ohshima, and E. Janzén, "Negative-U System of Carbon Vacancy in 4H-SiC," *Phys. Rev. Lett.*, vol. 109, no. 18, p. 187603, oct 2012.

[7] L. Storasta and H. Tsuchida, "Reduction of traps and improvement of carrier lifetime in 4 h-si c epilayers by ion implantation," *Appl. Phys. Lett.*, vol. 90, no. 6, p. 062116, 2007.

[8] K. Kawahara, X. Thang Trinh, N. Tien Son, E. Janz??n, J. Suda, and T. Kimoto, "Quantitative comparison between Z1/2 center and carbon vacancy in 4H-SiC," *J. Appl. Phys.*, vol. 115, no. 14, 2014.

[9] S. Ichikawa, K. Kawahara, J. Suda, and T. Kimoto, "Carrier Recombination in n-Type 4H-SiC Epilayers with Long Carrier Lifetimes," *Appl. Phys. Express*, vol. 5, no. 10, p. 101301, oct 2012.

[10] T. Kimoto, T. Hiyoshi, T. Hayashi, and J. Suda, "Impacts of recombination at the surface and in the substrate on carrier lifetimes of n-type 4h–sic epilayers," *J. Appl. Phys.*, vol. 108, no. 8, p. 083721, 2010.

[11] T. Zheleva, A. Lelis, G. Duscher, F. Liu, I. Levin, and M. Das, "Transition layers at the SiO_2/SiC interface," *Appl. Phys. Lett.*, vol. 93, no. 2, p. 022108, 2008.

[12] P. R. Briddon and R. Jones, "LDA Calculations Using a Basis of Gaussian Orbitals," *Phys. Status Solidi B*, vol. 131, pp. 131–171, 2000.

[13] C. Hartwigsen, S. Gœdecker, and J. Hutter, "Relativistic separable dual-space Gaussian pseudopotentials from H to Rn," *Physical Review B*, vol. 58, no. 7, p. 3641, 1998.

[14] J. P. Goss, M. J. Shaw, and P. R. Briddon, "Marker-method calculations for electrical levels using Gaussian-orbital basis sets," *Top. Appl. Phys.*, vol. 104, pp. 69–94, 2006.

[15] M. J. Rayson and P. R. Briddon, "Highly efficient method for kohn-sham density functional calculations of 500–10000 atom systems," *Physical Review B*, vol. 80, no. 20, p. 205104, 2009.

[16] H. J. Monkhorst and J. D. Pack, "Special points for Brillouin-zone integrations," *Phy. Rev. B*, vol. 13, no. 12, pp. 5188–5192, 1976.

[17] S. Salemi, "Electronic structure of sic/sio2 by density functional theory," Ph.D. dissertation, University of Maryland, USA, 2012.

[18] G. Henkelman and H. Jónsson, "Improved tangent estimate in the nudged elastic band method for finding minimum energy paths and saddle points," *J. Chem. Phys.*, vol. 113, no. 22, pp. 9978–9985, 2000.

[19] T. Hornos, A. Gali, and B. G. Svensson, "Large-Scale Electronic Structure Calculations of Vacancies in 4H-SiC Using the Heyd-Scuseria-Ernzerhof Screened Hybrid Density Functional," *Mater. Sci. Forum*, vol. 679-680, pp. 261–264, 2011.

[20] F. Gao, M. Posselt, V. Belko, Y. Zhang, and W. J. Weber, "Structures and energetics of defects: a comparative study of 3C- and 4H-SiC," *Nucl. Instrum. Methods B*, vol. 218, pp. 74–79, 2004.

[21] T. Liao and G. Roma, "Stability of neutral silicon interstitials in 3c- and 4h-sic: A first-principles study," *Defect and Diffusion Forum*, vol. 283, pp. 74–83, 2009.

[22] M. Bockstedte, A. Mattausch, and O. Pankratov, "Ab initio study of the annealing of vacancies and interstitials in cubic SiC: Vacancy-interstitial recombination and aggregation of carbon interstitials," *Phys. Rev. B*, vol. 69, no. 23, p. 235202, 2004.

[23] X. Shen, M. P. Oxley, Y. Puzyrev, B. R. Tuttle, G. Duscher, and S. T. Pantelides, "Excess carbon in silicon carbide," *J. Appl. Phys.*, vol. 108, no. 12, p. 123705, 2010.

Session N&MN

NANOSCIENCE; MICRO AND NANOPHOTONICS
(Poster session)

978-1-5386-4483-6/18 $31.00 © 2018 IEEE

Graphene and TiO₂ - PVDF nanocomposites for potential applications in triboelectronics

P. Pascariu[1], I. V. Tudose[2], C. Pachiu[3], M. Danila[3], O. Ionescu[3], M. Popescu[3], E. Koudoumas[2], M. Suchea[2,3]*

[1] "Petru Poni" Institute of Macromolecular Chemistry, Iaşi, Romania
[2] Center of Materials Technology and Laser, School of Engineering, Technological Educational Institute of Crete, Heraklion, Greece
[3] National Institute for Research and Development in Microtechnologies (IMT-Bucharest), Bucharest, Romania
* mira.suchea@imt.ro; mirasuchea@staff.teicrete.gr

Abstract— Novel graphene and titanium dioxide-polyvinylidene fluoride (PVDF) based nanocomposite materials (G-TiO2-PVDF) fabricated by electrospinning technique and their properties will be presented. The new composite materials show enhanced properties that make them candidates for large range of applications including triboelectronics.

Keywords—composites, graphene, electrospinning .

1. Introduction

PVDF is used in critical applications requiring an excellent chemical resistance, a high degree of purity and excellent mechanical properties. PVDF has a very good creep resistance that is superior to that of other fluoropolymers, it is UV resistant and has a high dielectric constant.

Nanocomposites represent an interesting class of materials because their applications are of multidisciplinary importance. Interactions at the interface of heterostructures, are leading to superior performance and sometimes to synergistic interactions. Graphene-based materials are among the most intriguing materials for researchers in recent times due to their exotic properties which find applications in various domains right from sensors to textiles, pharmaceuticals, intelligent coatings and biomedical applications. When such materials are incorporated in polymer matrices to form either composites or doped-polymers or simply carbon-reinforced-polymers, their applications broaden encompassing aerospace structures, sports equipments, auto vehicle parts, organic solar panels, supercapacitors, smart sensors, intelligent paints, electromagnetic interference shielding (EMS) and so on.

Charges induced in triboelectric effect are usually referred as a negative effect either in scientific research or technological applications, and they are wasted energy in many cases. Recently, the triboelectric nanogenerator based on coupling of a triboelectric effect and electrostatic induction has been extensively developed for harvesting mechanical energy from ambient environment. A triboelectric nanogenerator can directly translate mechanical energy into electrical energy and power the functional electronics instantly unlike a traditional chemical battery with limited lifespan and the need for frequent replacement or recharging. Many approaches based on different mechanisms have been demonstrated for effectively harvesting mechanical energy, such as electromagnetic generators [1, 2], piezoelectric nanogenerators [3−12] and electrostatic generators [13]. According to the literature, PVDF has a significant ability to gain electrons owing to the large electronegativity of the fluorine groups fact that makes it suitable for energy harvesting systems, self-powered anti-corrosion system and UV detectors among other applications [14, 15]. For fluorinated polymers, the charge density can be enhanced via controlling the polarization state by either using external electric field or via the use of suitable additives and processes to yield high polarization. The present communication report on properties of novel graphene and

978-1-5386-4483-6/18 $31.00 © 2018 IEEE

titanium dioxide-PVDF based nanocomposite materials fabricated by electrospinning technique and their potential use in electroactive applications.

2. Experimental

The novel graphene and titanium dioxide-PVDF based nanocomposite materials were fabricated by electrospinning technique using as precursor PVDF solution in which were dispersed the TiO2 nanoparticles (Degussa P25) and commercially available graphene nanoplatelets provided by EMFUTUR Technologies Ltd. Spain, 5 µm wide, with an average 5 nm thickness, a bulk density of 0.03 to 0.1 g/cc, a carbon content of >99.5 wt%, an oxygen content of <1% and a residual acid content of < 0.5 wt%. To study the properties of the G-TiO2-PVDF nanocomposites, the pure component materials were characterized and G-PVDF and TiO2-PVDF nanocomposites were fabricated in similar conditions. The G-TiO2-PVDF nanocomposites were studied by optical and Scanning Electron Microscopy (SEM), XRD and Raman spectroscopy and their electrical resistance were measured.

3. Results and discussion

As shown in Fig. 1 SEM characterization shows that the graphene platelets are very conductive (they are quite transparent for the electron beam) and quite flat, fact which prove their good quality.

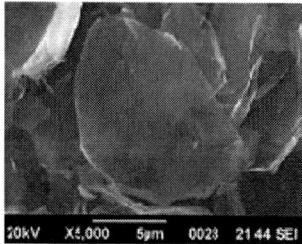

Figure 1. SEM characterization of graphene platelets.

SEM characterization of the G-PVDF, TiO2-PVDF and G-TiO2-PVDF nanocomposites obtained by electrospinning is presented in Fig. 2.

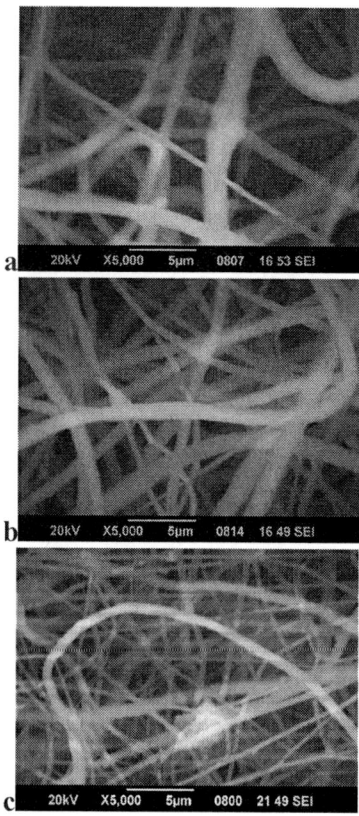

Figure 2. SEM characterization of nanocomposite materials a – 0.05g G-PVDF ; b – 0.05g TiO2-PVDF; c-.0.05g G- 0.05g TiO2-PVDF

It is obvious that nanocomposites containing only graphene or TiO2 are formed of fibres with an average diameter of about 500 µm (a and b) while the G-TiO2-PVDF nanocomposite is formed by thinner fibres with an average diameter of about 300-400 µm. The composites are quite conductive since the SEM characterization was performed in HV on uncoated samples at an acceleration voltage of 20kV.

XRD characterization of graphene platelets and TiO2 powders, electrospinned pure PVDF, 0.05g G-PVDF, 0.05g TiO2-PVDF, and 0.05g G- 0.05g TiO2-PVDF composite materials are presented in Fig. 3.

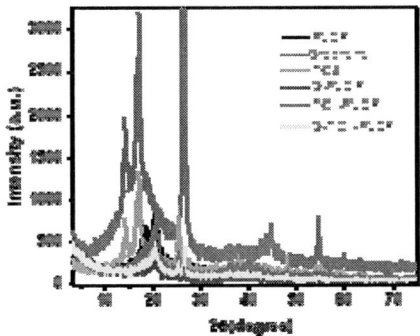

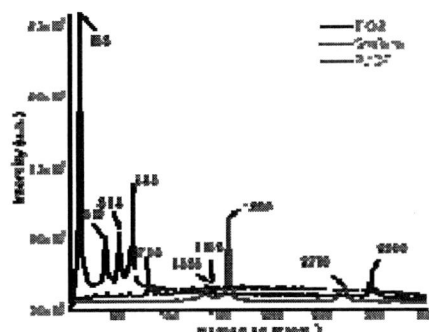

Figure 3. XRD characterisation of graphene platelets and TiO2 powders, electrospinned pure PVDF, 0.05g G-PVDF, 0.05g TiO2-PVDF, and 0.05g G- 0.05g TiO2-PVDF composite materials.

Figure 4. Raman spectra (@532 nm) of pure electrospinned PVDF, TiO2 and graphene powders.

It was observed that pure PVDF crystallizes as a mixture of α and β crystalline phases the α - phase peak at 19.9°(110) and (110) reflection of β-phase of crystal (@20.6°) being preferred. Forming the 0.05g G-PVDF, 0.05g TiO2-PVDF, and 0.05g G- 0.05g TiO2-PVDF nanocomposites leads to enhancement of (110) reflection of β-phase of crystal (@20.6°) and inhibition of α - phase peak at 19.9°(110). Graphene presence in composite seems to promote the preferential orientation of (021) plane (@26.6°) of α - phase peak while TiO2, the (110) reflection of β-phase of crystal (@20.6°). The mix composite appears as a mixture of both α and β crystalline phases. In this, the α - phase peak @26.6°(021) shows high, while the 19.9°(110) become almost invisible. It seems that the two nanomaterials presence in the PVDF matrix leads to synergistic interactions and the formation of a novel material with new properties. In order to deeper understand the 0.05g G- 0.05g TiO2-PVDF composite material structure, Raman spectroscopy was performed. Fig. 4 presents the Raman spectra (@532 nm) of pure electrospinned PVDF, TiO2 and graphene powders.

The Raman modes associated with graphene powder occurred at 1353 cm-1 and 1590 cm-1, reflecting the D and G band, respectively (Figure 4, the red spectrum). Similarly, the peak at 2710 cm-1 corresponded to the 2D band due to graphene. This modes were also reflected in the Raman spectra of the G-TiO2-PVDF composite (Fig. 5), but the intensity mode are decreased. The peak at 135 cm-1 confirms the widening of the interaction between the molecular chains of PVDF and the various functional groups of the graphene surface (Figure 5). Likewise, the TiO2 modes (Fig. 4: black line) Eg1 at 135 cm-1, B1g1 at 399 cm-1, A1g+B1g2 at 647 cm-1, Eg3 at 798 cm-1 does not cause any local structural changes. The structural properties of the TiO2, which covered the graphene, did not change after the composing process.

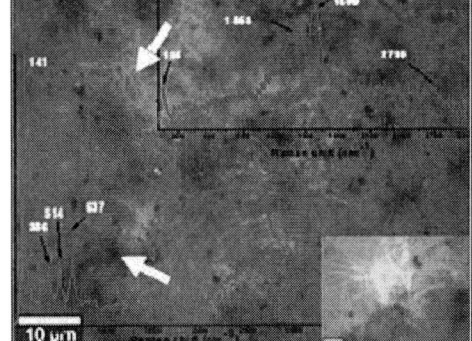

Figure 5. Raman spectra (@532 nm) of 0.05g G- 0.05g TiO2-PVDF composite in two different points. Inset in photo (right corner) shows a zoom optical microscopic of the composite.

Electrical resistance changes of pure PVDF, 0.05g G-PVDF, 0.05g TiO2-PVDF, and 0.05g G- 0.05g TiO2-PVDF composite materials in the 1 kHz- 1MHz frequencies range were measured using a 1cm x 1cm square sample. These results are presented in table 1.

Frequency	R PVDF	R G-PVDF	R TiO₂-PVDF	R G-TiO₂-PVDF
1 kHz	>10 GΩ	3 GΩ	3 MΩ	5 MΩ
10 kHz	1.1 GΩ	780 kΩ	354 kΩ	230 kΩ
50 kHz	43 MΩ	440 kΩ	370 kΩ	340 kΩ
100 kHz	9.5 MΩ	434 kΩ	357 kΩ	350 kΩ
500 kHz	640 kΩ	88 kΩ	71 kΩ	72 kΩ
1 MHz	331 kΩ	26 kΩ	20 kΩ	21 kΩ

Table 1 Electrical measurements of the composite materials.

4. Conclusions

PDVF based, 0.05g G-PVDF, 0.05g TiO2-PVDF, and 0.05g G- 0.05g TiO2-PVDF composite materials were obtained as fibrous membranes by electrospinning technique. SEM characterization showed that the fibers diameter is reduced by synergistic effect of composing PVDF with both nanomaterials, TiO2 and graphene flakes. XRD and Raman spectroscopy studies proved that the new multi-nanomaterial composites have an individual structure and are not just a mix of components. Further studies are necessary to better clarify these aspects. The novel composite materials are good insulator at low frequencies and very conductive at frequencies greater than 500 MHz. These make them, excellent candidates for potential use in various applications in electroactive applications, dissipative materials in electrostatic discharge (ESD) applications and electromagnetic shielding.

References

[1] Galchev, T.; McCullagh, J.; Peterson, R.; Najafi, K. Harvesting Traffic-Induced Vibrations for Structural Health Monitoring of Bridges. J. Micromech. Microeng. 2011, 21, 104005.

[2] Beeby, S. P.; Torah, R.; Tudor, M.; Glynne-Jones, P.;O'Donnell, T.; Saha, C.; Roy, S. A Micro Electromagnetic Generatorfor Vibration Energy Harvesting.J. Micromech. Microeng. 2007, 17, 1257.

[3] Cui, N.; Wu, W.; Zhao, Y.; Bai, S.; Meng, L.; Qin, Y.; Wang, Z. L. Magnetic Force Driven Nanogenerators as a Noncontact Energy Harvester and Sensor. Nano Lett. 2012, 12, 3701−3705.

[4] Park, K. I.; Son, J. H.; Hwang, G. T.; Jeong, C. K.; Ryu, J.; Koo, M.; Choi, I.; Lee, S. H.; Byun, M.; Wang, Z. L. Highly Efficient, Flexible Piezoelectric PZT Thin Film Nanogenerator on Plastic Substrates. Adv. Mater. 2014, 26, 2514−2520.

[5] Mao, Y.; Zhao, P.; McConohy, G.; Yang, H.; Tong, Y.; Wang, X. Sponge Like Piezoelectric Polymer Films for Scalable and Integratable Nanogenerators and Self-Powered Electronic Systems. Adv. Energy Mater. 2014,410.1002/aenm.201301624

[6] Jeong, C. K.; Park, K.-I.; Son, J. H.; Hwang, G.-T.; Lee, S. H.; Park, D. Y.; Lee, H. E.; Lee, H. K.; Byun, M.; Lee, K. J. Self-Powered Fully Flexible Light Emitting System Enabled by Flexible Energy Harvester. Energy Environ. Sci. 2014, 7, 4035−4043. Jeong, C. K.; Park, K. I.; Ryu, J.; Hwang, G. T.; Lee, K. J. Large Area and Flexible Lead Free Nanocomposite Generator Using Alkaline Niobate Particles and Metal Nanorod Filler. Adv. Funct. Mater. 2014, 24 , 2620−2629.

[7] Hwang, G. T.; Park, H.; Lee, J. H.; Oh, S.; Park, K. I.; Byun, M.; Park, H.; Ahn, G.; Jeong, C. K.; No, K. Self-Powered Cardiac Pacemaker Enabled by Flexible Single Crystalline PMN PT Piezoelectric Energy Harvester. Adv. Mater. 2014, 26, 4880−4887.

[8] Persano, L.; Dagdeviren, C.; Su, Y.; Zhang, Y.; Girardo, S.; Pisignano, D.; Huang, Y.; Rogers, J. A. High Performance Piezoelectric Devices Based on Aligned Arrays of Nanofibers of Poly (vinyl

[9] idenefluoride-co-trifluoroethylene). Nat. Commun. 2013, 4, 1633.

[10] Jeong, C. K.; Kim, I.; Park, K.-I.; Oh, M. H.; Paik, H.; Hwang, G.-T.; No, K.; Nam, Y. S.; Lee, K. J. Virus Directed Design of a Flexible BaTiO3 Nanogenerator. ACS Nano 2013, 7, 11016−11025.

[11] Ramadoss, A.; Saravanakumar, B.; Lee, S. W.; Kim, Y.-S.; Kim, S. J.; Wang, Z. L. Piezoelectric Driven Self Charging Supercapacitor Power Cell. ACS Nano 2015, 9, 4337−4345.

[12] Zi, Y.; Lin, L.; Wang, J.; Wang, S.; Chen, J.; Fan, X.; Yang, P. K.; Yi, F.; Wang, Z. L. Triboelectric Pyroelectric Piezoelectric Hybrid Cell for High Efficiency Energy Harvesting and Self Powered Sensing. Adv. Mater. 2015, 27, 2340−2347.

[13] Mitcheson, P. D.; Green, T. C.; Yeatman, E. M.; Holmes, A. S.Architectures for Vibration Driven Micropower Generators. J. Microelectromech. Syst. 2004, 13, 429-440.

[14] Z. Wang, L. Cheng, Y. Zheng, Y. Qin, Z.L. Wang Enhancing the performance of triboelectric nanogenerator through prior-charge injection and its application on self-powered anticorrosion, Nano Energy, 10 (2014), pp. 37-43

[15] Y. Zheng, L. Cheng, M. Yuan, Z. Wang, L. Zhang, Y. Qin, T. Jing An electrospun nanowire-based triboelectric nanogenerator and its application in a fully self-powered UV detector Nanoscale, 6 (2014), pp. 7842-7846

Kinetics of Lanthanum and Yttrium doped Zirconia crystallization by X-ray Powder Diffraction

D.V.Drăguț*, V.Bădiliță*, R.R.Piticesccu, A.Motoc

*. NATIONAL RESEARCH AND DEVELOPMENT INSTITUTE FOR NON-FERROUS AND RARE METALS –
INCDMNR- IMNR

dragutdumitruvalentin@yahoo.ro, viobadilita@yahoo.com, rpiticescu@imnr.ro, amotoc@imnr.ro

The need to develop high performance materials is increasing nowadays. Due to its extraordinary range of properties, the yttria-doped zirconia is placed in a special place among the ceramic oxide systems. Co-doping with other rare earths oxides, such as Lantania is required in order to obtain better properties for some applications. The aim of the following paper is to determine the kinetic growth parameters of the crystallite according to the nature and the content of dopant used. The ZrO2 – 3Y2O3 – nLa2O3 (n = 3, 6, 9) samples were obtained through a hydrothermal process.

Keywords—zirconia; hydrothermal; kinetics; lantania.

1. Introduction

Zirconia is one of the most promising multifunctional material, characterized by a set of unique composition combination of high strength and fracture toughness, corrosion resistance, low thermal conductivity, special electrophysical properties. This set of properties determines its application as a base for materials used in production of structural ceramics, fuel cells with ceramic oxide electrolyte (SOFC), bioceramics, catalyst carriers, heat-resistant coatings, etc. [1]

At normal pressure, zirconia can be found in three crystalografic forms depending on the temperature. At room temperature and up until 1170˚C, zirconia is found in a monoclinic state (P2₁/c). Between 1170˚C and 2370˚C we find a tetragonal structure (P4₂/nmc) and above 2370˚C zirconia is found in a cubic state (Fm3m). In some cases the tetragonal phase can be metastable. The transformation from the tetragonal phase to the monoclinic phase upon cooling is accompanied by a substantial increase in volume (~4.5%). A catastrophic failure can be observe with this increase. The transformation is reversible and begins at ~950˚C on cooling. By alloying zirconia with some stabilizing oxides, such as CaO, MgO, Y2O3 allows the retention of the tetragonal structure at room temperature. [3].

In order to investigate the kinetics of the reactions that take place during the formation of stabilised Zirconia, it is required to know a couple of information about the existing models in science. The Johnson-Mehl-Avrami is one of the most important equations that describes the way in which that phasic tranformations occur in solids at constant temperature. This equation describes the cristalization kinetics. [2]

$$X_T(t) = 1 - \exp(-bt^n) \quad (1)$$

,where n is an integer value between 1 and 4 which reflects the nature of the transformation, $X_T(t)$ represents the volume fraction crystallized in the conditions of maintaining the material at temperature T and for a duration t and b is a constant. [6]

All of the molecules have a certain amount of energy. This energy can be found as potential energy or as kinetic energy. When collisions between the molecules occurs, the kinetic energy can be used to bend, stretch or brake their bonds, thus the chemical reactions take place. [5]

The activation energy, Q, represents the minimum kinetic energy that the reactant particles must possess in order for a chemical reaction to take place. The higher the activation energy, the harder is for the reaction to take place and vice versa. One of the key parameters that controls the degree of transformation of the phases is the grain size. Thermal mechanisms responsible for transformation processes, crystallization, etc. are governed by diffusion, either at the grain boundary or grain volume. [4]

In order to find the value of activation energy the Arrhenius equation it is used.

978-1-5386-4483-6/18 $31.00 © 2018 IEEE

$$K = Ae^{-Q/RT} \quad (2)$$

Where K is rate constant, A is the Arrhenius factor, Q is the activation energy. [5]

2. Experimental

In order to prepare the samples, analytical grade zirconium chloride from Sigma Aldrich, as well as yttrium oxide and Lanthanum nitrate from Alfa Aesar was used. Via a hydrothermal process conducted in a 5-liter Berghoff autoclave at 200˚C for 2 hours the specimens were obtained. Afterwards the samples were subjected to a thermal treatment in a CARBOLITE furnace in order to observe the phase – structural behavior of doped zirconia according to the temperature and analysed through X-ray diffraction. All the samples were heated at 20˚C/min. , maintened for 2 hours and cooled in air. Considering the fact that through the hydrothermal synthesis homogenous solutions were obtained, it was considered that the mechanism described by the Arrhenius equation is the most suitable to characterize these materials. This mechanism being adequate for a homogenous nucleation and growth.

2. Results

By using a Bruker D8 Advance diffractometer, all of the samples were chacterized via x-ray powder diffraction at room temperature and pressure.

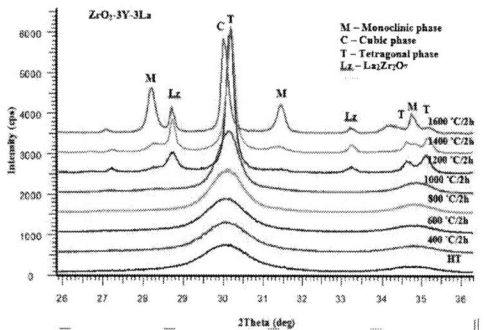

Fig.1 X-ray diffraction pattern of ZrO$_2$-3Y$_2$O$_3$-3La$_2$O$_3$ samples

In fig.1, fig.2 and fig.3 there are presented the x-ray diffraction patterns on the ZrO$_2$-3Y$_2$O$_3$-3La$_2$O$_3$, ZrO$_2$-3Y$_2$O$_3$-6La$_2$O$_3$ and ZrO$_2$-3Y$_2$O$_3$-9La$_2$O$_3$ performed for every temperature.

The crystallite sizes were calculated using the Scherrer equation and considering the instrumental broadening determined prior on a corundum sample (NIST– SRM 1976).

Table 1 Crystallite sizes of the ZrO$_2$-3Y$_2$O$_3$-3La$_2$O$_3$ samples

temp., °C	D for T-phase, nm	D for LZ-phase, nm	D for M-phase, nm	D for C-phase, nm
1600	-	56.7	28.0	33.1
1400	39.4	40.4	11.1	17.2
1200	36.4	25.6	6.8	8.1
1000	12.4	-	4.1	11.3
800	-	-	4.1	7.5
600	-	-	4.1	6.2
400	-	-	4.1	5.7
200	-	-	4.1	5.2

By adding 3%mol of La$_2$O$_3$ one can observe that above 1200˚C the dopant is segregating, leaving the stable cubic structure and forming a compound with a pyrochlore type structure – La$_2$Zr$_2$O$_7$. In table 1 there are presented the calculated crystallite sizes through X-ray diffraction of the stable phases detected.

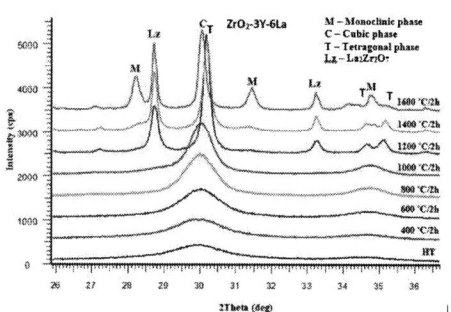

Fig.2 X-ray diffraction pattern of ZrO$_2$-3Y$_2$O$_3$-6La$_2$O$_3$ samples

By adding 6%mol of La$_2$O$_3$ it can be observed that the cubic phase is stable and at temperatures above 1200˚C the quantity of La$_2$Zr$_2$O$_7$ increases.

Table 2 Crystallite sizes of the ZrO$_2$-3Y$_2$O$_3$-6La$_2$O$_3$ samples

temp., °C	D for T-phase, nm	D for LZ-phase, nm	D for M-phase, nm	D for C-phase, nm
1600	-	59.0	27.0	30.1
1400	-	42.0	16.3	17.9
1200	28.6	30.1	11.6	-
1000	-	5.4	-	8.7
800	-	-	-	6.9
600	-	-	-	5.2
400	-	-	-	3.9
200	-	--	-	3.3

Table 2 presents the crystallite sizes of ZrO_2-$3Y_2O_3$-$6La_2O_3$ samples.

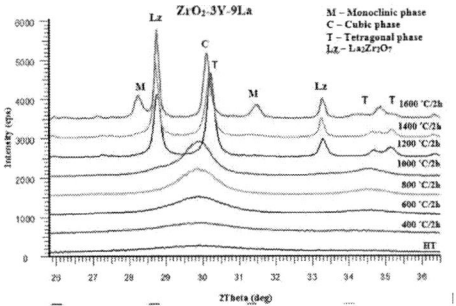

Fig.3 X-ray diffraction pattern of ZrO_2-$3Y_2O_3$-$9La_2O_3$ samples

In the case of doping with 9% mol La_2O_3 one can observe a higher increase in quantity of the compound, as well as the stabilization of the cubic phase. Table 3 shows the crystallite sizes calculated through X-ray diffraction of the stable phases.

Table 3 Crystallite sizes of the ZrO_2-$3Y_2O_3$-$9La_2O_3$ samples

temp., °C	D for T-phase, nm	D for LZ-phase, nm	D for M-phase, nm	D for C-phase, nm
1600	-	59.52	25.01	29.2
1400	36.78	43.82	10.17	17.5
1200	28.47	34.54	8.14	-
1000	-	5.81	-	7.3
800	-	-	-	5.5
600	-	-	-	3.9
400	-	-	-	2.1
200	-	-	-	1.7

3. Kinetic study

Keeping in mind the fact that the holding time of the thermal treatments at different temperatures was kept constant, the isocrone system study proposes to highlight how the temperature influenced the crystallite sizes of the present phases, at different compositions, thus determining the activating energies of the primary phases. In order to determine the values of the activation energies a power type kinetic model was used:

$$D^n - D0^n = kt \qquad (3)$$

Where D represents the medium crystallite size, D_0 represents the medium crystallite size at the initial moment and n represents the growth exponent (for a growth through diffusion at the grains interface n = 2, and for a growth through diffusion in volume n =3),

k is the kinetic coefficient and t is the time. The kinetic coefficient is temperature dependent and it can be expressed in an Arrhenius form:

$$K = k_0 \, e^{-Q/RT} \qquad (4)$$

Where k is the pre-exponential factor, which is constant, Q is the activation energy in Kj/mol and T is the temperature. Knowing that the time is constant and further calculating the equations resulted the following expression for the activation energy, expressed in the form of a first degree equation:

$$Q = -nR \cdot m \qquad (5)$$

Where m is the slope, R is the gas constant (8, 31445 KJ/molK), and n =1, 2, 3 according to the type of growth. In order to determine the activation energies of the identified phases, first the slopes have been calculated using graphs according to lnD and 1/T. Figures 4, 5, 6 presents the graphs created according to lnD and 1/T.

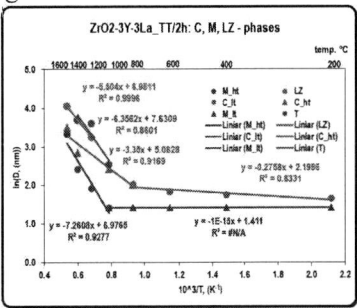

Fig. 4 Graphical reprezentation of lnD in relation to 1/T of the ZrO_2-$3Y_2O_3$-$3La_2O_3$ samples

It can be observed a development in two stages of cubic stabilised phase. The first stage characteristic to low temperatures is characterised by a light increase of crystallite sizes, from 5nm to 7.5nm, while the second stage, characteristic to high temperatures shows an increase of the sizes up to 33 nm al 1600°C.

When adding 6% of La_2O_3 (Fig.5) it can be observed that the crystallite size is smaller in the first stage as against adding 3%mol La_2O_3.

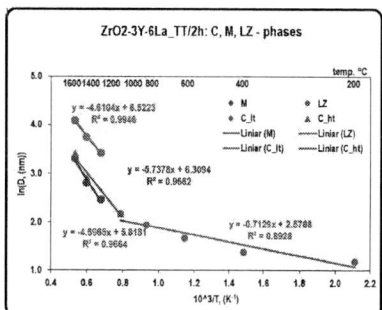

Fig. 5 Graphical reprezentation of lnD in relation to 1/T of the $ZrO_2-3Y_2O_3-6La_2O_3$ samples

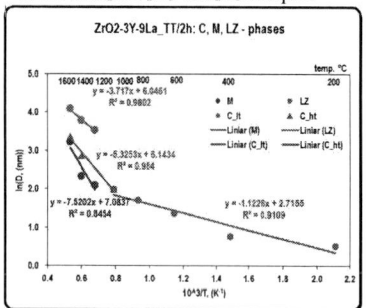

Fig. 6 Graphical reprezentation of lnD in relation to 1/T of the $ZrO_2-3Y_2O_3-9La_2O_3$ samples

In the case of adding 9%mol La_2O_3, (Fig.6) the crystallite size we can see a more pronounced increase, from 1.7 nm to 5.5 nm at 800°C. In order to calculate the activation energies the value of growth exponent was 1. Table 4 presents the calculated values of the activation energies of the phases identified in the samples through X-ray diffraction.

Table 4. Activation energies of the identified phases.

ZrO2-3Y (kJ/mol)	ZrO2-3Y-3La (kJ/mol)	ZrO2-3Y-6La (kJ/mol)	ZrO2-3Y-9La (kJ/mol)
QT_lt = 1.88	QC_lt = 2.29	QC_lt = 5.93	QC_lt = 9.34
QT_ht = 34.10	QC_ht = 27.85	QC_ht = 39.05	QC_ht = 44.28

Based on the calculated activation energies in Fig. 7 there is presented the forming mechanism of the main stabilised phase in zirconia doped with yttria, respectively lantania at different concentrations. It is noticed that as the content of the dopant is increased, the activation energy increases in both the low temperature stage, as well as in the high temperature stage.

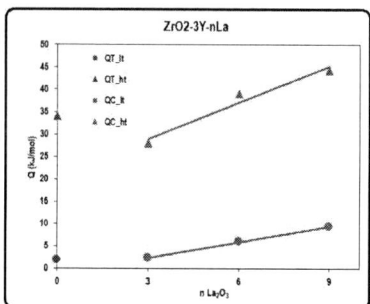

Fig. 7 The concentrations of La_2O_3 in relation to the activation energies of the cubic phase

Conclusions

By adding La_2O_3 the cubic phase is stabilised between 200°C– 800°C. As the concentration of La_2O_3 is increased, the crystallite size of the stabilised cubic phase is reduced. Also it can be observed that the value of the activation energy is increased as the concentration of dopant is higher, but without affecting the crystallite size of the cubic phase.

At temperatures above 1000°C, by adding La_2O_3 the formation of a pyrochlore like structure is favoured to form. As the quantity of the dopant is increased, a significant growth of the crystallite size of the $La_2Zr_2O_7$ compound is observed.

The researches continues in order to study the kinetics of the forming process using thermal analysis methods.

Acknowledgments.

The authors would like to acknowledge also the support from UEFISCDI in frame of the Grant 50/01.04.2018, Project ID 87-Monamix Call ERAMIN 2 Cofund and E.C.

References

[1] https://www.ariel.ac.il/sites/conf/mmt/ws2011/service%20files/papers/156-162.pdf

[2] https://en.wikipedia.org/wiki/Avrami_equation

[3] Isabelle Denrya, J. Robert Kelly, "State of the art of zirconia for dental applications", Dental Materials no. 24, 2008, pp.299–307

[4] Rachman Chaim, "Activation energy and grain growth in nanocrystalline Y-TZP ceramics", Materials Science and Engineering A 486 (2008) 439–446

[5] Arrhenius Equation - Chemistry LibreTexts, 13 Apr 2016

[6] J.J. Li, J.C. Wang, Q. Xu, G.C. Yang, " Comparison of Johnson–Mehl–Avrami–Kologoromov (JMAK) kinetics with a phase field simulation for polycrystalline solidification", Acta Materialia 55 (2007) 825–832

Comparative study of Sm and La doped ZnO properties

I. V. Tudose [1], P. Pascariu [2], C. Pachiu[3], F. Comanescu[3], M. Danila[3], R. Gavrila[3], E. Koudoumas[1], M. Suchea [1,3]*

*[1]Center of Materials Technology and Photonics, School of Engineering, Technological Educational Institute of Crete, Heraklion, Greece
[2]"Petru Poni" Institute of Macromolecular Chemistry, Aleea Grigore Ghica Voda, 41A, Iaşi 700487, Romania
[3]National Institute for Research and Development in Microtechnologies (IMT-Bucharest), 126 A, Erou Iancu Nicolae Street, P.O. Box 38-160, 023573 Bucharest, Romania
*E-mail: * mira.suchea@imt.ro; mirasuchea@staff.teicrete.gr

Abstract— Samarium and Lanthanum doped nanostructured ZnO thin films were grown onto glass substrates by spray deposition method. Influences of different concentrations (0% to 1%) of Sm and La on the ZnO structural and optical properties were investigated by scanning electron microscopy (SEM), X-ray diffraction (XRD), atomic force microscopy (AFM), Raman spectroscopy and by UV-VIS spectroscopy. X-ray diffraction studies revealed that the ZnO films have zincite crystalline structure and show a preferential growth orientation along (101) crystallographic orientation. Doping with Sm and La leads to changes of thin films crystallinity as well as their transparency in VIS region of electromagnetic spectrum. Increasing dopant concentration leads to slightly increased transparency.

Keywords—zinc oxide, Lanthan, samarium, dopant, optical properties.

1. Introduction

ZnO based nanostructured materials are especially attractive for nanoscience studies as well as for nanotechnology applications on photoluminescence, semiconductors, photocatalysists, gas sensors, UV photodetector, light emitting diodes (LEDs), solar cells etc. ZnO is a semiconductor material with band gap energy of ~ 3.37 eV that corresponds to the wavelengths of ~375 nm [1-4]. Zinc oxide has been regarded as an excellent UV shielding material which blocks wavelengths shorter than 375 nm. Size-dependent emission properties of ZnO nanostructured materials have been explored in many photochemical studies. The visible emission of ZnO, which usually arises from anionic vacancies, is very sensitive to hole scavengers. The emission is quantitatively quenched by hole scavengers such as iodide ions [5-8]. This particular work is dedicated to the study of the influence of Sm and La doping on the morphological, structural and optical properties of Sm and La ZnO thin films grown by spray technique.

2. Experimental

Sm and La doped nanostructured ZnO thin films with a thickness of about 100 nm were prepared by spray deposition method using as substrates glass sheets. These materials were synthesized from 0.1 M $Zn(NO_3)_2 \cdot 6H_2O$ and $La(NO_3)_3$ respectively $Sm(NO_3)_3$ solutions in various molar ratios of the rare earth metal (0.1, 0.5 and 1). The obtained solutions were deposited by spray technique onto heated glass substrates (about 250°C). Finally, the resulting thin films were annealed at 600°C under air atmosphere for 2h followed by slow cooling.

All films were characterized by SEM, XRD, AFM, Raman and UV-VIS spectroscopy.

3. Results and discussion

All, pure and doped ZnO films are uniform and transparent, crack free all over the coated area (microscope glass slides). Sm and La ZnO doping as well as their concentration has a very strong influence on films surface morphology and structuring. To illustrate this, low magnification (x100) SEM images of 0.5% Sm:ZnO, 1% Sm:ZnO; 0.5% La:ZnO; 1% La:ZnO are presented in **Figure 1.**

978-1-5386-4483-6/18 $31.00 © 2018 IEEE

Figure 1. *Low magnification SEM images 0.5% Sm:ZnO, 1% Sm:ZnO; 0.5% La:ZnO; 1% La:ZnO*

All films show a structured surface formed by grains grouped on large islands with uniform distribution onto the substrate. It was observed that, for similar doping concentrations, Sm doping tends to form larger islands than La doped ZnO. In Sm doping case, increasing concentration leads to larger islands formation while in the case of La is just opposite. To explore the intimate structure of the islands, AFM was used. Pure ZnO thin films grown in similar conditions were used for comparative purpose. AFM characterization shows that pure and La doped ZnO thin films have a granular structure with grains of ~30nm ZnO, ~20nm La:ZnO while in the case of Sm granular structure co-exist with the presence of micron long rods of ~20nm thickness. Illustrative AFM images of pure ZnO, 0.1%Sm:ZnO and 0.1%La:ZnO thin films are presented in **Figure 2**.

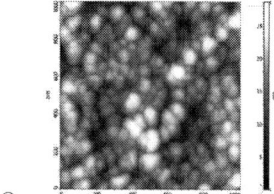

Pure ZnO

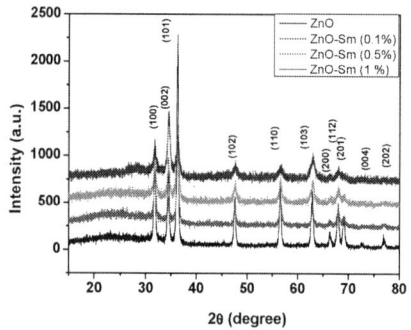

0.1%Sm:ZnO

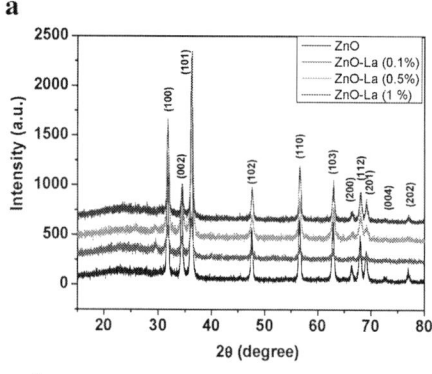

0.1%La:ZnO

Figure 2. *AFM images of pure ZnO, 0.1%Sm:ZnO and 0.1%La:ZnO thin films.*

XRD characterization of pure and doped thin films revealed that the ZnO films have zincite crystalline structure and show a preferential growth orientation along (101) crystallographic orientation. Film crystallinity slightly decreases when doping concentration increases. Sm doping shifts slightly the peaks position to larger angles while La doping does the opposite. XRD spectra of Sm and La doped ZnO are presented in **Figure 3 a and b**.

a

b

Figure 3. *XRD spectra of **a** Sm and **b** La doped ZnO*

Raman spectroscopy analysis was also performed. The studies were made using a Scanning Near-field Optical Microscope (SNOM) - Witec alpha 300S/Witec/2008 Witec Gmbh Germany with Nd-YAG - 532 nm laser. Raman band at 436 cm^{-1} corresponds to non-polar optical phonon E_{2H} and the band at 567 cm^{-1} is a typical band positioned A_{1L} mode of ZnO. The bands observed in the range 1050 - 1200 cm^{-1} are the second order A_{2L} scattering modes. The peaks can be assigned as second order of E_{1L} mode respectively as a consequence of defects such as O-vacancies and Zn interstitials. The additional peaks at 783 cm^{-1} and 2427 cm^{-1} appears in the Raman spectra of surface La-ZnO doped thick films. The peaks were conjectured to be associated with intrinsic host-lattice defects and was arises by doping with La. With an increase in the concentration of Lanthanum, increase the intensity of A_{1L} mode to A_{2L} mode and a it was observed a mode shifted slightly towards smaller values of the wave number by an amount of 1 to 6 cm^{-1}. There is a same trend for A_{1L} and A_{2L} mode for Sm-ZnO doped films. Especially, when the concentration of Samarium concentration increases, the intensity of modes increases and are shifted slightly with 5 to 15 cm^{-1}.

b

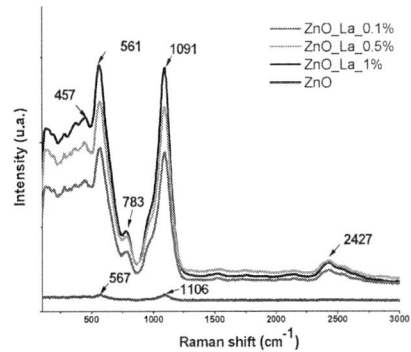

*Figure 4. a Raman spectra of the undoped ZnO thick films (blue line); Sm 0.1-ZnO (red line); Sm 0.5 – ZnO (green line); Sm 1.0 –ZnO (black line). ZnO excited by a 532 nm laser line. **b** Raman spectra of the undoped ZnO thick films (blue line); La 0.1-ZnO (red line); La 0.5 – ZnO (green line); La 1.0 –ZnO (black line). ZnO excited by a 532 nm laser line.*

Optical properties were investigated by UV-VIS spectroscopy. As observed in figure 5a and b, all films show high optical transmittance of ~90% on visible region of the electromagnetic spectrum.

a

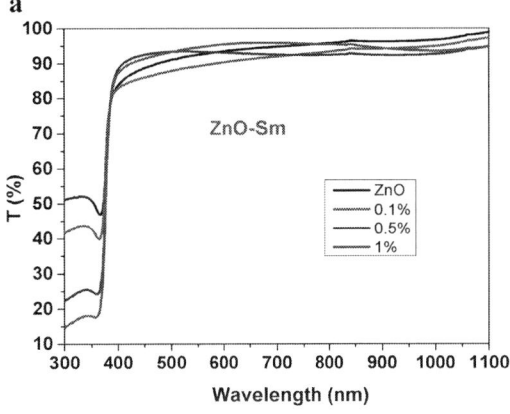

a

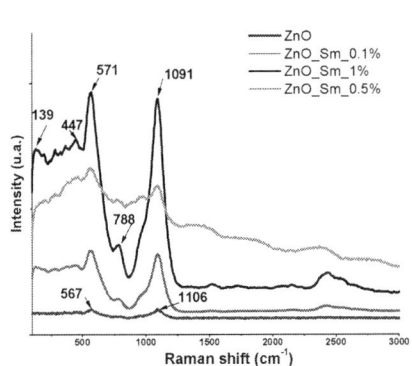

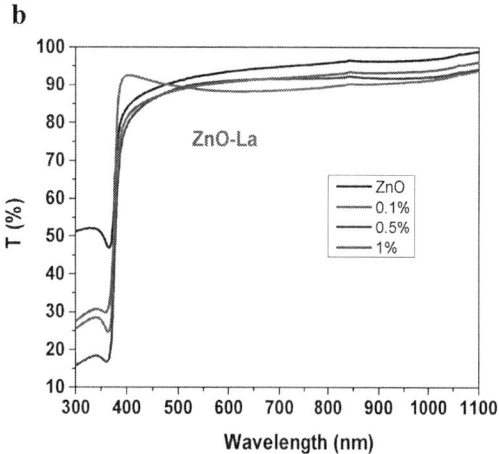

*Figure 5. UV-VIS transmittance spectra of **a** Sm and **b** La doped ZnO.*

Increase of dopant concentration slightly enhances optical transmittance of films. La:ZnO films shows slightly lower transmittance in the VIS region of the EM spectrum. Using Tauc plot, optical band gap energy was calculated. Pure ZnO has Eg~3.43eV. Doping determines optical bandgap narrowing. The Eg variations of Sm and La doped ZnO films are presented in **Table 1**.

concentration	Eg (eV) Sm:ZnO	Eg (eV) La:ZnO
0.1%	3.36	3.07
0.5%	3.20	3.05
1%	3.07	3.21

Table 1 Bandgap calculation results.

4. Conclusions

Transparent and uniform Sm and La doped ZnO thin films were grown by spray onto microscope glass slides. The effect of dopant nature and its concentration on film morphology, structure and optical properties were studied. It was fond out that Sm and La have opposite effects on surface structuring. Doping decreases crystallinity of ZnO films and as well as its optical band gap. La doping leads to stronger band gap narrowing than Sm doping.

References

[1] D.M. Alsebaie, W. Shirbeeny, A. Alshahrie, M. Sh. Abdel-Wahab, *"Ellipsometric study of optical properties of Sm-doped ZnO thin films Co-deposited by RF Magnetron sputtering"* Optik, 148, pp. 172–180, 2017.

[2] H.Y. He, J.F. Huang, J. Fei, J. Lu, *"La-doping content effect on the optical and electrical properties of La-doped ZnO thin films"*, J Mater Sci: Mater Electron, 26, pp. 1205–1211, 2015.

[3] N. N. Ilkhechi, N. Ghobadi, F. Yahyavi, *"Enhanced optical and hydrophilic properties of V and La co-doped ZnO thin films"*, Opt Quant Electron, 49:39, pp. 1–10, 2017.

[4] Y. Bouznit, Y. Beggah, F. Ynineb, *"Sprayed lanthanum doped zinc oxide thin films"*, Appl. Surf. Sci. 258, pp. 2967– 2971, 2012.

[5] M. Novotny, E. Maresova, P. Fitl, J. Vlcek, M. Bergmann, M. Vondracek, R. Yatskiv, J. Bulir, P. Hubik, P. Hruska, J. Drahokoupil, N. Abdellaoui, M. Vrnata, J. Lancok, *"The properties of samarium-doped zinc oxide/phthalocyanine structure for optoelectronics prepared by pulsed laser deposition and organic molecular evaporation"*, Appl. Phys. A, 122:225, pp. 1–8, 2016.

[6] T.P. Rao, S.G. Raj, M.C.S. Kumar, *"Optical Properties of Samarium Doped ZnO Thin Films"*, 2nd International Conference on Devices, Circuits and Systems (ICDCS), pp. 1–4, 2014.

[7] P. Velusamy, R. Ramesh Babu, K. T. Aparna, *"Effect of Sm doping on the physical properties of ZnO thin films deposited by spray pyrolysis technique"*, AIP Conf. Proc. 1832, pp. 080085-1–080085-3, 2016.

[8] A. Manikandan, E. Manikandan, B. Meenatchi, S. Vadivel, S.K. Jaganathan, R. Ladchumananandasivam, M. Henini, M. Maaza, J.S. Aanand, *"Rare earth element (REE) lanthanum doped zinc oxide (La: ZnO) nanomaterials: Synthesis structural optical and antibacterial studies"*, J. Alloys Compd., 723, pp. 1155–1161, 2017.

Carbon nanotube/polyaniline composite films prepared by hydrothermal- electrochemical method for biosensor applications

L. M. Cursaru (Popescu)*, A.G. Plaiasu**, C.M. Ducu**, R.M. Piticescu*, I.A. Tudor*

*National Research-Development Institute for Non-Ferrous and Rare Metals, 102 Biruintei blvd.,
077145, Pantelimon, Ilfov, Romania
E-mail: mpopescu@imnr.ro, roxana.piticescu@imnr.ro, atudor@imnr.ro
** University of Pitesti, 1 Targu din Vale Street, Pitesti, Romania
*E-mail: plaiasugabriela@yahoo.fr, catalinducu@yahoo.com

Abstract – In this study, CNT-PANI composites were prepared in soft chemical synthesis conditions using hydrothermal method. Our aim is to obtain CNT-PANI films with potential applications in VOC's detection, using an environmental friendly, low energy consumption technique: hydrothermal-electrochemical deposition of composite films. Thin films were characterized by AFM and FT-IR analyses.
Keywords— carbon nanotubes, poly(aniline), composites, electrochemical, biosensor.

1. Introduction

Carbon nanotubes (CNTs) have been the subject of numerous investigations in chemical, physical and material science research owing to their extraordinary structural, mechanical, chemical, thermal and electronic properties [1, 2]. CNTs can be used as electrode materials with useful properties for electrochemical and bioelectrochemical applications [1]. The ability of CNTs to promote electron-transfer reactions suggests interesting applications in the development of amperometric biosensors [1]. Carbon nanotubes have attracted considerable attention due to their potential application in electronic devices. They are attractive building blocks for the development of novel polymer-nanocomposite materials with enhanced functionality, especially if it comes to enhanced conductivity, thermal stability, and reinforcement properties [2]. Because of the high strength and stiffness of CNTs, they are ideal candidates for structural applications. For example, they may be used as

reinforcements in high strength, low weight and high performance composites. Presently there is a great interest in exploiting the exciting properties of these CNTs by incorporating them into some form of polymer matrix [3]. Composites of CNTs and conducting polymer are interesting and promising since they can combine two relatively cheap materials to gain the large pseudo-capacitance of the conducting polymers coupled with the conductivity and mechanical strength of the CNTs [4].

Polyaniline (PANI) is one of the most attractive conducting polymers due to its unique and controllable chemical and electrical properties; environmental, thermal, and electrochemical stability; and interesting electronic, optical, and electro-optical properties [1, 5]. PANI has a wide range of tunable properties emanating from its structural flexibility, leading to potential applications in many fields, such as battery electrodes, anticorrosive coatings, energy storage systems, gas sensors, and electro-catalytic devices [5]. Moreover, PANI has the highest environmental stability and is recognized as the only conducting polymer stable in air [5]. Composites consisting of CNTs and PANI have been developed for different applications in lithium ion batteries, supercapacitors, catalysts, solar cells, nanodevices, chemical sensors, and biosensors for vapor sensing [5, 6]. Composites of PANI and multi-walled carbon nanotubes (MWCNTs) have been

978-1-5386-4483-6/18 $31.00 © 2018 IEEE

widely exploited to sense a number of volatile organic compounds (VOCs) like ammonia, chloroform, explosive vapors like picric acid (PA), 2,6-dinitrotoluene (2,6-DNT) and 2,4,6-trinitrotoluene (TNT) and hydrogen gas. However, contrary to these reports of a positive influence of MWCNTs on PANI's electrical properties, P. Lobotka et al. reported that they didn't observe any increase in the sensitivity of their sensors when carbon nanotubes were incorporated into the PANI films and exposed to a range of organic vapors [6]. During the last decade PANI/CNT nanocomposites have been most frequently synthesized by aniline polymerization in the presence of CNTs, using APS as an oxidizing/polymerizing agent [7]. Different forms of nanotubes were used: oxidized B/N doped CNTs, single-walled carbon nanotubes (SWCNTs), MWCNT powders, aligned MWCNT films, oxidized MWCNTs or functionalized MWCNTs. Different modifications of the polymerization and functionalization procedures were applied, as well as non-covalent coating of a PANI shell on the wall of MWCNTs with the assistance of 1-pyrenesulfonic acid, and plasma induced grafting of PANI onto CNTs [7]. Lu et al. [8] fabricated a layer-by-layer PANI NPs-MWCNT film onto interdigitated electrodes for the fabrication of stable chemiresistive sensors for methanol (CH_3OH), toluene ($C_6H_5CH_3$), and chloroform ($CHCl_3$) detection with reproducible response upon chemical cycling. Double percolated conductive networks in PANI -MWCNT nanocomposite resulted in both higher sensitivity and selectivity than other formulations, demonstrating a positive synergy [8]. However, most of these studies normally require toxic organic solvents, which may be harmful to the environment [5]. To address this, it is important to develop a facile and efficient method without toxic organic solvents.

In the present study, CNT-PANI composites were prepared in soft chemical synthesis conditions (aqueous solution, low temperature T<100°C and moderate pressure (20-100 bar) using hydrothermal method. Our aim is to obtain CNT-PANI films with potential applications in VOC's detection, using an environmental friendly and low energy consumption technique: hydrothermal-electrochemical deposition of composite films from CNT -PANI aqueous suspension. The novelty of this study consists not only in synthesis and deposition method of composite films, but also in the formation of stable interactions between functionalized CNT and PANI, due to pressure effect. PANI, which is the only conducting polymer stable in air, was chosen due to its electrochemical and thermal stability. Moreover, a coating of CNTs with polyaniline enhances the conductivity, resulting in excellent performance of the new material as sensor for hydrocarbon vapors detection [9].

2. Experimental

2.1 Materials used

Multi-Walled Carbon Nanotubes, d=4.5-10 nm, L=4 μm, 6-8 walls, Sigma Aldrich; poly(aniline), emeraldine base, average Mw ~10,000, Sigma Aldrich; nitric acid (HNO_3); 5M KOH solution; distilled water; commercial Interdigitated Gold Electrodes on Alumina substrate (DROPSENS).

2.2 Thin films based on MWCNT and PANI obtained by the hydrothermal-electrochemical method in two stages:

a) The nanostructured hybrid material based on MWCNT and PANI was obtained by *hydrothermal process* at 60 bar and 40°C using Berghof autoclave (Germany); b) The aqueous suspension based on MWCNT and PANI resulting from hydrothermal synthesis was introduced into CORTEST (USA) autoclave, equipped with 3 electrodes, for *hydrothermal -electrochemical experiments* at 40°C and 60 bar. The three electrodes used are: i) the working electrode, commercial Interdigitated Gold Electrodes on Alumina substrate (DROPSENS). The interdigitated configuration typically enhances sensitivity and detection limits.

These types of substrates are suitable for decentralized assays, to develop specific (bio)sensors and other electrochemical studies; ii) reference electrode - Ag /AgCl fitted with a capillary tube containing 0.1 M KCl; iii) auxiliary electrode, platinum Nb. The three electrodes are connected to a VoltaLab 10 potentiostat with VoltaMaster 4 software.

2.3 Material characterization

Hydrothermally synthesized CNT-PANI composite materials which were further used for hydrothermal- electrochemical deposition of thin films were characterized by Fourier transform infrared spectroscopy (FTIR) and Scanning Electron Microscopy with Energy Dispersive Spectroscopy (SEM/EDS).

Surface topography of composite films was studied by atomic force microscopy (AFM) while FTIR analysis was carried out to study the chemical structure of the films.

3. Results and discussion

Experiments performed in this study were focused on hydrothermal-electrochemical deposition of CNT-PANI thin films starting from CNT-PANI stable aqueous suspension. Two electrochemical methods were applied to study the formation of CNT-PANI thin films: cyclic voltammetry and chronoamperometry.

3.1 SEM/EDS characterization of CNT-PANI composite powder revealed that CNT chains were coated by PANI (**Fig.1**).

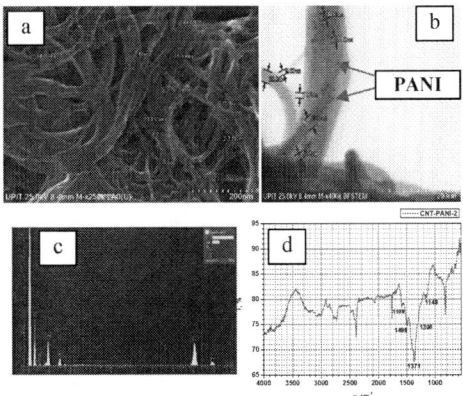

Fig.1. CNT-PANI composite powder: a) and b) SEM images; c) EDS spectrum; d) FTIR spectrum

Particle size of the organic phase (PANI) vary between 11-15 nm, while the diameter of CNT coated with PANI is about 22-25 nm. The interaction between PANI and CNT was confirmed by FT-IR spectra through the corresponding peaks of the two components.

3.2 Electrochemical methods

a) Cyclic voltammetry (CV)

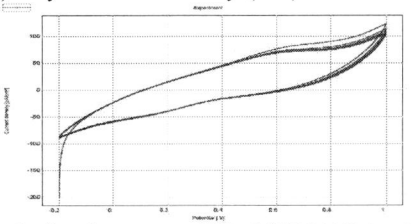

Fig.2. Cyclic voltammogram (CV) of CNT-PANI, 5 cycles, 10 mV/s, potential range:-0.2V ÷ +1V

The absence of cathodic or anodic peaks in **Fig. 2** indicates the existence of CNT-PANI thin film on the surface of DROPSENS substrate.

b) chronoamperometry (CA)

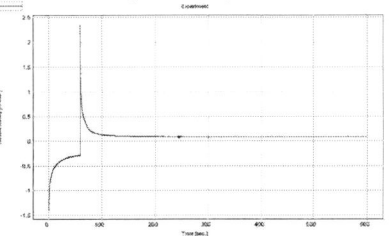

Fig.3. Chronoamperometry: -0.5V (1 min) ÷ +1V (9 min)

CA curve from **Fig.3** shows the electrochemical deposition of CNT-PANI film in hydrothermal conditions (60 bar and 40°C) at applied potential of +1V for 9 min.

3.3 Surface topography analysis (AFM)

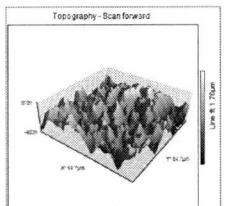 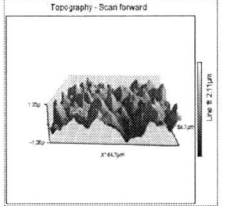

Fig.4. AFM images of commercial DROPSENS substrate (left) and one CNT-PANI thin film (right)

AFM images (**Fig.4**) showed the topography of the hydrothermal-

978-1-5386-4483-6/18 $31.00 © 2018 IEEE   251

electrochemical deposited thin films compared to the initial substrate prior to deposition. The structure of the commercial DROPSENS substrate was observed. The distance between interdigits is 5 μm.

The depicted substrate has a roughness of 1.76 μm, while CNT-PANI film roughness is higher (between 2.11 - 3.24 μm), indicating the hybrid deposition. The surface topography and film roughness are expected to favor the attachment of biomolecules and the detection of volatile organic compounds.

3.4 Structural Analysis by spectral methods (FT-IR) confirms the film deposition, in accordance with AFM results.

The characterization of CNT-PANI thin films deposited on commercial substrate (DROPSENS), by FT-IR revealed the presence of the following polyaniline bands: 1589 cm^{-1} attributed to the N = Q = N (quinoid) group, 1294 cm^{-1} corresponding to the CN (aromatic amine) stretching vibration and the 1146 cm^{-1} band attributed to the NH groups of the PANI chains or „in plane" vibration of CH group (**Fig.5**).

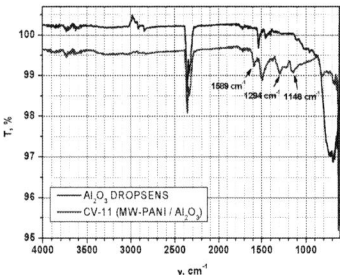

Fig.5. FT-IR spectra of DROPSENS substrate (black) and one CNT-PANI thin film (blue)

4. Conclusions

Thin films based on CNT-PANI hybrid nanostructures were obtained by hydrothermal-electrochemical method at 40°C and 60 bar on commercial substrates from DROPSENS.Prior to electrodeposition, morphology and structure of CNT-PANI nanopowders prepared by hydrothermal method were investigated, showing the presence of PANI on CNT surface by SEM analysis and the interaction between the two components by FTIR characterization. The

formation of thin films was studied using two electrochemical methods: cyclic voltammetry and chronoamperometry. Surface topography of the film compared to pristine substrate, as well as their roughness revealed the existence of a coating material. Chemical structure of the coating observed by AFM was studied using FTIR. The characterization of thin films through FT-IR revealed the presence of PANI through the vibration bands of this polymer. Testing of CNT-PANI thin films for VOCs detection is in progress.

Acknowledgments. The authors gratefully acknowledge the financial support of projects PN 16200301/2016 and MSCA-RISE-2014, Grant no. 645758/2014 (TROPSENSE).

References

[1] I. Cesarino, F. C. Moraes, M. R.V. Lanza, S.A.S. Machado, „*Electrochemical detection of carbamate pesticides in fruit and vegetables with a biosensor based on acetylcholinesterase immobilised on a composite of polyaniline–carbon nanotubes*", Food Chem. 135, pp. 873–879, 2012.

[2] S. B. Kondawar, M. D. Deshpande, S. P. Agrawal, „*Transport Properties of Conductive Polyaniline Nanocomposites Based on Carbon Nanotubes*", Int. J. Compos. Mater., 2(3), pp. 32-36, 2012.

[3] V. Choudhary, B.P. Singh and R.B. Mathur, „Carbon Nanotubes and Their Composites", Chapter 9 in „Syntheses and Applications of Carbon Nanotubes and Their Composites", pp. 193-222, 2013.

[4] M. Taki, F. Hekmat, B. Sohrabi, M. S. Rahmanifar, „*Carbon nanotube/polyaniline composite films prepared by in situ electrochemical polymerization for electrochemical supercapacitors*", Conference paper, d008. 10.3390/ecsoc-18-d008, 2014.

[5] V. H. Nguyen and J.-J. Shim, „*Green Synthesis and Characterization of Carbon Nanotubes/ Polyaniline Nanocomposites*", J. Spectroscopy, Article ID 297804, 2015.

[6] A. Bora, K. Mohan, D. Pegu, C. B. Gohain, S. K. Dolui, „*A room temperature methanol vapor sensor based on highlyconducting carboxylated multi-walled carbon nanotube/ polyaniline nanotube composite*", Sensors and Actuators B 253, pp. 977–986, 2017.

[7] G. Ciric-Marjanovic, „*Recent advances in polyaniline composites with metals, metalloids and nonmetals*", Synth Met. 170, pp. 31– 56, 2013.

[8] J. Lu, B.J. Park, B. Kumar, M. Castro, H.J. Choi, J.-F. Feller, "*Polyaniline nanoparticle carbon nanotube hybrid network vapour sensors with switchable chemo-electrical polarity*", Nanotechnology 21 255501, 2010.

[9] W. Li, D. Kim, "*Polyaniline/Multiwall carbon nanotube nanocomposite for detecting aromatic hydrocarbon vapors*", J. Mater. Sci. 45, pp. 1857-1861, 2011.

GeSi nanocrystals in SiO₂ matrix

with extended photoresponse in near infrared

I. Stavarache*, L. Nedelcu*,
V.S. Teodorescu*, V.A. Maraloiu*, I. Dascalescu*, M. L. Ciurea*,**

* National Institute of Materials Physics, 405A Atomistilor Street, 077125 Magurele, Romania
** Academy of Romanian Scientists, 050094 Bucuresti, Romania
E-mail: stavarache@infim.ro

Abstract— The films of SiGe nanocrystals in SiO₂ on Si substrate were obtained by co-sputtering Si, Ge, and SiO₂ followed by rapid thermal annealing. The films structure and morphology together with electrical and photoelectrical properties were studied by x-ray diffraction, transmission electron microscopy, current – voltage and spectral photocurrent measurements. The photocurrent spectra at 300, 200 and 100 K were correlated with results obtained from X-ray diffractograms and transmission electron microscopy. The photocurrent spectra show an extension in near infrared due to the enriching SiGe nanocrystals in Ge.
Keywords—SiGe nanocrystals; photoelectric; morphology; magnetron sputtering; near infrared.

1. Introduction

In recent years, the need for continued development of new materials and devices with improved optical and electrical properties compared to those of silicon based has attracted the attention of many research groups [1, 2]. The possibility to tune the electronic band structures of the binary alloys like SiGe between those of Si and Ge (bulk) and the SiGe compatibility with CMOS technology are the main arguments for alloying Ge with Si [3]. Furthermore, the alloying of Si with Ge has important advantages of a stronger quantum confinement (Ge exciton Bohr radius is four times bigger than that of Si), high absorption coefficient extended in near infrared (NIR) region, low crystallization temperature.

To date, various applications have been reported based on SiGe NCs from transistors [4], telecommunication [5], photo detectors [6] until to bio applications [7].

SiGe NCs with controlled size and composition embedded in different oxide matrices have been obtained by using different deposition methods like electron beam evaporation [8], low pressure chemical vapor deposition (PECVD) [9] and

magnetron sputtering (MS) [10, 11] followed or not by annealing [12, 13].

We report here on the synthesis of SiGe NCs embedded into SiO₂ deposited on Si substrate using MS method at room temperature followed by rapid thermal annealing (RTA). The co-deposited SiGe in SiO₂ films annealed at 800 °C for 15 minutes in N₂ was investigated. The spectral response at different temperatures (from 300 K to 100 K) correlated with structure and morphology of the films is discussed.

2. Experimental

In this study, thin film samples were prepared by co-deposition of Ge,Si and SiO₂ onto cleaned Si (100) substrates using a MS equipment (Gamma1000 tool from Surrey Nanosystems). Firstly, the deposition chamber was pumped down until the pressure reaching a base pressure of about 1×10^{-7} mTorr. During the deposition process the working pressure was maintained at 4 mtTorr by Ar (6N) gas. The substrate holder was rotated during deposition process to ensure a better film uniformity. After deposition, the obtained amorphous layers were annealed by RTA at 800 °C for 15 minutes in N₂ ambient for SiGe NCs formation.

In order to find out structural and morphological properties of the SiGe:SiO₂ thin films after annealing were performed grazing incidence x-ray diffraction (GIXRD) measurements using Bruker D8 ADVANCE diffractometer and transmission electron microscopy (TEM) using an equipment JEM-ARM200F, JEOL. The electrical characterization and spectral analysis were performed using a setup consisting of a

978-1-5386-4483-6/18 $31.00 © 2018 IEEE   253

Keithley 6517A electrometer, LakeShore 331 temperature controller, Stanford Research SR540 light modulator coupled with SR830 lock-in amplifier. The illumination of the investigated films was made with monochromatic light through a Newport monochromator and the illumination is provided by a 100 Watts Newport QTH lamp. The sample is placed into a Janis cryostat with optical windows. For the electrical and photoelectrical investigations, on the samples were deposited Al contacts by electron beam evaporation.

3. Results and Discussion

The structural, morphology, electrical and photoelectric results of the SiGe NCs embedded in SiO_2 are presented and discussed.

Details about the crystalline state of the SiGe embedded into SiO_2 films after annealing process at 800 °C for 15 min are presented in Fig. 1. The GIXRD measurements was made at 1.5 degree of incidence on angle dispersive x-ray diffraction beam line. For accurate peak assignment, the standard XRD tabulated patterns of Ge cubic (ICDD no. 004-0545) and Si cubic (ICDD no. 005-0565) are plotted. As it can be seen, the peaks are in between Ge and Si positions very close to Ge meaning that the film is Ge enriched.

In Fig. 2 is presented a low magnification cross section (XTEM) image of the SiGe:SiO_2 film annealed by RTA (800 °C,

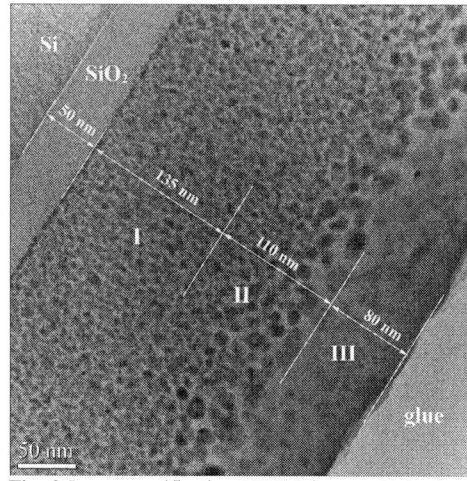

Fig. 2 Low magnification XTEM image obtained on SiGe:SiO_2 film annealed for 15 min in RTA.

15 min). As shown, the film thickness is of 325 nm and have three distinct zones. In the zone I (about 135 nm at the bottom of

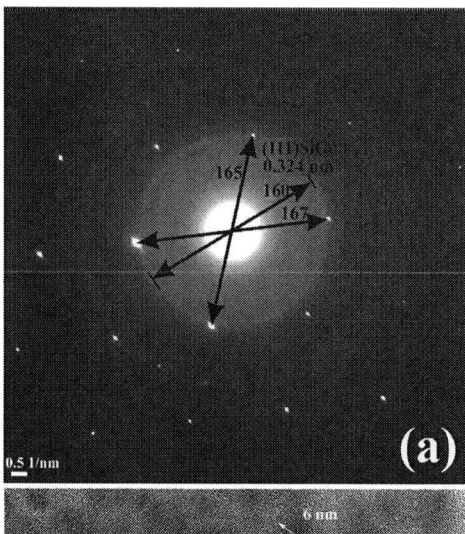

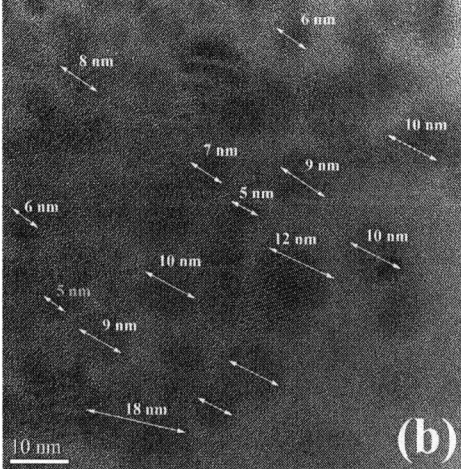

Fig. 3 SAED image on SiGe:SiO_2 film (a) and HRTEM image taken on zone II (b) respectively.

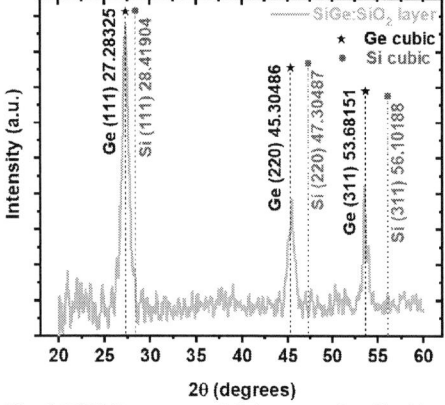

Fig. 1 GIXRD pattern at 1.5 degree angle of incidence on SiGe:SiO_2 film annealed for 15 min in RTA.

978-1-5386-4483-6/18 $31.00 © 2018 IEEE

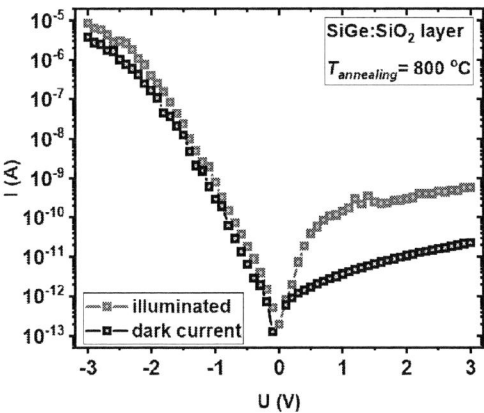

Fig. 4 *Current – voltage (I – V) characterisctics on SiGe:SiO₂ film annealed in RTA; dark current – dark line and under illumination – grey line.*

the samples), SiGe NCs with sizes of about 4-6 nm into SiO₂ are observed. In the middle of the films (zone II of about 110 nm) the crystallization degree increases with bigger SiGe NCs in the range 16-18 nm. A detailed analysis of the NCs fringes (111), observed in zone III in vicinity to zone II shows that the film is enriched in Ge (the lattice space is 0.328 nm that is close to Ge one). At the surface of the films no NCs were observed, and the film structure is in amorphous state.

If we take into consideration the increased sizes and change composition of the NCs observed to the surface, these results are in accordance with GIXRD investigations. Detailed high resolution (HR) TEM and SAED images obtained in zone II are presented in Fig. 3 (a) and Fig 3(b), respectively. In the SAED pattern the cubic SiGe is evidenced and in HRTEM image one can be seen that in zone II SiGe NCs size increase toward the free surface on the film.

Fig. 4 presents $I – V$ characteristics taken in dark and under illumination, plotted for a sample with SiGe NCs in SiO₂. The measurements were performed in co-planar configuration showing a strong rectifying behavior. Under illumination an increase of the current with more than one order of magnitude is evidenced.

The spectral dependence of the photocurrent ($I_{photo} – \lambda$) characteristics measured at different temperatures are shown

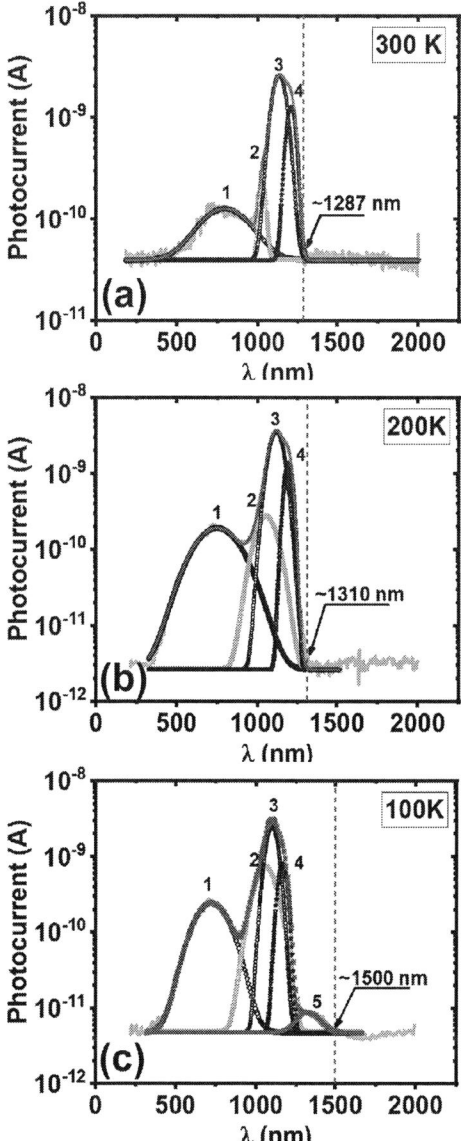

Fig. 5 $I_{photo} – \lambda$ *characterisctics on SiGe:SiO₂ film: (a) measured at 300 K, (b) at 200 K and (c) 100 K.*

in Fig.5. At 300 K, the photocurrent spectrum present two maxima located at ~ 790 nm and ~ 1140 nm, the one located at 1140 nm being dominant and two shoulders (peaks evidenced in deconvoluted curves) (Fig.5 (a)). At lower temperatures (200 K, Fig.5 (b) and 100 K, Fig.5 (c)), the peak 1 (~ 790 nm) is highlighted by increase of its intensity, the peak 4 (~ 1200 nm) is less pronounce in respect to peak 3 (1100 nm) and a new maximum located at ~ 1330 nm in NIR appears. We attribute the broad maximum located at ~ 720 – 780 nm is due to GeSi NCs contribution and the defects related to them while the peak at ~ 1090 - 1130 nm reflects the influence of Si substrate by capacity coupling through surface photovoltage and gating effects [14]. The most important peak from applications point of view is that located 1330 nm that shows the limit ~ 1500 nm. This result demonstrates that it is due to Ge NCs or highly enriched in Ge SiGe NCs and therefore we can conclude that inside the film Ge NCs and enriched SiGe NCs (in Ge).

4. Conclusions

NCs of SiGe embedded in SiO_2 were obtained by MS followed by annealing at 800 °C for 15 minutes in N_2. An extension in NIR of the spectral response was obtained on our samples suitable for optical sensor applications. This extension is obtained by preparing SiGe films with SiGe NCs enriched in Ge, the NCs being embedded in SiO_2 matrix.

Acknowledgments. This work was supported by the Ministry of Research and Innovation through NIMP Core Program PN18-11/2018 and M-ERA.NET PhotoNanoP 33/2016, TE 30/2018 and PCE 122/2017.

References

[1] T. Tah, Ch.K. Singh, S. Amirthapandian, K.K. Madapu, A. Sagdeo, S. Ilango, T. Mathews and S. Dash *"In-situ formation of Ge-rich SiGe alloy by electron beam evaporation and the effect of post deposition annealing on the energy band gap"*, Mat. Sci. Semicon. Proc., **80**, pp. 31–37, February 2018;

[2] J. Michel, J. Liu, and L.C. Kimerling, *"High-performance Ge-on-Si photodetectors"*, Nat.. Photonics, **4**, pp. 524-534, August 2010;

[3] H. Lafontaine, N.L. Rowell, G.C. Aers, D.C. Houghton, D. Labrie, R.L. Williams, S. Charbonneau, R.D. Goldberg and I.V. Mitchell, "*Band-gap tuning of SiGe/Si multiple quantum wells for waveguides and photodetectors*", Proc. SPIE Silicon-Based Monolithic and Hybrid Optoelectronic Devices, **3007**, pp. 48-54 , April 1997;

[4] E.Kasper, J. Eberhardt, H. Jorke, J.-F. Luy, H. Kibbel, M.W. Dashiell, O.G. Schmidt and M.Stoffel, *"SiGe resonance phase transistor:*

active transistor operation beyond the transit frequency f_T", Solid State Electron., **48** (5), pp. 837-840, May 2004;

[5] K. Hammani, M.A. Ettabib, A. Bogris, A. Kapsalis, D. Syvridis, M. Brun, P. Labeye, S. Nicoletti, D.J. Richardson and P. Petropoulos, *"Optical properties of silicon germanium waveguides at telecommunication wavelengths"*, Opt. Express, **21** (14), pp. 16690-16701, 2013;

[6] A. Yakimov, V. Kirienko, V. Armbrister and A. Dvurechenskii, *"Broadband Ge/SiGe quantum dot photodetector on pseudo substrate"*, Nanoscale Res. Lett., **8**:217, 2013;

[7] F. I. Jamal, S. Guha, M. H. Eissa, D. Kissinger, J. Wessel, *"A low-power 30 GHz complex dielectric chem-bio-sensor in a SiGe BiCMOS technology"*, 2017 First IEEE MTT-S International Microwave Bio Conference (IMBIOC), Gothenburg, Sweden ISBN: 978-1-5386-1713-7, DOI: 10.1109/IMBIOC.2017. 7965787

[8] V.A. Volodin, D.V. Marin, H. Rinnert and M. Vergnat, *"Formation of Ge and GeSi nanocrystals in GeO_x/SiO_2 multilayers"*, J. Phys. D: Appl. Phys., **46**, 275305, June 2013;

[9] M. Avella, A.C. Prieto, J. Jimenez, A. Rodrıguez, J. Sangrador, T. Rodrıguez, M.I. Ortiz and C. Ballesteros, *"Influence of the crystallization process on the luminescence of multilayers of SiGe nanocrystals embedded in SiO_2"*, Mat. Sci. Eng. B-Adv, **147**, pp. 200–204, 2008;

[10] N.N. Ha, N.T. Giang, T.T. T. Thuy, N. N. Trung, N.D. Dung, S. Saeed and T. Gregorkiewicz, *"Single phase $Si_{1-x}Ge_x$ nanocrystals and the shifting of the E_1 direct energy transition"*, Nanotechnology, **26**, 375701, August 2015;

[11] I. Stavarache, V.A. Maraloiu, C. Negrila, P. Prepelita, I. Gruia and G. Iordache, *"Photo-sensitive Ge nanocrystal based films controlled by substrate deposition temperature"*, Semicond. Sci. Technol., **32**, 105003, 2017;

[12] N.A.P. Mogaddam, A. S. Alagoz, S. Yerci, R. Turan, S. Foss, and T.G. Finstad, *"Phase separation in SiGe nanocrystals embedded in SiO_2 matrix during high temperature annealing"*, J. Appl. Phys., **104**, 124309, December 2008;

[13] C.-S. Lai, C.-Ming Yang, C.-Y. Wang, T.-C. Wang and D.G. Pijanowska, *"Chemical sensing properties of electrolyte/SiGe/SiO2/ Si/structure"*, J. J. Appl. Phys., **45** (8A), pp. 6192–6195, August 2006;

[14] A.M. Lepadatu, A. Slav, C. Palade, I. Dascalescu, M. Enculescu, S. Iftimie, S. Lazanu, V.S. Teodorescul, M.L. Ciurea, T. Stoica, "Dense Ge nanocrystals embedded in TiO_2 with exponentially increased photoconduction by field effect", Sci Rep-UK, **8**, 4898, March 2018.

The effect of H$_2$/Ar plasma treatment over photoconductivity of SiGe nanoparticles sandwiched between silicon oxide matrix

M.T. Sultan[1], J.T. Gudmundsson[2,3], A. Manolescu[1], M.L. Ciurea[4,5], H.G. Svavarsson[1]

[1]Reykjavik University, School of Science and Engineering, IS-101 Reykjavik, Iceland
muhammad16@ru.is, manoles@ru.is, halldorsv@ru.is
[2]Department of Space and Plasma Physics, School of Electrical Engineering and Computer Science,
KTH-Royal Institute of Technology, SE-100 44, Stockholm, Sweden
[3]Science Institute, University of Iceland, Dunhaga 3, IS-107 Reykjavik, Iceland
tumi@hi.is
[4]National Institute of Materials Physics, 077125 Magurele, Romania
[5]Academy of Romanian Scientists, 050094 Bucuresti, Romania
ciurea@infim.ro

Abstract—The effect of room temperature hydrogen plasma treatment on the photoconductive properties of the SiO$_2$ matrix containing SiGe nanoparticles is investigated. A considerable increase in photocurrent intensity is observed after plasma treatment. The increase is partly attributed to neutralization of dangling bonds around the nanoparticles and partly to passivation of non-radiative centers and defects in the matrix and at the nanoparticles-matrix interfaces.

Keywords—SiGe; SiO$_2$; hydrogenation; magnetron sputtering, HiPIMS; photoconductivity.

1. Introduction

Functional devices with targeted optoelectronics properties can be made by growing thin oxide films with embedded semiconducting nanoparticles. In the as-grown state, such films always have structural imperfections such as crystal defects and dangling bonds, present in various amounts. These imperfections reduce the optical and electrical performance of the structure, partly by acting as recombination centers and by trapping charge carriers.

It is well established that treatment with hydrogen (H$_2$) plasma (hydrogenation) can passivate deep level traps and defects in semiconductors [1, 2, 3, 4, 5, 6]. Moreover, hydrogenation is found to be particularly effective on structures containing embedded nanocrystals in an amorphous matrix [6].

Hydrogen plasma tends to passivates non-radiative centers (P$_b$) located at or in close proximity to nanoparticles. The aim of this study is to investigate the effect of hydrogenation on a structure composed of SiGe nanoparticles sandwiched between the SiO$_2$ matrix structure. The goal is to increase the efficiency of photoconductive devices by widening the spectral sensitivity and improving charge transport by suppressing defect concentration and/or neutralization of radiative defects and dangling bonds. Researches on the effect of hydrogenation on the photoluminescence of a similar system as here (Si and Ge NPs in SiO$_2$ matrix) have been published [6, 5, 7, 4] but few if any have reported on its effect on photocurrent as is done here. It is also to mention here that our sample treatment did not involve any external heating (neither annealing nor heating during hydrogenation).

2. Experimental method

Multilayers structure (MLs) of SiO$_2$ and SiGe were deposited on Si (001) substrates. The stacking order was a 200 nm of the SiO$_2$ buffer layer and SiO$_2$/SiGe/SiO$_2$ films with thicknesses of 40/20/40 nm, respectively, on

978-1-5386-4483-6/18 $31.00 © 2018 IEEE

top of that. The SiO_2 layer was deposited by reactive sputtering from a Si target by direct current magnetron sputtering (dcMS) using Advanced Energy MDX500 power supply. For the SiGe deposition, pure Si and Ge were co-sputtered from 99.9999% (6N) pure Si and Ge targets (with ratio 50:50) respectively. The Si was deposited using dcMS, while the Ge was deposited using a high impulse power magnetron sputtering (HiPIMS). For the HiPIMS, a square current-voltage waveform with a pulse length of 200 µs and frequency of 100 Hz was used. The average current density and the peak power density was 190 ± 10 mA/cm^2 and 84 ± 2 W/cm^2, respectively, at 470 V. The power for the HiPIMS was supplied by SPIK1000A pulse unit (Melec GmbH) operating in the unipolar negative mode at a constant voltage, which was charged by a DC power supply (ADL GS30).

The as-deposited samples (asd) were hydrogenated in 6N Ar/H$_2$ in 30:70 ratio in an inductively coupled discharge. The samples were kept inside a quartz tube that was placed in copper coil (see photograph in Fig. 1) connected to a radio frequency CESAR$^©$ 136 rf power generator (13.56 MHz), coupled with impedance matching unit to generate the plasma.

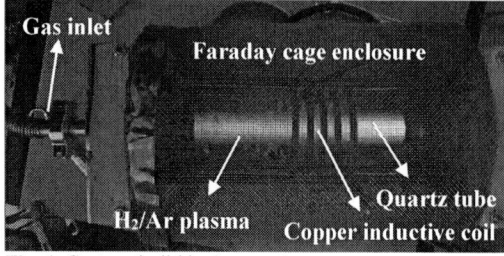

Fig. 1. Custom-build hydrogenation setup.

The hydrogenation was performed in cycles of 10 minutes. After each cycle, the samples were taken out of the quartz tube and measured. The rise in temperature during the cycle was determined using color-changing temperature stripes. For photocurrent measurement, two coplanar Al contacts were deposited on top of the MLs surfaces in the

asd state. The current passing through the contacts at a constant applied bias of 1 V in dark and under illumination from a tungsten-halogen lamp measured. The photocurrent spectra were obtained by extracting the dark current from the originally measured photocurrent and then normalizing the outcome with the light source spectrum.

3. Result and discussion

Fig. 2 shows the photocurrent spectra of asd sample. Three peaks are evident in the photocurrent spectra (in region A, B, and C). These peaks in corresponding regions can be attributed: to strain in MLs due to lattice mismatch between Si and Ge, to different density gradient between NCs and dielectric matrix, and to interface defects concentration in MLs (peak S$_t$); photo-effect from SiGe nanocrystals (peak NCs); and to surface photo-voltage (peak SPV) as explained by Lepadatu et al and Palade et al [8, 9].

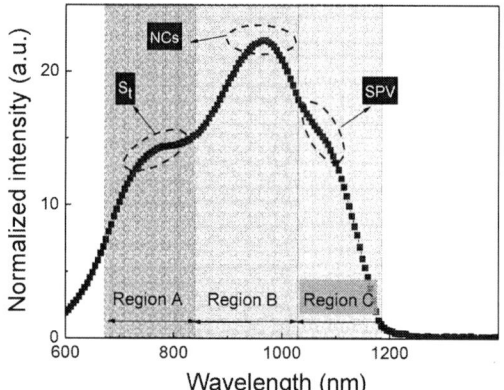

Fig. 2. Photocurrent spectra of asd MLs at room temperature.

The XRD curve (not shown here) reveals broad hump, which can be attributed to presence of nanoparticles (which may consist of large amorphous nanoclusters and/or small crystalline nanoparticles). The shape of the curve indicates that even in the asd contain a population of NPs (either amorphous or slightly crystalline at best) which does significantly contribute to the improved photosensitivity as also being discussed by E. G. Barbagiovanni, D. J. Lockwood, P. J. Simpson, L. V. Goncharova [10].

A photocurrent spectrum of asd MLs after hydrogenation of 0-50 min is shown in Fig. 3. The sample in asd condition underwent hydrogenation for several cycles of 10 min intervals. There is a clear increase in the photocurrent intensity with each hydrogenation cycle. During each cycle, the temperature inside the chamber reached a maximum of < 45 °C.

The observed increase in intensity can be understood on basis of interaction of H and H^+ (either directly from plasma or from dissociation of H_2^+ and H_3^+ into H and H^+ on the surface) with the SiGe-NCs/SiO_2 amorphous matrix interfaces and defects in matrix associated with nonstoichiometric SiO_2 [4]. These hydrogen products tend to passivate the dangling bonds.

role of non-radiative centers caused by Si-dangling bonds [6, 12, 5]. These dangling bonds can be passivated by hydrogenation, directly by exposure of the sample to plasma and introduction of atomic hydrogen in structure, which causes a reduction in dangling bond at the boundaries and at the surface via formation of Si-H bonds and annealing of defects [13, 14, 6, 4]. Peak decay in region C-dotted region (Fig. 3) after hydrogenation can be attributed to the reduction in charge trapping at SiO_2: Si-substrate interface.

Earlier studies [15, 16] have shown that H_2^+ and H_3^+ ions play a significant role in improving the efficiency of hydrogenated devices. It has also been established that electron density in plasma increases with increasing fractional Ar pressure and that $[H_2^+]/[H_3^+]$ ratio does not change when argon is added.

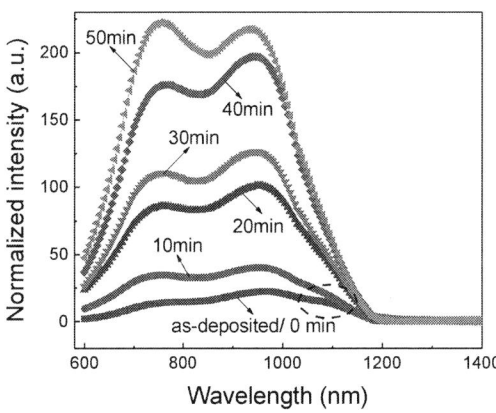

Fig. 3. Photocurrent spectra of asd MLs hydrogenated in intervals of 10 min (0-50 min in total).

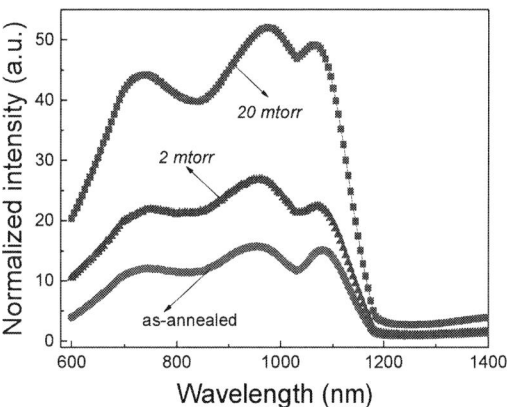

Fig. 4. Effect of hydrogen pressure on photocurrent of annealed MLs (600 °C, 1 min) treated for 10 min each.

Atomic hydrogen plays a vital role in the hydrogenation of SiO_2:SiGe interface, resulting in the reduction of positive fixed charge in dielectric matrices such as SiO_2 [11, 5]. Further, an increase in intensity during illumination of MLs after hydrogenation is partly related to the passivation/ decrease of non-radiative centers (P_b) which might be located in oxide matrix or at the sinterface between QD's and surrounding oxide matrix. Several researchers has studied the critical

Moreover, it was concluded that the ratio of H^+ ions increases with decreasing pressure while the H_2^+ and H_3^+ concentration increases with increased pressure. Fig. 4 shows the effect of increasing the gas pressure during hydrogenation. By increasing pressure, the H_2^+ and H_3^+ concentration increases which when encounters the surface, will dissociate into H^+. Thereby increasing the concentration of atomic hydrogen and the passivation level. However, in contrast to this E. S. Cielaszyk,

K. H. R. Kirmse, R. A. Stewart, A. E. Wendt [17] found H and H^+ in pure hydrogen plasma, present at low-pressure, results in higher degree of passivation. Thus, further work is required to determine the effect of pressure from as low as 0.2 mtorr to 20 mtorr or higher.

4. Conclusion

In summary, MLs underwent H_2 plasma treatment in intervals of 10 min at room temperature. Hydrogenation tend to increase the photocurrent intensity by passivation of P_b, dangling bonds and restructuring of oxide matrix without damaging the surface of MLs, that otherwise might have reduced the intensity due to surface scattering. Further work is required to compare and analyze the effect of long-term H_2 plasma treatment where MLs is being exposed for longer continuous time rather in short intervals.

Acknowledgments. This work was supported by M-ERA.NET projects PhotoNanoP UEFISCDI Contract no. 33/2016 and GESNAPHOTO UEFISCDI Contract no. 58/2016 and by Romanian Ministry of Research and Innovation through NIMP Core Program PN16-480102.

References

[1] J. Kassabov, E. Atanassova, D. Dimitrov, J. Vasileva, *"Effects of hydrogen plasma on thin-oxide Si-SiO₂ structures"*, Semiconductor Science and Technology, 3(7), pp. 686–690, July 1988.

[2] H. Olafsson, J. T. Gudmundsson, H. G. Svavarsson, H. P .Gislason, *"Hydrogen passivation of AlxGa1-xAs/GaAs studied by surface photovoltage spectroscopy"*, Physica B: Condensed Matter, **273-274**, pp. 689–692, December 1999.

[3] M. Mews, E. Conrad, S. Kirner, N. Mingirulli, L. Korte, *"Hydrogen Plasma Treatments of Amorphous/Crystalline Silicon Heterojunctions"*, Energy Procedia, **55**, pp. 827–833, 2014.

[4] A. N. Nazarov; V. S. Lysenko; T. M. Nazarova, *"Hydrogen plasma treatment of silicon thin-film structures and nanostructured layers"* Semiconductor Physics Quantum Electronics & Optoelectronics, **11** (2), pp. 101-123, July 2008.

[5] I.Z. Indutnyy; V.S. Lysenko; I.Yu. Maidanchuk; V.I. Min'ko; A.N. Nazarov; A.S. Tkachenko; P.E. Shepeliavyi; V.A. Dan'ko, *"Effect of chemical and radiofrequency plasma treatment on*

photoluminescence of SiOx films", Semiconductor Physics Quantum Electronics and Optoelectronics, **9**(1), pp. 9-13, January 2006.

[6] A. I. Yakimov, V. V. Kirienko, V. A. Armbrister, A. V. Dvurechenskii, *"Hydrogen passivation of self-assembled Ge/Si quantum dots"*, Semiconductor Science and Technology, **29** (8), 085011, June 2014.

[7] F. Mofidnakhaei, C. Mohammadizadeh, N. Refahati, *"Structural and Luminescence Properties of SiGe Nanostructures with Ge Quantum Dots"*, Research Journal of Environmental and Earth Sciences, **6** (9), pp. 466-468, September 2014.

[8] A. M. Lepadatu, A. Slav, C. Palade, I. Dascalescu, M. Enculescu, S. Iftimie, S. Lazanu, V. S. Teodorescu, M. L. Ciurea, T. Stoica, *"Dense Ge nanocrystals embedded in TiO2 with exponentially increased photoconduction by field effect"*, Scientific Reports, **8** (1), 4898, March 2018.

[9] C. Palade, I. Dascalescu, A. Slav, A. M. Lepadatu, S. Lazanu, T. Stoica, V. Teodorescu, M. L. Ciurea, F. Comanescu, R. Muller, A. Dinescu, A. Enuica, *"Photosensitive GeSi/TiO2 multilayers in VIS-NIR"*, 2017 International Semiconductor Conference (CAS), Sinaia, pp. 67-70, 2017.

[10] E. G. Barbagiovanni, D. J. Lockwood, P. J. Simpson, L. V. Goncharova, *"Quantum confinement in Si and Ge nanostructures"*,Journal of Applied Physics, **111** (3), 034307, February 2012.

[11] A. Szekeres, S. Alexandrova, *"Low-temperature treatment of Si/SiO2 structures in an RF hydrogen plasma"*,Vacuum, **47** (12), pp. 1483–1486, March 1996.

[12] A. P. Jacob, Q. X. Zhao, and M. Willander, *"Hydrogen passivation of self assembled InAs quantum dots"*, Journal of Applied Physics, **92** (11), 6794, August 2002.

[13] E. Cartier, J. H. Stathis, D. A. Buchanan, *"Passivation and depassivation of silicon dangling bonds at the Si/SiO2 interface by atomic hydrogen"*, Applied Physics Letters, 63 (11), 1510–1512, September 1993.

[14] L. P. Scheller, M. Weizman, P. Simon, M. Fehr, N. H. Nickel, *"Hydrogen passivation of polycrystalline silicon thin films"*, Journal of Applied Physics, **112** (6), 063711, August 2012.

[15] J. T. Gudmundsson, *"Experimental studies of Ar plasma in a planar inductive discharge"*, Plasma Sources Science and Technology, 7 (3), pp. 330–336, August 1998.

[16] C. F. Yeh, T. J. Chen, C. Liu, J. T. Gudmundsson, M. A. Lieberman, *"Hydrogenation of Polysilicon Thin-Film Transistor in a Planar Inductive H₂ /Ar Discharge"*, IEEE Electron Device Letters, **20** (5), 223–225, May 1999.

[17] E. S. Cielaszyk, K. H. R. Kirmse, R. A. Stewart, A. E. Wendt, *"Mechanisms for polycrystalline Silicon defect passivation by Hydrogenation in an Electron-Cyclotron-Resonance plasma"*, Applied Physics Letters, 67 (21), pp. 3099–3101, November 1995.

Direct Writing Patterns for Gold Thin Film with DPN technique

Carp Mihaela*, Pachiu Cristina*, Dediu Violeta *

* National Institute For Research And Development In Microtechnologies - IMT Bucharest

E-mail: mihaela.carp@imt.ro

Abstract— Dip-pen nanolithography combined with wet-chemical etching has been used to generate gold nanostructures with desired shapes and sizes. Self-assembled monolayers of 16-mercaptohexadecanoic acid have been patterned by DPN in different shapes: dots, lines and complex shapes, interdigits electrodes. AFM and LFM were used to measure the roughness of gold surface and to examine the thiol deposition and binding quality. These results show that DPN can be used as alternative method to generate different patterns used for complex devices, biosensor, and optoelectronic devices.

Keywords—dip pen nanolithography; direct-write; MHA; wet chemical etching.

1. Introduction

Dip - pen nanolithography (DPN) is a direct writing method that uses a tip of an atomic force microscope in order to transfer molecules or nanoparticles to the surface of a substrate in a controlled manner. In DPN, the material (the „ink") to be patterned is transferred from the coated tip (the „pen") to the substrate surface through a water meniscus (Figure 1).

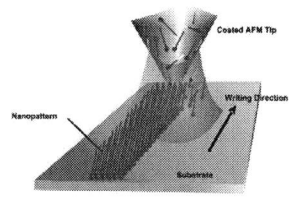

Fig. 1. The DPN process: a coated pen is drawing "nib." Molecules absorbed onto the pen became the "ink," and the sample substrate was the "paper."[1]

Comparing with micro/nano - contact printing, DPN provides a greater flexibility because it allows control over the size of the pattern by adjusting the tip speed dwell time and environmental control. Through this process features with sub - 100 nm resolution have been patterned [2-3]. In addition, DPN has more advantages over other nanolithography techniques: the equipment it

is not too expensive, doesn't need a restricted environment (as does electron beam lithography), and it can be carried out under ambient conditions. Another advantage is the possibility to immediately check the deposition quality by Lateral Force Microscopy (LFM). Disadvantages come from the diffusion characteristics of the ink in the substrate: limited speed and rounded corners. Diffusion of the ink influences, also, the resolution and the reproducibility of the pattern.

Dip - pen nanolithography can be used to create self - assembled monolayers (SAMs) on the substrate [4], with great selectivity, to deposit in a controlled manner: conductive polymers], biopolymers and macromolecules [5-6]. Also, hard materials can be deposited in a desired pattern: metals, inorganic sol precursors, nanoparticles catalyst, etc. [7-9]

In this paper we present fabrication of gold electrodes using DPN by direct deposition of self-assembled monolayers of molecular ink as positive resist on gold substrate, followed by wet chemical etching.

2. Experimental Section

A. Preparing MHA-inked AFM Tips for DPN

Dip Pen Nanolithography (DPN) experiments were carried out with a commercial Dip Pen writer called NSCRIPTOR™ DPN® System (NanoInk Inc., USA) situated in cleanroom class 100 environment. Single commercial pen with double side Probe A AFM cantilevers (Si_3N_4, NanoInk, Inc., USA) were used for patterning. The cantilever on side 1 has parallelepiped shape, 200 µm long and 40 µm wide with force constant of 0.04 N/m and the one on side 2 have triangular shape, 200 μm long and 200 μm base wide with force constant of 0.1 N/m. All DPN patterning experiments were carried out under ambient conditions (~ 50% relative humidity, 20-24 °C). To load the molecular ink (16-mercaptohexadecanoic acid MHA) onto the tips we used two methods[8]:

a) The cantilever was manually dip in the ink for 5 s then blow dry using a nitrogen gun with pressure fixed at the lowest pressure and the procedure is repeated again. This method is called the double dip method.

b) The cantilever was dip in the ink for 10 s, left to dry in air for 5 s followed by a quick dip in pure ethanol and the cantilever was then allowed to dry in ambient conditions.

Molecular ink MHA was purchased from NanoInk Inc.

B. Preparing gold substrates for DPN

Silicon substrates were freshly chemically cleaned of organic impurities using standard cleaning solutions: first piranha solution (3 parts of concentrated sulfuric acid and 1 part of 30% hydrogen peroxide solution) followed by insertion in deionized water until the acid is removed. The wafer is dried with a nitrogen gun. The second step of cleaning is a short immersion in a 1:50 solution of aqueous HF (hydrofluoric acid) at 25°C for about fifteen seconds, in order to remove the thin native oxide layer and some fraction of ionic contaminants. After drying in nitrogen the wafer is ready for metal deposition.

For obtaining a smooth gold surface, two different methods for metal deposition were perfomed:

1) A 1 nm thick adhesion layer of titanium followed by a 10 nm thick layer of gold were evaporated on these clean substrates

2) In the second method, the wafers were annealed at 105 °C for 30

minutes before metal deposition. After this 5 nm thick adhesion layer of titanium followed by a 50 nm thick layer of gold were then evaporated followed by a 1 h cooling step under vacuum.

After metallization the probes were portioned in 100 square microns to be fitted with working surface of DPN.

C. Procedure in InkCAD

InkCAD program represents main DPN application used to design nano-scale patterns and, also, it gives access to several other specialized modules absolutely necessary for the operations presented next.

Before drawing the desired pattern, the instrument was prepared going through the following steps: the substrate was placed on the puck, a laser (red dot) alignment was performed by reflecting the laser on the back of the cantilever, and then the cantilever was slowly approched the sample in the park coordinates.

In order to draw a dot with a desired surface area the ink was calibrated of which results will be discused in section 3.

After calibrating the ink for the layer we had designed three models - M1,M2 and M3. In Figure 2, was presented the M1 pattern, consisting of five elements: Van der Pauw structure, two intredigits electrodes, 2 sets of lines with vertical and horizontal orientation, and 4 dots with differnt diameters. On the left upper side of the image are presented all elements designed on the M1 mask and on the left lower side are the characteristics of the selected element on the mask (the yellow disk) where it was possible to modify the dwell time/ the speed time. The five lines sets are designed to be written with corresponding speed, from left to right, 0.5, 0.6, 0.7, 0.8, 1microm/s and corespondingly, from bottom to top for the horizontal ones. The elements are written in the same order in which they are drawn. The total writing time for M1 pattern was 5'30".

After designed each model mask, the drawings were selected and, accordingly, drawn them on the gold surface samples, and after that AFM scanning was performed. The AFM scannings has six figures: foward and reverse scanning for Lateral Force Microscopy (LFM), topography and error images. In Figure 4 – 5 were presented the LFM images of M2 and M3 masks.

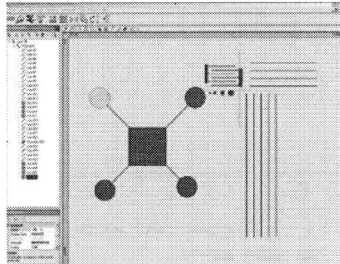

Fig. 2. M1 pattern preview:

D. Wet Chemical Etching of Gold

The gold from areas not modified with MHA were removed by treating with a ferri/ferrocyanide etchant aqueous mixture of 0.1 M $Na_2S_2O_3$, 1.0 M KOH, 0.01 M $K_3Fe(CN)_6$ and 0.001 M $K_4Fe(CN)_6$ [10]. The etching time varied from 4 to 20 minutes, depending on the thickness of the gold. A short immersion in a 1:50 solution of aqueous HF (hydrofluoric acid) at 25°C for about 5 seconds, in order to remove the thin adhesion layer of titanium. The substrates were afterwards rinsed with deionized water and dried with nitrogen.

3. Results and Discussions

The fabrication steps of microscale gold structures on Si substrate are described in the Experimental Section. After metallization, the substrate surfaces were scanned to determine the average roughness of gold and we obtain 1.77 nm for first method and 1.12 nm for the second one. The second one provided an ideal gold surface for DPN.

Figure 3 shows a line of Au microdots fabricated on a silicon surface at 50–60% relative humidity. The gold microdots were generated using 1 s, 2 s, 4 s, 6 s, 8 s and 10 s dwell times with the distance between the dots of 2 µm. The writing time was 0.55 s .As it can be seen in Figure 3A, the generated gold microdots are uniform, and the diameter

of the obtained Au microdots is progressively increase from 0.5 µm to 1.45 µm, with increasing dwell time. The longer the holding time, the higher the ink transport, and the microstructure diameter increases progressively. From Figure 3B the height of the profile reaches a maximum of 60 nm for 10 s DPN tip-holding time. From data fitting, the ink diffusion coefficient is 0.052 $\mu m^2/s$. The value was saved and used to draw lines with a designated width.

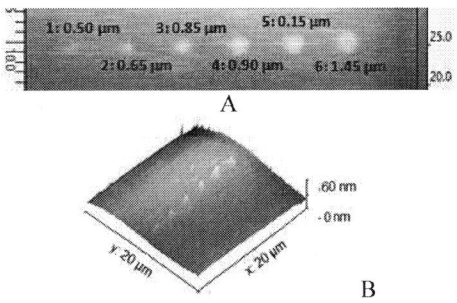

Fig.3. LFM image of the Au microdots on a Si surface generated by DPN; B) 3D view for height profile of the dot line.

Figure 4 shows 4 elements: two serpent lines, one vertical and another horizontal orientation, two horizontal parallel lines and array of 3x3 dots. The writing speed for first serpent line was 0.6 µm/s and 1 µm/s for the horizontal one, which correspond to respectively, 0.74 µm, and 0.43 µm width. The speed for parallel lines was 0.2 µm/s which correspond to 1.2 µm width. The dots were prepared with 4, 2, and 1 s dwell times, for each line resulting dots with diameters ranging from 0.54 to 1.36 µm. The writing time was 2 minutes.

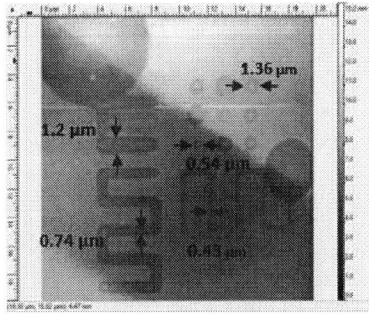

Fig. 4. LFM image of an array of M2 pattern generated on an gold substrate.

In Figure 5 we present the written pattern M3, representing few interdigitated electrods with different lines width and orientations. The left figure represents the LFM image of these elecrodes and the right figure represents an image from an optical microscope of the etched interdigited electrodes. It can be observed that the best results were obtained for the lines written in one model with the same writting speed.

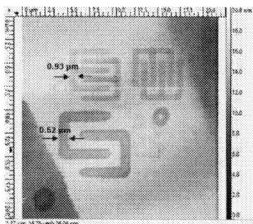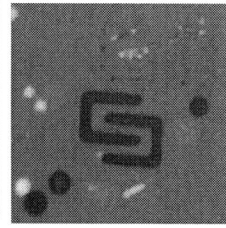

Fig.5 LFM image of interdigitated – test structures (left picture) and optical image (on the right) of the same structures after wet etching of gold.

In Figure 6 represents the LFM image of model M1(A) which was designed in Figure 2. The Figure 6B shows the position of laser spot during the writing process, and Figure 6C indicates the height profile of five lines. The z scale is 50 nm and x scale is about 0.3 µm for each line.

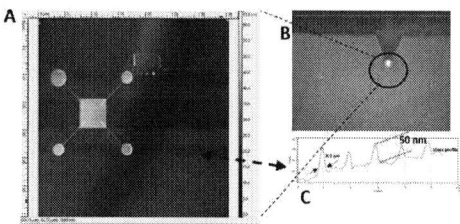

Fig.6. A) M1 patterning gold microstructures with MHA ink, **B)** cantilever with position of laser spot and **C)** height profile extracted from LFM image

4. Conclusions

DPN is a direct write, a direct deposit technology with major strengths such us insertion of materials at defined locations with nanoscale precision, flexibility on design of pattern and integration of a variety of atoms, molecules, and materials with accuracy at the nanometer length scale. Furthermore, DPN is a very useful tool to study the fundamental consequences of miniaturization and additional investigations

are expected in this field. Our results show that DPN can be used to generate different patterns used for complex devices.

Acknowledgments. We like to thanks to our colleges Antonio Radoi and Adina Bragaru Boldeiu for their help.

References

[1] NSCRIPTOR user manual, revision G, 4/2006,p. 14.
[2] C. A. Mirkin, S. Hong, and L. Demers, *"Dip-Pen Nanolithography: Controlling Surface Architecture on the Sub-100 Nanometer Length Scale"*, ChemPhysChem, **2**, pp. 37–39, 2001.
[3] S. H. Hong, J. Zhu, and C. A. Mirkin, *"Multiple ink nanolithography: Toward a multiple-pen nano-plotter"*. Science, **286**, pp. 523–525, 1999.
[4] O. A. Nafday, J. R. Haaheim, and F. Villagran, *"Site-Specific Dual Ink Dip Pen Nanolithography^{TM}"*, Scanning, **3**, pp. 122–126, 2009.
[5] M.Y. Yang, P.E.Sheehan, W.P.King, and L.J. Whitman, *"Direct writing of a conducting polymer with molecular-level control of physical dimensions and orientation"*, J. Am. Chem. Soc., **128**, pp. 6774 – 6775, 2006.
[6] K. Mitsakakis, S. Sekula-Neuner, S. Lenhert, H. Fuchs, and E. Gizeli, *"Convergence of dip-pen nanolithography and acoustic biosensors towards a rapid-analysis multi-sample microsystem"*, Analyst, **137**(13), pp. 3076--3082, 2012.
[7] N. S. John, G. Gundiah, P. J. Thomas and G. U. Kulkarni, *"Dip-pen nanolithography using colloidal inks"*, Int. J. Nanosci., **4**(05n06), pp. 921– 934, 2005.
[8] A. Calborean, I. Grosu, A. Colnit, D. Marconi, *"Design complexity of DPN patterning with Cr^{3+} and Co^{2+} metallic ions on Au (111) thin film"*, J. Alloys Compd., **747**, pp. 149–155, 2018.
[9] M. Su, X. Liu, S.Y, Li, V.P. Dravid, and C.A. Mirkin, *"Moving beyond molecules: patterning solid-state features via dip-pen nanolithography with sol-based inks"*, J. Am. Chem. Soc., **124**(8), 1560–1561, 2002.
[10] Hua Zhang, N. A. Amro, S. Disawal, R. Elghanian, R Shile, J. Fragala, *"Microstructure array on Si and SiOx generated by micro-contact printing, wet chemical etching and reactive ion etching"* App. Surf. Sc. 253 (2006) 1960–1963
[11] Y. Xia, X.M. Zhao, E. Kim, G.M. Whitesides, *"A selective etching solution for use with patterned self-assembled monolayers of alkanethiolates on gold"*, Chem. Mater.,7(12), pp. 2332–2337, 1995.

Substrate effect on the morphology and optical properties of ZnO nanorods layers grown by microwave-assisted hydrothermal method

Ana Filip*, Viorica Musat*, Nicolae Tigau*, Alina Cantaragiu*,
Cosmin Romanitan**, Munizer Purica**

*Centre of Nanostructures and Functional Materials, Faculty of Engineering, "Dunărea de Jos" University of Galati, Galati, Romania
*E-mail: viorica.musat@ugal.ro
**IMT Bucharest – National Institute for Research and Development in Microtechnologies, Bucharest, Romania
**E-mail: munizerp@imt.ro

Abstract — The substrate effect on the morphology and optical properties of zinc oxide nanorods synthesized by microwave-assisted hydrothermal method have been investigated by scanning electron microscopy (SEM), X-ray diffraction (XRD) and UV-VIS-NIR optical absorption and reflecting spectroscopy. The band gap energy of the investigated samples was calculated from absorbance spectra in the (200-1100) nm wavelengths range.

Keywords — ZnO nanorods; microwave-assisted hydrothermal synthesis; morphology; optical properties, band gap energy.

1. Introduction

Zinc oxide (ZnO) is among the most widely used nanomaterial, being a natural n-type semiconductor, inexpensive, simple to process. It has thermal and chemical stability at room temperature with high excitation energy of 60 m eV and the band gap of 3.37 eV [1][2].

ZnO 1D nanostructures have attracted much attention due to their high surface-to volume ratio and because they present deep defects that are responsible for self-assembly growth property. With unique properties, ZnO can be easily synthesized into different nanomorphologies, such as nanorods, nanotubes, nanoflowers, nanowalls, nanowires, nanocoral etc. [3][4][5][6]. For the fabrication of ZnO nanostructures various methods can be used (sol–gel, spray pyrolysis, sputtering, hydrothermal, physical vapor deposition, chemical bath deposition, electrodeposition and others) [1][7][8][9][11] Among the various synthesis methods, the hydrothermal technique is preferred due to several advantages including excellent control of particle morphology and size, high purity of the obtained nanostructures, high homogeneity and uniformity of nanostructured films.

Nowadays microwave-assisted (MW) hydrothermal synthesis has been widely used as a novel heating model in material science due to its various advantages such as short reaction time due to the fact that microwaves transfer energy directly to the reactive species, simplicity and large-scale production [1][9]. This method enable to create different 1D and 2D nanostructured materials with improved properties for application in optoelectronics [10][12].

This paper presents the substrate effect on the morphology and optical properties of zinc oxide nanorods (NRs) grown by MW-assisted method on glass substrates covered with ZnO thin layer have been investigated using X-ray diffraction (XRD), scanning electron microscopy (SEM) and optical transmittance-reflection spectra.

2. Experimental part

The nanorods layers were grown into an equimolar aqueous solution of 0.025M zinc nitrate hexahydrate ($Zn(NO_3)_2 \cdot 6H_2O$) and hexamethylenetetramine (HMTA, $(CH_2)_6N_4$) using a sealed Teflon-lined autoclave and the domestic Samsung M71A microwave oven. All the reagents (analytical grade purity) were purchased from Sigma Aldrich and used as received, without any further

978-1-5386-4483-6/18 $31.00 © 2018 IEEE

purification. The substrates were submerged in the growth solution inside a Parr 4782 Microwave Digestion Vessel and irradiated at 450 W for 3 minutes. After removing from solution, the obtained samples were washed, dried in air and heated to 90 °C for 60 minutes. The samples were grown on glass substrates covered with sol-gel ZnO thin films with thickness of ~40 nm (sample S1), ~70 nm (sample S2), ~100 nm (sample S3) and ZnO films deposited by physical vapor deposition (PVD) of ~120 nm thickness (sample S4).

The morphology of the synthesized samples was characterized using Scanning Electron Microscope (SEM) FEI Quanta 200 with a conventional tungsten electron source giving a resolution of 3.5 nm. The crystalline structure was identified by X-ray diffractometry using a Rigaku SmartLab X-ray diffractometer (Cu radiation). The crystallite sizes of the obtained nanostructures were calculated from the three main diffraction peaks using Scherer formula [9]. The optical transmission and reflectance spectra were acquired at room temperature with a Perkin Elmer Lambda 35 spectrometer, operated in air, at normal incidence, in the 200-1100 nm spectral range.

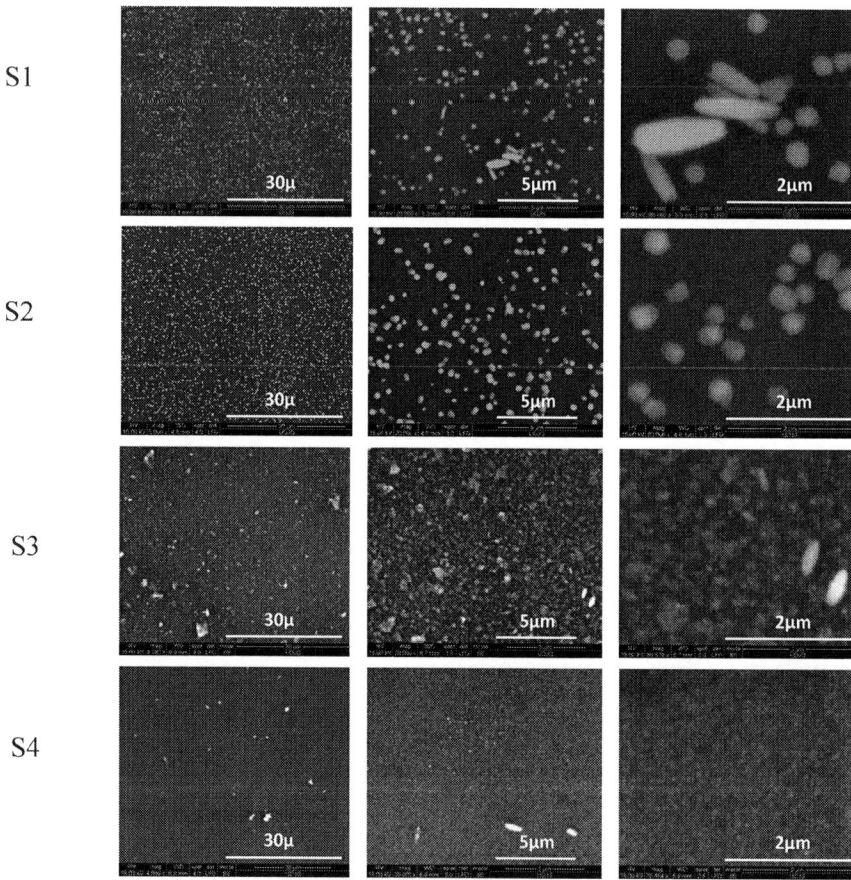

Fig. 1 SEM image at 30μm, 5μm, 2μm respectively of the obtained ZnO-based nanostructures

3. Results and discussion

The morphologies of the investigated samples at different magnifications are illustrated in Fig.1. As can be seen from these images, with the increase in the thickness of the ZnO layer deposited on these substrates, higher-density of the grown nanorods arrays can be observed. Also the rods diameter slightly increased from sample S1 to S2, and decreased consistently for S3 and S4 samples. So, the diameter of the well-defined nanorod-like structures increases from 330 nm (S1) to 400 nm (for S2) and their aspect ratio increases from 3 (S1) to 4 (S2). As concern the sample S3, the aspect ratio value increased at 6 and agglomerations of smaller diameter nanorods can be observed. This trend was also observed for sample S4, where the thicker (120 nm) PVD ZnO layer leads to a more compact layer of smaller diameter ZnO nanorods. Larger areas with higher density of smaller diameters nanorods arrays agglomerations are observed.

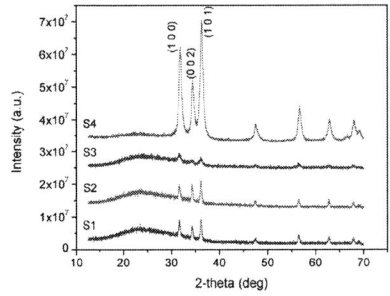

Fig. 2 XRD analyses of ZnO-based nanostructures obtained on thin films with different thicknesses

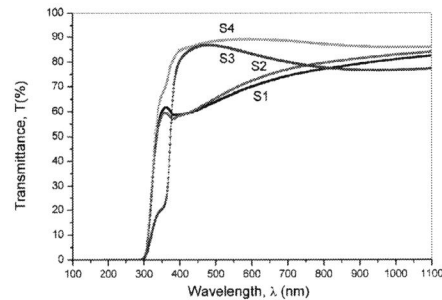

Fig. 3 The optical transmittance of ZnO-based nanostructures on thin films with different thicknesses

The XRD patterns of samples are shown in Fig.2. The X-ray diffraction peaks of the obtained ZnO nanostructures (Fig. 2) can be indexed to hexagonal wurtzite type ZnO structure with lattice constants, a and c ranging between 3.249-3.258Å and 5.212-5.220Å, respectively, consistent with the standard database (JCPDS file 36-1451).

The variation of intensity and the full-width-half-maximum of the most important three diffraction peaks, located at 2 theta of 30-40^0, highlight a decrease of the crystallite

size of ZnO nanostructures, when increases the thickness of ZnO layer on which nanorods grow (S1-S3). Values of 23.4 nm (S1), 23.2 nm (S2), 7.1nm (S3) and 10.0 nm (S4) have been obtained. Better crystallization can be observed for S4 sample.

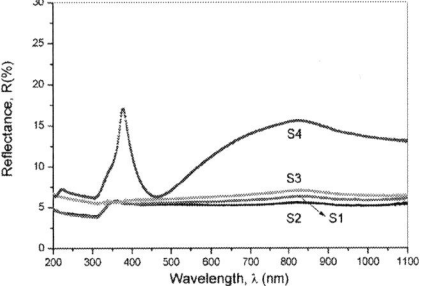

Fig. 4 The optical reflectance of ZnO-based nanostructures obtained on thin films with different thicknesses

Figure 3 shows optical spectra of the investigated samples. One can observe that samples S1 and S2, grown on thinner ZnO layers (40 and 70 nm) have VIS transmittance between 60-80%, while for samples S3 and S4 grown on thicker layers (100 and 120 nm) the transmittance increases up to 90%. The increased optical transmission values of samples S3 and S4, when compare with those of S1 and S2 samples, can be attributed to the significant decrease in crystallite size, from about 23 to 7-10 nm.

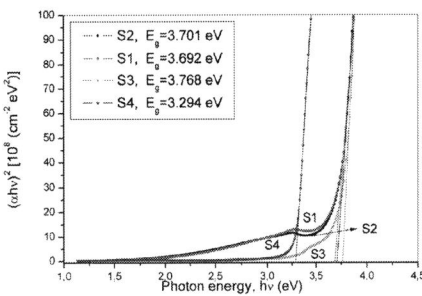

Fig. 5 The band gap energy calculated from optical absorbance spectra in UV-VIS-NIR range

The obtained ZnO nanostructured layers show low reflection between 4.0% - 7.0 % at normal incidence of light in both VIS and NIR spectra. An exception is the sample S4, which has a maximum reflection of about 17% at ~380 nm. The optical absorption edge

978-1-5386-4483-6/18 $31.00 © 2018 IEEE   267

of the obtained nanostructured layers (detail in Fig. 5) shows a continuous blue shift of band gap energy (Eg) values from S1 to S3 and then a red shift of S4 sample. According to quantum confinement effect [13], the increase in band gap energy (blue shift) of ZnO nanocrystalline films is associated to the decrease of the crystallites size and the red shift is associated with the increased size ZnO nanocrystallites, confirming XRD data.

4. Conclusions

ZnO nanorods layers with wurtzite type pure phase were grown by microwave–assisted hydrothermal method at 450W for 3 minutes span on glass substrates covered with zinc oxide thin layers.

The increase in the thickness of the ZnO layer deposited on the substrates increases the density of the grown nanorods arrays and their aspect ratio and decreases the nanorods diameter and crystallite size.

For nanostructured samples grown on sol-gel ZnO layers, high-density arrays of smaller diameter nanorods with smaller crystallite size (~7 nm) show higher transparency of 80-90% and low reflection between 4.0% - 7.0 % in the VIS-NIR range.

Concerning the nanostructures grown onto the PVD-deposited ZnO layer, the optical transmittance is similar to that of the sample S3, with which is similar in terms of ZnO layer thickness, the dimensions of nanorods and their crystallites size.

Acknowledgement:
This work was supported by project PNII-No.27/2014-NANOZON and PN-III -PED-No.223/2017.

References

[1] H. Yu, H. Fan, X. Wang and J. Wang, *"Synthesis and characterization of ZnO microstructures via microwave-assisted hydrothermal synthesis process"*, Optik, **125**, pp. 1461-1464, 2014.

[2] J. Cui, *"Zinc oxide nanowires"*, Mat. Charact., **64**, pp. 43-52, 2012.

[3] Z. Zhang, Y. Lv, J. Yan, D. Hui, J. Yun, C. Zhai, Wu Zhao, *"Uniform ZnO nanowire arrays: Hydrothermal synthesis, formation

mechanism and field emission performance"*, Journal of Alloys and Compounds, **650**, pp. 374-380, 2015.

[4] V. Musat, *"Filme subtiri multifunctionale"*, Cerni publishing house, Iasi, 2007.

[5] A. Pimentel, S. H. Ferreira, D. Nunes, T. Calmeiro, R. Martins and E. Fortunato, *"Microwave synthesized ZnO nanorod arrays for UV sensors: a seed layer annealing temperature study"*, Materials, **9**, pp. 299, april 2016.

[6] M. Y. Soomro, S. Hussain, N. Bano, I. Hussain, O. Nur and M. Willander, *"Hybrid organic zinc oxide white-light-emitting diodes on disposable paper substrate"*, Physica Status Solidi (a) Applications and Materials Science, **210(8)**, pp. 1600-1605, 2013.

[7] V. Musat, E. Fortunato, A. M. Botelho do Rego, M. Mazilu, B. Diaconu, T. Busani, *"Multifunctional zinc oxide nanostructures for a new generation of devices"*, Materials Chemistry and Physics, **132(2-3)**, pp. 339-346, febr. 2012.

[8] S. Pukird, W. Song, S. Noothongkaew, S. Ku Kim, B. Ki Min, S.J. Kim, K. W. Kim, S. Myung and Ki-Seok An, *"Synthesis and electrical characterization of vertically-aligned ZnO-CuO hybrid nanowire p-n junctions"*, Applied surface Science, **351**, pp. 456-549, 2015.

[9] K. Ocakoglu, Sh.A. Mansour, S. Yildirimcan, Ahmed A. Al-Ghamdi, F. El-Tantawy and F. Yakuphanoglu, *"Microwave-assisted hydrothermal synthesis and characterization of ZnO nanorods"*, Spectrochimica Acta Part A: Molecular and Biomolecular Spectroscopy, **148**, pp. 362–368, 2015.

[10] J.-Y Zhu, J.-X Zhang, H.-F Zhou, W.-Q. Qin, L.-Y Chai, Y.-H Chai, *"Microwave-assisted synthesis and characterization of ZnO-nanorod arrays"*, Trans. Nonferrous Met. Soc. China, **19**, pp. 1578-1582, 2009.

[11] A.I. Danciu, V. Musat, T. Busani, J.V. Pinto, R. Barros, A.M. Rego, A.M. Ferraria, P.A. Carvalho, R. Martins and E. Fortunato, *"1D nanostructured ZnO layers by microwave-assisted hydrothermal synthesis"*, J. of nanoscience and nanotechnology, **13**(10), pp.. 6701-6710, 2013.

[12] V. Musat, A. Filip, N. Tigau, R. Dinica, E. Herbei, F. Comanescu and M. Purica, *"Microwave-assisted hydrothermal synthesis of 1D nanostructured ZnO based layersfor optoelectronic applications"*, Revista de chimie, in Press, 2018.

[13] N.S. Pesika, Z. Hu, K.J. Stebe and P.C. Searson, *"Quenching of growth of ZnO nanoparticles by adsorption of octanethiol"*, J. Phys. Chem. B, **106(28)**, pp. 6985-6990, 2002.

Raman investigation of critical steps in monolayer graphene transfer form copper substrate to oxidized silicon by means of electrochemical delamination

Florin Comanescu, Anca Istrate, Munizer Purica

National Institute for Research and Development in Microtechnologies - IMT Bucharest, Romania

florin.comanescu@imt.ro

In this paper we present the results of the investigation by Raman spectroscopy of the monolayer graphene transfer on oxidized silicon substrates by highlighting the critical steps of the transfer using electrochemical delamination method. Characteristic Raman bands of graphene (D, G and 2D) are certifying if a transfer step has been performed successfully. The final step of Polymethyl methacrylate (PMMA) removal is the most critical one due to the partially folding of graphene. After transfer defects are induced in graphene which is confirmed by the increase of the ratio of the intensity of the D band to the G (I_D/I_G).

Keywords: Raman spectroscopy, electrochemical delamination, monolayer graphene,

1. Introduction

Current research and development of graphene and graphene based devices is driven due to its following properties [1- 4]: very good electron transport (depends only on the substrate on which was grown or transferred); scalability; it is the 2D building block for a multitude forms / derivatives of carbon materials (graphite, fullerenes, and carbon nanotubes).

Raman spectroscopy of graphene evidences its unique fingerprint given by its corresponding Raman bands: G band corresponds to E_{2g} phonon mode (~ 1580 cm^{-1}); 2D band (also called G' band) is the second order of D band (~ 2700 cm^{-1}); D band is related only to the defects (~1330 cm^{-1}); D' band is related only to the disorder in the crystalline structure of graphene (~1620 cm^{-1}); 2D' band is the second order of D' (~3240 cm^{-1}). 2D' and 2D bands do not require a defect for activation, [1, 2].

Graphene deposition on semiconductor substrate for electronic device applications is currently impossible. In consequence graphene is grown on transition metal substrates (Cu, Ni). Graphene can be synthesized by many methods but chemical vapor deposition (CVD) is currently considered the optimal method for producing graphene for high-performance electronics, while taking into account its cost, scalability and good crystalline quality [6,7]. Due to their low cost and good catalytic properties copper (Cu) and nickel (Ni) are currently the most suitable metal substrates for graphene deposition. Cu in particular can be easily removed after the growth by etching or can be recycled [8]. Graphene is later transferred on target substrate (silicon, oxidized silicon, glass, quartz). Due to increasing material cost with area most transfer techniques based on metal substrate etching are not suitable for large-scale fabrication. Low-cost, substrate recyclability and a reduced use of etching chemicals constitute key attributes of the electrochemical delamination process, [9-11].

In a previous study, we have investigated the transfer of graphene from copper to configured silicon substrate by electrochemical route [3, 4].

In this paper we are using Raman spectroscopy to determine the critical steps used in single layer graphene transfer from copper substrate to oxidized silicon by means of electrochemical delamination.

2. Experimental details

Samples of single layer graphene on copper substrate from Graphene Supermarket were used for graphene transfer by means of electrochemical delamination.

The GR on copper were transferred to SiO2/Si by electrochemical delamination method using a home-made system described elsewhere [3]. A brief description of the transfer process contains the following steps:

- 300 nm PMMA film is deposited on graphene/Cu, backed at 150 ℃ for 2 minutes and thereafter used as the cathode in the electrochemical cell (graphite |0.05 mM $K_2S_2O_8$| GR/Cu);
- An electrical potential of 40 V was applied to the cell until the entire film of PMMA/graphene was floating on the

---

978-1-5386-4483-6/18 $31.00 © 2018 IEEE

surface of the electrolyte solution;
- Collecting the PMMA /graphene from the electrolyte solution and placing it on the target substrates;
- Removal of PMMA in acetone.

Each of these steps is characterized by means of Raman spectroscopy in order to determine the critical one. Raman investigation has been performed by using LabRam HR800 Raman spectrometer in setup with 633 nm He-Ne laser.

3. Raman investigation

In figure 1 are presented Raman spectra acquired from CVD grown monolayer graphene on copper substrate. Here we find the characteristic Raman bands of graphene: G (~1587 cm^{-1}) and 2D (~2682 cm^{-1}).

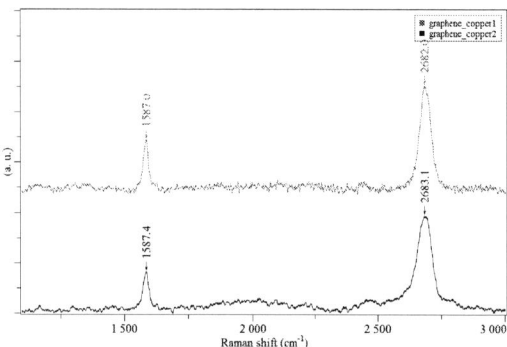

Fig.1 Raman spectra acquired from monolayer graphene / copper sample

Table 1: Average Raman bands parameters and standard deviation corresponding to graphene / Cu source for transfer

Graphene on copper before transfer			
Average values			
	Position (cm^{-1})	FWHM (cm^{-1})	ID/IG
D	1328,37	25,07	0,22
G	1584,02	8,87	I2D/IG
2D	2659,54	37,21	2,33
Standard deviation			
	D	G	2D
Position (cm^{-1})	10,94	3,38	26,24
FWHM (cm^{-1})	8,77	6,57	9,14
ID/IG	0,33	I2D/IG	0,8

In table 1 are presented average Raman band parameters extracted from 25 Raman spectra acquired from monolayer graphene / copper.

The first stage in graphene transfer is PMMA deposition on graphene / copper sample.

In figure 2 is presented a Raman spectrum acquired on PMMA deposited on graphene / copper. We find characteristic Raman bands of graphene: G (1590.8 cm^{-1}), 2D (2647.6 cm^{-1}) and the characteristic Raman bands of PMMA: 600.7, 813.7, 987.8, 1452.4, 1729.6, 2843.9, 2952.3, 2998.6 cm^{-1}, [5]. High intensity corresponding to PMMA Raman bands is due to the SERS effect on graphene surface.

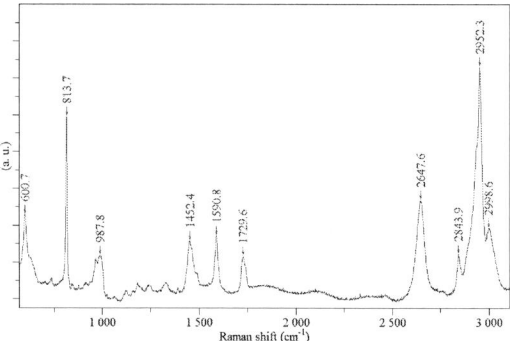

Fig.2 Raman spectrum acquired from PMMA / graphene / copper

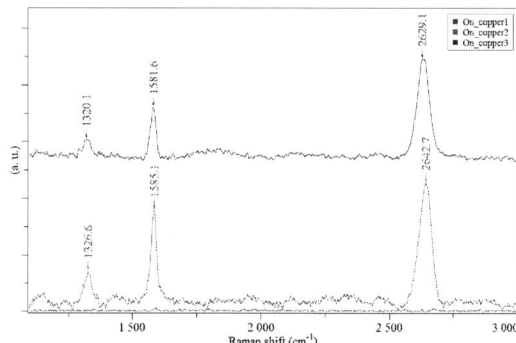

Fig. 3 Raman spectra acquired from copper foils after graphene transfer process

In figure 3 are presented Raman spectra acquired from copper foils after separation by means of electrochemical delamination. The Raman spectra are showing that after transfer there are areas without graphene and areas with graphene. In the areas with remnants of graphene the full with half maximum - FWHM of G and 2D bands are roughly 25 cm^{-1} and 50-55 cm^{-1}. Such values are much higher than the average values presented in table 1 for graphene on copper samples.

In figure 4 is presented a typical Raman spectrum acquired from graphene / PMMA sample extracted from the electrochemical delamination cell and placed on a glass substrate. The order of layers is the following: graphene /

978-1-5386-4483-6/18 $31.00 © 2018 IEEE

PMMA / glass substrate. We find the characteristic Raman bands of graphene D (1318 cm⁻¹), G (1582.2 cm⁻¹) and 2D (2638.8 cm⁻¹) and also the characteristic peaks of PMMA: 600.8, 810.9, 991.1, 1125,9, 1449.5, 1726, 2841.2, 2952.1, 3002 cm⁻¹, [5]. As in the previous case the high intensity corresponding to PMMA characteristic Raman bands is due to the SERS effect caused by the plasmons in graphene.

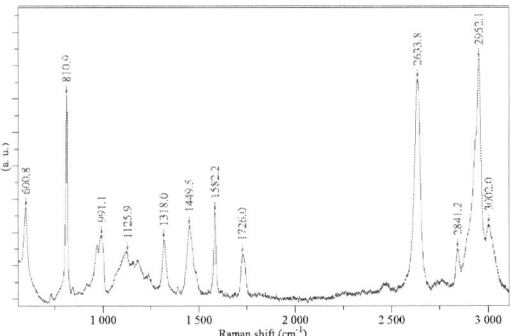

Fig. 4 Typical Raman spectrum acquired from graphene / PMMA /sample extracted from the electrochemical delamination cell and placed on a glass substrate

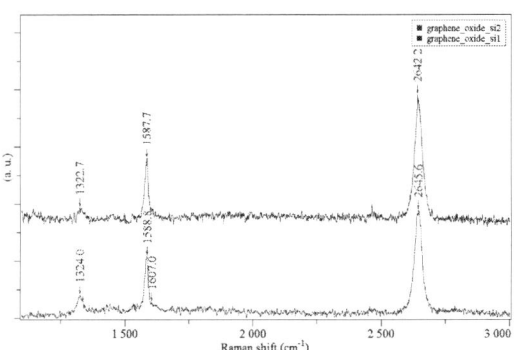

Fig. 5 Typical Raman spectra acquired from graphene / oxide / silicon samples

Fig. 6 Confocal microscope image acquired from graphene / oxidized silicon (2 μm scale)

Multiple spectra acquisitions have been performed on graphene / PMMA sample, almost all the Raman spectra are confirming the presence of graphene. The average value of G band FWHM is 13.26 cm⁻¹ which is close to stress free value – 15 cm⁻¹. The average value of 2D band FWHM is 35 cm⁻¹ a value that confirms the presence of monolayer graphene. By comparing these values with the corresponding ones in table 1 it results that graphene is relaxed on PMMA.

Almost all the spectra acquired from graphene / PMMA / glass are confirming the presence of graphene.

Tacking these into account it results that after introducing the sample in the electrochemical cell the remnants of graphene on copper foil have a high degree of crystalline disorder and therefore are not suited for further processing.

In figure 5 are presented 2 Raman spectra acquired from graphene transferred on oxidized silicon samples after PMMA removal. These spectra are confirming the presence of monolayer graphene due to the presence of the typical Raman bands D, G and 2D. In one spectrum we find D' band (~1607 cm⁻¹) which confirms a high concentration of defects in the investigated area.

In figure 6 is presented a confocal microscope image acquired from graphene / oxidized silicon. Due to the optical contrast we can determine the areas covered by graphene and also the areas where graphene was folded during PMMA removal. The black lines are areas containing folded graphene and remnants of PMMA.

Table 2: Average Raman bands parameters and standard deviation corresponding to graphene transferred by electrochemical delamination

Graphene on oxidized silicon after transfer			
Average values			
	Position (cm⁻¹)	FWHM (cm⁻¹)	ID/IG
D	1325,73	28,2	0,68
G	1587,42	17,11	I2D/IG
2D	2647,28	35,08	2,17
Standard deviation			
	D	G	2D
Position (cm⁻¹)	2,02	3,09	4,47
FWHM (cm⁻¹)	8,99	3,13	4,91
ID/IG	0,66	I2D/IG	1,04

In table 2 are presented average Raman band parameters extracted from 25 Raman spectra acquired from graphene transferred on oxidized silicon substrates.

By comparing the values obtained before and after the transfer it results that standard deviation for G and 2D band is reduced. Defects are induced after transfer as results from the increase of the defect ratio (I_D/I_G). The associated standard deviation for the defect ratio is larger after transfer. Graphene doping can occur as result of transfer.

4. Conclusions

In this paper we used Raman spectroscopy to investigate and to determine the critical steps used in monolayer graphene transfer from copper substrate to oxidized silicon by means of electrochemical delamination. Graphene withstood the first steps of pmma deposition and annealing. Graphene was separated in the electrochemical cell from the copper substrate. Graphene remnants on the copper substrate are having a high degree of crystalline disorder while also covering a much reduced area. Consequently the copper foil after transfer is not recommended for further processing. Almost all the spectra acquired on graphene / PMMA / glass are confirming the presence of graphene. Then we are attaching graphene / PMMA to oxidized silicon. The final step is PMMA removal in acetone. During removal of PMMA part of the graphene is folding. It results that PMMA removal is the most critical step due to the partially folding of graphene.

After transfer defects are induced in graphene which is confirmed by the increase of defect ratio (I_D/I_G) and the appearance of D' band in the Raman spectra.

Acknowledgments. The authors would like to acknowledge the financial support from the IMT Core Program.

References

[1]   A. C. Ferrari, D. M. Basko "Raman spectroscopy as a versatile tool for studying the properties of graphene", Nature Nanotechnology, VOL 8, APRIL 2013

[2]   Andrea C. Ferrari, "Raman spectroscopy of graphene and graphite: Disorder, electron–phonon coupling, doping and nonadiabatic effects", Solid State Communications 143 (2007) 47–57

[3]   Istrate, Anca-Ionela; Veca, L. Monica; Nastase, Florin; Baracu, Angela; Gavrila, Raluca; Comanescu, Florin; Tucureanu Vasilica; Dinescu Adrian; Sandu, Titus, „Scaling the graphene-silicon heterojunctions: fabrication and characterization", Romanian Journal of Information Science and Technology, vol. 19 (2016), No. 3, pp. 282-294

[4]   Constantin Florin Comanescu, Anca-Ionela Istrate, L. Monica Veca, Florin Nastase, Raluca Gavrila, Munizer Purica, "Micro-Raman Spectroscopy of graphene transferred by wet chemical methods",International semiconductor conference CAS 2015, Sinaia, Romania, pp. 49/52

[5]   C. Casiraghi, S. Pisana, K.S. Novoselov, A.K.Geim, A.C. Ferrari, "Raman fingerprint of charged impurities in graphene", Appl. Phys. Lett., Vol 91, 233108, 2007

[6]   K J Thomas, M Sheeba, V P N Nampoori, C P G Vallabhan, P Radhakrishnan, "Raman spectra of polymethyl methacrylate optical fibres excited by a 532 nm diode pumped solid state laser", JOURNAL OF OPTICS A: PURE AND APPLIED OPTICS, vol 10 (2008) 055303, doi:10.1088/1464-4258/10/5/055303

[7]   Keun Soo Kim, Yue Zhao, Houk Jang, Sang Yoon Lee, Jong Min Kim, Kwang S. Kim, Jong-Hyun Ahn, Philip Kim, Jae-Young Choi, Byung Hee Hong, "Large-scale pattern growth of graphene films for stretchable transparent electrodes", Nature, **457**, pp. 706–710, 2009.

[8]   Felipe Kessler, Pablo A.R. Muñoz, Ciaran Phelan, Eric C. Romani, Dunieskys R.G. Larrudé, Fernando L. Freire Júnior, Eunézio A. Thoroh de Souza, Christiano J.S. de Matos, Guilhermino J.M. Fechine, "Direct dry transfer of CVD graphene to an optical substrate by in situ photo-polymerization", Applied Surface Science, **440**, pp. 55–60, 2018.

[9]   Li X, Cai W, An J, Kim S, Nah J, Yang D, Piner R, Velamakanni A, Jung I, Tutuc E, Banerjee SK, Colombo L, Ruoff RS, "Large-area synthesis of high-quality and uniform graphene films on copper foils", Science, **324**, pp. 1312–1314, 2009.

[10] Yu Wang, Yi Zheng, Xiangfan Xu, Emilie Dubuisson, Qiaoliang Bao, Jiong Lu, Kian Ping Loh, "Electrochemical Delamination of CVD-Grown Graphene Film: Toward the Recyclable Use of Copper Catalyst", ACS Nano, **5**, pp. 9927–9933, 2011.

[11] Ahmed Ibrahim, Ghaith Nadhreen, Sultan Akhtar, Feras M. Kafiah, Tahar Laoui, "Study of the impact of chemical etching on Cu surface morphology, graphene growth and transfer on SiO/Si substrate", Carbon, **123**, pp. 402-414, 2017.

[12] Naoki Yoshihara, Masaru Noda, "Chemical etching of copper foils for single-layer graphene growth by chemical vapor deposition", Chemical Physics Letters, **685**, pp. 40–46, 2017.

Electron transfer and dye regeneration in Dye-Sensitized Solar Cells

Corneliu I. Oprea,[1] Atoumane Ndiaye,[2] Anamaria Trandafir,[1,3] Fanica Cimpoesu,[4] Mihai A. Gîrțu[1]*

[1]Department of Physics and Electronics, Ovidius University of Constanța, Constanța, Romania
[2]Department of Physics, Cheikh Anta Diop University of Dakar, Dakar, Senegal
[3]Department of Physics, University of Bath, Bath, United Kingdom
[4]Department of Theoretical Chemistry, Institute for Physical Chemistry, Bucharest, Romania
*E-mail: mihai.girtu@univ-ovidius.ro

Abstract—We report results of a computational study to understand the dye regeneration mechanism of organic dyes in conjunction with cobalt(II) complexes. We are able to determine the parameters of Marcus' theory for electron transfer by means of density functional theory calculations of the energy in various dye-electrolyte configurations.

Keywords—Dye-sensitized solar cells; photovoltaic conversion efficiency; organic dyes; cobalt-based complexes; electron transfer.

1. Introduction

Dye-sensitized solar cells are photovoltaic devices whose principle of operation is based on photoelectrochemical processes [1]. In such devices, the photoelectrode is a transparent conductive glass, such as fluorine doped SnO_2, covered with a mesoporous layer of a wide bandgap semiconductor, such as TiO_2, sensitized with a light absorbing system [2]. The electron in the dye absorbs the photon, and gains energy to the excited state of the sensitizer, from which is transferred to the semiconductor, through which diffuses to the conducting oxide and the external circuit. After performing work in the outer circuit the electron reaches the counter-electrode where, through redox reactions, is taken by the electrolyte back to the dye. The dye is thus regenerated and the process can restart [2].

Increasing energy demands and concerns over global warming have led to a greater focus on photovoltaic technologies [3]. In this context, dye sensitized solar cells (DSSC) may constitute a choice for affordable low power generation in urban areas particularly for producing power generating windows [4].

DSSCs have reached record efficiencies of over 13%, achieved through the molecular engineering of Zn-porphyrin sensitizers [5] or by using purely organic dyes [6]. Higher photovoltaic conversion efficiencies of about 15% [7] have been obtained with perovskite sensitizers with a solid-state electrolyte, which, more recently, have led to devices with over 23% efficiency [8].

The performance increase of DSSCs is possible, on one side by minimizing losses and increasing the fill factor of the devices, and, on the other side, by increasing the short-circuit current and the open-circuit voltage [9], as shown in Eq. 1:

$$\eta = \frac{I_{sc} \cdot V_{oc} \cdot FF}{P_o} \qquad (1)$$

The overall solar conversion efficiency, is a product of the short-circuit current, the open-circuit voltage, and the fill factor, divided by the total solar power incident on the cell, under standard (AM 1.5) conditions. Therefore, one clear way to improve the power efficiency is to increase the fill factor, which is the ratio of the maximum power from the solar cell to the product of V_{oc} and I_{sc}. This can be done by minimizing series and shunt losses through optimized fabrication processes. However, as the ranges of FF are already from 0.75 to 0.85, there is little room for further improvement.

The loss making processes are luminescence or nonradiative decay of the dye, without electron transfer to the semiconductor, back transfer of the electron

978-1-5386-4483-6/18 $31.00 © 2018 IEEE

from the oxide to the dye, and electron interception by electrolyte from the photoelectrode. The desirable processes are electron injection from the excited state of the dye into the semiconductor, the diffusion and charge collection at the photoelectrode and the regeneration of the dye through charge transfer from the electrolyte [9].

Thus, the improvement of device performance can be achieved by favoring the desirable processes, which would increase I_{sc} and V_{oc} [9]. The current can be increased by absorbing more light (by lowering the optical gap) and by injecting more charges (through better energy level alignment between the excited state of the dye and the conduction band edge of the semiconductor). The open-circuit voltage is the difference between the redox level of the electrolyte and the semiconductor's quasi-Fermi level (or the electron chemical potential). Since the electrochemical potential is difficult to change significantly in practice, the more effective way is to lower the redox level of the electrolyte [9], as shown in Fig. 1.

The main drawback of the commonly used triiodide/iodide redox (I_3^-/I^-) system is the high redox level, which limits the open circuit voltage to 0.7–0.8 V [10]. Therefore,

the strategy used has been to replace the iodide-based system with cobalt polypyridine complexes [10,11], as illustrated in Fig. 1.

Despite extensive studies, combining experimental and theoretical approaches, the electron transfer processes that take place between the electrolyte and the dye when the sensitizer is regenerated are still not well understood [12]. Progress is needed to study the electron transfer rate that describes the kinetics of the dye regeneration process [13,14,15], based on Marcus' theory [16,17].

Here we report computational studies that model the electron transfer using simple organic dye – cobalt electrolyte systems. We are able to determine the parameters of the Marcus theory by means of density functional theory calculations of the energy in various dye-electrolyte configurations.

2. Computational Details

A complex system comprising of a redox shuttle coordination complex, $[Co(terpy)_2]^{2+}$ (where terpy = terpyridine), a cationic dye molecule, L0 [18] and PF_6^- anions (for overall charge neutrality in the computation) was modeled with the aim to describe the dye regeneration process. The three constituents are represented in Fig. 2, as a result of geometry optimization calculations.

Optimized geometries of all components were computed for the initial case of low spin Co^{2+} state combined with the cationic dye molecule of the opposite spin -½ into a total singlet spin state, but also for the final case of

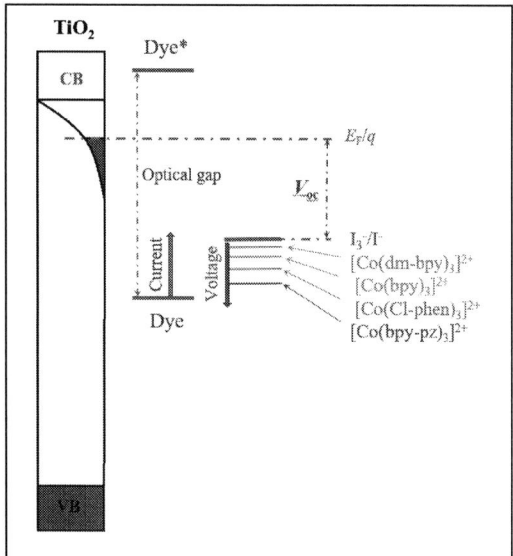

Fig. 1 Schematic energy diagram for illustrating the two strategies for improving DSSC efficiency: increasing current by improving light absorption and/or increasing voltage by shifting down the redox level of the electrolyte [9].

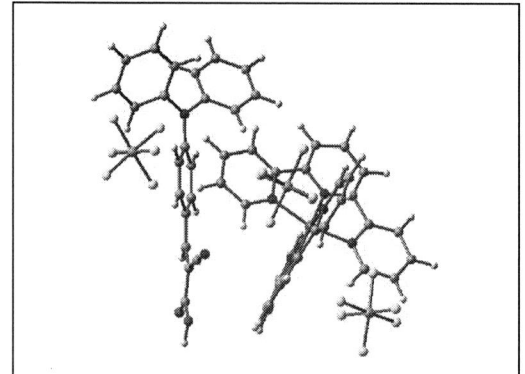

Fig. 2 Optimized geometries of the $[Co(terpy)_2]^{2+}$ redox complex and the L0 dye molecule computed before the regeneration process by DFT/UB3LYP/LANL2DZ method.

zero spin Co^{3+} ion and neutral, regenerated dye molecule. The method employed to obtain the electronic density is the density functional theory (DFT) [19] with the spin-unrestricted B3LYP exchange-correlation functional [20,21], and effective core potential LANL2DZ basis set [22], available in the GAUSSIAN09 quantum chemistry program package [23].

3. Results and Discussion

According to Marcus' electron transfer theory, the rate of the transfer is given by an exponential activation expression, in which the energy barrier, ΔG^*, can be expressed in terms of the driving force for the transfer, ΔG^0, and the reorganization energy, λ, [16]:

$$\Delta G^* = \frac{(\Delta G^0 + \lambda)^2}{4\lambda} \qquad (2)$$

The three quantities are represented schematically in Fig. 3. To determine these quantities we calculated the variation of the total electronic energy with respect to the displacement of nuclear coordinates, by considering intermediate structures obtained by simultaneous changes of all internal coordinates (bonds, angles, dihedrals) in ten

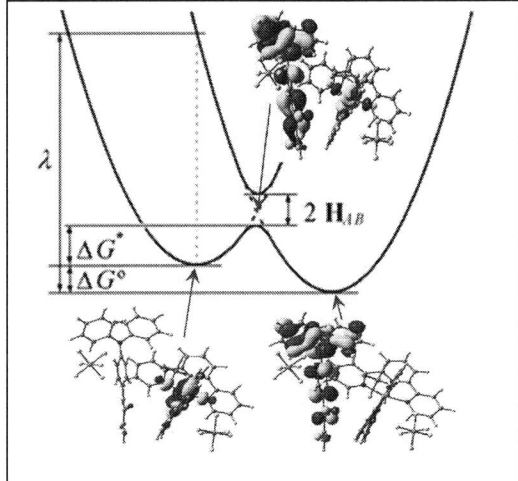

Fig. 3. Schematic illustration of Marcus' theory parameters with highest singly occupied electronic states of the broken symmetry solution showing the transition from $[Co(terpy)_2]^{2+}$ & $L0^+$ to $[Co(terpy)_2]^{3+}$ & $L0$, due to te electron transfer from the electrolyte to the dye. Isodensity surfaces are at 0.02 e/bohr³.

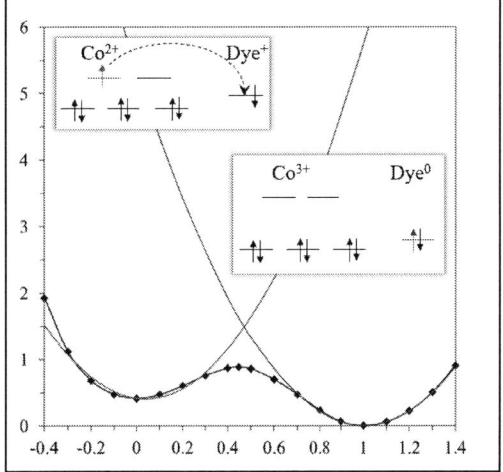

Fig. 4 Total electronic energy calculated for configurations with the electron on the electrolyte, 0 ($[Co(terpy)_2]^{2+}$ & $L0^+$), with the electron on the dye, 1 ($[Co(terpy)_2]^{3+}$ & L0), 10 interpolated and 8 extrapolated structures. Continuous curves represent fits of the calculated values to parabolas.

equal steps between the state with the electron on the electrolyte and the state with the electron transferred to the dye. Four additional calculations were performed by extrapolating for other geometries on both sides of the interval, in order to properly fit the resulting potential energy curve to second order polynomials. The results are displayed in Fig. 4. The calculated parameters are: $\Delta G^* = 1.077$ eV, $\Delta G^0 = 0.417$ eV, $\lambda = 5.347$ eV and the energy matrix element $H_{AB} = 0.610$ eV. An indication of the reliability of these results is the typical error of DFT energy calculations, which for these systems is within 0.2 eV [21].

4. Conclusions

We were able to determine by means of DFT calculations of several dye-electrolyte configurations, the key parameters describing the electron transfer based on Marcus' theory. The calculations applied here to simple models can be extended to other dye-electrolyte systems.

Acknowledgments. M.A.G acknowledges the financial support received from SNSF and UEFISCDI under the Romanian-Swiss Research Programme, through the grant

RSRP #IZERO-142144/1, and A.N. is thankful to the Romanian Ministry of Foreign Affairs and AUF for the E. Ionescu grant.

References

[1] M. Grätzel, Photoelectrochemical cells. *Nature 414*, 338-344, 2001.

[2] A. Hagfeldt, G. Boschloo, L. Sun, L. Kloo, and H. Pettersson, Dye-sensitized solar cells, *Chem. Rev.* **110,** 6595-6663, 2010.

[3] M.A. Green, Y. Hishikawa, E.D. Dunlop, D.H. Levi, J. Hohl□Ebinger, A.W.Y. Ho□Baillie, Solar cell efficiency tables (version 51), Prog. Photovolt. Res. Appl. 26, 3–12, 2018.

[4] R. Baetens, B.P. Jelle, A. Gustavsen, Properties, requirements and possibilities of smart windows for dynamic daylight and solar energy control in buildings: A state-of-the-art review. Sol. Energy Mater. Sol. Cells, 94, 87–105, 2010.

[5] S. Mathew, A. Yella, P. Gao, R. Humphry-Baker, B.F.E. Curchod, N. Ashari-Astani, I. Tavernelli, U. Rothlisberger, M.K. Nazeeruddin, M. Grätzel, *Nature Chem.* **6**, 242-247, 2014.

[6] K. Kakiage, Y. Aoyama, T. Yano, K. Oya, J.-i. Fujisawa, M. Hanaya, Highly-efficient dye-sensitized solar cells with collaborative sensitization by silyl-anchor and carboxy-anchor dyes, Chem. Commun. 51, 15894-15897, 2015.

[7] J. Burschka, N. Pellet, S.-J. Moon, R. Humphry-Baker, P. Gao, M.K. Nazeeruddin, M. Grätzel, *Nature* **499**, 316-319, 2013.

[8] K.A. Bush, A.F. Palmstrom, Z.J. Yu, M. Boccard, R. Cheacharoen, J.P. Mailoa, D.P. McMeekin, R.L.Z. Hoye, C.D. Bailie, T. Leijtens, I.M. Peters, M.C. Minichetti, N. Rolston, R. Prasanna, S. Sofia, D. Harwood, W. Ma, F. Moghadam, H.J. Snaith, T. Buonassisi, Z.C. Holman, S.F. Bent, M.D. McGehee, 23.6%-Efficient Monolithic Perovskite/Silicon Tandem Solar Cells with Improved Stability, Nature Energy 2, 1-7, 2017.

[9] T.W. Hamann, R.A. Jensen, A.B.F. Martinson, H. van Ryswyk, J.T. Hupp, Advancing beyond current generation dye-sensitized solar cells. Energy Environ. Sci. 1, 66–78 2008.

[10] J.H. Yum, E. Baranoff, F. Kessler, T. Moehl, S. Ahmad, T. Bessho, A. Marchioro, E. Ghadiri, J.E. Moser, C. Yi, M.D. Nazeeruddin, M. Grätzel, A cobalt complex redox shuttle for dye-sensitized solar cells with high open-circuit potentials. Nature Communications 3, 631 (2012).

[11] M. Wang, C. Gratzel, S.M. Zakeeruddin, M. Gratzel, Recent developments in redox electrolytes for dye-sensitized solar cells, Energy Environ. Sci. 5, 9394-9405, 2012.

[12] Y. Saygili, M. Söderberg, N. Pellet, F. Giordano, Y. Cao, A. Belen Muñoz-García, S.M. Zakeeruddin, N. Vlachopoulos, M. Pavone, G. Boschloo, L. Kavan, J.-E. Moser, M. Grätzel, A. Hagfeldt, M. Freitag, Copper Bipyridyl Redox Mediators for Dye-Sensitized Solar Cells with High Photovoltage, J. Am. Chem. Soc. 138, 15087−15096, 2016.

[13] S.M. Feldt, P.W. Lohse, F. Kessler, M.K. Nazeeruddin, M. Grätzel, G. Boschloo, A. Hagfeldt, Regeneration and recombination kinetics in cobalt polypyridine based dye-sensitized solar cells, explained using Marcus theory, Phys. Chem. Chem. Phys. 15, 7087-7097, 2013.

[14] Y. Xie, J. Baillargeon, T.W. Hamann, Kinetics of Regeneration and Recombination Reactions in Dye-Sensitized Solar Cells Employing Cobalt Redox Shuttles, J. Phys. Chem. C 119, 28155−28166, 2015.

[15] H.-H. Chou, C.-H. Yang, J.T. Lin, C.-P. Hsu, First-Principle Determination of Electronic Coupling and Prediction of Charge Recombination Rates in Dye-Sensitized Solar Cells, J. Phys. Chem. C 121, 983−992, 2017.

[16] R.A. Marcus, Chemical and Electrochemical Electron-Transfer Theory, Annu. Rev. Phys. Chem., 15, 155−196, 1964.

[17] R.A. Marcus, On the Theory of Electron-Transfer Reactions. VI. Unified Treatment for Homogeneous and Electrode Reactions. J. Chem. Phys. 43, 679−701, 1965.

[18] D.P. Hagberg, T. Marinado, K.M. Karlsson, K. Nonomura, P. Qin, G. Boschloo, T. Brinck, A. Hagfeldt, L. Sun, Tuning the HOMO and LUMO Energy Levels of Organic Chromophores for Dye Sensitized Solar Cells J. Org. Chem. 72, 9550–9556, 2007.

[19] W. Kohn, L.J. Sham, Self-consistent equations including exchange and correlation effects. Phys. Rev. 140, A1133–A1138, 1965

[20] A.D. Becke, Density-functional exchange-energy approximation with correct asymptotic behavior. Phys. Rev. A38, 3098–3100, 1988.

[21] C. Lee, W. Yang, R.G. Parr, Development of the Colle-Salvetti correlation-energy formula into a functional of the electron density. Phys. Rev. B37, 785–789, 1988.

[22] P.J. Hay, W.R. Wadt, Ab initio effective core potentials for molecular calculations. Potentials for K to Au including the outermost core orbitals. J. Chem. Phys. 82, 299–311, 1985.

[23] GAUSSIAN 09, Revision A.02, M. J. Frisch, et al., Gaussian, Inc., Wallingford CT, 2016, citation available online at http://gaussian.com/g09citation/ (accessed on 22 June 2018).

Session **SD&Ms**

SEMICONDUCTOR DEVICES AND MICROSYSTEMS
(Poster session)

978-1-5386-4483-6/18 $31.00 © 2018 IEEE

Efficiency and Total Harmonic Distorsion in Composite Right-/Left-Handed Distributed Oscillators

Giancarlo Bartolucci*, Lucio Scucchia* and Stefan Simion**

* Dept. of Electronics Engineering, University of Roma Tor Vergata, Roma, Italy
E-mails: bartolucci@eln.uniroma2.it, scucchia@uniroma2.it
** Dept. of Electronics and Communications Engineering, MTA – Bucharest, Romania
E-mail: stefan.simion@yahoo.com

Abstract— In this paper the performance of two configurations of Composite Right-/Left-Handed (CRLH) distributed oscillators is investigated. The analysis of both circuits is carried out by means of a numerical simulator. A comparison between the two oscillators is presented in terms of output power, efficiency, and total harmonic distortion.

Keywords—distributed oscillator; composite right-/left-handed circuit; output power; efficiency; total harmonic distortion.

1. Introduction

The travelling wave configuration is a well known topology, adopted in many high frequency circuits such as distributed amplifiers [1-4], phase shifters [5-8] and oscillators [9-11]. For the latter application, a remarkable modification has been presented in [12], where the low-pass basic cell has been replaced by a Composite Right-/Left-Handed (CRLH) two-port network. This oscillator is shown in Fig. 1, it has two output ports, thus providing two sinusoidal signals with almost the same power level. This circuit can be transformed in a single output generator, by summing the two output signals by means of a Wilkinson combiner [13]. The so obtained structure is also depicted in Fig. 1, where the gate line and the drain line must be connected to the network which is drawn in the rectangular dashed box. The main drawback of the proposed single output oscillator is the large die area of the Wilkinson power combiner when the working frequency is low, but this disadvantage may be solved by using lumped element combiner. The single output oscillator has been fabricated, and its experimental characterization was presented

in [13]. This paper is focused on the evaluation of the performance of the two distributed oscillators as a function of the number of the utilized transistors. To this end, an analysis of both configurations is carried out. Three electric parameters are chosen for comparing the two topologies: the output power computed at the oscillation frequency, the efficiency as sinusoidal generator and the distortion due to the presence of higher order harmonics. The numerical values of these parameters are obtained by means of a widely used microwave simulator. Starting from the so obtained data, a comparison between the two configurations is presented. Finally, design considerations for reducing the number of employed transistors and for increasing the performance of the oscillator are discussed.

2. Oscillator Design and Parameters Definition for the Comparison

Both the double output and the single output CRLH distributed oscillators are composed by active and passive components. In particular, the latter are essentially the CRLH cells of the gate artificial line and of the drain artificial line. According to [14], the topology of these cells should consist of an inductor in series with a capacitor and a shunt capacitor in parallel with an inductor. However, for the practical implementation of the cells, the microstrip configuration depicted in Fig. 2 has been chosen. It is designed on a RT/duroid 5870 substrate with thickness 1.575 mm. The values of its distributed and lumped elements are different

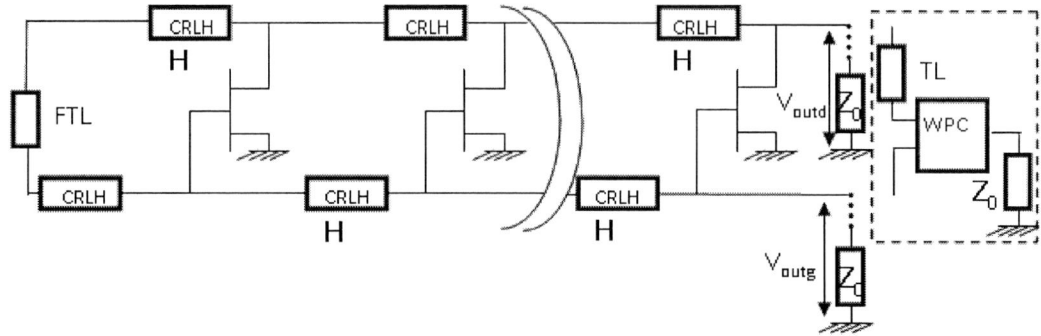

Fig. 1 The single and double output CRLH distributed oscillators. The FTL and TL elements are transmission lines, the first one being used as a positive feedback circuit element, whereas WPC is the Wilkinson power combiner.

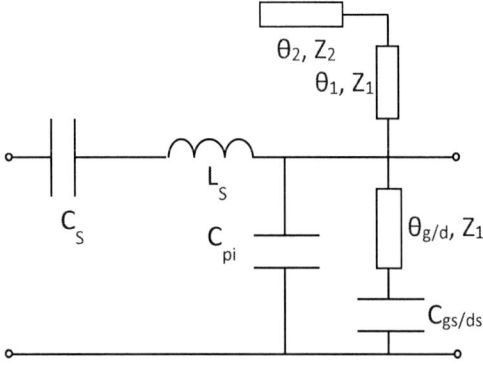

Fig. 2 The basic cell topology for the CRLH gate and drain lines.

for the gate line and for the drain line, and have been provided in [13]. The inductor L_s and the capacitors C_s and C_{pic} are realized by means of a series connected interdigital capacitor. The basic cells are characterized by the electromagnetic simulator Momentum, implemented in the Advanced Design System (ADS) - Keysight software package.

For the realization of the active devices, the MGF 4941AL InGaAs HEMT is utilized. For the characterization of this transistor the non-linear model given by the Mitsubishi company is adopted. The analysis of both distributed oscillators is performed by means of the non-linear simulator included in ADS. As a result, an oscillation frequency of 2.8 GHz is obtained. The following parameters are used for investigating the performance of the two configurations:

• the power of the first harmonic from the

output ports of the gate and drain lines, and from the combiner;
• the efficiency Eff, defined for each output port as the ratio of the power of the first harmonic to the DC power of the oscillator;
• the total harmonic distortion THD, defined for each output port as follows:

$$THD = \sqrt{\frac{\sum_{i=2}^{hm} P_i}{P_1}}$$

where P_i is the power of the i-th harmonic for the chosen output port, and *hm* is the number of harmonics used for the non-linear simulation of the oscillators.

The above defined parameters will be used to develop a comparison between the two topologies of CRLH distributed oscillators.

3. Numerical Results and Performance Comparison

In this section we shall examine the influence of the number of active elements on the performance of the distributed oscillators. To this end, the ADS software package is used, with *hm*=13.

The first parameter to be considered is the output power at the oscillation frequency (first harmonic). In Fig. 3, the powers of the output signals from the gate line and the drain line, and from the Wilkinson combiner, are shown versus number of transistors (N). As expected, the single output oscillator exhibits the highest values for the power. Therefore, for a required power value, it represents a useful solution for reducing the number of active devices. For both the configurations, Fig. 3 suggests that the output power cannot be increased indefinitely by adding active devices. A similar property was found for the gain of distributed amplifiers, and it was explained in [1] by the presence of losses, and therefore of attenuation, on the artificial gate and drain lines.

Another important parameter is the efficiency Eff, whose behavior versus N is illustrated in Fig. 4. A remarkable increase of Eff can be obtained by choosing the single output topology. It is worth noting that, for the analyzed oscillator, the efficiency is maximized for N = 3.

The last parameter utilized for the comparison is the previously defined total harmonic distortion. The THD as a function of N is plotted in Fig. 5. It can be inferred that in the output signal from the Wilkinson combiner, the presence of higher order harmonics is almost the same that in the gate line signal. Conversely, the THD for the single output oscillator is much lower than the THD for the drain line signal.

For sake of completeness, the three output waveforms in time domain are depicted in Fig. 6 for the two configurations, assuming N = 3, when the efficiency is optimized for the single output oscillator. We can note that the drain line signal shows a deviation from a sinusoidal function higher than that one

presented by the combiner output signal. This result is in full agreement with the THD behavior in Fig. 5.

4. Conclusion

In this paper, we have presented a comparison between two topologies of

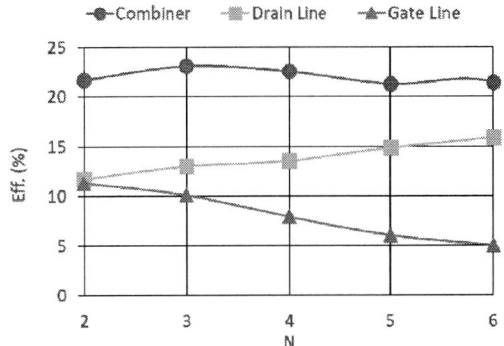

Fig. 4 Plots of the efficiency for the distributed oscillators, as a function of the transistors number.

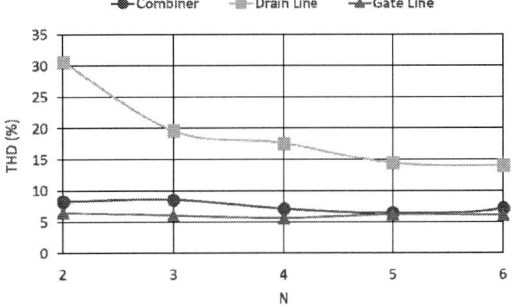

Fig. 5 Plots of the total harmonic distortion for the distributed oscillators, as a function of the transistors number.

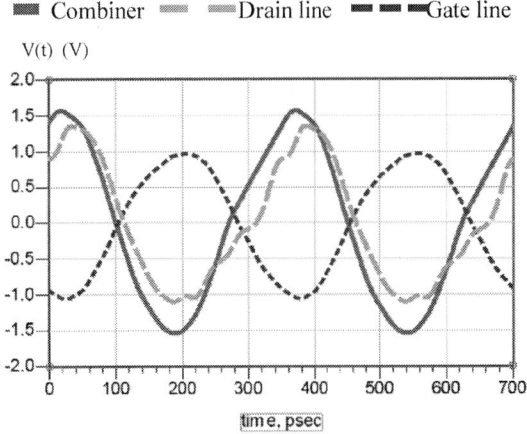

Fig. 6 Output waveforms for the two distributed oscillators, assuming N = 3.

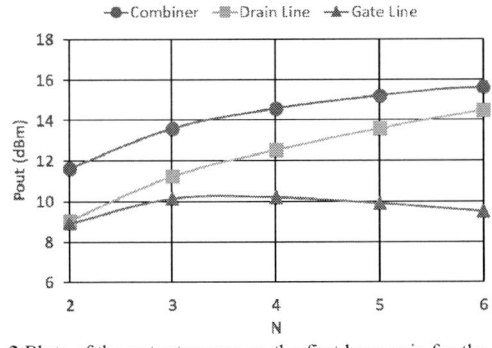

Fig. 3 Plots of the output power on the first harmonic for the distributed oscillators, as a function of the transistors number.

distributed oscillators, composed by CRLH cells. Three parameters have been defined and utilized, related to the output power at the fundamental and higher order harmonics. The behavior of these parameters has been studied as a function of the number of transistors. Starting from these numerical results, provided by a non-linear simulator, design considerations have been proposed. In conclusion, the noticeable advantages of the single output configuration are clearly demonstrated in terms of output power, efficiency, and distortion optimization.

References

[1] J.B. Beyer, S.N. Prasad, R.C. Becker, J.E. Nordman, G.K. Hohenwarter, "*MESFET distributed amplifiers design guidelines*", IEEE Trans. Microwave Theory Tech., **32** (3), pp. 268–275, March 1984.

[2] B. M. Ballweber, R. Gupta, D.J. Allstot, '*A fully integrated 0.5 – 5.5-GHz CMOS distributed amplifier*', IEEE Trans. Solid-State Circuits, **35** (2), pp. 231–239, February 2000.

[3] G. Bartolucci, F. Giannini, L. Scucchia, "*Design considerations for the gate circuit in distributed amplifiers*", IET Circuits Devices Systems, **4** (3), pp. 181-187, 2010.

[4] X. Guan, C. Nguyen, "*Low-power-consumption and high-gain CMOS distributed amplifiers using cascade of inductively coupled common-source gain cells for UWB systems*", IEEE Trans. Microw. Theory Tech., **54** (8), pp. 3278–3283, August 2006.

[5] J.J Hung, L. Dussopt, G.M. Rebeiz, " *Distributed 2- and 3-bit W-band MEMS phase shifters on glass substrates*", IEEE Trans. Microw. Theory Tech., **52** (2), pp. 600–606, February 2004.

[6] G. Bartolucci, "*Image parameter modeling of analog traveling-wave phase shifters*" IEEE Trans. Circuits and Systems I: Fundamental Theory and Applications, **49** (10), pp. 1505–1509, October 2002.

[7] D.W. Kang, S. Hong, "*A 4-bit CMOS phase shifter using distributed active switches*", IEEE Trans. Microw. Theory Tech., **55** (7), pp. 1476–1483, July 2007.

[8] G. Bartolucci, S. Catoni, F. Giacomozzi, R. Marcelli, B. Margesin, and D. Pochesci, "*Realisation of distributed RF MEMS phase shifter with very low number of switches*" Electronics Letters, **43**, (23), pp. 1290–1292, November 2007.

[9] Z. Skvor, S.R. Saunders, C.S. Aitchison, "*Novel decade electronically tunable microwave oscillator based on the distributed amplifier*", Electronics Letters, **28**, (17), pp. 1647-1648, 1992.

[10] L. Divina, Z. Skvor, "*The distributed oscillator at 4 GHz,*" IEEE Trans. Microwave Theory Techn., **46**, (12), pp. 2240-2243, 1998.

[11] H., Wu, A. Hajimiri, "*Silicon-based distributed voltage-controlled oscillators*", IEEE Journal of Solid-State Circuits, **36**, (3), pp. 493-502, March 2001.

[12] S. Simion, G. Bartolucci, "*High power efficiency distributed oscillator based on composite-right-/left-handed unit cells*", Appl. Phys. Lett., **107**, 104102, 2015.

[13] G. Bartolucci, S. Simion, L. Scucchia, "*Power performance and spurious frequencies analysis of Composite Right-/Left-Handed distributed oscillators*", Progress In Electromagnetics Research Letters, **75**, pp. 67–73, 2018

[14] A. Lai, T. Itoh, C. Caloz, "*Composite Right/Left-handed transmission line metamaterials*", IEEE Microwave Magazine, **5** (3), pp. 34–50, September 2004.

Effect of Degeneration on a Millimeter Wave LNA: Application of Microstrip Transmission Lines

M. Fanoro *, S. S. Olokede**, S. Sinha ***

* Department of Electrical & Electronic Engineering Science, Faculty of Engineering & the Built Environment, University of Johannesburg, Kingsway Campus, Auckland Park, Johannesburg, South Africa

mfanoro@uj.ac.za

** Department of Electrical & Electronic Engineering Technology, Faculty of Engineering & Built Environment, University of Johannesburg, Doornfontein Campus, Beit Street, Johannesburg, South Africa

solokede@uj.ac.za

** Deputy Vice-Chancellor, Research and Internalization, University of Johannesburg, Johannesburg, South Africa

ssinha@uj.ac.za

Abstract— This paper presents the impact of a singlewire degeneration microstrip transmission line (MTL) on the gain and noise figure of a millimeter wave low noise amplifier (LNA) at 60 GHz, designed using 0.13 μm SiGe BiCMOS technology. To accomplish this, the performance of the designed LNA is varied with and without the presence of singlewire degeneration MTL in a setup at the first stage of the LNA, using a common emitter transistor topology. Initial results show that the introduction of the singlewire degeneration MTL in the schematic resulted in a decrease in the gain and an increased noise figure of the LNA, while without the presence of a singlewire MTL, the common emitter transistor of the cascode configuration gave rise to a satisfactory increase in gain and reduced noise figure for the LNA. A maximum gain of 15.91 dB and a minimum noise figure of 6.74 dB were recorded when MTL was added to the LNA circuit, while a maximum gain of 20.84 dB and a minimum noise figure of 6.16 dB was recorded when the singlewire MTL was disconnected.

Keywords — singlewire Microstrip transmission line; degeneration; common emitter; low noise amplifier; noise figure; gain, millimeter wave; silicon germaniμm.

1. Introduction

A microstrip transmission line (MTL) is a wideband transmission line system that facilitates the propagation of an electromagnetic wave and offers a higher effective area per length. It consists of conductive strips of a definite thickness and a wider group plane with isolation permitting many dielectric layers, which can be used in power amplifier, low noise amplifier and antenna design, among others. Likewise, MTL has lower capacitance per unit length and high impedance for the design of compact circuitry [1]. Another variation of transmission lines is the coplanar wave (CPW), which helps in simple matching and is useful in significantly reducing substrate loss through the concentration of the electromagnetic field around the slot region. Despite the fact that CPW has a higher quality factor (Q-factor) in comparison to MTL, this inadvertently diminishes the operational bandwidth of the LNA.

2. Theory of degeneration using transmission line at millimeter wave

Van der Ziel and Strutt [2] formulated a method to improve the output signal-to-noise ratio, and to decouple the input impedance from the noise factor while introducing design flexibility. Inductor degeneration in common source amplifiers was introduced for this purpose. Emitter inductance offers a high Q-factor and low noise contribution [3] to the LNA. Fig. 1 shows a simple representation of the inductance at the emitter and the circuit representation. This concept was initially designed for a narrowband LNA, but was extended to ultra-wideband (UWB).

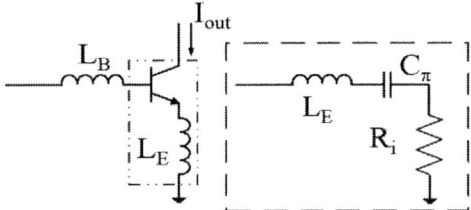

Fig. 1 Equivalent circuit section for the inductively degenerated transistor

In UWB, the fundamental principle of the degeneration inductor was extended. However, it was established that this approach was itself not sufficient, as the frequency response rolls off at higher frequencies. Thus, the redeployment of the feedback resistors, used at the emitter of the transistor, gives an optimized performance of the LNA, improving linearity and noise performance.

L_E is the common emitter (CE) degenerative inductor inserted at the emitter, shown in Fig. 1. Its small-signal model is used mainly to achieve simultaneous gain and noise matching by bringing gain circles and noise circles on the Smith chart closer together. The degeneration introduces negative feedback in the transistor, enhancing stability. The presumed noiseless inductor at the emitter ensures the decoupling of the input impedance from the input noise, ensuring optimized power consumption. The input impedance Z_{in} can be expressed as:

$$Z_{in} = \omega_T L_E + j\left(\omega L_E - \frac{1}{\omega C_\pi} \right) \qquad (1)$$

where C_π is the capacitance across the collector and the base of the CE transistor. The real part required for matching the input impedance, R_{in}, provided the unity current gain frequency, ω_T, is known, can be used in calculating the value L_E. The expression is derived in (2).

$$R_{in} = \frac{g_m}{C_\pi} L_E = \omega_T L_E \qquad (2)$$

where g_m is the transconductance of the transistor. In order to use a singlewire degeneration MTL in a wideband circuit, the optimum source reflect coefficient, Γ_{sopt}, must move towards the center of the Smith chart. This is also expected to cause the minimum noise figure to approach the maximum gain point. Based on Haus's theory [4], it can be expressed in (3) as:

$$\Gamma_{sopt} = S_{11}^* \qquad (3)$$

where S_{11}^* is the conjugate of the input impedance of the LNA. The principle described has been applied in different LNA circuits from the GHz to the THz range.

A. Associated passive elements in degeneration circuits

Passive components in the CE transistor have

been widely used in the narrowband LNA. In the last decade, these have been widely used for wideband and ultra-wideband LNA. Examples of passive components include the resistor [5], spiral inductor [6] and singlewire MTL [7]. Transformer feedback [6] has been used as a degenerative agent on the CE transistor. As obtained from the literature, Table 1 provides the list of passive components, their strengths and known weaknesses.

Table 1. Component deployment of emitter degeneration in LNA circuit

#	Passive Elements	Strengths	Demerits
1	Resistor	Smaller area; insensitivity to process variation when precision resistor is used	Stability issues; low gain and poor noise figure
2	Inductor	Superior noise performance; higher gain; high linearity; power noise matching; flexibility for input impedance and source impedance	Occupies large chip area
3	Transformer	Simultaneous matching of the source impedance to the imaginary and real part of the input impedance in orthogonal way. Any type of inductor can be used (lossy or lossless). Power matching is independent of the cutoff frequency of the transistor.	Occupies larger chip area
4	MTL	Higher gain-bandwidth product; better linearity and noise figure; more accurate model for simulations	Circuit complexity; presence of interconnect parasitics; inductance per unit length of MTL is very low; high transmission losses; high power consumption

3. Theory of degeneration using transmission line at millimeter wave

The mm-wave LNA was designed at the 60 GHz frequency, with the expected bandwidth ranging from 56 – 64 GHz. The 130 nm SiGe HBT

BiCMOS process design kit was used. A series-shunt-series MTL (T-topology) was adopted to ensure a wide bandwidth. While introducing the singlewire MTL degeneration, conditions stated in (3) were observed.

To find the optimal size of the transistor, especially the length, the chosen cascode (CC) topology was set up for testing and estimating the correct width and current density. The circuit consisted of a DC blocking capacitor at the input and the output and singlewire MTL for the input and the output matching respectively. The bias for the base of the transistor was also included. Table 2 gives the result of the process of picking an optimal value of the transistor that must be used for the CE of the CC topology.

Table 2. Optimal value for testing and estimating the size of the transistor

Transistor width/length		10/ 0.12 μm	7/ 0.12 μm	5/ 0.12 μm	3.8/ 0.12 μm
CC w degeneration	NF	4.9	4.6	4.4	4.4
	Gain	3.4	4.0	4.4	4.6
CC w/o degeneration	NF	4.9	4.6	4.4	4.4
	Gain	3.7	4.3	4.8	4.9

Though the CC with degeneration (CC w degeneration) has a transistor of length 5 um that cancels out the effect of the noise in the CC topology at the first stage, the performance at the LNA circuit with two stages was poor. Even when the 3.8 μm transistor, which had the highest gain in the CC w/o degeneration, was deployed, the gain dropped significantly owing to the common base (CB) transistor. Therefore, a compromise was reached in using two transistors yielding the best performance in terms of gain and NF. These transistors were 10 μm and 3.8 μm in length with a constant length of 0.12 μm.

For optimum performance of the designed LNA, transistor sizes used at the CE in the cascode topology were 10 μm in length, with a constant width of 0.12 μm for all transistors. The length was the same for all the transistors used, except the transistor at the CB, which was

3.8 μm in length. The input matching network was built with a T-topology and a CE degeneration using MTLs of quarter-wave length each. The T-topology also acted as the bias line for the circuit.

All the transmission lines and metal insulator-metal (MIM) capacitors were made of three terminals, $\lambda/3$ in length. TL_1, TL_2, TL_3, TL_6, TL_7 and TL_8 had a metal layer AM and ground layer MQ. The MIM capacitor was made up of three terminals, with the last connected to the substrate. The MIM capacitors used were of metal layer metal AM and ground layer LY. TL_4 was used as the singlewire degeneration MTL. Table 3 and Table 4 show the value of the components and the circuit diagram can be seen in Fig. 2:

Table 3. Transmission line specifications

Transmission Line (TL)	Length (μm)	Width (μm)
TL1	200	6.11
TL2	400	6.11
TL3	400	4.00
TL4	80	4.00
TL5	300	4.00
TL6	300	6.11
TL7	180	6.11
TL8	400	6.11
TL9	400	4.00
TL10	300	4.00
TL11	380	6.11

Table 4. MIM capacitor specifications

Transmission Line	Length (μm)	Width (μm)	Value (fF)
CM1	8	8	66
CM2	6	4	26
CM3	16	9	150.05
CM4	7	7	50
CM5	7	4	29
CM6	16	9	150.05
CM7	7	4	29
CM8	7	6	44
CM9	4	5	22

An extra MIM capacitor was introduced to cater for the bondpad in the circuit. *CM2, CM3, CM5, CM6 and CM8* were used for this purpose.

4. Results and discussion

The circuit diagram of the millimeter wave LNA, including the singlewire degenerative MTL in the first stage, is depicted in Fig. 2. In Fig. 3, simulation results are plotted. In the result, the effect of introducing the degeneration MTL in the first stage of the LNA with a cascode topology shows that there was an increased reduction in the gain of the millimeter wave during the process. A maximum gain of 15.91 dB was recorded at 62.40 GHz and a minimum noise figure of 6.74 dB was recorded at 58.20 GHz through simulation when the singlewire degeneration MTL was connected.

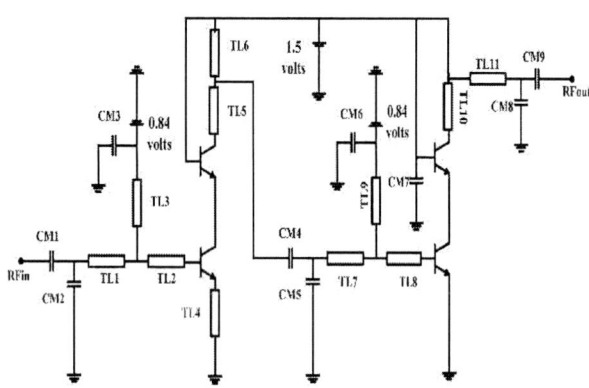

Fig. 2 Circuit diagram of the millimeter wave LNA including the singlewire degenerated MTL transistor in the first stage

However, once the singlewire degeneration MTL was removed, there was a considerable increase in the gain and a further decrease in the NF of the LNA. A maximum gain of 20.84 dB was recorded at 62 GHz and a minimum noise figure of 6.16 dB was recorded at 55 GHz through simulation when the singlewire degeneration MTL was disconnected. The degradation in the NF and gain was due to the capacitance across the base-emitter junction, the MTL at the base and the emitter respectively.

However, the absence of the singlewire degeneration MTL did not affect the stability factor and the linearity of the millimeter-wave LNA. The presence of the degeneration MTL increased the input impedance, while ensuring the matching of the noise figure and the input matching.

4. Conclusion

In this paper, a review of a component utilized in degeneration in a CE at the first stage of a millimeter wave was reported. Thereafter, we established the effect that the degenerative component at the CE of a cascode transistor has on the gain and the noise figure of a 60 GHz LNA. An MTL was introduced for this purpose. From simulations, it was observed that the noise figure was reduced to a minimal level when MTL degeneration was disconnected compared to when the MTL was connected at the CE. Similarly, when the MTL was removed, the gain of the LNA increased to an appreciable extent.

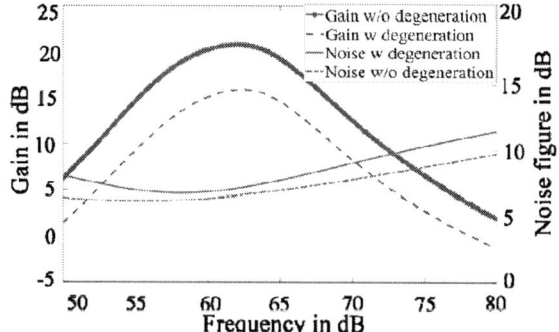

Fig. 3 Simulated gain and noise figure plotted against the frequency in GHz

References

[1] P. Song, A. C. Ulusoy, R. L. Schmid, and J. D. Cressler, "A high gain, W-band SiGe LNA with sub-4.0 dB noise figure," *IEEE MTT-S Int. Microw. Symp. Dig.*, pp. 4–6, 2014.

[2] M. J. Strutt and A. Van der Ziel, "Suppression of spontaneous fluctuations in amplifiers and receivers for electrical communication and for measurement devices," *Physica*, vol. 9, no. 6, pp. 556–562, 1942.

[3] M. Fanoro, S. S. Olokede, and S. Sinha, "Investigation of the effect of input matching network on 60 GHz low noise amplifier," in *2016 International Semiconductor Conference*, 2016, pp. 71–74.

[4] R. C. H. Li, *RF Circuit Design*, 1st ed. New Jersey: John Wiley and Sons,Inc., 2009.

[5] J. Lee and J. D. Cressler, "A 3 - 10 GHz SiGe resistive feedback low noise amplifier for UWB applications," in *Digest of Papers - IEEE Radio Frequency Integrated Circuits Symposium*, 2005, pp. 545–548.

[6] A. Tasic, W. A. Serdijn, and J. R. Long, "Concept of transformer-feedback degeneration of low-noise amplifiers," in *Proceedings of the 2003 International Symposium on Circuits and Systems, 2003. ISCAS '03.*, 2003, vol. 1, pp. 421–424.

[7] S. Jang and C. Nguyen, "High-gain power-efficient wideband V-band LNA in 0.18μm SiGe BiCMOS," *IEEE Microw. Wirel. Components Lett.*, vol. 26, no. 4, pp. 276–278, 2016.

Millimeter wave and Terahertz investigations on some dielectric materials

M. G. Banciu*, T. Furuya**, D. C. Geambasu*, L. Nedelcu*, D. Pantelica***, M.-D. Dracea***, P. Ionescu***, A. Iuga*, C. Chirila*, L. Hrib*, L. Trupina*, M. Tani**

*National Institute of Materials Physics, Magurele, Jud. Ilfov
E-mail: gbanciu@infim.ro
**Research Center for Development of Far-Infrared Region, Fukui University, Fukui, Japan.
E-mail: tani@fir.u-fukui.ac.jp
***Horia Hulubei National Institute for Research and Development in Physics and Nuclear Engineering
E-mail: pantel@nipne.ro

Abstract—Investigations of barium strontium titanate (BST) layers deposited on MgO and Si substrates are presented. Since the Sr content determines the dielectric and optical properties of the BST layers at room temperature, accurate compositional analysis was performed by using Rutherford Backscattering technique at 3.041 Mev.

Keywords—dielectric materials, millimeter waves, THz Time-domain spectroscopy.

1. Introduction

There is a continuous demand for a more efficient use of the electromagnetic spectrum. Therefore, systems using millimeter waves and Terahertz radiation are increasingly attractive to the developers of new solutions.

Dielectric and ferroelectric materials play an important role in device development for very high frequencies. As a consequence, previous research was carried on dielectrics in millimeter waves [1,2].

On the other side, Terahertz time-domain spectroscopy was developed as a robust method of material investigations. The method measures both the amplitude and the phase of the signal, therefore is more convenient compared to other methods as Fourier Transform Infrared Spectroscopy (FTIR).

The Time-Domain Spectroscopy in Terahertz continuously evolved. The used femtolasers with pulses shorter than 10 fs allowed the increasing of the investigation band up to 7 THz in special conditions by using bow tie H shaped photoconductive GaAs antennas [3]. Moreover, very fast TDS method was also proposed [4].

2. $(Ba_{1-x}Sr_x)TiO_3$ films deposited on different substrates

Barium strontium titanate $(Ba_{1-x}Sr_x)TiO_3$ or BST proved to be a very attractive material at lower frequencies (microwaves) especially for tunable devices [5]. In this paper, dielectric substrates and BST films deposited on such transparent substrates were investigated at higher frequencies. As substrates, we measured high resistive Si, (100) MgO and alumina.

The BST films were deposited by using Pulsed Laser Deposition or RF sputtering deposition technique from in-house developed ceramic targets. The X ray diffraction patterns proved a good crystallinity the BST films as showed in Fig.1. Moreover, for the layers deposited by PLD, the Atomic Force Microscopy investigations showed an acceptable roughness as shown in Fig. 2.

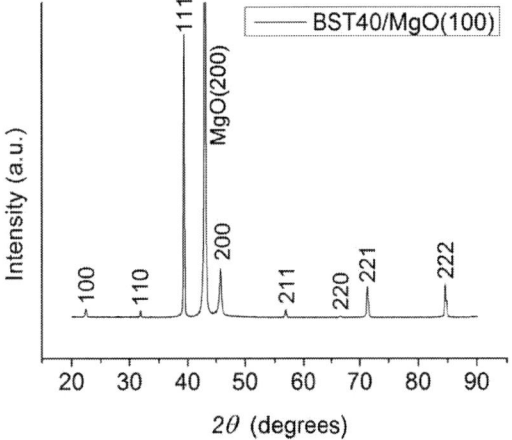

Fig. 1. X-ray diffraction pattern of a BST film with 40% Sr content.

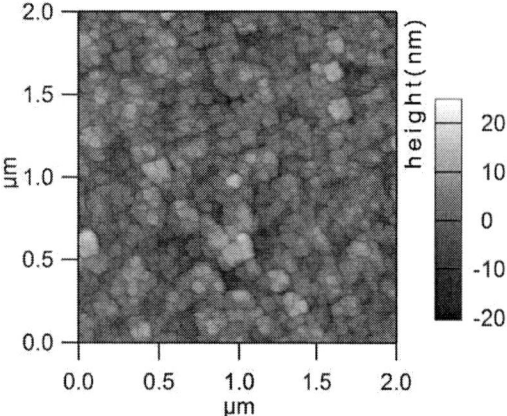

Fig. 2. Atomic Force Microscopy of the BST film obtained by Pulsed Laser Deposition.

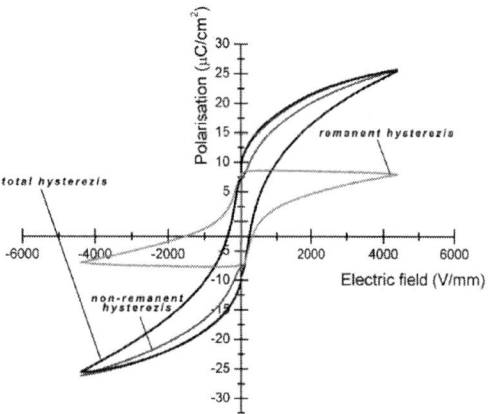

Fig. 3 . Hysteresis loop of BST sample for x=0.2.

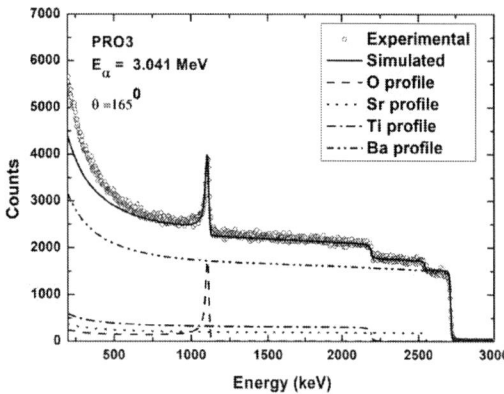

Fig. 4. Rutherford Back-Scattering spectra on a BST single crystal sample.

The Curie temperature of BST sample increases with the Sr content. Therefore, for a small Sr content as x=0.2, at room temperature the BST is in ferroelectric state. The ferroelectric BST samples show a clear hysteresis as showed

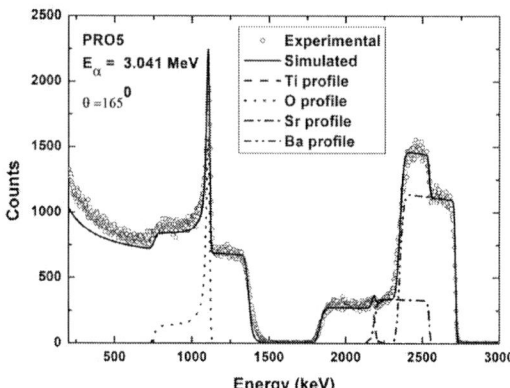

Fig. 5. Rutherford Back-Scattering spectra on a BST film.

in Fig. 3 for x-0.2 Sr content.

The dielectric and optical properties of the BST samples at room temperature very much depend on the Sr content. Hence, Rutherford Backstattering (RBS) technique using an alpha beam accelerated to 3.041 MeV was employed for precise compositional analysis. In Fig. 4 are presented the RBS spectra for a BST crystaline sample (x=0.2) and In Fig. 5 are shown the RBS spectra for a BST film deposited on Si. The RBS technique found a stoichiometry very closed to the expected values used in target preparation.

3. Millimeter wave investigations

For investigations in millimeter wave domain, two methods were employed. The first method consisted in a transmission setup in which the dielectric sample is connected just between the flanges of two rectangular guides as shown in Fig. 6 and Fig. 7.

Fig. 6. Transmission measurements in millimeter waves..

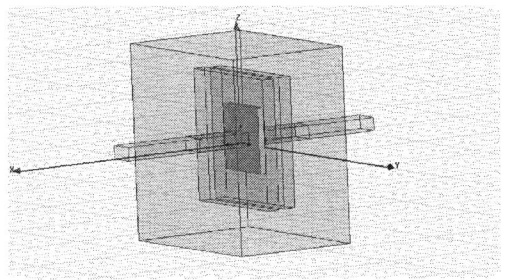

Fig. 7. Position of the sample between the flanges in the transmission setup.

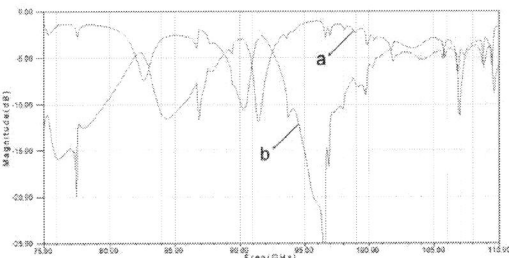

Fig. 8. Scattering parameters for a Si square sample in a transmission setup. a) magnitude of S_{21}, b) magnitude of and S_{11}.

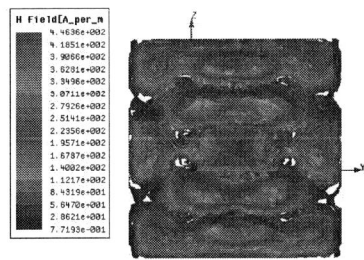

Fig.9. H field pattern in a square 5mm x 5 mm x 0.5mm Si sample in transmission setup at 96.5 GHz.

The measurements were performed in two frequency ranges 75-110 GHz and 110-140 GHz by using a system from AB Millimetre Vector Network Analyzer. The measurements on square transparent samples showed significant resonances as shown in Fig. 8. The simulations performed by using a commercial full-wave simulation software (HFSS) showed that, for some frequencies specific electromagnetic modes are excited as for a rectangular dielectric resonator. Such modes are clearly present in the H pattern configuration in the square Si sample shown in Fig. 9.

A second method used in millimeter wave domain was the free space method. The corrugated horn antennas are placed at a certain distance. Two Teflon lenses focus the waves on

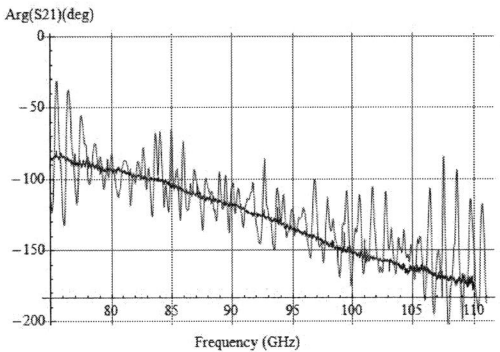

Fig. 10. Phase of the transmitted signal through an alumina sample in free space measurements for two distances between the horn antenna (light curve 400mm, dark curve 760 mm)

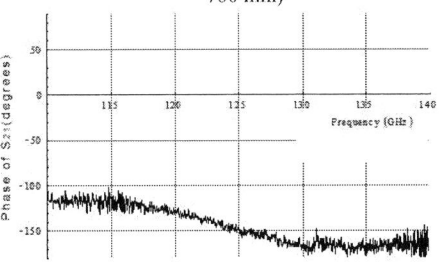

Fig. 11. Phase of the transmitted signal through a Si in free space measurements in 110-140 GHz band.

the sample. The measured phase in the presence of sample showed noise due to the multiple reflections between the antennas. This noise was drastically reduced as shown in Fig. 10 and Fig. 11 by increasing the distance between the corrugated horn antennas to 760 mm.

3. Terahertz Time Domain Spectroscopy

THz TDS method [3] is a very attractive method for low loss high dielectric constant materials [6, 7]. In Fig. 12 are shown the time-domain waveforms for the reference signal and the signal, which passed through the sample consisting of a 2 μm thick BST layer deposited on a (100) MgO substrate. The delay of the main peak and the multiple reflections showed by the signal through the sample are mainly due to MgO substrate. The MgO substrate exhibits a refraction index in THz around 3.2, much smaller than for the BST, but the substrate thickness is around 500 μm, mach greater than the film thickness. However, in terms of

978-1-5386-4483-6/18 $31.00 © 2018 IEEE

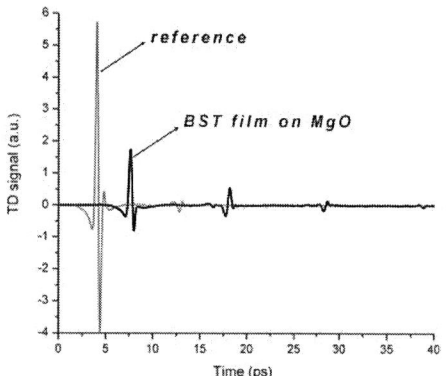

Fig. 12. Time domain pulses during the measurements of BST film on (100) MgO substrate

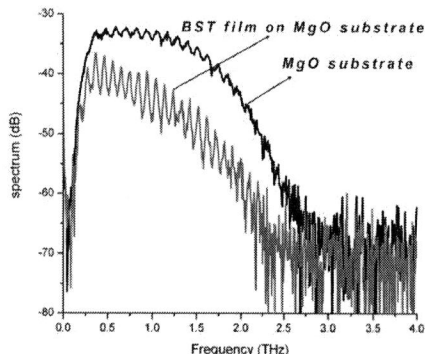

Fig. 13. THz spectrum for the BST film on MgO substrate compared to the spectrum for MgO substrate without deposition.

amplitude, the absorption of BST film can easily put in evidence as in Fig. 13. For an accurate determination of the dielectric constant of the BST, it is necessary the comparison with the very same MgO substrate, since the variation in thickness provided by the manufactured is about +- 2 μm, as the film thickness.

4. Conclusion

Several dielectric substrates and $(Ba_{1-x}Sr_x)TiO_3$ films deposited on those substrates were investigated in millimeter wave domain and THz.

Two methods were applied in millimeter wave domain. The direct transmission method, for which the sample is connected between the flanges of the rectangular waveguides put in evidence the excitation of dielectric resonator modes, which depend on the shape of the sample. For the second method, the free space technique, better results consisting in a significant reduction

of the phase noise were achieved for a long distance of 760 mm between the ends of the corrugated wave antennas.

The time domain spectroscopy put easily in evidence the absorbance of the BST layer. However, for an accurate determination of the dielectric constant of the BST layer, a greater thickness of the BST layer and the previous measurement of exactly the same substrate sample in is required.

Acknoledgments. This work was partially supported by a grant of the Romanian National Authority for Scientific Research and Innovation, Programme for research- Space Technology and Advanced Research - STAR, project no. 630/2016 (MCOATANT) and by CCCDI - UEFISCDI project number 61/2016 (acronym MASTERS).

References

[1]   E. M. Nanni, S. K. Jawla, M. A. Shapiro, P. P. Woskov, R. J. Temkin, "Low-loss transmission lines for high-power Terahertz radiation", J. Infrared. Milli. Terahz. Waves 33, 695-714, 2012.

[2]   G. Annino, M. Cassettari, M. Martinelli, "Study on Planar Whispring Gallery Dielectric Resonators. I. General Properties", International, Journal of Infrared and Millimeter Waves, vol. 23, pp. 597-615, 2002.

[3]   ***, AiSPEC , IRS 200 pro, THz Time-Domain spectrometer - user manual, 2010.

[4]   T. Furuya, E.S. Estacio, K. Horita, C T. Que, K. Yamamoto, F. Miyamaru, S. Nishizawa, M. Tani, "Fast-scan Terahertz Time-Domain Spectrometer Based on Laser Repetition Frequency Modulation",, Japanese Journal of Applied Physics, Vol. 52, Number 2R, 022401, 2013.

[5]   A, Ghalem, M. Rammal, L. Huitema, A, Crunteanu, V. Madrangeas, P. Dutheirl, F. Dumas-Bouchiat, P. Marchet, C. Champeaux, L. Trupina, L. Nedelcu, M. G. Banciu, "Ultra-high tunability of Ba(2/3)Sr(1/3)TiO3-based capacitors under low electric fields", IEEE Microwave and Wireless Components Letters, vol. 26, issue 7, pp. 504-506, 2016.

[6]   M.G. Banciu, L. Nedelcu, K. Yamamoto, S. Tsuzuki, M. Tani, "THz TDS investigations on dielectrics for microwave applications", invited paper to the 5th International Workshop on Far-Infrared Technologies, Fukui, Japan,. March 5-7, 2014

[7]   M. G. Banciu, L. Nedelcu, H. V. Alexandru, T. Furuya and M. Tani, „Investigations on dielectric parameters of some ferroelectric materials in Terahertz waves", published in the Symposium Digest of the Second International Symposium on Frontiers in THz Technology FTT 2015, 30Aug – 02 Sept 2015.

978-1-5386-4483-6/18 $31.00 © 2018 IEEE

Methods for Art Preservation and Restauration. Identification of parameters for potential monitoring the temporal evolution of putties

Ioana-Maria Giura*, Cristina Pachiu**, Marian Popescu, **
Bogdan Bita**, Octavian Narcis Ionescu** and Mirela Suchea**,***

*Restorer - Romanian Patriarchy, Bucharest, Romania
E-mail: ioanamariagiura@yahoo.com
**National Institute for Research and Development in Microtechnologies (IMT-Bucharest), Bucharest, Romania
***Center of Materials Technology and Laser, School of Engineering, Technological Educational Institute of Crete, Heraklion, Greece
E-mail: ** mira.suchea@imt.ro; mirasuchea@staff.teicrete.gr

Abstract—Art preservation, conservation and restauration is a very important niche field that may beneficiate of recent advances in sensor technology. Identification of parameters for potential monitoring the temporal evolution of artworks (wood, stone, textiles, paints, putties etc.) is a first step in designing novel specialized and personalized sensing and detection devices. This work regards a particular comparative study of a new putty material for restauration and old unknown putty from the beginning of XIXth century for preliminary identification of parameters for potential monitoring the temporal evolution of putties used in artworks restauration process.

Keywords—sensor, aging, artworks restauration, conservation.

1. Introduction

Conservation is a blanket term which covers all aspects of preserving and caring for artworks. Restoration is a conservation process which generally involves making physical additions to artworks as a way of preserving them. A variety of methods are used to repair damage and stabilize artworks and restoration can be approached in several ways. Conservation also involves preventing the deterioration of an artwork, and this is a large part of the conservators' work. Activities that might be considered preventive conservation range from adding backboards to paintings to advising on packing and display conditions. The monitoring of paintings, both on canvas and wooden support, is a crucial issue for the preservation and conservation of this kind of artworks. Many environmental factors (e.g. humidity, temperature, illumination, etc.), as well as bad conservation practices (e.g.

wrong restorations, inappropriate locations, etc.), can compromise the material conditions over time and deteriorate an artwork. The main factors impacting on the conservation state of an artwork are recognized to be, as far as microclimate is concerned, relative humidity, temperature and light. Biological aspects such as insect attacks are also considered relevant and therefore should be studied and monitored. Another important identified factor is air quality and air dispersed pollutants which are deposited on the artwork's surface and may cause a noticeable decay even in a short time. In present, there are very few customized sensors and devices to serve this purpose although the present technology advancement may successfully help. In the last years this issue came into attention. In 2006 the expression "Heritage Science" appeared in the horizon of sciences but, today has not yet a final definition. According to Matija Strlic who has published a significant editorial entitled "Heritage Science: A Future Oriented Cross Disciplinary Field" in a virtual special issue of Angewandte Chemie in May 2018, it should be seen as a new cross-disciplinary field of scientific research and presents the challenges that this should face. It includes applications of modern characterization techniques of materials, novel intervention methods that base on nanotechnologies and advanced materials science and devices including real time monitoring of artworks [1]. Also in the last years, various basic temperature/humidity/air quality sensor systems were implemented in some museums in the world [2, 3] and even appeared commercially available products such as "SensorPush" that is a device working as a temperature and humidity sensor, collecting data that can be retrieved via iPhone, iPad or Android smartphone or tablet. [4] and even Internet-of-Things (IoT) Approach for Remote Sensing in Museums [5, 6]. Since 2016 "NANORESTART" Project is under development. The NANORESTART project consortium comprises 27 museums, academic institutions, private conservators and private companies working together to find new methods for the conservation and preservation of contemporary art www.nanorestart.eu . In this context, the present work regards a particular comparative study of a new putty material for restauration and an old unknown putty from the beginning of XIXth century for preliminary identification of parameters for potential monitoring the temporal evolution of putties used in artworks restauration process.

2. Experimental premises
Nowadays the installation of climate control units to provide human comfort as well as good conservation conditions inside museums is facilitating conservation and preservation actions. On the other hand museums' depots and storage rooms which host large part of collections are sometimes not maintained in a stable environment al conditions due to lack of funding and bad practice. This inappropriate practice is crucial for all artworks, particularly for paintings conservation (especially for those realized on wooden supports). These kinds of artworks indeed are very sensitive to the variation of some parameters like light, temperature and humidity. In Romania -a Christian orthodox country- it is traditional to have the wooden painted icons in churches and monasteries and these constitute an invaluable patrimonial possession. Unfortunately, there are many old such items that were improperly stored during

communist years and after due to lack of founding as well as ignorance of their keepers that were not trained personnel but clerics. The present studies were performed on old putty used during a restoration in 1930's on a wood icon from Antim Monastery-Bucharest Collection and compared with one of the most popular today commercially available putty. Scanning electron microscopy (SEM), Raman confocal microscopy characterization and mechanical hardness measurements were performed on various samples as preliminary attempt to identify parameters that may become the subject of permanent monitoring using a kind of sensor that can be customized for the purpose.

3. Experimental
SEM characterization was performed using a (FE-SEM) Nova NanoSEM 630 (FEI Company, SUA), equipped with an EDX detector (EDAX TEAM™, SUA) FEI Nova NanoSEM 630 field emission microscope. Raman studies were made using a Scanning Near-field Optical Microscope (SNOM) - Witec alpha 300S/Witec/2008 Witec Gmbh Germany with Nd-YAG - 532 nm laser. For hardness measurements evaluation was made using a Shore durometer and coverted in Vickers hardeness equivalent units using atbles.

4. Results and discussions
Optical microscopy and SEM characterization were performed to identify the putties surface morphologies. Some examples of the two compared materials are presented in **Figure 1.**

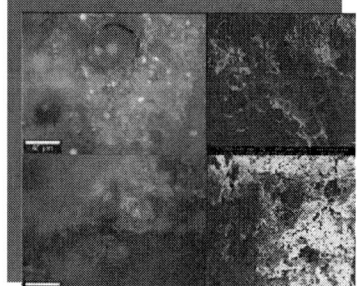

Figure 1 *Optical and SEM images of new putty upper line and old putty –down images.*

As it can be observed, the two materials have a completely different appearance. The new putty has a foamy structure with bubbles that, at longer electron beam exposure start to collapse (see **Figure 2**) while the old putty has a compact structure formed by compact packing of small (perhaps inorganic) grains. It presents also large voids and cracks that may be attributed to aging and low quality of material.

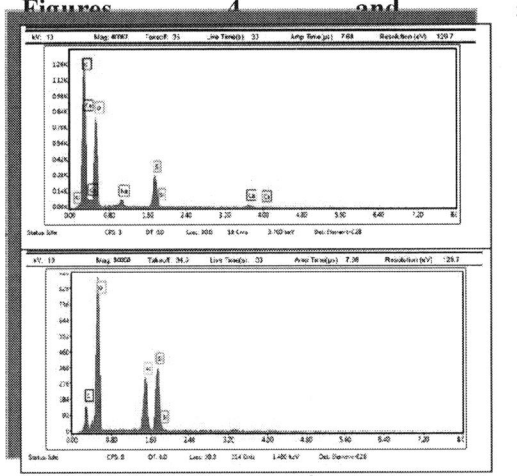

Figure 2 *SEM image of exploded bubbles due to electron beam exposure in the new putty.*

EDX analysis of the materials shows that the two materials have a very different composition. In the new material were identified C, O, Na, Si, and Ca while the old material contains C, O, Al and Si. **Figure 3** shows an example of EDX specta of the two materials. To resolve the composition of the two putties Raman spectroscopy analysis was performed in various points of the optical microscopy image and a large number of spectra were collected and compared. Some examples are presented in Figures 4 and 5.

Figure 3 *EDX spectra of new putty – up image and new putty –down image.*

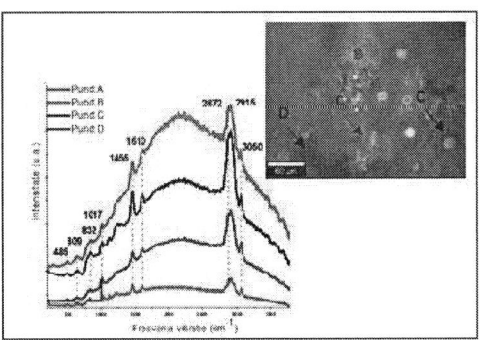

Figure 4. *Raman spectra and optical microscopy images (insets) of the new putty. Identified composite within the material through Raman analyses applied in four different points/areas that are spotted in the optical images.*

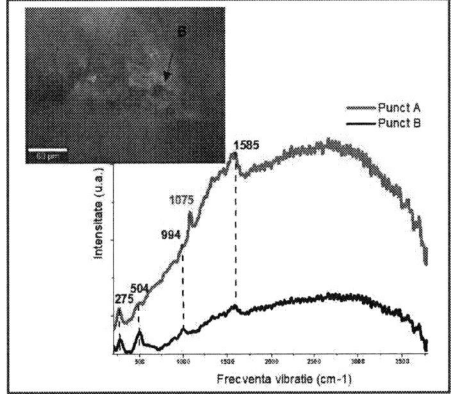

Figure 5. *Raman spectra and optical microscopy images (insets) of the old putty. Identified composite within the material through Raman analyses applied in two different points/areas that are spotted in the optical images.*

For the new putty presence of Al, C, Na, Si, Ca and polypropylene was identified while in the old putty Al, SiO_2 and C are present. Calcium is frequently found as calcium carbonate and calcium phosphate. Calcium phosphate is present in many alive organisms, for example in bone minerals, tooth enamel, while the calcium carbonates are frequently used by the pharmaceutical industry, constructions and agriculture. The fact it was found in the putties is not random as the calcium carbonate is the basic composite of the putties and the colors used in conservation and restoration of paintings. [7]. Progressive decay in time of these composites are caused by bacteria attacks and accelerated by the environmental conditions (temperature and humidity). Thus calcium was not found by Raman spectra analysis for the old putty. Polypropylene o is a thermoplastic polymer frequently used in constructions due to is technical features (energy efficiency, weather resistance). As it has been recently introduced in the industry, it could not be found in the old restoration putties by the Raman spectra analyses.

Sodium is present in minerals like natrosilite or nahcolite [8] that are used in constructions for preventing from early aging and microbe attacks [9].

In the new putty optical microscopy images was observed the presence of various colors regions. Raman spectroscopy of these regions showed compositional variations. These observations create premises of further studies of composition-color correlation and its evolution during artificial aging. This may lead to a simple optical way of aging evolution monitoring of this new and very popular putty for paints and sculptures restoration. Regarding this material the information regarding its behavior in time and its aging in various conditions are yet unknown but just predicted by provider based on its composition. Except the provider documentation there is no available information yet and, when invaluable historical artifacts are the subject of intervention, a poor knowledge of the material aging behavior may lead to disasters as in the old putty case. The old putty compact and dense material used during 1930's restoration seems to be the main cause for the irreversible damage of the wood paint icon from Antim Monastery collection.

Hardness measurements show that new putty is much softer than the old one. Old putty had a hardness of ~537 N/mm^2 while the new putty has ~250 N/mm^2.

5. Conclusions

Modern material characterization and analysis methods can be used to characterize compounds use din restoration and conservation works and identify specific characteristic and parameters that can be used for future monitoring. The identified parameters can act as stimuli for specially designed and fabricated sensors and detectors. Preliminary results of old and new putties for wood artworks restoration leads to the conclusion that bade on a detailed compositional analysis correlated with optical and electron microscopy, some parameters optical detection of the material condition may be elaborated.

References

[1] M. Stirlic "Heritage Science: A Future⁪Oriented Cross⁪Disciplinary Field" Angew.Chem.Int. Ed., 57,, pp. 7260 –7261, May 2018.

[2] M. Zarazo, A Fernández-Navajas, and F-J García-Diego, " *Long-Term Monitoring of Fresco Paintings in the Cathedral of Valencia (Spain) Through Humidity and Temperature Sensors in Various Locations for Preventive Conservation*", Sensors, 11(9), pp. 8685-8710; 2011.

[3] FJ Mesas-Carrascosa, D, Verdú Santano JE,Meroño de Larriva R, Ortíz Cordero RE Hidalgo Fernández, A García-Ferrer. "*Monitoring Heritage Buildings with Open Source Hardware Sensors: A Case Study of the Mosque-Cathedral of Córdoba*". Sensors.; 16(10):1620, 2016.

[4] SensorPush Staff, "*Environmental Conditions for Art Collection and Preservation*" 2018 http://www.sensorpush.com/articles/environmental-conditions-for-art-collection-and-preservation

[5] P, Londero, T. Fairbanks-Harris and P. M. Whitmore "*An Open-Source, Internet-of-Things Approach for Remote Sensing in Museums*", J. of the American Institute for Conservation, 55:3, 166-175, 2016.

[6] J. Shah and B. Mishra, "*Customized IoT enabled Wireless Sensing and Monitoring Platform for preservation of artwork in heritage buildings*," International Conference on Wireless Communications, Signal Processing and Networking (WiSPNET), Chennai, 2016, pp. 361-366., 2016

[7] 15th Triennial conference New Delhi. 22-26 September 2008, 978-81-8424-344-4, 2008

[8] http://rruff.info/chem/Na/display=default/R070237

[9] J. McCutcheon, "Microbially Accelerated Carbonate Mineral Precipitation as a Strategy for in Situ Carbon Sequestration and Rehabilitation of Asbestos Mine Sites", Environmental Science & Technology 50(3):1419–1427,2016.

Temperature measurements with four-resistor sensor patterned on golden layer

Milija Sarajlić*, Miloš Frantlović*, Predrag Poljak*, Katarina Radulović*, Dana Vasiljević Radović*

*University of Belgrade, Institute of Chemistry, Technology and Metallurgy,
Center of Microelectronic Technologies, Njegoševa 12, 11000 Belgrade, Serbia
{milijas, frant, predrag.poljak, kacar, dana}@nanosys.ihtm.bg.ac.rs

Abstract—Test of a sensor with four resistors structure on its sensitivity as a temperature probe is described. The chip is sensitive on a temperature difference between Si substrate and air and less sensitive on the temperature of the whole chip. The sensor consists out of four resistors realized as metal meanders on a silicon chip patterned in 150 nm thick gold layer whose lateral dimensions are 0.94 mm by 0.6 mm and the length 14.1 mm. Width of meander line is 0.02 mm with clearance 0.02 mm. The resistors form Wheatstone bridge configuration. Current through resistors was kept constant on 5 mA. Offset of the bridge on no temperature difference was 1.5 mV. In the case of temperature difference on the sensor surface, sensor output is changing with linear dependence. This has a potential for the use in temperature stabilization systems.

Keywords—sensor; temperature sensor; temperature measurement.

1. Introduction

Temperature measurement, monitoring and temperature stabilization is a very important part of industrial processes. [1]. Miniaturized sensors for this purpose are developed in many different forms and with various operating principles [2,3,4,5]. Here, an attempt to use a chip with four resistors made out of golden layer is shown.

At ICTM, Center of Microelectronic Technologies, (ICTM–CMT), Belgrade, Serbia this type of sensor was developed for the primary use as a Mercury vapor sensor [6,7]. It is also possible to apply the same structure as a temperature sensor. Even though this sensor is not optimized for the use as a temperature sensor it was worth testing its performance before trying to optimize it. Details of the structure and fabrication are given elsewhere [6,7]. Basic

Fig. 1 Photograph of the sensor connected to the measurement setup.

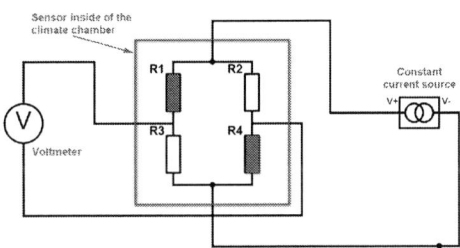

Fig. 2 Schematic diagram of the experimental setup. Two of the resistors shown in red color are covered with photoresist on the chip.

testing and principle of operation as a temperature sensor are shown in this paper.

2. Experiment

The sensor resistors are connected in the Wheatstone bridge configuration, Fig. 1 and Fig. 2. Through the bridge a constant current of 5 mA is applied from the constant current source, (Keithley 220, Programmable Current

978-1-5386-4483-6/18 $31.00 © 2018 IEEE

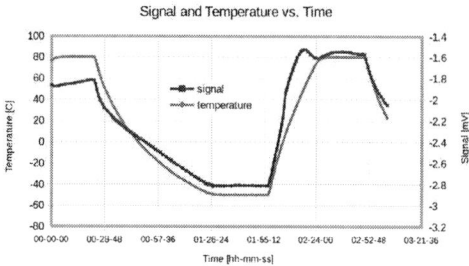

Fig. 3 Output from the sensor (blue) and temperature in the climate chamber (red) as recorded by the built-in thermometer.

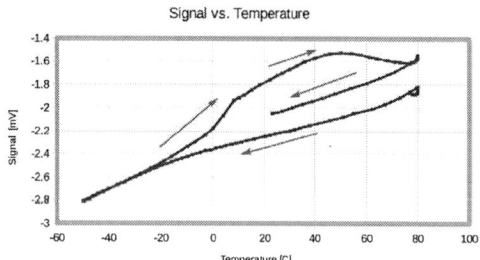

Fig. 4 Thermal hysteresis of the sensor in the climate chamber. Red arrows show time flow.

Source). Two of the resistors are covered with photoresist material, AZ1505, 0.5 μm thickness while the other two are open and exposed to the outer air, Fig 2. Therefore it is to be expected that covered resistors would have a different response to the temperature change of the air on the surface of the chip than resistors which are not covered.

Output from the sensor is monitored by voltmeter Agilent 34410A and recorded on the computer. Initial offset of the sensor on the room temperature is around 1.5 mV. The sensor was placed in a climate chamber, Heraeus VÖTSCH, and heated up from the room temperature of 20°C up to the 80°C. When the temperature 80°C was reached, sensor was left to stabilize for about 20 min. After this period the sensor was cooled to the temperature -50°C in order to observe the change of the output signal, Fig. 3. It is visible that the sensor signal is following the temperature drop inside of the climate chamber. It took around one hour for the climate chamber to reach -50°C. After this temperature has been reached the sensor was

left again to stabilize for about 30 min. During this period temperature was kept constant and the sensor also gave the constant output. After the period of constant temperature the climate chamber was set to the warming up sequence up to 80°C. Again, the sensor output was able to follow increase of the temperature in the climate chamber but with the higher slope and at the end before temperature stabilization it showed a pronounced overshoot and afterwards it came back to the stable value. It is possible that this sensor is very sensitive to the small temperature variations what can be the cause of the signal overshoot. Climate chamber applies simultaneous heating and cooling in order to control the temperature but it is possible that there are temperature variations inside of the chamber which are not visible by the built-in thermometer. After approximately 30 min of the temperature stabilization the climate chamber was cooled down to 20°C and the sensor again was able to follow this change.

Another issue with this sensor is thermal hysteresis. Fig. 4 depicts this behavior as diagram signal vs. temperature. Starting point is at 80°C and the signal 1.8 mV. The sensor was cooled from 80°C to -50°C and it followed the path shown in the diagram. As it was warmed up it initially followed the same path as it was during cooling but at the temperature around -20°C it starts to deviate from the previous values. The reason for this is not clear and it would take more thermal cycling to clarify the origin of the hysteresis. In the region of temperatures between 20°C and 60°C the sensor shows a large discrepancy between the current and the previous measurements. As it reached steady temperature of 80°C this discrepancy becomes less severe. When the sensor is exposed to the cooling once more it starts to follow the previous slope of the cooling curve but with the values shifted by about 0.2 mV upwards which represents 15% of the total signal span in the measurements.

3. Discussion

Based on the behavior of the sensor at stable temperatures and during the temperature changes, we conclude that the sensor like this would be more suitable for measurements of temperature oscillations in the system. This sensor could be used for the systems where temperature stabilization is of the primary importance. For instance in the typical climate chamber one should know when the temperature is stabilized in the case that the door is opened and closed quickly, typically during the sample change inside. In this case the sensor that can indicate that the temperature oscillations are not present anymore would be of the great help and would reduce waiting time for the system stabilization.

What was tested here is a four resistor sensor originally developed for the mercury vapor detection in air. It turned out that it is also sensitive to the temperature variations but for this use it should be optimized. First of all, heater which is present in the Hg sensor would not be needed for temperature sensor. The resistors could be made out of thinner lines of metal thus increasing their resistance and making the system more sensitive in the Wheatstone bridge configuration.

4. Conclusion

A novel field of application is considered for the sensor which was originally developed as a mercury (Hg) vapor sensor. The sensor was tested to the temperature change in the range from +80 C to -50 C. In this region it shows good response to the temperature change and relatively good response in the periods when the temperature was kept constant. With this result it is possible to conclude that this type of the sensor can be used for temperature measurements. A major problem discovered so far is thermal hysteresis where the output of the sensor depends on the previous thermal history. More experiments that include temperature cycling, as well as sensor optimization will be needed to overcome this issue.

Acknowledgments. This work was supported by The Ministry of Education, Science and Technological Development of the Republic of Serbia, within the Projects TR 32008 and TR32019.

References

[1] Peng Zhang, "*Advanced Industrial Control Technology*", pp. 85-86, eBook ISBN: 9781437778083, Hardcover ISBN: 9781437778076, William Andrew, August 2010.

[2] T.P. Nguyen, E. Lemaire, S. Euphrasie, L. Thiery, D. Teyssieux, D. Briand and P. Vairac, "*Microfabricated high temperature sensing platform dedicated to scanning thermal microscopy (SThM)*", Sensors and Actuators A, **275**, pp. 109–118, 2018.

[3] Jung-Soo Kim, Kyoung-Yong Chun and Chang-Soo Han, "*Ion channel-based flexible temperature sensor with humidity insensitivity*", Sensors and Actuators A, **271**, pp. 139–145, 2018.

[4] Ying Huang, Xiao Zeng, Wendong Wang, Xiaohui Guo, Chao Hao, Weidong Pan, Ping Liu, Caixia Liu, Yuanming Ma, Yugang Zhang and Xiaoming Yang, "*High-resolution flexible temperature sensor based graphite-filled polyethylene oxide and polyvinylidene fluoride composites for body temperature monitoring*", Sensors and Actuators A, **278**, pp. 1–10, 2018.

[5] Ran Zhao, Gang Shao, Yejie Cao, Linan An and Chengying Xu, "*Temperature sensor made of polymer-derived ceramics for high-temperature applications*", Sensors and Actuators A, **219**, pp. 58–64, 2014.

[6] Milija Sarajlić, Zoran G. Đurić, Vesna B. Jović, Srđan P. Petrović and Dragana S. Đorđević, "*An adsorption-based mercury sensor with continuous readout*", Microsyst. Technol., **19**, pp. 749-755, 2013.

[7] Milija Sarajlić, Zoran Đurić, Vesna Jović, Srđan Petrović and Dragana Đorđević, "*Detection limit for an adsorption-based mercury sensor*", Microelectronic Engineering, **103**, pp. 118-122, March 2013.

978-1-5386-4483-6/18 $31.00 © 2018 IEEE

Manufacture and investigation of a Vertical MEMS switch

A.Baracu*, R. Müller*, R. C. Voicu *, Marius Pustan **, Corina Birleanu**, Adrian Dinescu *

National Institute for Research and Development in Microtechnologies – IMT Bucharest, Romania
** Micro & Nano Systems Laboratory, Technical University of Cluj-Napoca, Cluj-Napoca, Romania

*E-mail: angela.baracu@imt.ro

Abstract — This paper presents a switch with vertical movement, based on surface micromachining processes. The central thermal driving structure contains a suspended rectangular shape connected by 5 pairs of symmetric mobile beams. SEM investigations and COVENTOR simulations were performed in order to investigate the behavior of the switch.

Keywords: MEMS, electro- thermal actuator, micrcmachining, sacrificial layer

1. Introduction

There are many taypes of MEMS actuataors: magnetic, electrostatic or thermal. They are used in diffeent applications as automatisations for microassembling tools or are incorporated in MEMS Systems like diffraction gratings or gear motors [1]. It is important to obtain small size actuators, compatible with CMOS technology, with low actuation voltages.

Between the electro- thermal actuator, based on Joule effect, Chevron type is mostly studied and used [2-4]. By applying an actuation voltage to a suspended structure with a pre-bent angle, an in-plane movement will be produced.

In our case we propose an electro-thermal actuator with vertical movement, and with parallel suspending beams, perpendicular to the central driving structure, using for manufacturing similar surface micro-machining processes [5].

The paper is focused on the investigation of the critical technological dimensions (gap, thickness of the arms) in order to obtain a released, stable mechanical structure and to avoid stickness of the metallic shapes. Technological experiments, SEM analyses and COVENTOR simulations were performed.

2. Design

The proposed MEMS vertical switch is driven by a metal electro-thermal actuator. Due to the thermal Joule effect induced in the central mobile part, a movement in the vertical plane will occur and commute the switch from the Off stage to On.

The thermal driving structure contains a suspended central, rectangular shape connected by 5 pairs of symmetric mobile beams. The gap between the bottom Aluminum electrodes and the moving part of the switch is equal with the thickness of the sacrificial layer (in our case photoresist). When applying a voltage to the actuation electrodes, an electric current passes and due to the electro-thermal effect the driving structure and the beams will heat and deform, moving down and contacting the bottom electrode, signal line (a metal-metal DC contact). The switch will commute form the Off into On stage. When the voltage is removed, the thermal driving structure will go back to the initial position.
We design an actuator with vertical displacement.

The geometrical dimensions are as following: central rectangular shape: length 500 μm and width 30 μm, lateral beams: length 260 μm and width 5 μm, distance between the beams 90 μm and the gap between the two upper and bottom electrodes: 0.5.-1 μm.

3. Fabrication

The vertical switch was fabricated using standard microelectronic processes Figure1. We start with a n type 525 μm <111> Silicon wafer. A 300 nm Si₃N₄ layer, obtained by LPCVD, was used in order to passivate the substrate.

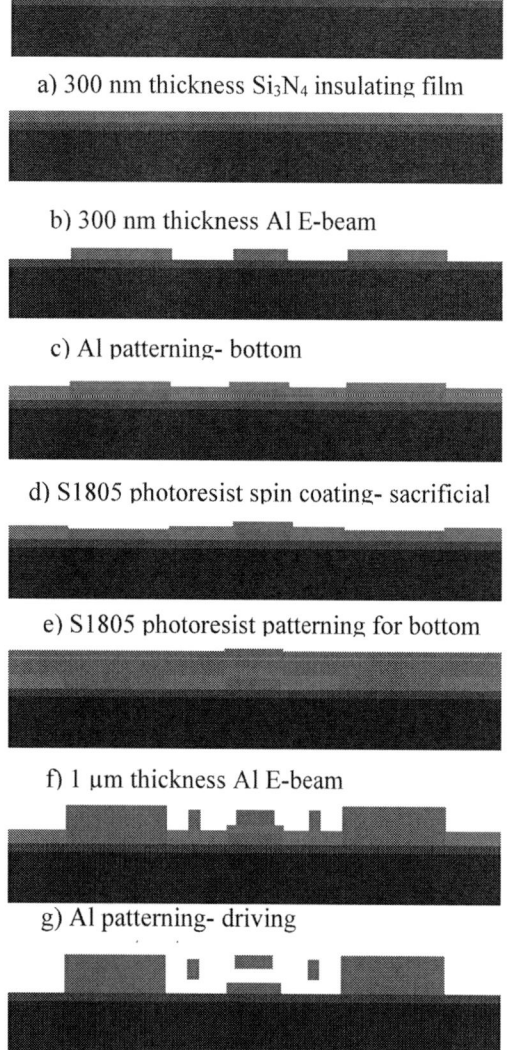

a) 300 nm thickness Si₃N₄ insulating film

b) 300 nm thickness Al E-beam

c) Al patterning- bottom

d) S1805 photoresist spin coating- sacrificial

e) S1805 photoresist patterning for bottom

f) 1 μm thickness Al E-beam

g) Al patterning- driving

h) S1805 photoresist removing- structure

Figure 1: Fabrication steps of the switch

The first metal deposition was Aluminum, thickness 300 nm, deposited by E-beam evaporation. It was configured using photolithographic process and wet chemical etching.
 An optical photo can be seen in Figure 2. We can observe the two rectangular electrodes were will be applied the actuation voltage and the other two electrodes (the signal line) used for the on/off contacts of the switch.

The next step was the deposition of a positive photoresist, S1805, with a thickness of 500 nm.

This will be used as a sacrificial layer in order to release the final mechanical MEMS switch structure. After the photoresist was patterned, we removed the unwanted sacrificial layer, in the region of the actuation electrode. A second Al layer of 1 μm was deposited by E-beam evaporation.

Using the last photolithographic mask we defined the vertical switch structure, using again wet etching of the top metallic layer.

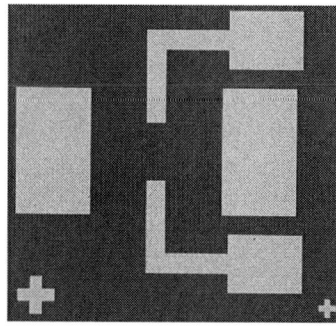

Figure 2: Optical image of the configured Al bottom electrodes

The optical image is presented in Figure 3. The last step was the removal of the photoresist sacrificial layer and the released of the mobile-driven structure of the switch.

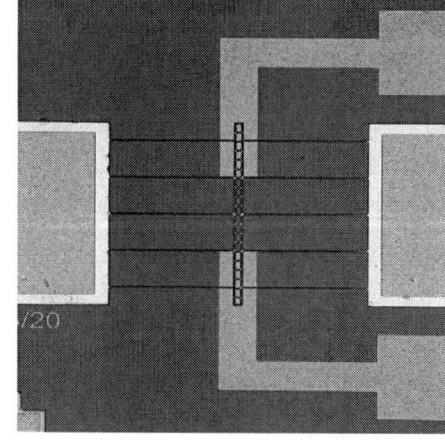

Figure 3: The top Al configured structure of the vertical switch (before the removal of the sacrificial layer- optical microscope image) [4]

This process was realized using oxygen plasma etching, which allowed us to avoid the unwanted stiction phenomena which can occur during classical wet removal between the metals.

4. Simulations

FEM simulation were performed by COVENTORWARE software in order to investigate the electro-thermo-mechanical behavior of the structure, for an actuation (pull-in) voltage.

The results are presented in Figures 4 – 7. The simulated 3D model was simplified and was meshed using hexahedral parabolic elements (Extruded bricks).

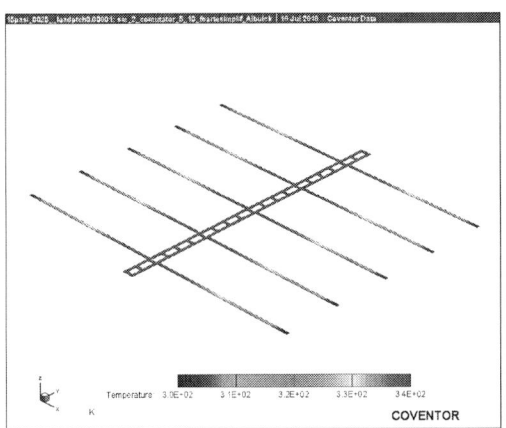

Figure 4: Temperature distribution in the structure for 50 mV applied (COVENTOR simulation)

The boundary conditions were set for simulations with an initial environment temperature of 27°C and a convection coefficient of 200 W/m²K.

The melting temperature of Aluminum is 655°C [6], so that the Al structure can be actuated for more than 200 mV.

In order to close the switch an applied voltage of less than 50 mV and a current of less than 80 mA are needed. In this case the maximal values of the temperature reached in the central part of the structure is less than 70°C (Figure 6 and Figure 7). As a first study, the out-of-plane displacement obtained for 50 mV applied and an initial

pressure of 0.0001 μN/μm² is estimated by simulations, at around 9μm (Figure 6).

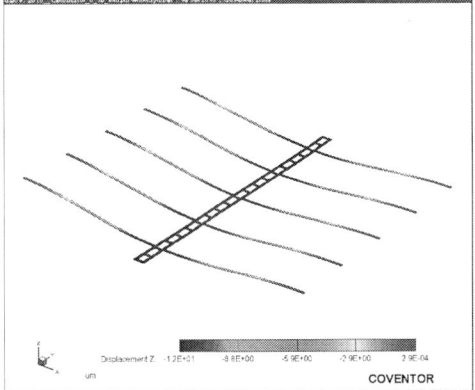

Figure 5: Out-of-plane displacement for 50 mV applied (COVENTOR simulation)

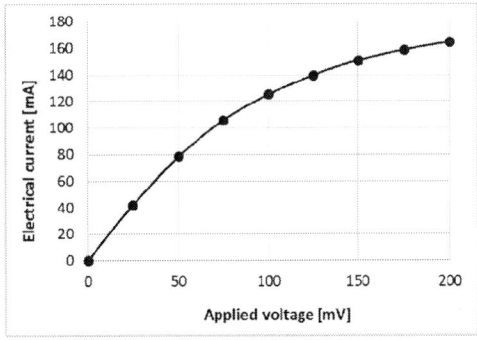

Figure 6: Electrical current versus the applied voltage (COVENTOR simulation)

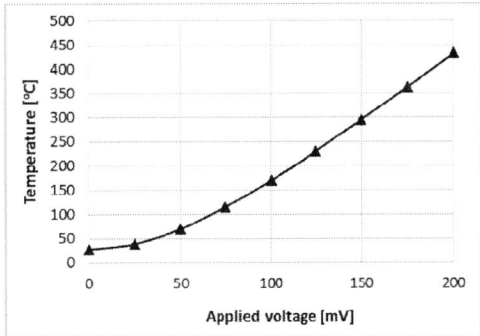

Figure 7: Maximal values of the temperature reached in the structure for an applied voltage (COVENTOR simulation)

5. Results and Discussions

We manufactured two different structures, by changing the depth of the gap (respective photoresist: 500 nm and 1 μm). We investigated the experimental switch using optical and Scanning Electron Microscopy (SEM) and we compared the two structures.

978-1-5386-4483-6/18 $31.00 © 2018 IEEE

As can be seen from Figure 8 and 9 in both cases were obtained good results: free standing structures which were completely released.

Figure 8: SEM image of the released beam of the switch for a 500 nm gap

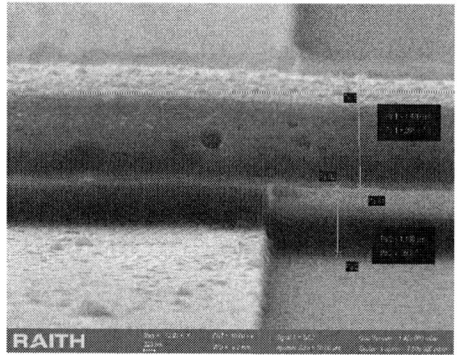

Figure 9: SEM image of the released beam of the switch for a 1 μm gap

Figure 10: SEM image of the released central shape with holes and the suspended beam of the switch

In addition, in the central rectangular shape we designed regular, rectangular holes that facilitate a fast and entire removing of the S1805 photoresist sacrificial layer in order to obtain a free, released structure (Figure 10).

Conclusions

Using Aluminum as structural material we fabricated a vertical MEMS switch, by surface micromachining and a photoresist as sacrificial layer. By varying the gap dimensions we achieved released MEMS switch. The process is relatively simple with good reproducibility. We simulated by COVENTOR the electro-thermo-mechanical behavior of the switch. This structure can be used as electro-thermal-mechanical switch and as test device for investigation of mechanical stiffness and the adhesion force between two metals.

Different stiffness of investigated switch structure can be obtained if the number of the suspended beams is modified with effect on the acting signal. Moreover, as the stiffness increases, the restoring force of the mobile electrode from the fixed one increases and the stiction decreases, respectively. The presented switch structure can also be integrated in MEMS applications where a thermal gradient occurs.

Acknowledgment: This work was supported by a Grant of the Romanian Agency by Scientific Research and innovation CNCS/CCDI UEFISCDI, project PNIII-P2-2.1-PED-2016-1727.

References

[1] Alex Man Ho Kwan, et al, *"Improved design for an Electro-Thermal IN-Plane Microactuato";* Journal of Electromechanical Systems, pp.586-595, vol. 21, No.3, June 2012

[2] L. Xiuhan, L. Leijie, et.al, *"Electro-thermally actuated RF MEMS switch for wireless communication"*, Proc. of the IEEE Int. Conf. on Nano/Micro Eng. And Mol. Syst., pp. 497-500, 2010.

[3] J. Sun, Z Li, et.al, *"Design of DC-contact RF MEMS switch with temperature stability*, AIP Advances 5, 041313, pp. 041313-1 - 041313-8, 2015

[4] M. Pustan, C. Birleanu, Dudescu, R. Muller, A. Baracu, *"Integrated thermally actuated MEMS switch with the signal line for the out-of-plane actuation"*, Proc. DTIP 2018, pp. 91-94, May 2018.

[5] A. Baracu, R. Voicu, R. Müller, A. Avram, M. Pustan, R. Chiorean, C. Birleanu, C. Dudescu, *"Design and fabrication of a MEMS chevron-type thermal actuator"*, AIP Conference Proceedings 1646, 25 (2015), pp. 25-30.

[6] www.asminternational.org

The Gate Current in MOSFETs versus planar-NOI Devices

Cristian Ravariu[1], Elena Manea[2], Catalin Parvulescu[2], Florin Babarada[1], Alina Popescu[2] and Avireni Srinivasulu[3]

[1] UPB-University "Politehnica" of Bucharest, Faculty of Electronics ETTI, Splaiul Independentei 313, Sect.6, 060042, Bucharest, Romania; E-mail: cristian.ravariu@upb.ro

[2] IMT-National Institute for Development and Research of Microtechnology, Str. Erou Iancu Nicolae 126A, 077190, Bucharest, Romania; E-mail: elena.manea@imt.ro; catalin.parvulescu@imt.ro; alina.popescu@imt.ro

[3] JECRC University, Dept. of E.C.E, Jaipur-303905, Rajasthan State, India, E-mail: avireni@ieee.org

Abstract—Recently reported, the Nothing On Insulator (NOI) device is based on the tunneling through a ultra-thin insulator placed between two semiconductors. A direct implementation of the NOI transistor that requires a vertical cavity etching in Si of 2nm width is a difficult technological task. Therefore, this paper proposes a simpler structure, based on the planar Si-technology. Rotating the NOI structure by 90⁰, the width of the cavity becomes the thickness of the cavity. If the vacuum is replaced by oxide, results a MOS capacitor without lateral junction but with lateral drain that is called p-NOI (planar-NOI variant). The p-NOI structure is simulated in Atlas and the results are compared with measured currents through the gate of fabricated MOSFETs. The main conduction mechanism is Fowler-Nordheim and secondary is quantum tunneling. The tunneling currents of the p-NOI structures obeys to the exponential law and are similar to the gate MOSFET currents. The currents are dominated by the insulator thickness and the gate voltage.

Keywords—Si-technology, oxide tunneling, simulation

1. Introduction

In the last decade, a lot of alternative devices are developed to fulfill the CMOS co-integration desiderate: CNT-FET, vacuum-FET, FinFET, [1-3].

A NASA research group recently claimed in few papers that a vacuum-channel transistor closely resembles an ordinary MOSFET and discussed the vacuum transistor as a device made of "Nothing". The same research group fabricated for the first time a vacuum transistor with 10nm cavity gap that experimentally presented a swing of SS=4V/dec and operation voltages of 20...30V [4], considerably lower versus the traditional vacuum devices.

In this context, the device abbreviated by NOI (Nothing On Insulator), conceptually proposed in 2005 [5], patented in 2013 [6], and periodically optimized [7], enters in this vacuum devices class of international interest. A direct implementation of the NOI transistor that requires a vertical cavity etching in Si of 10nm depth and 2nm width, is a difficult technological task. Therefore, this paper is searching for implementation solutions for a planar p-NOI variant based on the Si-technology. Rotating the NOI structure by 90⁰, the width problem is transferred into the thickness problem, within a p-NOI variant, Fig. 1.

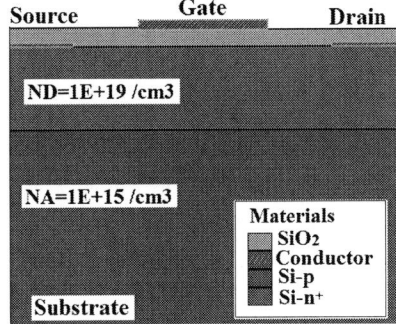

Fig. 1 The p-NOI conceptual structure.

The insulator of 2...90nm thickness can be vacuum or rather oxide. The oxide can be easily grown by the Si-planar technology. However, the Fowler-Nordheim tunneling through 2nm - oxide or vacuum was proved by simulations [8] and was also measured for the MOSFET gate currents, [9].

The main objective of this paper is to

978-1-5386-4483-6/18 $31.00 © 2018 IEEE

produce virtual tests by Atlas simulations, before a planar variant p-NOI fabrication. By these simulations, we intend to demonstrate the tunneling through the "oxide" region instead the "nothing" region, which is expected to offer a non-linear I-V dependence too, dominated by exponential laws, as in the NOI device case.

2. First p-NOI-1 variant

Firstly, a p-NOI-1 variant with an oxide thickness of 10nm is considered. In fact, this first structure is a MOS capacitor with lateral drain contact on the n^+-type semiconductor at entire surface. Consequently, the simulations can be easily compared to similar MOS capacitors from MOSFETs.

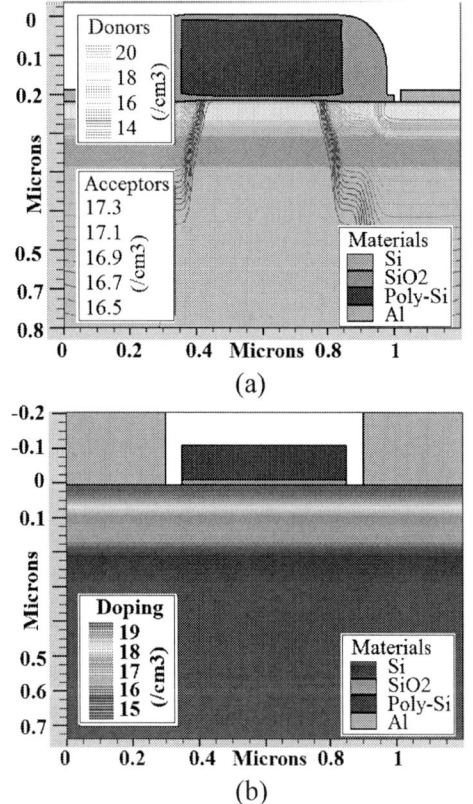

(a)

(b)

Fig. 2 The simulated structures of: (a) standard MOSFET structure, (b) p-NOI-1 variant with similar features excepting the n+ diffusions at entire surface.

Fig. 2a presents a standard MOSFET structure, available on Silvaco examples

[10], while Fig. 2b presents the p-NOI-1 variant with similar features, excepting the n^+ Source/Drain diffusions that spread to the entire surface of the Si-wafer in the p-NOI-1 case. This MOSFET configuration presented by Silvaco, as a standard example, gets static characteristics anchored in experiments. Both devices have: Si-substrate with doping profiles as in Fig. 2, SiO_2 insulator of 10nm thickness, poly-Si as gate contact and Aluminum as Source/Drain contacts.

Both MOSFET and p-NOI-1 structures are initially biased to $V_{Sb}=V_S=V_D=0V$ and V_G increases from 0V to 30V. The main tunneling mechanism for the gate current is Fowler-Nordheim model, activated by the FNORD parameter in the model statement, both for the standard MOSFET and p-NOI-1 device, Fig. 3a.

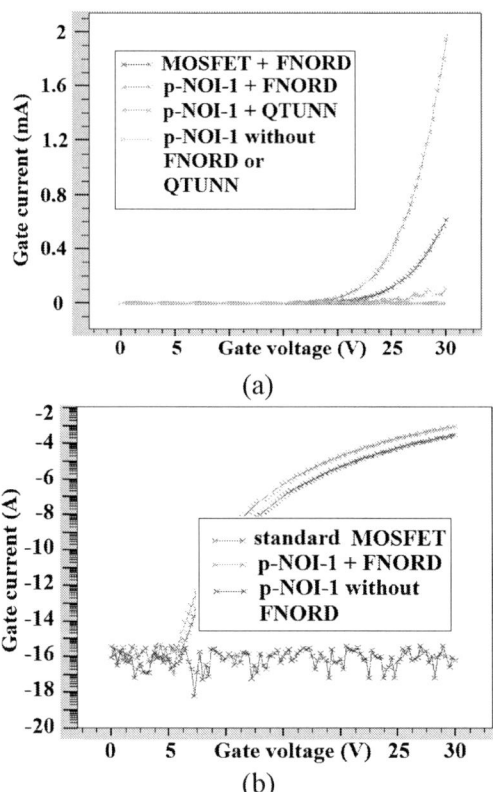

(a)

(b)

Fig. 3 The I_G-V_G curves for MOS and p-NOI-1 with different models at: (a) linear scale; (b) log scale.

An alternative phenomenon that can lead to a gate current increasing is the direct quantum tunneling. This event occurs if the oxide is thin enough to create a quasi-rectangular barrier potential between the gate electrode and semiconductor. Subsequently, the gate current of p-NOI-1 is simulated using sole QTUNN parameter, Fig. 3a. The simulations proved an insignificant quantum tunnelling components, 20 times lower than FNORD component. Hence, in next simulations QTUNN parameter is ignored. The Fowler-Nordheim mechanism offers the maximum current excursion till 2mA for p-NOI-1 with donor doping of 10^{19}cm^{-3} under the gate oxide and till 0.6mA for MOSFET with acceptor doping of 10^{17}cm^{-3} under the gate oxide.

Fig. 3b comparatively presents the I_G-V_G characteristics at logarithmic scale for: (i) p-NOI-1 without FNORD parameter; (ii) p-NOI-1 with FNORD model; (iii) standard MOSFET with FNORD model. If FNORD parameter is absent, a noisy almost-null current is captured through p-NOI-1 structure, Fig. 3b.

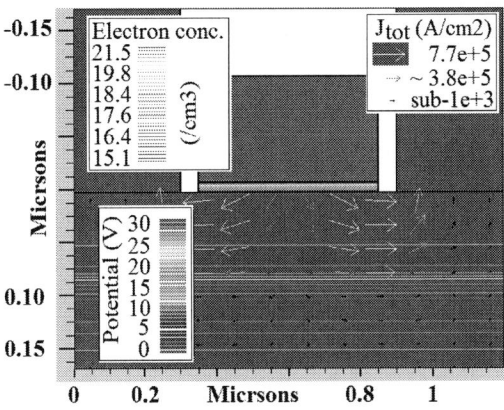

Fig. 4 The current vectors flow validation through the p-NOI-1 structure simulated at V_G=30V.

In the case of p-NOI-1 structure, the current flow starts from the gate and laterally occurs toward source and drain electrodes that are grounded, Fig. 4. This is possible for the p-NOI-1 structure, because the included MOS capacitor works in accumulation and the conduction occurs through a n-type film.

3. Second p-NOI-2 variant

Taking into account the NOI-transistor requirements with 2-5nm gap for tunnelling [8], we re-simulated the p-NOI structure, selecting as insulator few nm of HfO$_2$ over SiO$_2$, to respect the same structure of a fabricated MOSFET for comparisons, [11]. In all cases, the mesh gets 5nm on Ox axis and 2nm till 0.2μm on Oy, then 100nm. This p-NOI-2 structure is presented in Fig. 5.

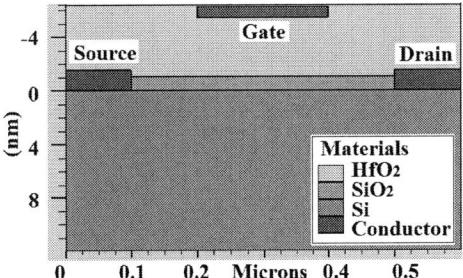

Fig. 5 The simulated p-NOI-2 structure.

Fig. 6 presents the current vectors inside the p-NOI-2 structure with 1nm-SiO$_2$/4nm-HfO$_2$, finally biased at V_G=10V. The maximum total current density of 2.6×10^7A/cm^2 in n$^+$-film toward source and drain, is extracted by simulations, Fig. 6.

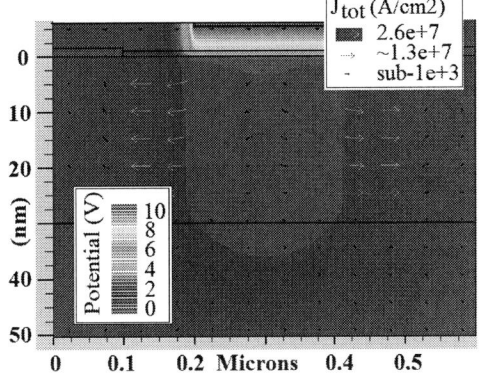

Fig. 6 The simulated current vectors through the p-NOI-2 structure biased at V_G=10V.

Fig. 7 comparatively presents the simulated I_G-V_G characteristics of p-NOI-2 variants with 1nm-SiO$_2$/4nm-HfO$_2$, 1nm-SiO$_2$/8nm-HfO$_2$, and sole 1.39nm-SiO$_2$ besides to some experimental I_G-V_G picked points from a fabricated MOSFET with 1nm-SiO_2/4nm-HfO_2 [11]. The shape and current variation are quite similar for p-NOI-2 and experimental MOSFET. For MOSFET the experimental gate current density ranges from 10^{-6} to $10^2 A/m^2$ that means 10^{-18} to $10^{-10}A$, Fig. 7b. In fact, the atto-Amperes currents are extrapolated from the pico-Amperes measurements, [11].

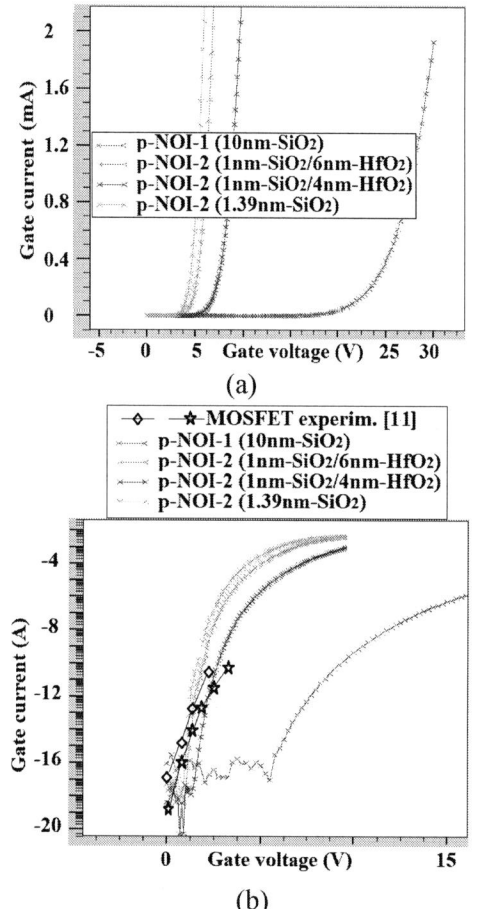

Fig. 7 Comparisons of I_G-V_G curves among different p-NOI-2 variants and real MOSFETs from ref. [11].

4. Conclusions

In conclusion, the Fowler-Nordheim mechanism was firmly demonstrated to be the main tunnelling current component through the p-NOI structure, being in agreement with the gate current simulated through a standard MOSFET and a real MOSFET of 5nm oxide. The gate current mainly depends on the insulator type and thickness and gate voltage.

Acknowledgments. This work is supported by grants of the Romanian National Authority for Scientific Research and Innovation, CNCS/CCCDI UEFISCDI: PN-III-P2-2.1-PED-2016-0427 project number 205PED/2017 and partially supported by PN-III-P4-ID-PCE-2016-0480 project number PCE no. 4/2017.

References

[1] J. Singh, J. Ciavatti, et al, "14-nm FinFET Technology for Analog and RF Applications, " *IEEE Transactions on Electron Devices*, vol 65, no.1, pp. 31-37, 2018.

[2] P.Kavitha, Sarada Musala, K.V.Vardhan, Y.S. Vani and Avireni Srinivasulu,"Carbon nano tube field effect transistors based ternary Ex-OR and Ex-NOR gates", *Journal of Current Nanoscience*, Vol 12, Issue 4, pp. 520-526, Aug 2016.

[3] J.W. Han, J.S. Oh, M. Meyyappan, "Cofabrication of Vacuum Field Emission Transistor (VFET) and MOSFET," *IEEE Transactions on Nanotechnology*, vol. 13, nr. 3, pp. 464 - 468, May 2014.

[4] J-W. Han, M. Meyyappan, "Introducing the vacuum transistor: a device made of Nothing," *IEEE Spectrum*, nr. 7, pp. 25-29, Jul 2014.

[5] C. Ravariu. "A NOI – nanotransistor," *in Proc. IEEE Int. Conf. of Semiconductors Proceedings*, Sinaia, Romania, 2005, pp. 65-68.

[6] C. Ravariu, "Field effect transistors with a cavity on insulator, NOI and a-NOI," Romanian Patent Number: RO126811-A0, OSIM, Aug. 2013.

[7] C. Ravariu, Deeper Insights of the Conduction Mechanisms in a Vacuum SOI Nanotransistor, *IEEE Transactions on Electron Devices*, vol. 63, no. 8, pp. 3278 - 3283, 2016.

[8] C. Ravariu, "Semiconductor Materials Optimization for A TFET Device with Nothing Region On Insulator," *IEEE Trans. on Semiconductor Manufacturing*, vol. 26, no. 3, pp. 406-413, Aug. 2013.

[9] TA Karatsori, CG Theodorou, E Josse, C. Dimitriadis, G. Ghibaudo, All Operation Region Characterization and Modeling of Drain and Gate Current Mismatch in 14-nm Fully Depleted SOI MOSFETs, *IEEE Transactions on Electron Devices* Vol 64, Issue: 5, pp. 2080 - 2085, 2017.

[10] ***, Silvaco/Examples, MOS1 : MOS Application Examples, 2018, available at: https://www.silvaco.com/examples/tcad/section34/exampl e1/index.html

[11] R. Basak, B. Maiti, A. Mallik, Analytical model of gate leakage current through bilayer oxide stack in advanced MOSFET, *Superlattices and Microstructures*, vol. 80, no. 1, pp. 20–31, 2015.

Improved Ti/Pt/Au - n-type Si contacts by post-metallization annealing in nitrogen atmosphere

Razvan Pascu[1], Mihai Danila[1], Pericle Varasteanu[1], Mihaela Kusko[1], Gheorghe Pristavu[2], Gheorghe Brezeanu[2], Florin Draghici[2]

[1] National Institute for R&D in Microtechnology - IMT Bucharest, Romania
razvan.pascu@imt.ro
[2] University "Politehnica" of Bucharest, Romania

Abstract — A metallic sandwich (Ti/Pt/Au) is deposited on n-type Si and subsequently subjected to post-metallization annealing at temperatures ranging from 500°C to 950°C, with a step of 50°C. XRD microstructural investigations evince the effect of post-metallization annealing, focusing on the formation of the silicide interface layer at each of the annealing temperatures. The electrical quality of the contacts was analyzed based on the linear transfer length method using test structures with different gaps between the pads. Accordingly, the sheet resistance, the contact resistance, the transfer length and the specific contact resistivity are determined. It is demonstrated that sheet resistance improves with more than three orders of magnitude at 900°C, compared the reference sample. The contact resistance also improves with more than one order of magnitude, reaching a minimum value of 1.68 Ω at 950°C. The transfer length reaches a maximum value of 10.8 µm at 750°C, corresponding to a specific contact resistivity of $4.82 \cdot 10^{-5}$ $\Omega \cdot cm^2$. Finally, an effective resistivity for the fabricated ohmic contact of 0.033 Ωcm is obtained for PMA at 950°C.

Keywords — *ohmic contacts, post-metallization annealing, silicide, specific contact resistance.*

1. Introduction

Smart sensors, embedded systems and overall interconnected electronic devices are the task force behind the technological revolution driven by cloud computing and artificial intelligence. As these "information and resource gatherers" grow in both number and level of integration, minimizing technology-induced parasitics has become a concern [1]. Since ohmic contacts are ubiquitously present in the structure of any devices, the minimization of their associated Specific Contact Resistance (SCR) is an essential research focus. Low SCR values are very desirable for junction power-loss reduction and better heat management. Furthermore, in sensing and harvesting applications, the output performances are strongly dependent on SCR [2]. In particular, recent advancements in interdigitated back-contact silicon (Si) solar cells have re-attracted interest in optimizing Si-based ohmic contacts [3].

The preferred approach in obtaining a nearly ideal ohmic contact consists of determining the best post-metallization annealing (PMA) conditions, specific to the metal-semiconductor composition. The aim of this paper is to investigate the parameter - variation of Ti/Pt/Au – Si ohmic contacts for different post-metallization annealing conditions. Microstructural analyses are performed in order to assess interface homogeneity and standard Transfer Length Method (TLM) [4] measurements are carried out on experimental structures, for annealing temperatures up to 950°C.

2. Experimental

Ohmic contact investigations were carried out on a test structure shown in **Fig. 1**. The ohmic contacts were fabricated on (100) n-type Si wafers with resistivity of 5 mΩcm, where, after a standard RCA cleaning [5], a metallic sandwich formed by Ti (30 nm) / Pt (50 nm) / Au (100 nm) was deposited by e-beam evaporation on the front side of the wafer. The Ti layer ensures good adherence on the Si wafer. The Au layer is deposited to facilitate the wire bonding for further device packaging. Pt acts as a barrier against Au diffusion at high temperatures. In order to guarantee both the ohmic contact performances and the adherence of the deposited metals, classical PMAs have been performed between 500°C and 950°C, in nitrogen atmosphere (with a step of 50°C), time for 30 min with a heating-ramp of 30 min. The contact electrodes have been defined using a photolithographic process, with different gaps (d = 1.5, 2, 3 and 4 µm) between them (see **Fig. 1**). Microphysical investigations of the ohmic contacts were performed with a 9 kW Rigaku SmartLab rotating anode diffraction system from Rigaku Corporation, Japan, operating at 45 kV,

200 mA, equipped with an X-ray multilayer mirror (set in parallel beam mode – PB) and a flat diffracted beam monochromator (graphite (002)) at the detector side, configured for the standard wide-angle diffraction method (coupled θ–2θ scan).

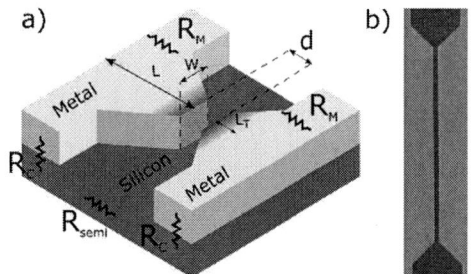

Fig. 1 a) 3D picture of the designed test structure; b) optical image of the experimental test structure.

3. Results and discussions

3.1 Microstructural analyses

The X-ray measurements were performed from 10° to 90°, using a step of 0.01°, a speed of 12°/min., 5° Soller slits and standard 1 mm wide incident and receiving slits.

Fig. 2 shows the changes in the X-ray diffraction pattern of the Ti/Pt/Au metallic stack by annealing at different elevated temperatures.

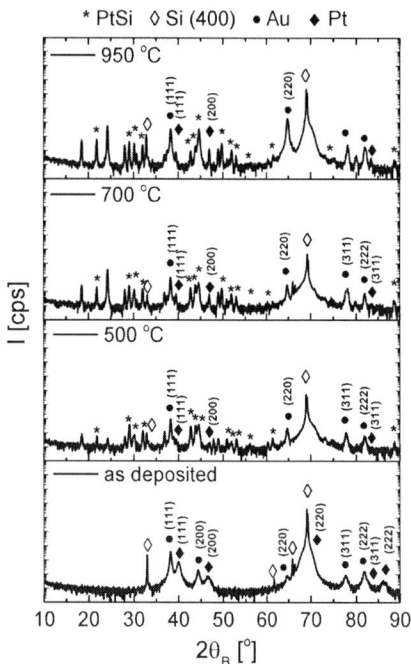

Fig. 2 X-ray diffraction patterns of Ti / Pt / Au layers on Si (400) annealed at various temperatures.

Firstly, the as-deposited Ti/Pt/Au layers on Si (400) have a strong (111) texture (Au and Pt) while titanium layer is fully amorphous. The presence of two large amorphous humps around 26 - 52° (centered at ~ 43°) and 55 - 90° (centered at ~76°) reveals that the as-deposited Au and Pt layers are also not completely crystalline. The most relevant effect of thermal annealing is formation of the orthorhombic platinum silicide (PtSi) phase (ICDD PDF card number 01-071-0523, space group Pbnm (62), a = 5.916 Å, b=5.577 Å, c = 3.587 Å) even starting from 500°C. The Pt related peaks are absent in the diffraction spectra after the annealing processes above 500°C, while the strong Au (111) peak decreases to 9400 cps (T = 500°C) from 40700 cps with no apparent texture (relative intensities match the ASTM/ ICDD/JPCSD values of Au, 00-004-0784 card number, s.g. 225). This decrease of the X-ray Au (111) diffraction peak point to the formation of either an amorphous Ti (Au) or TiSi$_x$ (Au) interface layer (x=1.2). This interface layer is close to the Si (400) substrate and, together with the PtSi interface layer, act as an effective barrier for the diffusion of both Au and Pt into silicon substrate even at higher temperatures of 900 and 950°C (no X-ray diffraction lines of Au-Pt / Au$_3$Pt alloys are present). Although some TiSi$_2$ and/or TiSi diffraction lines are present in the XRD spectra (especially at high annealing temperatures of 900°C and 950°C) a clear match is not possible (overlapping and/ or missing lines).

3.2 Electrical characterization

The electrical performances of the fabricated ohmic contacts were evaluated using the TLM method. This technique, which requires performing electrical measurements between two adjacent metal contacts with different gaps between them, uses the test structure from **Fig. 1**. I-V characteristics were acquired for structures subjected to each of the PMA using a Keithley 4200 semiconductor characterization system; all the measurements were performed at room temperature, in dark conditions. Thus, **Fig. 3** presents electrical curves of the as deposited sample (**Ref**) and those annealed at 500°C, 700°C and 950°C, respectively. Each graph contains plots associated to the different gap distances (d).

978-1-5386-4483-6/18 $31.00 © 2018 IEEE

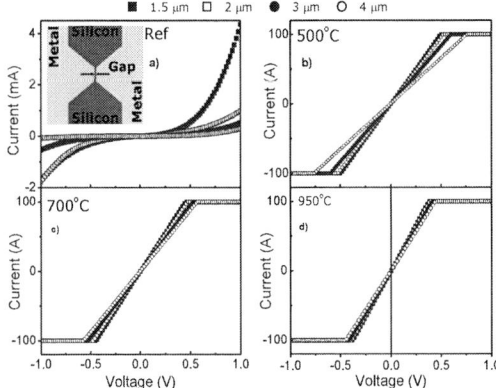

Fig. 3 I-V characteristics for the fabricated ohmic contacts: a) Ref; b) 500 °C; c) 700 °C; d) 950 °C.

The linear variation of the current with applied voltage clearly demonstrates that the ohmic contact is significantly improved from **Ref** to the PMA samples. Furthermore, with increasing annealing temperature, the total resistance ($\mathbf{R_T}$) decreases, while the influence of gap-size (1.5 μm – 4 μm) is strongly reduced, indicating a tendency towards an ideal ohmic contact.

Plotting the total resistance (inverse slope of I-V characteristics in **Fig. 3**) as a function of gap size, the sheet resistance ($\mathbf{R_{sh}}$), the contact resistance ($\mathbf{R_c}$) and the transfer length ($\mathbf{L_T}$), can be extracted according to [6]:

$$R_T = \frac{R_{sh}d}{W} + 2R_c \approx \frac{R_{sh}}{W}(d + 2L_T) \qquad (1)$$

where $\mathbf{R_T}$ is the total resistance measured, $\mathbf{W}$ the contact width, $\mathbf{d}$ the distance between the two measured contacts and $\frac{\mathbf{R_{sh}d}}{\mathbf{W}}$ is the semiconductor resistance (R_{semi} – **Fig. 1**). $\mathbf{L_T}$ represents the effective contact length used for the dominant current flow over the metal-semiconductor interface.

Fig. 4 illustrates $\mathbf{R_T}$ as a function of $\mathbf{d}$ for each experimental test structure. As can be observed, the quality of the ohmic contacts is significantly improved when a PMA is used, the total resistance decreasing with almost three orders of magnitude for all annealed samples in comparison with the reference one. Moreover, a slow decrease in $\mathbf{R_T}$ with increasing annealing temperature is observed. The more uniform ohmic contact is obtained due to formation of PtSi during the PMAs, in accordance with XRD measurements.

$\mathbf{R_{sh}}$, $\mathbf{R_c}$ and $\mathbf{L_T}$ were estimated using the slope and the intercept resulted from the linear fits of the curves from **Fig. 4** ($\mathbf{R_T}$ as a function of gap distance).

$\mathbf{L_T}$ is necessary in order to obtain specific contact resistance ($\mathbf{\rho_c}$) [6]:

$$\rho_c = R_c \cdot W \cdot L_T \qquad (2)$$

These parameters, determined using the TLM method as a function of annealing temperature, are depicted in **Fig. 5**.

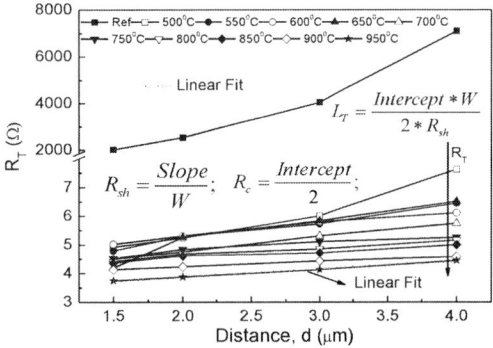

Fig. 4 The total resistance measured as a function of the distance between the contact electrodes.

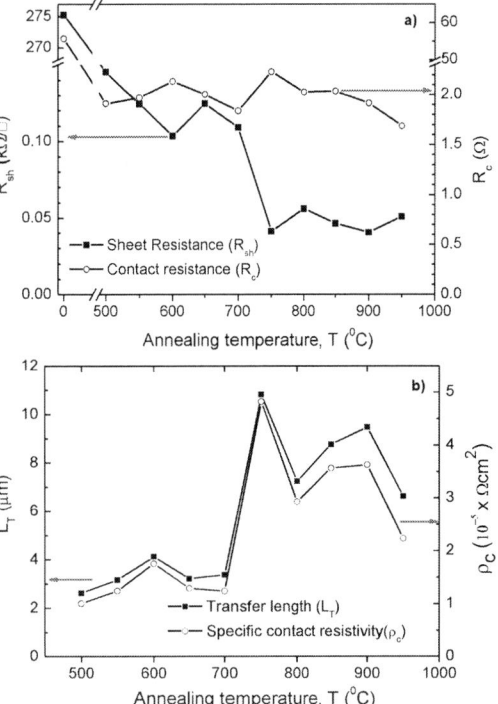

Fig. 5 Electrical parameters as a function of PMA temperature: a) sheet and contact resistances; b) specific contact resistance and transfer length variation.

As can be seen in **Fig. 5a**, the sheet resistance has a value of 275.27 kΩ/□ for the **Ref** sample and it is significantly improved when the PMA temperature is increased, reaching a minimum value of 40.42 Ω/□ at 900°C, with more than three orders of magnitude lower than the initial one. In fact, a continuous diminishing with PMA temperature is observed up to 700°C, followed by a plateau, around 50 $\Omega/\square$, up to 950°C. On the other hand, the contact resistance has a value of 55.7 Ω for the **Ref** sample that decreases by one order of magnitude even at the lowest annealing temperature (500°C), to 1.9 Ω. It stays approximately constant with PMA temperature up to 900°C and reaches a minimum value of 1.68 Ω at 950°C.

A very small transfer length value for the **Ref** sample (around 0.04 µm), was calculated which means that the dominant current crosses a very small portion of the total length of the contact pads, $L_T \ll L$ ($L = 3$ mm). The L_T results for annealed structures, given in **Fig. 5b**, show relatively stable values, around 3 µm, up to 700°C, followed by a maximum of 10.8 µm at 750°C. The high L_T values demonstrate a significant conductivity improvement.

The specific contact resistance determined using eq. 2 presents a similar variation as the transfer length with PMAs (**Fig. 5b**). Thus, a maximum value of $4.82 \cdot 10^{-5}$ Ωcm^2 was obtained, corresponding to the maximum value of L_T. At 950°C, the specific contact resistance is $2.23 \cdot 10^{-5}$ Ωcm^2 which corresponds to a transfer length of 6.6 µm. For a better illustration of the effective resistance of the ohmic contacts, the ratio between the specific contact resistance and the transfer length as a function of annealing temperature is plotted in **Fig. 6**. It can be seen that the effective resistance is substantially improved for the annealed samples by almost two orders of magnitude, reaching a minimum value of 0.033 Ωcm at 950°C.

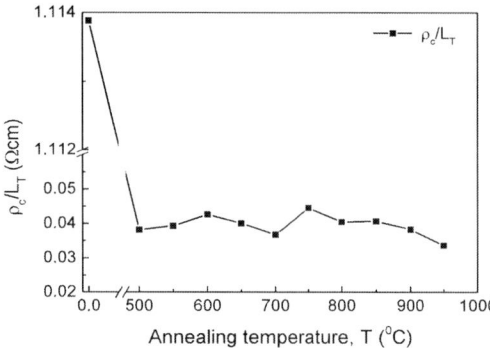

Fig. 6 The ratio between the specific contact resistivity and the transfer length as a function of PMA temperature.

These results demonstrate the necessity of PMAs, even at a moderate temperature (500°C), for fabrication of good quality ohmic contacts.

4. Conclusions

We investigated the effects of post metallization annealing on the electrical parameters of the ohmic contacts realized on n-type Si wafer, using a metallic sandwich consisting of Ti/Pt/Au. The microstructural XRD analyses showed formation of a silicide layer at the metal/semiconductor interface. The electrical measurements performed on fabricated structures with different gaps between the contact pads allowed estimation of the sheet and contact resistance, the transfer length and the specific contact resistivity, all of them being determined using the transfer length measurements method. The results clearly demonstrate the beneficial influence of the used PMAs, all the parameters being substantially improved, even for treatments performed at moderate temperatures (500°C).

Acknowledgments. This work was supported by a grant of the Romanian Ministry of Research and Innovation, CCCDI – UEFISCDI, project number PN-III-P1-1.2-PCCDI-2017-0419/ SENSIS, within PNCDI III.

References

[1] M. Abraham, *"Engineering Ohmic Contacts to III-V, III-N and 2D Dichalcogenides: The Impact of Annealing and Surface Preparation on Contact Resistance"*, Pensylvania State University, 2017.
[2] R. S. Okojie and D. Lukco, *"Pt:Ti Diffusion Barrier, Interconnect, and Simultaneous Ohmic Contacts to n- and p-Type 4H-SiC"*, Materials Science Forum, 924, pp. 381-384, 2018.
[3] N. Song et al., *"Metallization Method for Interdigitated Back-Contact Silicon Solar Cells Employing an Insulating Resin Layer and a Ti/Ag/Cu Metal Stack"* IEEE Journal of Photovoltaics, PP (99), pp. 1-7, 2018.
[4] W. Shockley, A. Goetzberger, R.M. Scarlett, R. Finch, R. Gereth, V. Williams and N Zetterquist, *"Research and Investigation of Inverse Epitaxial UHF Power Transistors,"* Rep. No. AFAL-TDR-64-207, Air Force Avionics Lab., Wright-Patterson Air Force Base, OH, Sept. 1964.
[5] W. Kern, D.A. Puotinen, *"Cleaning solutions based on hydrogen peroxide for use in silicon semiconductor technology"* RCA Rev., 31, pp. 187, 1970.
[6] D. Schroder, *"Semiconductor material and device characterization"*, Third edition, John Wiley & Sons Inc., New York 2006.

Comparative study of the electrical properties of CZTS-TiO$_2$ and CZTS-ZnO heterojunctions for PV applications

M. Covei*, C. Bogatu*, D. Perniu*, S. Cisse**, A. Duta*

* R&D Centre: Renewable Energy Systems and Recycling, Transilvania University of Brasov, Romania
E-mail: maria.covei@unitbv.ro; cristina.bogatu@unitbv.ro; d.perniu@unitbv.ro; a.duta@unitbv.ro
** Cheikh Anta Diop University of Dakar, Sénégal
E-mail: salifcisse22@yahoo.com

Abstract—Two heterojunctions based on Cu$_2$ZnSnS$_4$ were obtained as thin films, on FTO glass, by spray pyrolysis deposition. The choice of the n-type layer (TiO$_2$ or ZnO) has impacted the nucleation and growth of the CZTS layers deposited on top. Crystalline thin films were obtained with drastically different morphologies. Small spherical grains promoted a better match at the interface between TiO$_2$ and CZTS, while the hexagonal plates of ZnO promoted discontinuations. Both junctions proved photosensitive and gave good rectifying behavior but efficiencies lower than 0.01% suggest charge recombination at the interface or in the CZTS layer.

Keywords—CZTS-based heterojunctions; spray pyrolysis deposition; curent-voltage variation; interface.

1. Introduction

Thin film photovoltaics gained attention due to their ease of construction, low material requirements and good efficiency. Several ternary and quaternary compounds are currently used and among these CuInS(Se)$_2$ and CuInGaS(Se)$_2$ are the most well-known however, since In and Ga are critical materials [1], alternatives are investigated. By replacing In with Zn and Ga with Sn, Cu$_2$ZnSnS$_4$ (CZTS) layers with similar opto-electrical properties could be obtained [2]. Coupled with high-performance n-type layers (TiO$_2$ or ZnO), efficient PVs could be reached. The most efficient CZTS cells only partially preserve the sustainability features, due to the deposition conditions and the use of toxic materials (e.g. CdS used as buffer layer). Therefore, focus is set on obtaining thin layered heterojunctions through low-cost, less-energy intensive deposition methods, like spray pyrolysis deposition (SPD).

This work presents a comparative study of two types of SPD-deposited layer structures, based on TiO$_2$-CZTS and ZnO-CZTS heterojunctions with the aim of identifying the best electrical response and the possible bottlenecks.

2. Materials and methods

The thin films were deposited on commercial FTO covered glass (TEC 15/ 3mm Pilkington) by the spray pyrolysis deposition technique.

(i) The TiO$_2$ layer was deposited using ethanol solution of TiCl$_4$ (0.05M), at 350°C, using 20 spraying sequences with 60 seconds break between them, followed by annealing at 500°C for 1hour to increase crystallinity.

(ii) The ZnO layer was obtained using ZnCl$_2$ in a 30% water-70% ethanol solution, using similar deposition and annealing conditions to those employed for the TiO$_2$ layer.

(iii) The Cu$_2$ZnSnS$_4$ layer was obtained from cationic chlorides and thiourea mixed in a water-ethanol solution (1:1) to give an atomic ratio of 2:1:1:16. The excess sulfur in the precursor solution was chosen to compensate the loss that occurs during spraying and the subsequent annealing stages, due to the sulfides instability in the ambient atmosphere that also limits the deposition and annealing temperatures. To obtain an up-scalable deposition method, no reducing atmosphere was used and the layer was deposited using 15 spraying sequences with 60 seconds break between them, at 300°C followed by 1h annealing at 400°C.

The samples' crystallinity was studied by X-Ray Diffraction (XRD) using a Bruker Advanced D8 Discover Diffractometer with locked coupled continuous scan, 2s/step and

---

Acknowledgments: The financial supports of the projects: M-ERA.net WaterSafe, contract no. 39/2016 and AUF Eugen Ionesco are gratefully acknowledged.

CuKα1 radiation (1.5406Å), while Raman analysis (LabRAM HR800 Horiba) was used to confirm the CZTS formation.

The surface morphology was investigated using Scanning Electron Microscopy (SEM, Hitachi model S-3400N) and Atomic Force Microscopy (AFM, NT-MDT), in semicontact mode, after the deposition and annealing of the CZTS layer.

The bandgap energy of the component layers was obtained using the Tauc formula from the transmittance and reflectance spectra recorded using a UV–Vis spectrophotometer (Perkin-Elmer Lambda 25UV/Vis), in the 250–2500 nm range.

To investigate the electrical properties of the n-p heterojunctions, the current density – voltage (J-V) curves were recorded in dark and under irradiation. A Newport Oriel Apex Illuminator 1000 W/m^2 was used, located at 30 cm from the device for opto-electrical characterization of materials (patent request no. A/01147), with the working electrode in contact with the CZTS film and the counter electrode on a thin carbon film on the TCO substrate (reference). The voltage was varied between -3.5 and 3.5 V for both samples.

3. Results and discussion

The structural properties of the two layered structures were investigated by XRD. The spectra presented in Fig. 1 show that the oxides (TiO$_2$ or ZnO) and the sulfide (CZTS - kesterite) compounds are identified in the samples. However, the kesterite CZTS peaks partially overlap those of zinc or copper sulfide, as previously outlined [3].

The multiple sulfide phases in the film could be detrimental to the charge carrier transport promoting charge recombination or scattering at the boundaries. Moreover, zinc sulfide is a wide bandgap semiconductor with a conduction band that prevents the electrons from traveling to the n-type semiconductor. Therefore, to confirm the existence of the kesterite in the film, Raman spectra were obtained for the CZTS thin film on FTO glass (Fig. 2), as it is able to distinguish between the different sulfide phases. The presence of the characteristic peak close to 331 cm^{-1} indicates that the targeted quaternary compound was indeed obtained in the p-type layer, while the absence of any peaks at 475 cm^{-1}, 356 cm^{-1} indicated that no ZnS or Cu$_2$S are present [4].

This was also supported by the bandgap measurement of the CZTS thin film deposited on the substrate as a single layer (Fig. 2 inset). As illustrated, there are multiple fringes corresponding to multiple bandgap energy values, suggesting that additional phases are in the CZTS film. The E$_g$=1.5 eV could be attributed to CZTS [5], but the other two do not match those of zinc or copper sulfides (2.2 eV for Cu$_2$S and 3.9 eV for ZnS) and are more likely matched to some non-stoichiometric compounds.

The crystallite sizes, D, of the kesterite, wutzite and anatase phases were calculated using Scherrer's formula and it was observed that relatively small crystallites are present in all three cases: 4.2 nm for kesterite CZTS, 5.2 nm for anatase TiO$_2$ and 8.9 nm for wurtzite ZnO.

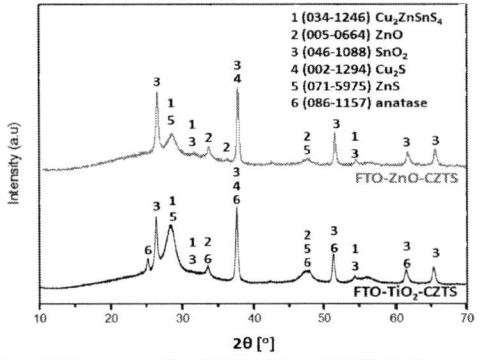

Fig. 1 XRD spectra of the CZTS-TiO$_2$ and CZTS-ZnO structures

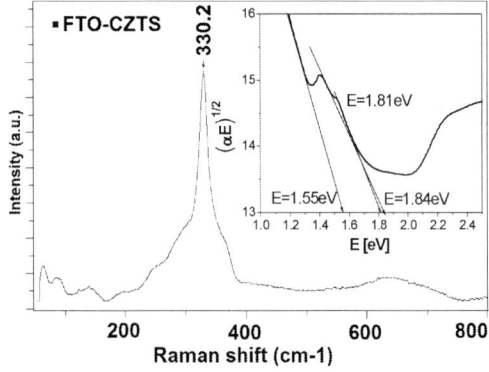

Fig. 2 Raman spectra of CZTS film deposited on FTO glass. Inset: Tauc plot of the CZTS layer

Small crystallites imply larger grain boundaries that can act as scattering centers for charge carriers, which would suggest that the wurtzite phase will promote improved charge transfer due to its slightly higher crystallite size compared to that of anatase. However, comparing the CZTS crystallite size with the n-type semiconductors, the small crystallite size of the kesterite phase better matches TiO_2 compared to ZnO. This could imply a better contact between CZTS and TiO_2 at the heterojunction interface.

The morphology of the two structures was investigated by SEM. As the images in Fig. 3 show, the n-type semiconductor has an important effect on the growth of the CZTS films on top. The titanium dioxide thin film was not perfectly homogeneous and contains agglomerations on the surface (Fig. 3a inset) which acted as growth centers for the CZTS layer deposited on top. Therefore the surface consists of two types of morphologies (Fig. 3a): one with lower roughness (11.8 nm, as measured by AFM) that appears to be a better electron conductor (the darker shades in the micrographs) and one that has a higher roughness (29.4 nm) which could induce charge recombination and scattering and thus appears lighter in the SEM image.

Similarly, in the case of the FTO-ZnO-CZTS structure, the surface exhibited two morphology types (Fig. 3b), with different roughness (14.8 nm compared to 48.5 nm). However, due to the growth of ZnO as hexagonal plates (Fig. 3b inset), further covered by the CZTS, the surface morphology of this structure substantially differs from the spherical, cluster grained one of the FTO-TiO_2-CZTS structure.

To better assess the impact of the n-type layer choice on the morphology of the structures, cross-section SEM images were taken at the interface (see Fig. 4).

The interface between ZnO and CZTS (Fig. 4b) shows some areas with imperfect contact between the layers, (indicated with red circles) that has a detrimental effect on the charge transport within the heterostructure. By comparison, a much better match appears between the TiO_2 and CZTS layers at the interface. Also, it is worth remarking that while the TiO_2 and ZnO layers have different thicknesses (with the titania layer almost twice the height of the ZnO one), the CZTS thin film thickness was not affected and remains mostly constant in both investigated cases.

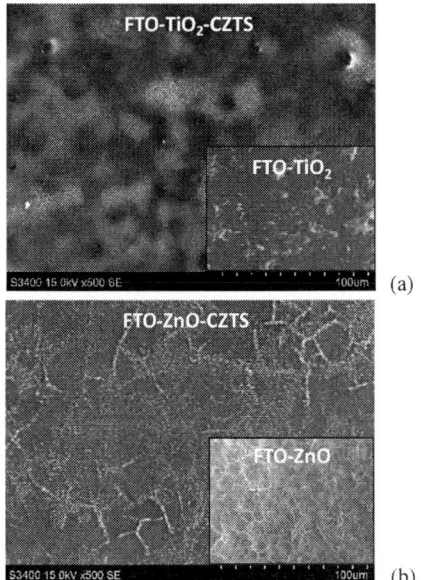

Fig. 3 SEM images of the (a) FTO-TiO_2-CZTS and (b) FTO-ZnO-CZTS structures. Insets of the FTO-TiO_2 and FTO-ZnO layers, respectively

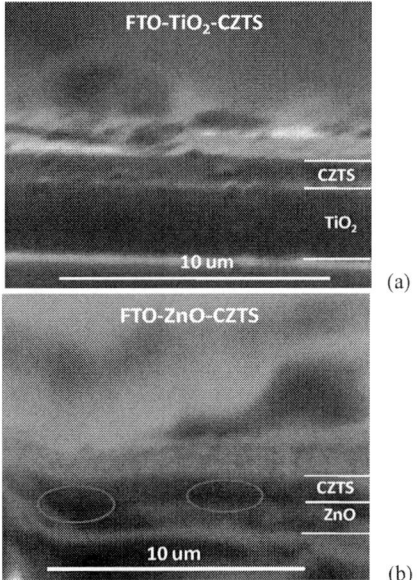

Fig. 4 SEM cross-section images of (a) FTO-TiO_2-CZTS and (b) FTO-ZnO-CZTS structures

The J-V curves obtained in dark and under irradiation are presented in Fig. 5 and show rectifying behavior for both heterostructures.

The fill factor (FF) is the ratio between the PV cell maximum power and the ideal power, as described in (1):

$$FF = (V_{max} \times I_{max}) \times 100 / (V_{OC} \times I_{SC})[\%] \quad (1)$$

Where: V_{max} and I_{max} are the maximum voltage and current, V_{OC} is the open circuit voltage and I_{SC} is the short-circuit current.

The fill factor in both cases is quite low, at about 22% for the TiO$_2$-CZTS and 25% for the ZnO-CZTS junctions. This is in good agreement with other results reporting on SPD CZTS solar cells, with FF lower than 36% due to high series resistance and low shunt resistance [6], leading to efficiencies raging between 0.3 - 1.1% [1, 7].

The TiO$_2$-CZTS increased photosensitivity (illustrated by the different shapes of the light and dark curves) could be due to the reduced charge recombination at the interface.

For the structures hereby analyzed, the efficiency values are lower than 0.01% in both cases, as a possible result of the multiple phases in the CZTS layer, but also of the imperfect interface between the two semiconducting thin films.

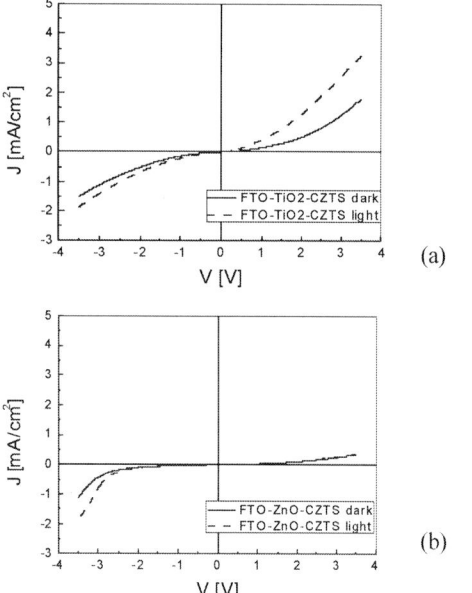

Fig. 5 I-V curves in the dark and under irradiation of the (a) FTO-TiO$_2$-CZTS and (b) FTO-ZnO-CZTS junctions

4. Conclusions

Spray deposited CZTS-TiO$_2$ and CZTS-ZnO hetero-junctions were obtained. The choice of the n-type layer affects the morphology of the heterostructure both at the surface as well as at the interface, with consequences on the charge transport. Although the TiO$_2$-CZTS interface seems more continuous than the ZnO-CZTS one, the latter shows a slightly better fill factor. This can be the consequence of the additional non-stoichiometric phases in the CZTS layer that may affect the bands alignment between the n and p types of semiconductors.

References

[1] S.M. Bhosale, M.P.Suryawanshi, J.H.Kim, A.V.Moholkar, *"Influence of copper concentrationon sprayed CZTS thin films deposited at high temperature"*, Ceram Int, **41**, pp. 8299–8304, 2015.

[2] E.M. Mkawi, K. Ibrahim, M.K. Ali, A.S. Mohamed, *"Dependence of copper concentrationon the properties of CZTS thin films prepared by electrochemical method"*, Int J Electrochem Sci, **8**, pp. 359–368, 2013.

[3] M.Covei, D.Perniu, C.Bogatu, A.Duta, *"CZTS-TiO$_2$ thin film heterostructures for advanced photocatalytic wastewater treatment"*, Catal. Today, 2017, DOI:10.1016/j.cattod.2017.12.003

[4] K. Diwate et. al, *"Synthesis and characterization of chemical spray pyrolysed CZTS thin films for solar cell applications"*, Energ. Proced., **110**, pp. 180–187, 2017.

[5] S.A. Khalate, R.S. Kate, J.H. Kim, S.M. Pawar, R.J. Deokate, *"Effect of deposition temperature on the propertie of Cu$_2$ZnSnS$_4$ (CZTS) thin films"*, Superlattices Microstruct, **103**, pp. 335-342, 2017.

[6] M. Courel, E. Valencia-Resendiz, F.A. Pulgarin-Agudelo, O. Vigil-Galan, *"Determination of minority carrier diffusion length of sprayed-Cu$_2$ZnSnS$_4$ thin films"*, Solid State Electron, **118**, pp. 1-3, 2016.

[7] S.M. Bhosale et al, *"Influence of growth temperatures on the properties of photoactive CZTS thin films using a spray pyrolysis technique"*, Mater Lett, **129**, pp. 153–155, 2014.

Three Phase Synchronous Boost Rectifier

V. Trifa[*], Gh. Brezeanu[**], E. Ceuca[***]

[*] PhD. Student, Politehnica University of Bucharest, Romania

vasi.trifa@gmail.com

[**] Department of Applied Electronics, Politehnica University of Bucharest, Romania

gheorghe.brezeanu@dce.pub.ro

[***] Department of Applied Electronics, "1 December 1918" University of Alba Iulia, Romania

emilian.ceuca@uab.ro

Abstract: In this paper we will analyze two systems, one synchronous and one nonsynchronous, for drive and recover energy from a three-phase motor (a three phase motor can be used as an electric generator according to the mechanical power applied to its shaft) and we will propose a more efficient circuit to drive a three-phase motor and recover energy from the motor.

The paper is a survey and represent the preliminary work in achieving the knowledge for developing a practical solution for to make the system more efficient.

Keywords: three phase inverter, synchronous boost, synchronous rectifier, recovering energy, optimized dead time.

1. Introduction

Today's global trend is to reduce carbon dioxide (CO_2) emissions. Power electronics is the electronics sector where most energy is consumed. That is why, in this paper, we want to improve the performance of electronic circuits. [1]

Reducing CO_2 emissions can be done in two ways: (1) increasing the efficiency of electronic circuits, or (2) recovering some of the energy consumed. By increasing the efficiency of electronic circuits, their lifetime will be higher and implicitly the risk of failure due to aging components decreases.

2. The Schematic Proposed and Preliminary Simulations

A. Three phase rectifier

The first case for rectifier circuit is a nonsynchronous rectifier. In this case we use the three-phase inverter like a three-phase rectifier. In the *Fig. 1* the gate is connected to

source for each transistor. We are used only the NMOS transistors because the NMOS transistors has the R_{dsON} small then a PMOS transistors for the same package.

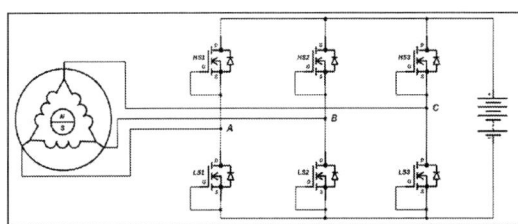

Fig. 1 The diagram for the three-phase typical rectifier

The simulated result for first case is shown in the *Fig 2*. All simulations for this paper were performed in LTSpice.

The rectified voltage is lower with two voltage forward on the diode. In this case is around 1.8V.

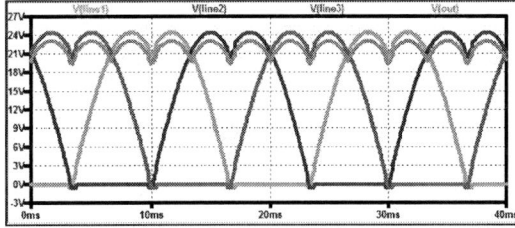

Fig. 2 Simulated result for three phase rectifier

The second case for rectifier circuit is a synchronous rectifier (*Fig. 3*). In this case the gate from transistors is controlled by comparators. When the potential of the transistor source is higher than the potential of the drain, the comparator turn on the transistor. When the potential of the transistor drain of is higher than the potential of the source, the comparator will turn off

978-1-5386-4483-6/18 $31.00 © 2018 IEEE

the transistor. All the time the voltage drain - source (v_{GS}) is monitored by the comparators for each transistor.

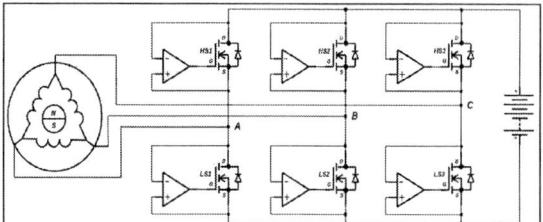

Fig. 3 Three phase synchronous rectifier

The advantage of these method result from the voltage drops across the transistors (when the transistors is turn on), much smaller than the forward voltage on the diode.

The simulated result for synchronous rectifier is shown in the *Fig 4*.

The rectified voltage is smaller with two voltages drop on the transistor. The voltage drop in this case is around 0.4V, but the most important aspect in this case is the very large influence given by R_{dsON} of transistors.

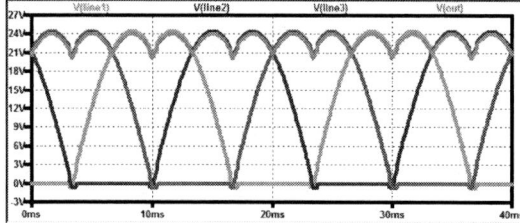

Fig. 4 Simulated result for synchronous rectifier

B. Three phase boost rectifier

The first case for boost rectifier circuit is a nonsynchronous boost rectifier based on of the three half-bridge as three boost converters. In the *Fig. 5* the high side transistors: HS1, HS2 and HS3 are used as diodes (the transistors are the gate connected to the source) and the low side transistors: LS1, LS2 and LS3 are controlled with pulse width modulation (PWM) signals in gate. [2]

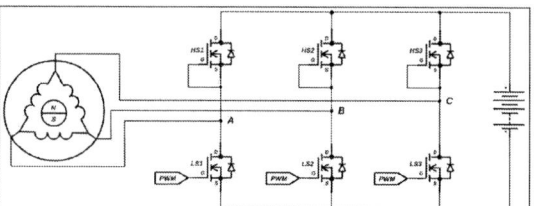

Fig. 5 The diagram for the three phase boost rectifier

The simulated result for nonsynchronous boost rectifier is in the next two figures (*Fig. 6* and *Fig. 7*). The rectified voltage is lower with two voltage forward on the diode. In this case is around 1.8V.

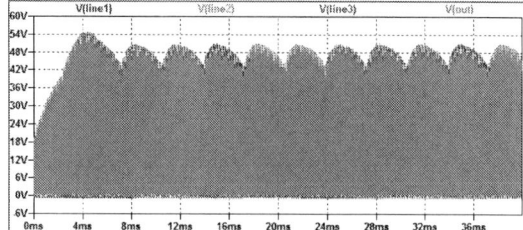

Fig. 6 Simulated result for three phase boost rectifier

In the next figure (*Fig. 7*) it shown the voltage drop on the two diodes in conduction. This voltage drop causes a loss of energy and contributes to the inverter heating, lowering efficiency and accelerating the aging of components.

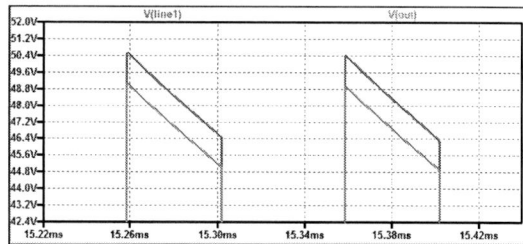

Fig. 7 Simulated result for three phase boost rectifier

The second case for boost rectifier circuit is a synchronous boost rectifier (*Fig. 8*). In this case the gate from the high side transistors is controlled by comparators and the gate from the low side transistors is controlled by comparators and PWM signals generated by microcontroller.

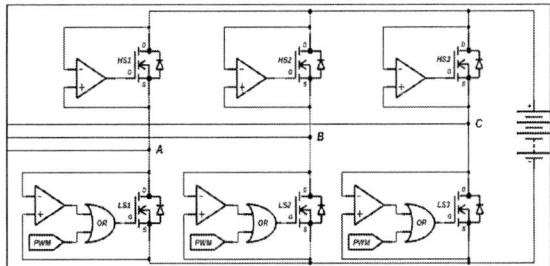

Fig. 8 Three phase synchronous boost rectifier

The rectified voltage is lower with two voltage drop on the transistor in conduction. The voltage drop in this case is around 0.8V, but the Rds$_{ON}$ of this transistor used in simulation is a little big.

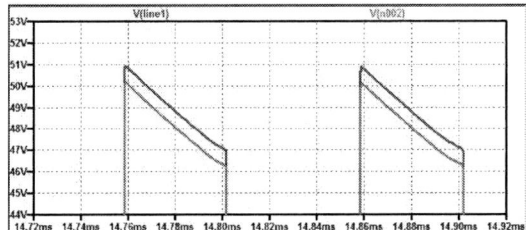

Fig. 9 Simulated result for synchronous boost rectifier

C. Three phase synchronous inverter

Our original work proposed in the paper is the synchronous inverter which also include the synchronous rectifier function (*Fig. 10*).

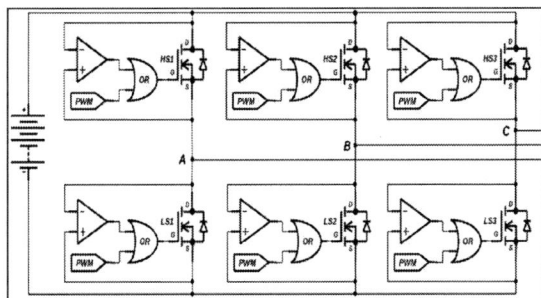

Fig. 10 Three phase synchronous inverter

This synchronous inverter efficiency is higher than a classic inverter because during operation like the conventional inverter, the inactive time is generated hardware by comparators and with a good strategy is optimized to the minimal value for each PWM pulse. This hardware reaction (made by comparators) is much faster than a software reaction (made software by micro-controller).

The advantage when the inverter works in synchronous rectifier mode is that the command is performed automatically and is faster due the hardware strategy.

When the inverter works in synchronous boost mode, the logic function to assure the boost operation (adapting duty cycle for PWM according to the voltage generated by each phase of the generator) is generate software by microcontroller, but some of the „synchronous" operations (dead time optimization and transistor conduction instead of diode conduction) are generated by hardware blocks.

3. The Results of Experimental Measurements

Most importantly, is needing to know at what specific voltage within this range we get the maximum peak current. As mentioned, the peak is critical from the standpoint of ensuring there is no inductor saturation.

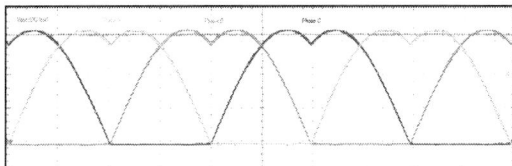

Fig 11. The experimental result for the three-phase typical rectifier.

The Vout (green signal in oscillogram) is without ripple because is connected to battery.

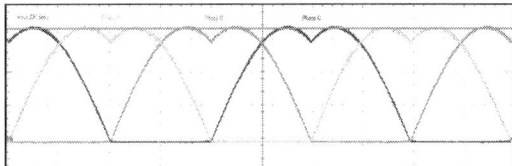

Fig 12. The experimental result for the three phase synchronous rectifier.

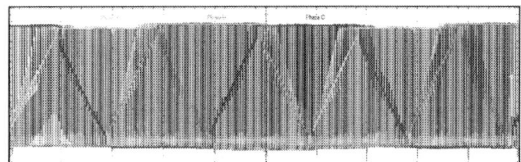

Fig 13. The experimental result for the three phase boost rectifier.

In the next two figures (*Fig. 14* and *Fig. 15*) is detailed the voltage drop for nonsynchronous and synchronous circuit.

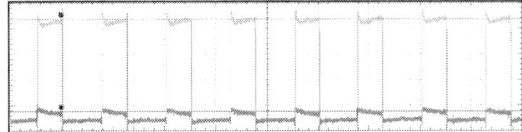

Fig 14. The experimental result for the three phase boost rectifier.

The rectified voltage is lower with two voltage forward on the diode. In this case is: 3.6div x 0.5V/div = 1.8V.

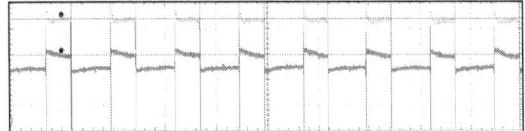

Fig 15. The experimental result for the three phase synchronous boost rectifier.

The rectified voltage is lower with two voltage on the Rds_{ON}. In this case is: 1.4div x 0.5V/div = 0.7V.

4. Conclusions

This paper is a survey of the two effective regenerative braking methods (synchronous and nonsynchronous) based on a general full-bridge inverter. Theoretical analysis and experimental results have been simulated and verified by practical measurements.

The objective of the paper is to identify and substantially reduce time in design using the customized interfaces that will be created by the authors.

By developing the customizable interfaces, the experience of working with simulator it improved and allow to feel the accuracy of using the real equations of circuits which is very close to the exact behavior unlike some computational models which can't substitute the proper operation.

References

[1] M. Toyota, Z. L. Liang, Y. Akita, H. Miyata, S. Kato, and T. Kurosu, *"Application of Power Electronics Technology to Energy Efficiency and CO₂ Reduction"* Hitachi Rev., vol. 59, no. 4, p. 141, 2010.

[2] V. Trifa, G. Brezeanu, E. Ceuca, „*Analyzing the Efficiency of Different Ways Recovering Energy*", International Spring Seminar on Electronics Technology (ISSE), Zlatibor, Serbia, May 16–20, 2018

[3] S. Maniktala, *Switching Power Supplies A-Z, Second Edition*, Elsevier, 2012.

[4] A. Emadi, Y. J. Lee, and K. Rajashekara, *"Power Electronics and Motor Drives in Electric, Hybrid Electric, and Plug-In Hybrid Electric Vehicles,"* Trans. Ind. Electron., vol. 55, no. 6, pp. 2237-2245, Jun. 2008.

[5] A. Kawahashi, "A New Generation Hybrid Electric Vehicle and Its Supporting Power Semiconductor Devices," Proc. 16th Int. Symp. Power Semicond. Devices and ICs, Kitakyushu, Japan, pp. 23-29, 2004.

[6] C. H. Chen and M. Y. Cheng "Design and Implementation of a High-Performance Bidirectional DC/AC Converter for Advanced EVs/HEVs," Proc. Inst. Electr. Eng.-Electr. Power Appl., vol. 153, no. 1, pp.140-148, Jan. 2006.

[7] M. Duoba, T. Bohn and H. Lohse-Busch *"Investigating Possible Fuel Economy Bias Due to Regenerative Braking in Testing HEVs on 2WD and 4WD Chassis"*, SAE 2005 Word Congress, pp. 11-14, Apr. 2005.

[8] H. Seki, K. Ishihara, and S. Tadakuma, *"Novel Regenerative Braking Control of Electric Power-Assisted Wheelchair for Safety Downhill Road Driving,"* IEEE Trans. Ind. Electron., vol. 56, no. 5, pp. 1393-1400, May 2009.

Session **WS**

WORKSHOP "MICROSYSTEMS FOR ENERGY HARVESTING AND ENVIRONMENT MONITORING"

978-1-5386-4483-6/18 $31.00 © 2018 IEEE

Power harvesting and storage circuit for a double array of lead-free piezoelectric cantilevers

George Muscalu[1,2], Bogdan Firtat[1,2], Silviu Dinulescu[1]
Carmen Moldovan[1], Adrian Anghelescu[1], Ion Stan[3]

1. IMT-Bucharest, 126A Erou Iancu Nicolae, 077190, Voluntari, Romania
george.muscalu@imt.ro
2. Faculty of Electronics, Telecommunications and Information Technology, UPB, 1-3 Iuliu Maniu, Bucharest, Romania
3. ROMELGEN SRL, 82 Baicului, Bucharest, Romania

Abstract—The purpose of this paper is the studying of an off the shelf solution for power harvesting and storage circuitry for a previously designed lead-free energy harvester. The energy harvester is designed to work at a resonant frequency of 460Hz and consists in a double array of piezoelectric cantilevers (2 x 10) using zinc oxide as a lead-free piezoelectric material. The energy from the harvester is processed by LTC3588, a nanopower energy harvesting power supply (Abstract).

Keywords— energy harvester; lead-free; ZnO; power harvesting circuit.

1. Introduction

The rapid development of microsensors and microsystems call forth the importance of their power consumption. Energy harvesting comes as a solution to this problem, because, by harvesting the environmental energy, we should be able to respond to the energy shortage or to replace polluting power supplies, as batteries. In particular, the piezoelectric energy harvesters can offer the ability to meet application requirements such as compatibility for miniaturization and high efficiency in energy conversion.

The purpose of piezoelectric energy harvesters is to "harvest" mechanical energy, in our case environmental vibrations, which induces strain in the piezoelectric layer, and, due the direct piezoelectric effect, it is converted into electrical energy. Because of the environmental friendly requirement, we chose lead-free materials for the piezoelectric layer, such as zinc oxide (ZnO), aluminum nitrate (AlN) or potassium sodium niobite (KNN), even if the best piezoelectric material used in energy harvesting is PZT (lead zirconate titanate).

Usually, energy harvesters are developed to be used with ultra-low power circuits [1] and that implies a power harvesting and storage circuitry.

For efficient harvesting of electric energy, from device to load, different strategies have been used to increase efficiency of the storage circuitry, including the use of active diodes for the rectifier [2], resistive load matching with maximum power point tracking, synchronized switch harvesting on inductor [3], synchronous electric charge extraction [4], or a combination of these.

The energy harvester studied in this article consists in a double array of piezoelectric cantilevers, using ZnO as a lead-free piezoelectric material, and we have considered using LTC3588 as a power harvesting and storage circuit, an off the shelf solution which integrates a low-loss bridge rectifier with a DC to DC converter to obtain a useful stabilized voltage level output. An intermediary capacitor placed between the bridge rectifier and converter allows for the accumulation of energy up to the point where input voltage levels to the converter allow efficient energy transfer to the stabilized output [5].

2. Design of the energy harvester

A. Geometric modeling

A piezoelectric cantilever converts a mechanical vibration into electrical energy via direct piezoelectric effect.

The structure of the proposed cantilever is shown in Fig. 1. It is an unimorph structure, designed to work for the first mode of vibration (flexure) and with the 3-1 transversal mode. As a piezoelectric material it is used a lead-free material, ZnO.

978-1-5386-4483-6/18 $31.00 © 2018 IEEE

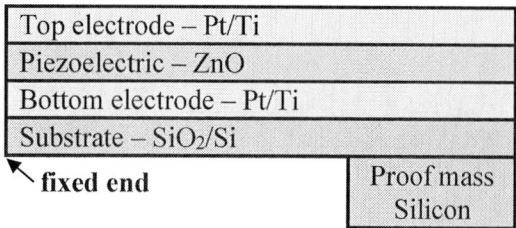

| Top electrode – Pt/Ti |
| Piezoelectric – ZnO |
| Bottom electrode – Pt/Ti |
| Substrate – SiO₂/Si |

fixed end

Proof mass
Silicon

Fig. 1 The structure of the proposed cantilever

Table 1. Material properties for the cantilever structure

Material	Thickness [μm]	Density [kg/m³]	Young's modulus [Gpa]
Si	10	2331	130
SiO₂	0.5	2200	70
Pt/Ti	0.2	21400	145
ZnO	1	5680	155
Pt/Ti	0.2	21400	145
Si	400	2331	130

These dimensions result in a resonant frequency of around 460Hz, frequency which keeps us in the environmental frequencies domain.

B. Electrical circuit simulation - FEA

Using finite element analysis (FEA) in COMSOL Multiphysics 5.2, we determine the electrical parameters of the cantilever structure, parameters which are further used in the simulation of the electrical circuit.

The maximum power transfer which can be obtained from a harvester is when the load impedance is equal with the intern impedance of the piezoelectric harvester and its value is given by the expression [6]:

$$R = \frac{1}{2\pi f C_0} \qquad (1)$$

Where f is the resonant frequency and C_0 is the intern capacitance of the piezoelectric harvester and is given in the expression:

$$C_0 = \frac{\varepsilon W L}{t_p} \qquad (2)$$

W, L and t_p are the width, length and thickness of the piezoelectric layer and ε is its dielectric constant.

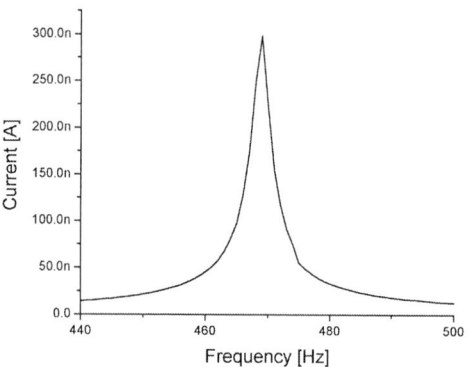

Fig. 2 Current vs frequency curve

Using expressions (1) and (2), we obtain a value of 9.28MΩ for the resistance and 39.32pF for the capacitance.

The electrical circuit analysis is done by connecting a load resistance between the two electrodes of the piezoelectric cantilever. As piezoelectric material is used zinc oxide from the software database and the simulated acceleration is 1g. The results are shown in Fig. 2.

The output current of the piezoelectric cantilever has a maximum, of approximately 300nA, at the resonant frequency.

3. Electrical simulations of the circuit

LTC3588 offer four output voltages: 1.8V, 2.5V, 3.3V, 3.6V [5]. Because of the low power of our energy harvester, we chose 1.8V as the regulated output power, value which is enough to supply an ultra-low power microcontroller [7]. The circuit is shown in Fig. 4.

C_{in} and C_{out} are chosen using the expressions below [5]. The scenario for this circuit was with a load current of 100μA (I_{LOAD}) [7] for 1 second (t_{LOAD}). The other values are chosen from the LTC3588 datasheet.

$$P_{LOAD} t_{LOAD} = \frac{1}{2}\eta C_{IN}(V_{IN}^2 - V_{UVLOFALLING}^2) \quad (3)$$

$$C_{OUT} = (I_{LOAD} - I_{BUCK})\frac{t_{LOAD}}{V_{OUT}^+ - V_{OUT}^-} \qquad (4)$$

978-1-5386-4483-6/18 $31.00 © 2018 IEEE

LTC3588 integrates a low-loss bridge rectifier with a DC to DC converter to obtain a useful stabilized voltage level output of 1.8V in this case. An intermediary capacitor placed between the bridge rectifier and converter allows for the accumulation of energy up to the point where input voltage levels to the converter allow efficient energy transfer to the stabilized output.

The proposed energy harvester consists in a double array (2 x 10) of piezoelectric cantilevers in order to increase the output power. Using the data from the FEA analysis, one cantilever can be modeling into a simplified reduced model [6], as it's shown in Fig. 3, which consists in a sine wave current source with the amplitude of 300nA and the frequency of 469Hz. In parallel with the current source are connected the internal resistance and capacitance of the piezoelectric layer. Those 20 cantilevers are connected and form an equivalent harvester, from the electrical point of view, *PZeq*.

The cantilevers were connected (in series and parallel) in different configurations such that on every branch there is an equal number on cantilevers because, otherwise, there will be power losses. We tested six configurations (described in Table 2) to see which one is suitable for our need. All the circuit simulations were made using LTspice XVII. The results are shown in Fig. 5.

In this case an initial condition, $V_{IN} = 4\ V$, was set in order to reduce the simulation time. When V_{IN} reach a value around 4.05V, the energy from C_{IN} is transfer at the output of the circuit, on C_{OUT}, and V_{IN} drops to 2.8V. The cycle is repeated until output voltage is regulated at $V_{OUT} = 1.8V$.

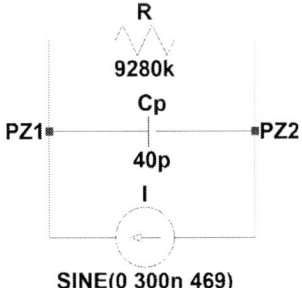

Fig. 3 The simplified reduced model of the piezoelectric cantilever

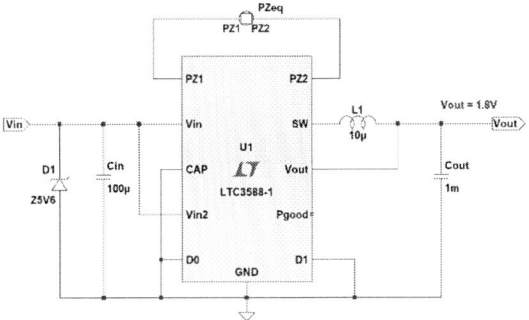

Fig. 4 LTC3588 circuit with a regulated output voltage of 1.8V

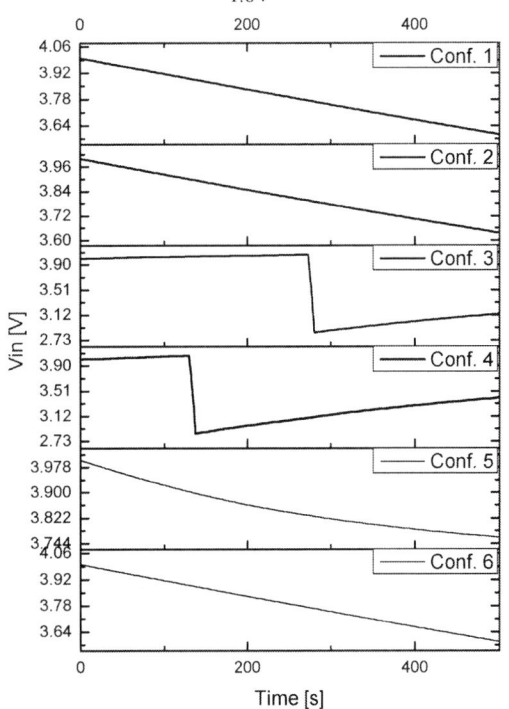

Fig. 5 V_{IN} for the six configurations tested in this article

Table 2. The six configurations of the cantilevers

Configuration	Description
1	20 cantilevers connected in series
2	2 branches connected in parallel with 10 cantilevers connected in series each
3	4 branches connected in parallel with 5 cantilevers connected in series each
4	5 branches connected in parallel with 4 cantilevers connected in series each
5	10 branches connected in parallel with 2 cantilevers connected in series each

6	20 cantilevers connected in parallel

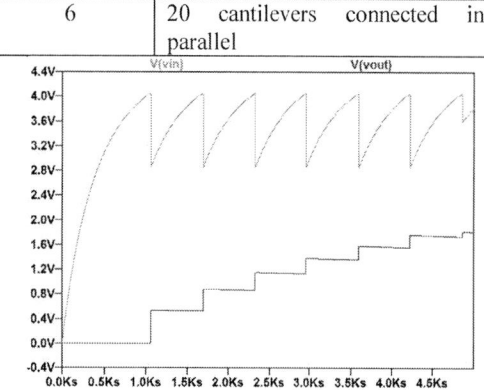

Fig. 6 V_{IN} and V_{OUT} for the configuration 4

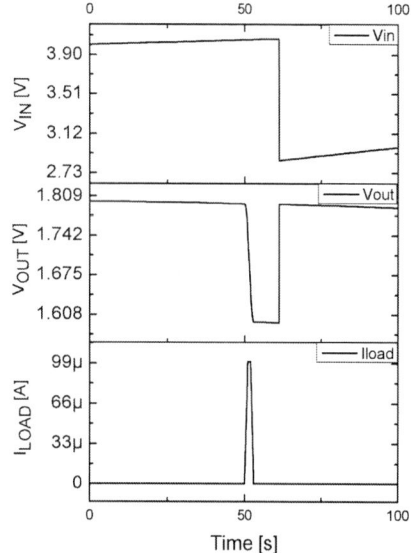

Fig. 7 V_{IN} and V_{OUT} with a load current of 100μA

As we can observe from Fig. 5, only in configurations 3 and 4 the V_{IN} is increased and this implies that only in these two configurations V_{OUT} will pe stabilized at 1.8V. Configuration 4 is faster and it needs around 90 minutes, in open load, for V_{OUT} to reach 1.8V (Fig. 6).

Fig. 7 shows V_{IN} and V_{OUT} when a current, $I_{LOAD} = 100$μA, is connected on the load of the circuit for 1 second. The standby load current is 120nA and some initial conditions are necessary in order to reduce the simulation time ($V_{IN} = 4$V and $V_{OUT} = 1.8$V).

4. Conclusions

A FEA analysis has been made in order to see the current capabilities of a lead-free piezoelectric cantilever structure with a resonant frequency around 460Hz.

With the FEA analysis results, output current of 300nA and resonant frequency of 469Hz, a simplified reduced model has been made in order to simulate a power harvesting and storage circuit.

Using LTspice XVII, six configurations of 20 cantilevers have been tested in order to find the most suitable configuration for our needs.

The output voltage reached the 1.8V only in two configurations, but configuration 4, 5 branches connected in parallel with 4 cantilevers each, was the fastest. V_{OUT} reach the value of 1.8V in 90 minutes in open load.

The system, after it reaches stabilization, is able to sustain a load current of 100μA for 1 second and a standby current of 120nA but the periods between the load current spikes must be long enough for the system to stabilize again.

Acknowledgments: The research work has been supported by The European Project M-ERA.NET, "PiezoMEMS", contract 12/2015.

References

[1] J. J. Estrada-López, A. Abuellil, Z. Zeng and E. Sánchez-Sinencio, "Multiple Input Energy Harvesting Systems for Autonomous IoT End-Nodes," *Low Power Electronics and Applications*, vol. 8, no. 6, 2018.

[2] M. Nielsen-Lönn, P. Harikumar, J. J. Wikner and A. Alvandpour, "Design of efficient CMOS rectifiers for integrated piezo-MEMS energy-harvesting power management systems," in *2015 European Conference on Circuit Theory and Design (ECCTD)*, Trondheim, Norway, 2015.

[3] E. E. Aktakka and K. Najafi, "A Micro Inertial Energy Harvesting Platform With Self-Supplied Power Management Circuit for Autonomous Wireless Sensor Nodes," *IEEE Journal of Solid-State Circuits*, vol. 49, no. 9, 2014.

[4] E. Lefeuvre, A. Badel, C. Richard, L. Petit and D. Guyomar, "Optimization of piezoelectric electrical generators," in *Dans Symposium on Design, Test, Integration and Packaging of MEMS/MOEMS - DTIP 2006*, Stresa, Italy, 2006.

[5] Linear Technology, "LTC3588-1 Datasheet".

[6] A. Townley, Vibrational Energy Harvesting Using MEMS Piezoelectric Generators.

[7] STMicroelectronics, "STM32 Ultra Low Power MCUs," [Online]. Available: http://www.st.com/en/microcontrollers/stm32-ultra-low-power-mcus.html?querycriteria=productId=SC2157.

978-1-5386-4483-6/18 $31.00 © 2018 IEEE

Design and simulation of piezoelectric energy harvester for aerospace applications

George Muscalu[1,2], Bogdan Firtat[1,2], Silviu Dinulescu[1]
Carmen Moldovan[1], Adrian Anghelescu[1], Ciprian Vasile[3,2], Daniela Ciobotaru[3], Cristian Hutanu[3]

1. IMT-Bucharest, 126A Erou Iancu Nicolae, 077190, Voluntari, Romania
E-mail: george.muscalu@imt.ro
2. Faculty of Electronics, Telecommunications and Information Technology, UPB-Bucharest
3. Advanced Technologies Institute, 10 Dinu Vintila, Bucharest, Romania

Abstract—*The focus of our study is to design piezoelectric harvesting cantilevers in order to convert environmental mechanical vibrations of low frequencies into electrical energy via direct piezoelectric effect. The final device consists of two arrays of cantilevers with a silicon substrate, a PZT-5H piezoelectric layer and a tungsten proof mass in order to obtain low resonant frequencies. The cantilevers are designed to work around 30, 45 and 90Hz and with an acceleration of 15.4, 8.6 and 1.5 m/s2 respectively. All the models are simulated using COMSOL Multiphysics 5.2 from the point of view of resonant frequencies and von Mises stress. (Abstract).*

Keywords— energy harvester; cantilevers; low frequencies, PZT-5H.

1. Introduction

With the rapid development of microsensors comes the importance of their power consumption. Therefore, a new field becomes more and more important, that of energy harvesting. Energy harvesting is the most effective way to respond to the energy shortage and to produce sustainable power sources from the surrounding environment. It is of interest in wireless and mobile devices as an alternative energy supply since the feature size of the devices becomes smaller and power consumption increases with diversified functions. In particular, the piezoelectric energy harvesters can offer the ability to meet application requirements such as compatibility for miniaturization and high efficiency in energy conversion. The energy source for piezoelectric energy harvesters is mechanical energy, mostly vibration, which induces strain in the piezoelectric layer, and, due to the direct piezoelectric effect, it is converted into electrical energy.

Lead zirconate titanate (PZT) is one of the most popular piezoelectric materials due to its high piezoelectric constant. Therefore, many microfabrication processes of PZT film for energy harvesters have been reported, like sol-gel process [1], spin-coating [2] or screen printing [3].

Liu H. et al. developed a piezoelectric MEMS harvester for low frequencies using a sol-gel deposited PZT as piezoelectric material and silicon as substrate for the vibrational structure. This device generated an output power between 19.4 and 51.3 nW with an operating frequency range between 30 and 47Hz at 1g [1].

In 2013, S.B. Kim et al. developed PZT energy harvesters on silicon substrate with a resonant frequency of 243Hz and an output power up to 2.33μW for an acceleration of 0.5g [2].

Also, B. Yang et. Al developed a high performance PZT thick films harvester with a resonant frequency around 500Hz and an output power up to 57.6μW for an acceleration of 1g [4].

More recently, Y. Tian et al. developed a low-frequency MEMS piezoelectric energy harvester on bulk PZT film. The resonant frequency was of 34.3Hz and the output power was 216.66μW at 1.5g [5].

Regarding the piezoelectric material, PZT-5H is the best material for energy harvesting due to high piezoelectric constant and high coupling factor [6].

The purpose of this paper is to develop a device capable to harvest vibration energy at three different frequencies, 30, 45 and 90Hz, and acceleration of 15.4, 8.6 and 1.5m/s^2, respectively. These frequencies have been met at the vertical stabilizer of a PZL SW-4 helicopter [7]. The device consists of two arrays with three cantilevers each. The piezoelectric cantilevers have a silicon substrate, a screen-printed PZT-5H piezoelectric layer and a proof mass of tungsten.

978-1-5386-4483-6/18 $31.00 © 2018 IEEE

2. Design of cantilevers

A. Theoretical considerations

The piezoelectric effect is a direct transformation of mechanical energy into electrical energy. When piezoelectric materials are deformed or stressed, voltage appears across the materials.

The cantilever structure, simulated in this paper, is the one shown in Fig. 1. It is an unimorph structure, designed to work for the first mode of vibration (flexure) and with the 3-1 transversal mode as the electromechanical coupling mode, in which the electric field is produced on an axis orthogonal to the axis of applied strain. This type of structure is easier to obtain from the technological point of view. The first mode of vibration has the lowest resonant frequency and provides the most deflection and therefore electrical energy. Also, by adding a proof mass, the resonant frequency becomes lower. A lower resonant frequency is closer to physical vibration sources and generally more power is produced at lower frequencies.

Assuming the added mass on the tip is much larger than the mass of the beam itself and the stiffness is unaffected, the resonant frequency is given by the expression [8]:

$$f = \frac{1}{2\pi} \sqrt{\frac{Y_{eq} W t^3}{4 L^3 (m_t + 0.24 m_c)}} \quad (1)$$

Where Y_{eq} is the equivalent Young's modulus, t, L and W are the thickness, the length and the width of the cantilever beam and m_i and m_c are the mass of the proof mass and the cantilever mass, respectively.

B. Geometric modeling

As it can be seen, the resonant frequency is given by the materials properties and the geometric parameters of the cantilever: length, thickness and the proof mass. The width of the cantilever is not relevant in the resonant frequency calculation because it is properties are fixed as soon as we choose a material, the resonant frequency is obtained through variation of geometric factors: length, thickness and proof mass.also included in the m_i and m_c parameters and is therefore simplified. Because the material

Table 1. Material parameters for the cantilever structure

Material	Thickness [μm]		Density [kg/m³]	Young's modulus [Gpa]
Si	15	10	2331	130
SiO₂	0.5		2200	70
Pt/Ti	0.2		21400	145
PZT-5H	5		7500	63
Pt/Ti	0.2		21400	145
Tungsten	10		19350	411

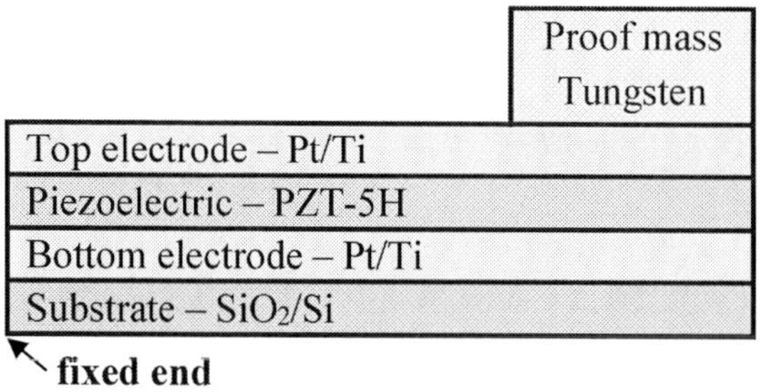

	Proof mass Tungsten
Top electrode – Pt/Ti	
Piezoelectric – PZT-5H	
Bottom electrode – Pt/Ti	
Substrate – SiO₂/Si	

↖ **fixed end**

Fig. 1. The structure of piezoelectric unimorph cantilever

Considering the expression (1) and the values from Table 1, the resulted geometry for the three frequencies is shown in Table 2. These geometries were simulated in order to check the resonant frequency and von Mises stress.

3. Results and discussions

The structures were designed and simulated using COMSOL Multiphysics 5.2. In order to lower the resonant frequency, different proof mass geometries were studied, as shown in Fig. 2 and in Table 2.

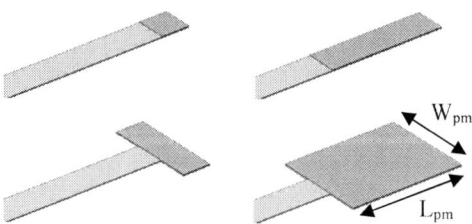

Fig. 2. Different proof mass geometries

Table 2. Results of modal and frequency domain analyzes

Frequency [Hz]	Simulated Frequency [Hz]	Frequency error [%]	Substrate thickness [μm]	Proof mass width (W_{pm}) [μm]	Proof mass length (L_{pm}) [μm]	Cantilever length [μm]	von Mises stress [MPa]
90	89,89	-0,11	15	500	500	13180	-
90	97,82	8,69	15	500	2000	11020	-
90	83,57	-7,14	15	1500	500	11600	-
90	94,68	5,20	15	1500	2000	8480	-
90	90,07	0,07	10	500	500	10720	-
90	101,38	12,64	10	500	2000	8600	-
90	86,49	-3,89	10	1500	500	9140	-
90	102,1	13,44	10	1500	2000	6440	-
45	45,08	0,19	15	500	500	19060	-
45	47,29	5,10	15	500	2000	16580	-
45	43,25	-3,88	15	1500	500	17280	-
45	44,98	-0,04	15	1500	2000	13160	-
45	46,86	4,13	10	500	500	15600	-
45	48,46	7,69	10	500	2000	13080	-
45	43,54	-3,23	10	1500	500	13760	-
45	46,94	4,31	10	1500	2000	10060	-
30	30,35	1,16	15	500	500	23580	159
30	31,43	4,78	15	500	2000	20920	213
30	29,88	-0,38	15	1500	500	21700	209
30	29,77	-0,76	15	1500	2000	16960	263
30	31,01	3,36	10	500	500	19380	180
30	32,04	6,82	10	500	2000	16620	239
30	28,84	-3,85	10	1500	500	17380	233
30	30,77	2,59	10	1500	2000	13020	369

A. Modal analyzes

Modal analyzes represent a fast way to check the resonant frequency and the geometry of the structures. The dimensions of the simulated structures are given by Tables 1 and 2. The width of the proof mass was varied in order to concentrate a larger mass at the tip of the cantilever. For these simulations the width of the cantilevers was 500μm. The results are shown in Table 2.

As was expecting, the smallest dimensions were obtained for a substrate thickness of 10μm and maximum proof mass (width of 1500μm and length of 2000μm).

B. Frequency domain analyzes for von Mises stress

Frequency domain analyzes are used to check the von Mises stress which gives information about the equivalent tensile stress from a material. When the von Mises stress exceeds the yield strength, the material yields which bring strong deformation in the structure.

Because this type of analysis is time consuming and the purpose of the work is to obtain all the three structures (for 30, 45 and 90 Hz), we run frequency domain analyzes only for the 30Hz structures. These have the largest dimensions and exert the largest stress. One of the simulated structures is shown in Fig. 3 and the results are given in Table 2. The simulated acceleration was 15.4m/s².

Bulk silicon (hundreds of μm) has a yield strength of 7GPa but in thin layers (tens of μm), the yield strength of silicon is close to 200MPa [9].

According to the results from Table 2, only the structures with a proof mass of 500μm by 500μm, for both the substrate thickness, have the von Mises stress lower than 200MPa.

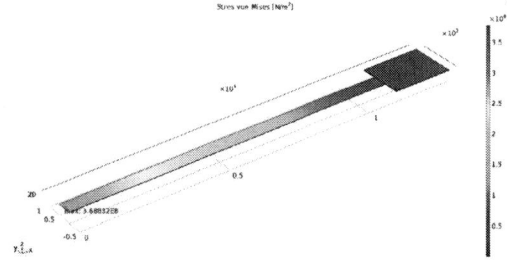

Fig. 3 The stress distribution for the last structure from Table2

Between the two 30Hz structures with a von Mises stress below 200MPa, we choose the structure with a substrate thickness of 15μm. This one has the lowest von Mises stress and it can give us an error margin. Because the three cantilevers are part of the same device, the other two cantilevers must have the same substrate thickness.

Therefore, in Fig. 4 there are a few possible solutions for the position of the two cantilevers arrays. By choosing a different geometry for the proof mass of 45 and 90Hz cantilevers, we were able to reduce the area with almost 30mm² but the frequency error is above 5% for these structures (as seen in Table 2).

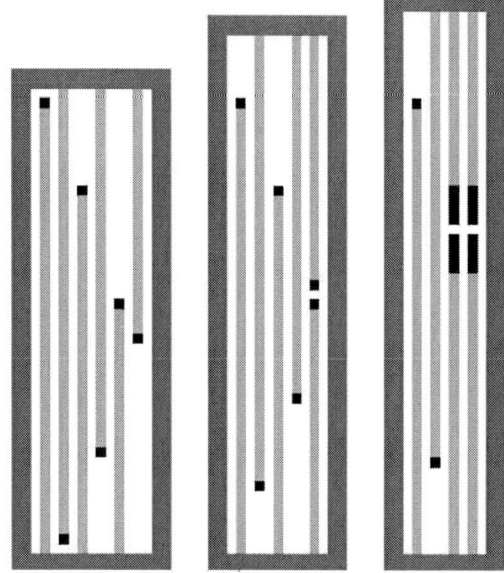

Fig. 4. Three possible solutions for the proposed device.

4. Conclusions

Different structures of cantilevers were simulated in order to obtain a device for energy harvesting at three frequencies, 30, 45 and 90Hz, frequencies which are met on a vertical stabilizer of a PZL SW-4 helicopter.

In order to reduce the resonant frequency, all the cantilevers were designed with different types of proof mass.

Modal and frequency domain analyzes have been run in order to check the geometry and von Mises stress.

From the proposed 30Hz cantilever structures only two had a von Mises stress lower than yield strength of thin silicon.

The resulted structures were used to propose different device layouts for the future technological process.

Acknowledgments: The research work has been supported by The National Project STAR, "PiezoHARV", contract 164/2017.

References

[1] H. Liu, C. Tay, C. Quan, T. Kobayashi and C. Lee, " Piezoelectric MEMS energy harvester for low-frequency vibrations with wideband operation range and steadily increased output power," *Journal of Microelectromechanical Systems,* vol. 20, no. 5, pp. 1131-1142, 2011.

[2] S. Kim, H. Park, S. Kim, H. Ill, J. Park and D. Kim, "Comparison of MEMS PZT Cantilevers Based on d31 and d33 Modes for Vibration Energy Harvesting," *Journal of Microelectromechanical Systems ,* vol. 22, no. 1, pp. 26-33, 2013.

[3] S. Beeby, A. Blackburn and N. White, "Processing of PZT piezoelectric thick films on silicon for microelectromechancial systems," *Journal of Micromechanics and Microengineering,* vol. 9, no. 3, 1999.

[4] B. Yang, Y. Zhu, X. Wang, J.-q. Liu, X. Chen and C. Yang, "High performance PZT thick films based on bonding technique for d31 mode harvester with integrated proof mass," *Sensors and Actuators A: Physical,* vol. 214, no. 88-94, p. 2014.

[5] Y. Tian, G. Li, Z. Yi, J. Liu and B. Yang, "A low-requency MEMS piezoelectric energy harvester with a rectangular hole based on bulk PZT film," *Journal of Physics and Chemistry of Solids,* vol. 117, pp. 21-27, 2018.

[6] K. V. Allamraju and S. Korla, "Design and performance of a prototype clamped free beam novel energy harvester for low power applications," *Materials Today: Proceedings 4,* pp. 3542-3548, 2017.

[7] M. Stamos, C. Nicoleau, R. Torah, J. Tudor, N. R. Harris, A. Niewiadomski and S. P. Beeby, "Screen-Printed Piezoelectric Generator for Helicopter Health and Usage Monitoring Systems," in *8th International Workshop on Micro and Nanotechnology for Power Generation and Energy Conversion Applications (PowerMEMS 2008),* Sendai, Japan, 2008.

[8] C. Williams and R. Yates, "Analysis of a micro-electric generator for microsystems," vol. 52, no. 1-3, 1996.

[9] AZO Materials, "Properties: Silicon," AZO Materials, [Online]. Available: https://www.azom.com/properties.aspx?ArticleID=599.

Electrochemical sensors for detection of different ionic species (nitrites/nitrates and heavy metals) in natural water sources

M. Gartner[1], C. Lete[1], M. Chelu[1], H. Stroescu[1], M. Zaharescu[1],
C. Moldovan[2], C. Brasoveanu[2], M. Gheorghe[3], S. Gheorghe[3], A. Duta[4],
Z. Labadi[5], B. Kalas[5], A. Saftics[5], M. Fried[5,6], P. Petrik[5],
E. Tóth[7], H. Jankovics[7], F. Vonderviszt[7]

[1] Institute of Physical Chemistry, "Ilie Murgulescu" of the Romanian Academy, Bucharest, Romania
mgartner@icf.ro
[2] National Institute for Research and Development in Microtechnologies, 077190 Bucharest, Voluntari, Romania
[3] NANOM MEMS SRL, Râşnov, Romania
[4] Transilvania University of Braşov, Braşov, Romania
[5] Institute for Technical Physics and Materials Science, Centre for Energy Research (MFA), Budapest, Hungary
[6] Institute of Microelectronics and Technology, Óbuda University, Budapest, Hungary
[7] Research Institute of Biomolecular and Chemical Engineering, University of Pannonia, Veszprém, Hungary

Abstract — Thin films prepared by different technologies (electrodeposition, screen printing) were electrochemically investigated as sensitive layers for nitrates/nitrites. Bacterial flagellar filaments (special protein molecules) were engineered as sensitive biolayers for heavy metal detection.

Keywords — Electrochemical sensors; nitrites/nitrates; heavy metals

1. Introduction

The quantitative evaluation of the nitrites and nitrates became a very important issue in different fields as wastewater treatment, food industry and drinking water control. Both anions must be carefully controlled because of their own toxicity, as e.g. nitrites could be transformed into dangerous substances. In the soil, fertilizers containing inorganic nitrogen and wastes containing organic nitrogen are firstly decomposed to give ammonia, which is then oxidized to nitrite and nitrate. The nitrate anion is metabolized by plants during growth and used in the synthesis of organic nitrogenous compounds.

Heavy metals exist in natural waters in colloidal, particulate and/or dissolved phases. Their toxic effects are different, but all of them (e.g. mercury, cadmium, nickel, arsenic, chromium, thallium, lead) are dangerous, especially due to their potential bioaccumulation. Heavy metals can enter in a water supply by industrial or consumer waste, or due to the acidic rain that breaks down the soil and releases heavy metals into streams, lakes, rivers and groundwater.

The aim of this paper is the investigation of a broad range of materials, sensitive to nitrites/nitrates and heavy metals. The sensitive material is deposited by physical and chemical methods as thin layer on the working electrode of a microsensor specially designed for this purpose. The deposited layers are electrochemically tested to assess the sensitivity and the detection limit for each dangerous component.

2. Electrochemical microsensors

Two types of electrochemical sensor structures were designed (**Fig. 1**) as active components of a portable device for the detection of different ionic species (nitrites/nitrates and heavy metals) in natural water sources. The first one (**Fig. 1a**) is designed to be integrated on a Si wafer (**Fig. 1b**) covered with Au/Ti/SiO$_2$ or Pt/Ti/SiO$_2$ while the second one (**Fig. 1c**) was designed to be built on a ceramic support (Al$_2$O$_3$).

Different films based on a combination of metal oxides, Cu nanoparticles and polymers, were deposited on the working electrode of these transducers and tested as

---

Acknowledgments - The financial supports of the projects:M-ERA.net WaterSafe / Nr. 39/2016 and EU (ERDF) and Romanian Government, that allowed for acquisition of the research infrastructure under POS-CCE O 2.2.1 project INFRANANOCHEM - Nr. 19/01.03.2009 are gratefully acknowledged

978-1-5386-4483-6/18 $31.00 © 2018 IEEE

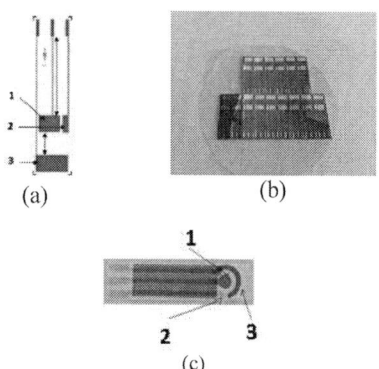

Fig. 1. (a) Electrochemical structure (first type) with 3 eletrodes on Pt/Ti/SiO₂/Si substrate; (b) Integrated sensors on Si wafer; (c) Electrochemical structure (second type) with 3 electrodes on ceramic substrate;1-working electrode; 2-reference electrode; 3-counter electrode

sensors to nitrites/nitrates or heavy metals by cyclic voltammetry (CV), Square Wave Voltammetry (SWV) or linear sweep voltammograms (LSVs). The best results are further presented.

3. Sensing films for nitrates (NO_3^-)
A. CuO-NPs/PANI

PANI films were prepared in two steps: the first step was the potentiostatic deposition of the PANI film at 1V for 150s from a mixture of 1M HCl and 50mM aniline followed by the over-oxidation of PANI film at 0.1V for 5 minutes in 0.1 M NaOH solution. Then, the CuO-NPs were also electrodeposited on the PANI coating in two steps: firstly, CuNPs film was prepared in CV from a solution containing 0.1M KNO_3 and 0.01 M $Cu(NO_3)_2$ sweeping the potential between -0.7 to 0.8 V for 25 scans with 25 mV/s. Afterwards, the CuNPs were oxidized to CuONPs in 0.05 M NaOH, scanning the potential from 0 to 0.8 V for 40 scans with 100 mV/s.

The voltammetric sensor based on CuO-NPs/PANI showed a narrow linear range (0.8–2 mM) and a limit of detection (LOD) of 80 mg/L.

B. Cu-NPs/PEDOT

There were two approaches in the preparation of these coatings. The first approach in the *electrodeposition* of the

composite films based on PEDOT and CuNPs was based on modifying the electrode layer by layer: the PEDOT film was electrodeposited at 0,95V, for 3 minutes using a solution containing 10 mM EDOT and 0,1 M $LiClO_4$, followed by the electrodeposition of CuNPs from a mixture of 10 mM $CuSO_4$ and 10 mM H_2SO_4 at $E_{applied} = -0,7$ V for a deposition duration of 200 seconds. The second approach was a galvanostatic ($0.4mA/cm^2$ for 120 seconds) or potentiostatic (0.95V for 180 seconds) electrodeposition from a sonicated solution containing 10 mM EDOT, 0,1M $LiClO_4$, and 1 mg/ml CuNPs (prepared through a modified protocol of the alkaline polyol method). The solution was sonicated for 5 h.

The oxidation current of NO_3^- at CuNPs/PEDOT/Au electrode in 0.1M Na_2SO_4 + 0.1M HCl solution (pH=3) is linear to its concentration in the range of 0.4–2mM and LOD was found to be 7.1 mg/L with a sensitivity of 2.5 μA/mM. The analytical parameters, such as good linear range, low detection limit, good sensitivity and short response time, are comparable or superior to other previously reported NO_3^- sensors. The linear sweep voltammograms (LSVs) of the CuNPs-PEDOT coating prepared by *galvanostatic* method recorded for additions of different NO_3^- concentrations are presented in **Fig. 2**.

C. Copper hexacyanoferrate (CuHCF) coatings preparation

CuHCF is a polymeric inorganic compound with outstanding properties in electroanalysis applications. The coating was

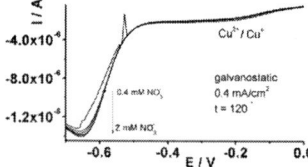

Fig. 2. LSVs for CuNPs-PEDOT/Au electrode in 0.1M Na_2SO_4 + 0.1M HCl solution containing different KNO_3 concentration, pH=3

prepared by scanning the potential between 0.4 and 0.8V with 50mV/s for 20 scans. The electrodeposition used a solution of 0.1M

KCl, 0.1M HCl, 1mM CuCl$_2$ and 1mM K$_3$[Fe(CN)$_6$]. During the characterization of the CuHCF coating in transfer solution (0.1M KCl and 0.1M HCl) one redox process is observed with an oxidation potential at 0.67V and a reduction potential at 0.63V, as the cyclic voltammogram recorded for the characterization of the prepared CuHCF shows in **Fig. 3**.

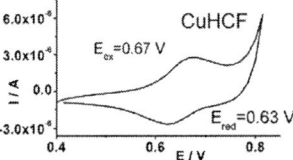

Fig. 3. Cyclic voltammogram (CV) of CuHCF/Au electrode in transfer solution, 0.1M KCl + 0.1M HCl; 50 mV/s

The redox process can be described by the reaction (1):

$$Cu3[Fe^{III}(CN)_6]_2 + 2e^- + 2K^+ \rightarrow$$
$$K_2Cu_3[Fe^{II}(CN)_6]_2 \qquad (1)$$

$K_2Cu_3[Fe^{II}(CN)_6]_2$ exists in the solution as $Cu_2[Fe^{II}(CN)_6]_2$ or as $K_2Cu[Fe^{III}(CN)_6]_2$.

The electrochemical sensor based on CuHCF showed a linear response in the NO$_3^-$ concentration range 0.4 … 4.6mM and a good limit of detection (LOD) of 30 mg/L, lower than 50mg/L, as required by the Council Directive 98/83/EC of 3 November 1998 on the quality of water intended for human consumption. CuHCF coating also exhibited good electrocatalytic properties toward NO$_2^-$ oxidation in phosphate buffer solution, pH=7. The linearity of the NO$_2^-$ sensor response was from 0.4 to 2.8mM and the LOD was 40.8 mg/L. The behaviour of the CuHCF coating for the NO$_3^-$ reduction and for the NO$_2^-$ oxidation in 0.1M acetate buffer, pH=4.1 solution is presented in **Fig. 4** and **5**.

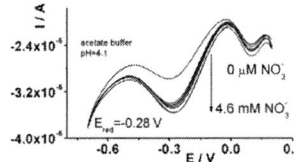

Fig. 4. SWVs for CuHCF/Au electrode in 0.1 M Acetate buffer, pH=4.1 with different KNO$_3$ concentrations

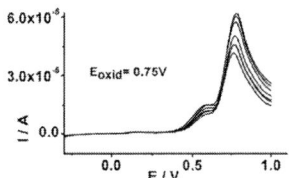

Fig. 5. SWVs for NO2- determination on CuHCF/Au electrode in acetate buffer pH=4 containing 0.8, 1.2, 1.6, 2, 2.4, and 2.8 mM KNO$_2$

4. Sensing films for nitrites (NO$_2^-$)

A very sensitive material to nitrite is the Pt-Au alloy deposited as SPE (screen printed electrode) on ceramic (Al$_2$O$_3$) substrate. The voltammograms obtained by SWV analysis in acetate buffer solution, with pH=4, containing 0.4, 0.8, 1.2, 1.6, 2, 3 and 4mM NO$_2^-$ are presented in **Fig. 6**. A good calibration curve was obtained as presented in the inset of **Fig. 6**).

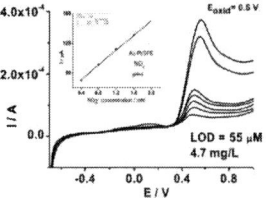

Fig. 6. SWVs on Pt-Au alloy electrode for NO$_2^-$ determination in acetate buffer solution with pH=4 containing 0.4, 0.8, 1.2, 1.6, 2, 3 and 4mM NO$_2^-$ Inset: calibration curve

From the calibration curve, the lower detection limit obtained was 4.1 mg/L, which recommend it for the NO$_2^-$ sensor.

5. Sensing films for heavy metals

A 3-electrode sensor structure on Si substrate having flagellin as sensing layer, **Fig. 7**, was used. A good (EIS) based impedance correlation was observed for different Ni^{2+} concentrations in PBS buffer (**Fig. 8**).

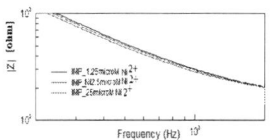

Fig. 7. Experimental setup for electrochemical measurements.

Fig. 8. Bode diagram of EIS measurement for different Ni^{2+} concentration using 3 electrode structure.

978-1-5386-4483-6/18 $31.00 © 2018 IEEE

Ni(II)- and As(III)-binding flagellin variants

Heavy metal binding flagellin variants were engineered by the appropriate modification of the D3 domain of flagellin or by replacing it with different metal binding polypeptide motifs using computer modeling and conventional genetic engineering techniques. One of the nickel(II) binding flagellin variants was constructed by mutating four amino acids of the D3 domain to histidines situated at an adequate distance and position to each other being able to form a buried nickel(II) binding site. In another Ni-binding flagellin mutant, the whole D3 domain was replaced with a His-rich oligopeptide. Arsenic (III) binding flagellin variants were constructed by substituting the D3 domain of flagellin with strong As-binding polypeptide motifs derived from the ArsR arsenical resistance operon repressor proteins of different organisms. Proper folding of the binding motif and the flagellin part of the fusion protein is crucial for retaining both binding and polymerization ability. It was ensured by appropriate choosing the linker peptides (**Fig. 9**) for insertion.

Various genetically engineered flagellin mutants capable of heavy metal (preferentially Ni and Co) binding were successfully prepared. Heavy metal binding flagellar filaments were polymerized from the modified flagellin monomers [1].

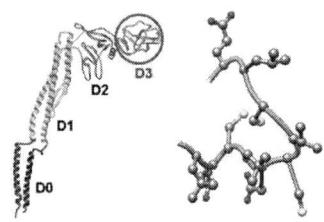

Fig. 9. Replacement of D3 by an As (III)-binding motif

Their binding ability both in monomeric and polymeric forms was characterized by isothermal titration calorimetry.

Immobilization of these metal-binding filaments on a gold surface was followed by new optical methods investigations [1].

Si wafer integrated sensor structures (**Fig. 1b**) were used to test the Ni binding flagellar filaments. The flagellin layers were deposited onto the gold surface using DSP (dithiobis(succinimidyl propionate)) cross-linking agent.

The chips covered with filaments were tested in buffer solution (pH=7.4) by CV when 2, 10, 50 and 100 µmol/l Ni^{2+} ions were added to the buffer respectively. The resulting voltammograms are presented in **Fig. 10**.

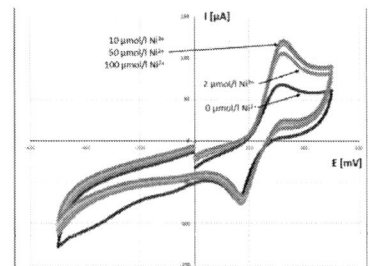

Fig. 10. CV curves of Ni sensing filament layers with different Ni concentrations

Fig. 10 shows that Ni samples exhibit a characteristic signal at + 300 mV. The effect is observable at 2 µmol/L Ni^{2+} concentration and saturates above this level. The detection limit using this method is at least 2 µmol/L.

CONCLUSIONS

All the obtained films exhibited good electrocatalytic activity toward reduction of NO_3^- and respectively to NO_2^- and a good stability. The genetically engineered flagellin mutant proteins were capable of heavy metal sensing as confirmed in the case of Ni.

References

[1]. H. Jankovics, E. Toth, A. Klein, A. Sebestyen, B. Toth, F. Vonderviszt, *"Engineering heavy metal binding flagellin-based nanostructures for sensor elements monitoring natural waters"*, Hungarian Chemical Society (MKE), 13th European Biological Inorganic Chemistry Conference (EuroBIC 13), Budapest, 2016.08.26 - 09.01, p. 110, (ISBN: 9789639970670).

AUTHOR INDEX

AbdelAal R.M.	17	Cimpoesu F.	273	Enache A.	179
Al-Ani O.	231	Ciobotaru D.	325	Faccio F.	133
Alata E.	215	Cisse S.	311	Fanoro M.	283
Albu A.	69	Ciurea M.L.	61	Filip A.	265
Aldrigo M.	101		73	Fiorenza P.	7
Alsnani H.	231		87	Firtat B.	321
Anghel V.	205		253		325
Anghelescu A.	321		257	Frantlović M.	295
	325	Cobianu C.	69	Freitag M.	43
Antonescu A.M.	183	Cojocariu I.N.	197	Fried M.	329
Arnold B.	43	Cojocaru O.	73	Fu Y.	123
Babarada F.	303		87	Furuya T.	287
Bădiliță V.	241	Coll P.	209	Garabagiu S.	83
Banciu M.G.	287	Comanescu C.F.	245	Gartner M.	329
Banu M.	143		269	Gavrila R.	245
Banu V.	155	Cotofana S.D.	51	Geambasu D.C.	287
Baracu A.	299	Covei M.	311	Gheorghe M.	329
Barbu-Tudoran L.	83	Cracan A.	161	Gheorghe S.	329
Bartolucci G.	279		173	Gîrțu M.A.	17
Bawedin M.	143	Cristoloveanu S.	35		273
Benamrouche B.	169		143	Giura I.-M.	291
Benea L.	143	Cursaru (Popescu) L.M.	249	Gkougkousis E.L.	227
Birleanu C.	83	Damianos D.	35	Godignon P.	155
	299	Dan C.	191	Gologanu M.	137
Bita B.	291	Danila M.	69	Goras L.	165
Bîzîitu F.	165		237		187
Blanc-Pelissier D.	35		245	Goss J.P.	231
Bogatu C.	311		307	Greco G.	7
Boianceanu C.M.	219	Dascalescu I.	87	Gri M.	35
Bonteanu G.	161		253	Grosa G.	35
	173	De Martino G.	147	Gudmundsson J.T.	61
Borghello G.	133	Dediu V.	261		257
Bouchard A.	35	Delacour C.	143	Horsfall A.B.	231
Brasoveanu C.	329	Della Corte F.G.	147	Hrib L.	287
Brezeanu Gh.	179	Dicianu M.	191	Hurez I.	205
	205	Dinescu M.A.	97	Hutanu C.	325
	209		299	Ionescu O.	69
	307	Dinulescu S.	321		237
	315		325		291
Briddon P.R.	231	Djamai D.	227	Ionescu P.	287
Bucher M.	133	Dobre M.-D.	209	Ionescu V.	191
Buiculescu V.	119	Dobrescu D.	183	Ionica I.	35
Buiu O.	69	Dobrescu L.	183		143
	79	Dracea M.-D.	287	Iordanescu S.	105
Bunea A.-C.	97	Dragan A.M.	179	Isac L.	79
Buzo A.	201	Draghici F.	209	Istrate A.	269
Cantaragiu A.	265		307	Istrati D.	65
Carp M.	261	Dragoman M.	101	Iuga A.	287
Ceuca E.	315	Dragomirescu D.	169	Jankovics H.	329
Chahdi M.	227		215	Jorda X.	155
Chan K.Y.	93	Drăguț D.V.	241	Kain C.	201
	123	Ducu C.M.	249	Kalas B.	329
Changala J.	35	Dumbravescu N.	69	Kaminski-Cachopo A.	35
Chelu M.	329	Dupont F.	115	Kang H.	151
Chen T.	205	Duta A.	79	Kawanago T.	3
Chevas L.	133		311	Kirscher J.	201
Chirila C.	287		329	Koch H.D.	133

Koudoumas E.	237	Pachiu C.	237	Scucchia L.	279
	245		245	Serban B.	69
Kraft M.	115		261	Simion M.	143
Kriebel D.	43		291	Simion S.	105
Kusko M.	143	Palade C.	61		279
	307		73	Sinha S.	283
Labadi Z.	329		87	Slav A.	73
Laurenciu N.C.	51	Panait P.	17		87
Lazanu S.	73	Pantelica D.	287	Srinivasulu A.	303
	87	Papadopoulou A.	133	Stan I.	321
Lei M.	35	Parvulescu C.	303	Stanca M.	65
Lepadatu A.M.	87	Pascariu P.	237	Stavarache I.	87
Lete C.	329		245		253
Li X.	93	Pascu R.	307	Stoica T.	73
	123	Pelz G.	201		87
Lounis A.	227	Perniu D.	311	Stoukatch S.	115
Makris N.	133	Petrica R.-V.	209	Stricker J.	201
Manea E.	303	Petrik P.	329	Stroescu H.	329
Manolescu A.	61	Pezzimenti F.	147	Suchea M.	237
	257	Piticesccu R.R.	241		245
Maraloiu A.V.	61	Piticescu R.M.	249		291
	253	Placinta V.M.	197	Sultan M.T.	61
Marconi D.	83	Plaiasu A.G.	249		73
Masotti D.	101	Plesa C.-S.	219		257
Maurer L.	201	Pobegen G.	223	Svavarsson H.G.	61
Mehner J.E.	43	Poljak P.	295		73
	129	Popescu A.	303		257
Mescot X.	35	Popescu M.	237	Tache A.M.	179
Mihaiescu D.E.	65		291	Takacs A.	215
Moldovan C.	321	Pristavu G.	307	Tani M.	287
	325	Prohinig J.	223	Teodorescu V.S.	73
	329	Purica M.	265		253
Monti A.	215		269	Tibeica C.	137
Montserrat J.	155	Pustan M.	83	Tigau N.	265
Motoc A.	241		299	Tismanar I.	79
Müller R.	299	Radnovic I.	109	Tóth E.	329
Musat V.	265	Radović D.V.	295	Trandafir A.	273
Muscalu G.	321	Radulović K.	295	Trif A.	83
	325	Ramer R.	93	Trifa V.	315
Nastase F.	69		123	Trupina L.	287
Naumann M.	43	Randjelović D.V.	27	Tudor I.A.	249
Ndiaye A.	273	Rasinger F.	223	Tudose I.V.	237
Neag M.	219	Ravariu C.	65		245
Neculoiu D.	97		197	Udrea F.	151
Nedelcu L.	253		303	Varachiu N.	169
	287	Rayson M.J.	231	Varasteanu P.	307
Nedelcu O.T.	137	Rebigan R.	119	Vasile C.	325
Negut A.	179	Roccaforte F.	7	Vitrant G.	35
Nesic D.A.	109	Romanitan C.	69	Vlădoianu F.	205
Nica I.-A.	187		265	Voicu R.C.	299
Nikolaou A.	133	Rumeau A.	169	Vonderviszt F.	329
Noullet J.-L.	169	Saftics A.	329	Wakabayashi H.	3
Obreja A.C.	79	Sandu T.	137	Zaharescu M.	329
Oda S.	3	Sarajlić M.	295		
Olokede S.S.	283	Savinescu V.-S.	187		
Olsen S.H.	231	Schiebold M.	43		
Oprea C.I.	17		129		
	273	Schmidt H.	43		
Oussalah S.	227	Schulze H.-J.	223		

IEEE
445 Hoes Lane
Piscataway, NJ 08854-4141

ISBN 978-1-5386-4483-6